SPRINGER HANDBOOK OF AUDITORY RESEARCH

Series Editors: Richard R. Fay and Arthur N. Popper

SPRINGER HANDBOOK OF AUDITORY RESEARCH

Volume 1: The Mammalian Auditory Pathway: Neuroanatomy
Edited by Douglas B. Webster, Arthur N. Popper, and Richard R. Fay

Volume 2: The Mammalian Auditory Pathway: Neurophysiology
Edited by Arthur N. Popper and Richard R. Fay

Volume 3: Human Psychophysics
Edited by William Yost, Arthur N. Popper, and Richard R. Fay

Volume 4: Comparative Hearing: Mammals
Edited by Richard R. Fay and Arthur N. Popper

Volume 5: Hearing by Bats
Edited by Arthur N. Popper and Richard R. Fay

Forthcoming volumes (partial list)

Auditory Computation
Edited by Harold L. Hawkins, Theresa A. McMullen, Arthur N. Popper, and Richard R. Fay

Clinical Aspects of Hearing
Edited by Thomas R. Van de Water, Arthur N. Popper, and Richard R. Fay

Arthur N. Popper
Richard R. Fay
Editors

Hearing by Bats

With 138 Illustrations

Springer-Verlag
New York Berlin Heidelberg London Paris
Tokyo Hong Kong Barcelona Budapest

Arthur N. Popper
Department of Zoology
University of Maryland
College Park, MD 20742, USA

Richard R. Fay
Parmly Hearing Institute and
Department of Psychology
Loyola University of Chicago
Chicago, IL 60626, USA

Series Editors: Richard R. Fay and Arthur N. Popper

Cover illustration: A big brown bat, *Eptesicus fuscus,* is about to capture a mealworm hanging on a fine filament. Photograph by S. P. Dear and P. A. Saillant. This figure appears on p. 148 of the text.

Library of Congress Cataloging in Publication Data
Hearing by bats / Arthur N. Popper, Richard R. Fay, editors.
    p.   cm. — (Springer handbook of auditory research : v. 5)
    Includes bibliographical rcferences and index.
    ISBN 0-387-97844-5.
    1. Bats—Physiology.   2. Echolocation (Physiology)   3. Hearing—
Behavior. I. Popper, Arthur N. II. Fay, Richard R. III. Series.
QL737.C5H435   1995
599.4′041825—dc20   94-41860
                                                           CIP

Printed on acid-free paper.

Production managed by Terry Kornak; manufacturing supervised by Jeffrey Taub.
Typeset by TechType Inc., Upper Saddle River, NJ.
Printed and bound by Braun-Brumfield, Ann Arbor, MI.
Printed in the United States of America.

9 8 7 6 5 4 3 2 1

ISBN 0-387-97844-5 Springer-Verlag New York Berlin Heidelberg

Series Preface

The *Springer Handbook of Auditory Research* presents a series of comprehensive and synthetic reviews of the fundamental topics in modern auditory research. It is aimed at all individuals with interests in hearing research including advanced graduate students, postdoctoral researchers, and clinical investigators. The volumes will introduce new investigators to important aspects of hearing science and will help established investigators to better understand the fundamental theories and data in fields of hearing that they may not normally follow closely.

Each volume is intended to present a particular topic comprehensively, and each chapter will serve as a synthetic overview and guide to the literature. As such, the chapters present neither exhaustive data reviews nor original research that has not yet appeared in peer-reviewed journals. The series focuses on topics that have developed a solid data and conceptual foundation rather than on those for which a literature is only beginning to develop. New research areas will be covered on a timely basis in the series as they begin to mature.

Each volume in the series consists of five to eight substantial chapters on a particular topic. In some cases, the topics will be ones of traditional interest for which there is a solid body of data and theory, such as auditory neuroanatomy (Vol. 1) and neurophysiology (Vol. 2). Other volumes in the series will deal with topics which have begun to mature more recently, such as development, plasticity, and computational models of neural processing. In many cases, the series editors will be joined by a co-editor having special expertise in the topic of the volume.

Richard R. Fay
Arthur N. Popper

Preface

Of all vertebrate groups, the echolocating mammals have probably provided the most intrigue to auditory researchers. The biological significance of echolocation sounds is obvious, yet the sounds are generally inaudible to humans and are used in ways that seem alien to our auditory experience.

Of the echolocators, bats are the most widely and intensively studied, and are the best understood. In fact, we probably know and understand more about the acoustic neuroethology of bats than of any other vertebrate group.

Our decision to include a volume on bats in the *Springer Handbook of Auditory Research* was motivated by several considerations. First, to our knowledge, there has been no modern volume that addresses hearing and the neuroethology of echolocation in bats in a systematic and comprehensive way. Thus, this volume provides detailed insights into the current state of our knowledge on these fascinating species, as well as a guide to areas for future research. Second, and as pointed out by George Pollak, Jeffery Winer, and William O'Neill in Chapter 10, bats are exceptionally interesting and useful model systems for the study of mammalian hearing in general, and as such, this volume provides investigators interested in other species with detailed insights into the auditory mechanisms and capabilities of this model system. Finally, as pointed out elegantly by Alan Grinnell in Chapter 1, our understanding of hearing and echolocation by bats is " . . . one of the triumphs of neuroethology . . . " and, as such, bats provide one of the few comprehensive "stories" about the uses of sound in the auditory literature.

Grinnell (Chapter 1) provides an overview of what is known about hearing and echolocation by bats, and provides a valuable perspective on the questions and controversies of greatest interest. In Chapter 2, Fenton discusses the natural history and evolution of echolocation and provides an introduction to the myriad species that are discussed in the rest of this volume. The behavioral capabilities of bats are discussed in Chapter 3 by Moss and Schnitzler, and discrimination is treated in detail by Simmons and his colleagues in Chapter 4. The structure and function of the auditory system is dealt with in detail in Chapters 5 through 9. In Chapter 5, Kössl and Vater describe the inner ear. The brainstem and lower auditory pathways are discussed by Covey and Casseday in Chapter 6, while Pollak and Park describe the inferior colliculus in Chapter 7. The thalamus is treated in Chapter 8 by Wenstrup and the auditory cortex in Chapter 9 by O'Neill. Finally, Pollak, Winer, and O'Neill give an overview of the mammalian auditory system and provide compelling arguments for the use of bats as a model system for mammalian hearing.

In inviting the authors to write chapters for this volume we asked them to not only consider the species upon which they work, but also to provide a comparative overview of the auditory system of the wide variety of bat species. Thus, the volume not only deals with bat hearing per se, but also provides a perspective that shows the wide range of variation in detection and processing mechanisms even within this one amazing group of mammals.

The editors would like to express their gratitude to George Pollak for his guidance, insight, and help in developing this volume. Dr. Pollak has a unique and invaluable perspective on bats, and he shared his thoughts with us freely and generously when we invited him to contribute chapters to this volume.

Arthur N. Popper
Richard R. Fay

Contents

Contributors

John H. Casseday
Department of Neurobiology, Duke University Medical Center, Durham, NC 27710, USA

Ellen Covey
Department of Neurobiology, Duke University Medical Center, Durham, NC 27710, USA

Steven P. Dear
Department of Neuroscience, Brown University, Providence, RI 02912, USA

M. Brock Fenton
Department of Biology, York University, North York, Ontario, Canada M3J 1P3

Michael J. Ferragamo
Department of Neuroscience, Brown University, Providence, RI 02912, USA

Alan D. Grinnell
Department of Physiology, UCLA School of Medicine, Jerry Lewis Neuromuscular Research Center, Los Angeles, CA 90024, USA

Tim Haresign
Department of Neuroscience, Brown University, Providence, RI 02912, USA

Manfred Kössl
Zoologisches Institut, Universität Munchen, D8000 Munchen 2, Germany

David N. Lee
University of Edinburgh, Edinburgh, Scotland

Cynthia F. Moss
Department of Psychology and Program in Neuroscience, Harvard University, Cambridge, MA 02138, USA

William E. O'Neill
Department of Physiology, University of Rochester School of Medicine and Dentistry, Rochester, NY 14642, USA

Thomas J. Park
Department of Zoology, The University of Texas at Austin, Austin, TX 78712, USA

George D. Pollak
Department of Zoology, The University of Texas at Austin, Austin, TX 78712, USA

Prestor A. Saillant
Department of Neuroscience, Brown University, Providence, RI 02912, USA

Hans-Ulrich Schnitzler
Department of Animal Physiology, University of Tübingen, Tübingen, Germany

James A. Simmons
Department of Neuroscience, Brown University, Providence, RI 02912, USA

Marianne Vater
Universität Regensburg, Fachbeireich Biologie, Zoologisches Institut, 93040 Regensburg, Germany

Jeffrey J. Wenstrup
Department of Neurobiology, Northeastern Ohio Universities College of Medicine, Rootstown, Ohio 44272, USA

Jeffery A. Winer
Department of Molecular and Cell Biology, Division of Neurobiology, University of California, Berkeley, CA 94720-2097, USA

Janine M. Wotton
Department of Neuroscience, Brown University, Providence, RI 02912, USA

1

Hearing in Bats: An Overview

ALAN D. GRINNELL

1. Introduction

As predominantly visual animals, we have great difficulty imagining how sound can be used to orient precisely in a complex natural environment. Indeed, most animals, even nocturnal animals, share our dependence on vision. Comparatively recently evolved, hearing has reached a high degree of sophistication in birds and mammals. No wonder, then, that bats, with their abilities to orient, find food, and lead active lives in the dark, have long fascinated humans. This volume summarizes the current understanding of hearing in bats — the behavioral skill with which bats obtain information about their environment and the adaptations of the mammalian auditory nervous system that make this possible.

Our understanding of how bats use hearing began in the last years of the eighteenth century, when the great Italian scientist, L. Spallanzani, noted that blinded bats could fly, avoid obstacles, land on walls and ceiling, and survive in nature as well as bats with sight. He and his Swiss counterpart, C. Jurine, established that hearing was the sense bats used in orientation (see Galambos 1942a; Griffin 1958; and Dijkgraff, 1946 for accounts of these experiments). However, it was not until 140 years later that technological advances enabled Donald R. Griffin, then an undergraduate at Harvard, to demonstrate that bats emitted trains of high-frequency sounds during flight and could use the echoes of these sounds to detect objects and orient in a complex environment. In the next few years, Griffin and his collaborators showed that bats use "echolocation" to do almost everything a bird such as a swift or flycatcher can do with vision (Griffin 1958; see also Dijkgraaf 1946 for accounts of independent experiments leading to similar conclusions). At the same time, Griffin's friend and colleague, Robert Galambos, recorded cochlear microphonics from bat ears to at least 98 kHz (Galambos 1942b).

From the beginning, the study of hearing in bats has been characterized by a synergistic partnership between behavioral and physiological approaches. The results have been nothing short of spectacular, as this

1

volume, written by leaders in the field, testifies. Echolocation, for all of its unanswered questions, is one of the triumphs of neuroethology—better understood, perhaps, than any other complex mammalian behavior, certainly any auditory behavior.

The reasons are straightforward. It is not that bats are intrinsically easier to work with or that their nervous systems are more simple. On the contrary, the size of bats, the difficulty of obtaining and maintaining them in captivity, the inaccessibility of their native habitats, and their nocturnal activity in a territory that covers square miles all combine to make the study of bats daunting. What has made bats such superb subjects for the study of acoustic behavior and neural processing is that they rely almost exclusively on hearing for the information they need about their environments, at least while flying. We can construct test situations in which we know the information they need as well as the signals they use to obtain that information. Moreover, these signals tend to be simple in structure and are employed in reproducible ways under similar conditions. Although bats use much more complex sounds to communicate with one another, and in many cases take advantage of prey-generated sounds, the overwhelming emphasis is on extracting information from echoes of their relatively simple and stereotyped echolocation signals. Moreover, echolocation sounds are consistent within a given species and differ in species-specific ways. Comparative studies show that these differences correlate with preferred habitats and hunting strategies. The differences also help establish the information-gathering value of different components of the echolocation sounds. Finally, the behavioral skills of bats are so remarkable, even unbelievable, that investigators find them fascinating subjects.

2. The Natural History of Echolocating Bats

Bats evolved early in the history of mammals (Jepsen 1970). Enlargements in fossil bat skulls from the Eocene reflect enlarged auditory neural structures. Furthermore, still older fossils of noctuid moths—a favorite prey of bats—demonstrate the existence of tympanic organs, suggesting that the animals that preyed on them (presumably bats) used echolocation. As Fenton points out in Chapter 2 echolocating bats have radiated widely into seemingly all conceivable niches except the polar cap. There are more than 800 species in the suborder Microchiroptera, all of which apparently can echolocate. (One of the puzzles of the bat world is the existence of another suborder of Chiroptera, the Megachiroptera, restricted to the Old World, all with excellent night vision and, with the exception of one genus, all lacking echolocation. That exception, *Rousettus*, has independently evolved a primitive form of echolocation using tongue clicks similar to those used by some cave-dwelling birds [Griffin 1958]). Little study has been devoted to hearing in nonecholocating bats, which probably are like most other small animals in their hearing capabilities (Grinnell and Hagiwara 1972b). This volume is devoted almost exclusively to echolocating bats and their orientation skills. In this we also reflect the field's general neglect of social communication sounds and their processing. A good introduction to that subject can be obtained from Fenton (1985).

Echolocating bats range in size from 2 g to more than a kilogram, with wing spans from 8 in. to more than 2 m. Different species feed exclusively or in various combinations on pollen, nectar, fruit, blood, on flying, crawling, or floating insects or other arthropods, and on frogs, fish, and small birds and mammals. All bats must be able to avoid obstacles such as tree branches, wires, buildings, cave walls, and other flying bats. Typically, they land by approaching a potential landing site and, at the last moment, flipping around and grabbing a surface irregularity with their hind claws. Many drink by swooping down over a body of water and dipping their lower jaw. These maneuvers require not only superb flight skills but a precise knowledge of where objects are located. In most cases, this information is obtained from echoes of emitted sounds. It is not that other cues are unavailable. Many bats probably have much better eyesight than is typically assumed, and some find their food by prey-generated sounds or olfaction. In most species, however, echolocation appears to be the necessary and adequate source of information.

2.1 Echolocation Signal Characteristics

2.1.1 FM Bats

As mentioned, bats have evolved a variety of echolocation sounds that can, to some degree, be correlated with different hunting strategies and mechanisms of information processing. Most families of bats use relatively short, frequency-modulated (FM) sounds that sweep through about an octave ("FM bats"). One or more harmonics increase the bandwidth of these signals. With few exceptions, the frequencies employed are ultrasonic (i.e., above the human range of hearing). Typically these sounds last several milliseconds and are emitted 2–10 times per second while a bat searches for prey. After the bat detects a target of interest, it increases the repetition rate and decreases the duration of its signals. As it nears the target, the bat produces a terminal "buzz," emitting pulses as rapidly as 100–200/sec (Fig. 1.1A). At this time, the intensity and the starting frequency of the sweeps typically decrease. Representative sonograms of a number of species are shown in Fenton, Chapter 2, Figure 2.2. Depending on the target size, a bat first detects an object at distances as close as a few centimeters to as distant as 3–5 m. With these echoes, the bats determine the direction and distance of targets and discriminate the nature of the target.

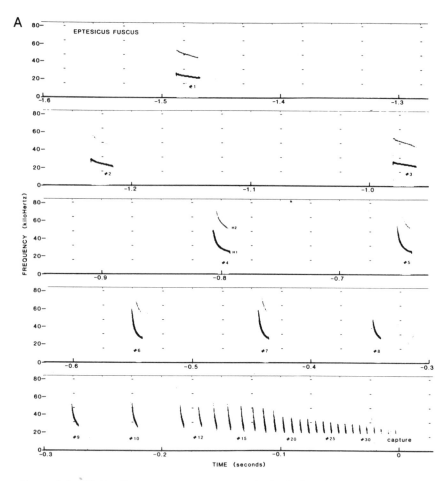

FIGURE 1.1. (A) Continuous spectrogram record of echolocation sounds emitted by a big brown bat, *Eptesicus fuscus*, during pursuit of an insect in the field. Time is shown with respect to the moment of capture. Note the very shallow sweep of the initial search pulses. Courtesy of J. Simmons based on data from O.W. Henson, Jr.

2.1.2 Long CF/FM Bats and Doppler Shift Compensation

A much smaller number of species belonging to three different families have evolved a different type of signal, with a 10- to 100-msec-long, constant-frequency (CF) component preceding an FM sweep ("long CF/FM bats"). In the two best-studied species, the CF component is approximately 83 kHz in the horseshoe bat, *Rhinolophus ferrumequinum*, and 60–62 kHz in the mustached bat, *Pteronotus parnellii*. Individual bats exhibit slight differences, but these signals are remarkably consistent in any given bat.

Long CF/FM bats are strikingly specialized for detailed analysis of sounds in the range of the emitted CF. A large fraction of the inner ear and

B

FIGURE 1.1. (B) (*cont'd*.) Continuous record of pulse emission pattern in a long constant-frequency/frequency-modulated (CF/FM) bat, *Rhinolophus ferrumequinum*, during pursuit and capture of a moth (shown) in [B]). Numbers below the sonograms indicate the time of each image. Capture occurred between images 3 and 4. The CF frequency was 82 kHz; the vertical calibration is 10 kHz. (From Neuweiler, Bruns, and Schuller, 1980.)

most of the neurons in all auditory neural centers are devoted to this narrow range of frequencies. This neural configuration has been termed an "acoustic fovea" (Schuller and Pollak 1979). The importance of this narrow frequency band to the bats is demonstrated by their ability to compensate for Doppler shifts of returning echoes by lowering the emitted CF just enough to restrict the echo CF within the narrow range of frequencies where their sensitivity, and change in sensitivity with frequency, are greatest (see Pollak and Park, Chapter 7, Fig. 7.14). This frequency regulation is accurate to within 50 Hz in 83 kHz, or within 0.06% (Schuller, Beuter, and Schnitzler 1974). Given the concentration of the signal energy within this frequency range, these adaptations optimize the bat's detection of echoes that are slightly displaced in frequency from background echoes, as would result from a difference in the relative velocity of target and bat. The Doppler shift changes systematically with angle relative to the bat's flight direction. But near the axis of flight, or to a bat emitting CF/FM signals while perched on a branch, the echo of an approaching insect would stand out clearly. Indeed, the bats appear to be exquisitely sensitive to the small frequency and amplitude oscillations caused by insect wingbeats and able to use these oscillations not only to detect but also to discriminate prey (Schnitzler and Flieger 1983).

CF/FM bats tend to hunt in cluttered environments where prey detection is probably harder for bats that use exclusively FM signals. CF/FM bats also emit FM signal components, however. During pursuit of prey, these bats progressively shorten the CF portion of their pulses and increase the repetition rate until, in some species, the pulses in the final buzz are composed of almost pure FM signals (Fig. 1.1B). In all bats, it is likely that broadband, sharply timed FM components are used for distance detection and target localization.

2.1.3 Short CF/FM Bats

More numerous than the long CF/FM bats are bats that employ an intermediate pulse design, with pulses containing a short CF component (up to 8–10 msec) and terminating in an FM sweep. These species have been comparatively little studied. Short CF/FM bats may use Doppler shift information to some degree, but they are less specialized for analysis of the CF frequency band. Instead, they probably use the CF component to enhance target detection, but then obtain most information about the target from the FM sweep. Some bats that emit pure FM pulses when close to vegetation prolong the pulse and reduce the amount of sweep in uncluttered environments. Presumably this modification helps them detect faint echoes from relatively distant targets (see Fig. 1A, and Chapter 4, Fig. 2, this volume).

Thus there is considerable facultative flexibility in the design of echolocation pulses. The duration, amount of CF and FM sweep, overall frequency range, intensity, and repetition rate can be varied in ways that optimize the type and amount of information obtainable in a given echolocation situation. The major aims of research in echolocation have been directed at determining the accuracy with which bats obtain information through echolocation and the probable mechanisms that underlie this behavior. Most of this work is discussed in detail in the chapters by Moss and Schnitzler (Chapter 3) and by Simmons and his colleagues (Chapter 4) (this volume). I mention a few of the major findings and issues here.

3. Echolocation Abilities

The ability of bats to detect and capture small insects close to shrubbery or among tree branches involves remarkable echolocation skills. However, our understanding of these skills has grown only gradually as the technology of recording and analyzing ultrasonic sounds has improved and as ever more ingenious ways of explaining their capabilities have been developed. The most important initial studies were done by Griffin and his collaborators. In careful laboratory experiments, they documented the changes in orientation sound characteristics and emission patterns as bats detected and avoided arrays of fine wires. In these studies, the distance at which the pulse emission rate changed were used as the criterion for the distance of detection (Griffin 1958). Parallel field observations established that bats detect and capture flying insects in the wild with a similar pattern of search, pursuit or approach, and terminal buzz pulses. In the late 1950s, Griffin and his colleagues, in collaboration with Fred Webster, succeeded in moving insect capture into the lab (a large Quonset hut) where small FM bats, *Myotis lucifigus*, were strobe photographed and recorded simultaneously as they caught small insects or mealworms thrown into the air by a "mealworm gun." Not only could these bats localize airborne targets accurately enough to catch them with the tail membrane or the tip of a purposely outstretched wing, but they also identified and caught mealworms in the presence of a number of other targets of similar size but different shape (e.g., disks or spheres) that contributed interfering signals (clutter) (Webster 1967; Webster and Brazier 1965; also see Chapter 4 by Simmons et al. [this volume] for a detailed description of these and similar recent experiments).

In short, bats can detect and identify individual small targets in the presence of other targets that return similar echoes that overlap in time; they can determine the distance and direction of each target, and they can avoid unwanted targets while catching the "insect" (see Simmons et al., Chapter 4, Fig. 4.8). It is difficult to escape the conclusion that bats are able to form an acoustic image of the world in front of them. It appears that the echoes of each pulse, returning from different directions and delays corresponding to the distance of different targets, are processed by the brain to provide a three-dimensional map. This map is comparable to the visual image produced by the flash of a strobe light and tells the bats simultaneously about the location of potential prey and all other objects within hearing range, such as tree branches, wires, buildings, and the ground. Indeed, one of the most inexplicable capabilities of bats is that of memorizing a home territory of several square miles, within which they fly along regular flight paths, use the same night roosts, and always find their way back to the colony site. Bats also migrate long distances (e.g., from summer colony sites to overwintering caves), a related skill that is even less well understood (Griffin 1958).

How do bats construct this acoustic image, and how accurately can they do it? Moss and Schnitzler (Moss and Schnitzler, Chapter 3) discuss the experimental results that help answer, or at least define, these questions. The most illuminating experiments have been done with the behavioral conditioning techniques introduced by James Simmons. Simmons showed that bats could readily be trained to discriminate accurately between targets presented simultaneously or sequentially in an automatic forced choice (AFC) or yes/no paradigm. The bats indicate their choice by crawling to one side or the other of a holding platform. Real targets can be used, or "phantom" targets produced by recording the emitted pulse with a microphone near the bat and playing back either a faithful or altered "echo" at the desired delay and angle from a loudspeaker (see Moss and Schnitzler, Chapter 3, Fig. 3.2). This technique makes it is possible to test how sensitively a bat can detect echoes, measure echo delays and discriminate various targets at different distances, resolve target directions in both horizontal and vertical axes, discriminate target shapes, dimensions, and surface textures, and make all these judgments in the presence of different amounts of acoustic clutter.

3.1 Detection

In a laboratory setting, the detectability of an echo is subject to forward and backward masking by echoes from platform, microphone, loudspeaker, walls, and floor. Consequently, no adequate measure of absolute threshold of echo detection has been made. There is no reason to believe, however, that bats are absolutely more sensitive than any other mammal at its best frequency. Behavioral (and neurophysiological) audiograms show that bats tend to be sensitive at the range of their emitted orientation pulses, a frequency range far above that used by most other mammals. FM bats tend to be broadly sensitive throughout a frequency range that often covers several harmonics. CF/FM bats have a sharp peak in sensitivity near the most prominent CF frequency in their emitted sounds (which is usually the second harmonic), with lesser peaks at other harmonics of this frequency (Fig 1.2) (see also Kössl and Vater, Chapter 5, Figs 5.2, 5.3, and Moss and Schnitzler, Chapter 3, Fig. 3.1).

The CF signal used by CF/FM bats is thought to increase their ability to detect distant targets by concentrating signal energy into a narrow frequency band. This is premised on the assumption that information returning at that frequency can be integrated for the duration of the CF component, up to 100 msec. Because any given frequency in an FM signal typically lasts only a fraction of a millisecond, an equivalent ability to use FM signals for detection would require that the auditory system act as a matched filter receiver, programmed by the outgoing signal to sum all the energy in returning sounds with the same frequency-time structure. Although this capability cannot be ruled out (and evidence from ranging experiments suggests that such an operation does occur for detection of pulse–echo delay), two observations argue against detection involving such matched filter performance. First, field observations show that FM bats which hunt in open space tend to elongate their search pulses and reduce

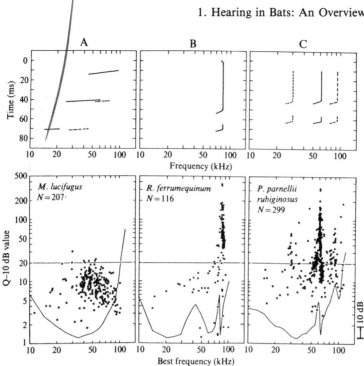

FIGURE 1.2. Characteristic frequency–time plots (sonograms) of the emitted echolocation sounds of a typical FM bat, *Myotis lucifugus*, and the two most-studied long CF/FM bats, *Rhinolophus ferrumequinum* and *Pteronotus parnellii* (*top panels*) as well as behavioral or neurophysiological audiograms of each species (*solid lines* in *lower panels*, with 10 dB calibration on right) and the $Q_{10\ dB}$ values of single auditory nerve fibers (*dots*). (From Suga and Jen 1977, including data from Suga 1973 (A), Suga, Neuweiler, and Möller, 1976 (B), and Long and Schnitzler 1975 [audiogram in B].)

their frequency sweep, concentrating most of the signal energy into a narrow frequency band. Second, FM bats are almost equally capable of detecting artificial "echoes" that sweep upward instead of downward (Møhl 1986; Masters and Jacobs 1989). Further research is needed to determine whether the integration time differs for different echo frequencies and how bats alter their emitted signals to optimize target detection under different echolocation conditions.

The effective limit to the range of echolocation may be that imposed by the emission of another orientation sound. In most cases, pulse duration appears to be reduced progressively during the approach to a target so the emitted sound and its returning echo never overlap. On the other hand, with FM pulses and sharp frequency resolution, it is unclear why some overlap should not be tolerated. And there is no convincing evidence that the echoes of one pulse, returning to a bat after a subsequent emission, are not useful, as might be the case for echoes from large obstacles in the environment or from the ground. Moreover, the acoustic reaction to a target, whether pursuit or avoidance, appears to be a facultative response. Bats show such a response when they are pursuing a target or avoiding a particular obstacle or landing, but they still obtain useful information about other objects in the environment — branches, other insects, etc. — that they do not respond to with pursuit sequences and buzzes.

3.2 Temporal Resolution

3.2.1 Ranging

Sound travels at 34.4 cm/msec in air; consequently, the interval between an emitted sound and a returning echo is an accurate measure of target distance. To capture a flying insect, even in its tail or wing membranes, a bat must be able to localize it to within about 1 cm at a distance of 10 cm or more (a minimal reaction time at a bat's typical flight speed). Many behavioral measurements have been made of the ability of different bats to discriminate target distance in the laboratory: identification of the closer of two real or phantom targets presented simultaneously or sequentially, or a yes/no decision whether a target is at a given trained absolute distance. Interestingly, all such tests using any of the experimental paradigms outlined here, including recognition of absolute distance, have yielded a maximum accuracy of about 10–11 mm (50–55 μsec time separation) for all species studied (see Moss and Schnitzler, Chapter 3, Fig. 3.4).

3.2.2 "Jitter" Resolution

Initially this 50- to 55-μsec value was accepted as a realistic limit because it could explain the bats' behavioral ability to localize a target in space. The fact that similar values were obtained when bats were tested at long or short target distances was bothersome. Then it was realized that as a bat moves around on its holding platform, it shifts the position of its head and body from one side to the other as it emits pulses, and the head movement might well introduce this much variability in the distance being measured. Hence this type of experiment was supplanted by one in which there are two phantom targets, one of which returns "echoes" at a constant delay, while the other shifts back and forth in time, from one pulse to the next, by variable amounts (Simmons 1979). In such jitter experiments, the bat holds its head in one position as it scans one target with its pulses, then holds its head relatively constant in a different position as it directs pulses toward the other. An intensity comparator is used to restrict "echo" production to only one target at a time (jittering or nonjittering), the one toward which the bat is directing its emission. The results of these experiments have been so dramatic that they have dominated behavioral research in the field for the last several years.

Initial results with *Eptesicus* yielded a jitter threshold of less than 5 μsec, a value that was soon lowered to about 1 μsec (Simmons 1979). Further training with the same bats eventually pushed the value down to the incredible value of 10 nsec (corresponding to a range sensitivity of less than 2 μm) (see Moss and Schnitzler, Chapter 3, Fig. 3.5) (Simmons et al. 1990b).!! This finding has been received with understandable skepticism (Pollak 1993; Simmons 1993; also see Moss and Schnitzler, Chapter 3; Simmons et al., Chapter 4). Besides the difficulty of explaining how the nervous system could encode an event with 10-ns accuracy, two major problems argue against interpretation of the jitter experiments as showing micrometer range resolution.

First, even in the jitter situation, the bat moves its head. Not as much as a centimeter, surely, but even while facing in one direction, a bat's entire body jerks with each emission. Its head movement is likely to vary by at least a fraction of a millimeter between pulses (and the jitter is being measured between pulses). Variations in head positions could be minimized if the bat emits its pulses and hears the echoes with its head at approximately the same phase of each pulse-induced vibration. Its seems inevitable, however, that residual changes in the bat's head position (not to mention the complications of ear movements) would be much greater than the resolution implied by 10-nsec threshold jitter detection. Perhaps even more troubling is that bats consistently perform in behavioral experiments as if they interpret a 1-dB reduction in echo intensity as a 13- to 17-μsec increase in delay (distance) (Simmons et al. 1990a) (see Chapter 3, Fig. 3.6). A 1-dB difference in emitted sound or reflected echo strength of two successive (jittered) echoes would yield an intensity/latency shift about 1000-fold greater than their apparent jitter resolution.

Nevertheless, reproduction of the jitter experiment in a different lab, with instrumentation that allowed jitter down to 400 nsec, showed no decrement of performance (Menne et al. 1989, Moss and Schnitzler 1989). Moreover, variants on these experiments showed that bats are sensitive to 90°–180° phase shifts in phantom echoes, indicating that they can resolve the phase of sounds at these frequencies. So despite reluctance to believe the 10-nsec finding and the widespread feeling that there must be some qualitative difference between the jittering and nonjittering targets, no one has identified an artifactual origin or a flaw in the experimental procedure that could yield this result. In fact, one is tempted to suspend disbelief in this result for the simple reason that temporal resolution of almost this magnitude seems to be necessary for bats to resolve one target from others nearby (as Griffin and Webster and their colleagues showed so clearly) and to discriminate the dimensions of individual targets (see Simmons et al., Chapter 4).

3.2.3 Multidimensional Targets and Two-Wavefront Range Discrimination Resolution

Targets of the size caught by most insectivorous bats reflect echoes that consist of two or more "glints," echoes from different parts of the target that reflect stronger echo components than other parts. Each glint is an echo of the emitted pulse, displaced in arrival time by a few tens of microseconds at most. Hence, the glints overlap for most of their duration. Interference and reinforcement between overlapping echoes produces troughs and peaks in echo intensity during the frequency sweep. To simulate such natural sounds, echoes that contain two wavefronts a few microseconds apart have been used to test discriminability of such overlapping glints. For example, two different hole depths in targets can be detected with thresholds of 0.6–0.9 mm, corresponding to 4-Δ to 6-μsec time resolution. Initially, these experiments were interpreted as showing detection of spectral cues, but in fact they can be interpreted equally well as a time difference (Mogdans, Schnitzler, and Oswald 1993, and see Moss and Schnitzler, Chapter 3). Indeed, in a particularly intriguing experiment, Simmons and his collaborators (1990a) reported that such overlapping echoes can be mimicked by echoes that jitter back and forth by the same Δt but are always heard separately. In summary, although the specific cues that bats use are hotly debated, they clearly are exquisitely sensitive to time or spectral cues that result from multiple glints from targets. Furthermore, they may be capable of range discrimination on the order of a millimeter or less. Is this an exciting field or what?!

3.2.4 Some Other Puzzles in Ranging

The ranging, jitter, and two-wavefront discrimination experiments pose an interesting dilemma. The accuracy of resolution of range of two different targets, presented either simultaneously or sequentially on either side of the animal or sequentially from the same direction, is always about 1 cm (50 μsec) or more. In contrast, jitter or two-wavefront discrimination appears to be well under 1 mm. Why this difference? The answer is not clear. However, both figures could be correct. Greater head movement in two-target range resolution experiments might explain the difference, but that explanation is not particularly compelling. If the bats can process all available and useful information, then almost inevitably echoes from other objects in the environment (the microphone or loudspeaker, for example) at a constant distance would serve as an accurate reference against which target distance could be compared, independent of head position. Also, experiments with two real targets yield the same 50- to 60-μsec Δt value when echoes from bats would be heard overlapping in time (but from different angles), where again head position should have been unimportant.

Thus, the difference in range thresholds for the two experimental situations very likely is real. One possible explanation for this might be that echoes returning from different targets, at significantly different angles, are kept separated in the brain (a separation that seems necessary for an effective auditory "map" of space, which bats appear to enjoy). Different populations of neurons would process the information from different targets, or at least those targets separated sufficiently in space. In contrast, glints (or jittered echoes) from the same target would be analyzed by the same population of neurons. It may be that temporal comparisons (or spectral notches) are measured much more accurately within one population than between different populations. (The 50-μsec ranging limit for sequential presentation of targets at different Δt were done with a minimal time separation of several seconds between trials at different apparent distance. This measurement might be no more accurate than that between different populations.)

Another interesting question is whether bats can detect range information from single pulse–echo pairs. This ability seems necessary if bats are to adjust their flight path to intercept fast-moving insects, as some bats do (see Simmons et al., Chapter 4). On the other hand, it is possible that range determination requires the integration of several echoes. This is suggested by behavioral evidence from the lesser fishing bat, *Noctilio albiventris*, a short CF/FM bat. When these bats were required to perform an easy range discrimination task in the presence of loud white noise interrupted by brief gaps of silence, they could do so only if at least two pulse–echo pairs fell within a single silent gap. Difficult discriminations required as many as seven pulse–echo pairs. Simple detection required only a single pulse–echo pair (Roverud and Grinnell 1985a). This might be explained simply as forward and backward masking, but related experiments show that free-running loud CF/FM pulses simulating the bat's own sounds could also interfere with range discrimination. This interference occurred even when the sounds appeared well after or before a natural pulse–echo pair, if one such artificial pulse fell between any two natural pulse–echo pairs. If two emitted sounds and their echoes occurred without such an artificial pulse, the bats could discriminate range (Roverud and Grinnell 1985a).

These experiments also support the idea that an echo (or a sound simulating the bat's emission) is interpreted as bearing range information if it occurs during a restricted time window following each emitted pulse. In *Noctilio albiventris* a loud artificial FM sweep disrupted range discrimination for about 27 msec after the onset of the emitted pulse (corresponding to a target distance of approximately 5 m). Outside that window, the FM sweep by itself was ineffective; only a more complete CF/FM pulse could degrade the bat's performance (Roverud and Grinnell 1985b). These findings suggest that a loud CF signal at the right frequency resets the system, activating a time window of approximately 25–30 msec during which a returning FM sweep is interpreted as carrying distance information. If the artificial FM sweep is sufficiently loud during this time window, it can interfere with range discrimination. Outside this time window, it is effective only if the gate is again opened by a loud CF component.

Other laboratory experiments with *Noctilio* have been aimed at discovering which features of the FM sweep are critical to range determination. *Noctilio* can be induced to increase the repetition rate of its pulses as it would to an approaching target if echoes of its emitted pulses are returned at progressively shorter delays. The acoustic situation mimics that of an object moving toward the bat, and clearly gets the bat's attention. By manipulating the properties of returning echoes, R. Roverud (personal communication) has shown that the direction of the FM sweep is important for this response (in contrast to detection). An upward sweeping signal does not induce the response. With this technique, Roverud has also shown that a series of brief CF tone components, electronically merged seamlessly in phase with each other, can replace the FM setup if there are more than 90 steps and none is closer together than about 100 Hz (Roverud 1993). A 10-kHz series of such steps (50-μsec duration and 100-Hz separation) is as effective as a 40-kHz sweep (50-μsec duration, 400-Hz separation), and the tonal steps can be as short as 20 μsec without degrading the recognition of that component as one of the steps. This finding not only implies extremely rapid and accurate discrimination of different frequencies in an echo, but also that the FM component of an echo is recognized by the sequential activation of a required number of separate channels of information.

3.3 Angular Resolution and Target Localization

Bats appear able to determine the direction of an echo's source to within ±2°–5°. This conclusion reflects several lines of evidence: their ability to capture a flying insect by spreading their tail membrane like a catcher's mitt or reaching out to haul it in with a tip of a wing membrane, their ability to point their head directly at an insect with that degree of accuracy even if the flight path is on a different line (Masters, Moffat, and Simmons 1985), and their ability to resolve and discriminate one target from several others in the same approximate direction (see Simmons et al., Chapter 4). Laboratory behavioral experiments confirm this degree of accuracy. In the azimuth, *Eptesicus* can discriminate targets at a separation as small as 1.5° (Simmons et al. 1983). Most mammals are much less accurate in resolving the direction of a sound source in the vertical plane. In bats, however, forced-choice behavior experiments show vertical localization to be almost as accurate, about 3° (Lawrence and Simmons 1982).

How is this done? More importantly, how are many targets at different directions and distances localized simultaneously with this accuracy, which seems necessary to explain the ability to catch an insect in the midst of branches and foliage? To a remarkable degree, bats act as if this were no problem. Indeed, as mentioned earlier, multiple targets seem not to be interpreted as interfering signals, but instead are analyzed accurately in their own right to construct an acoustic map, or image, of the world in front of the bats. How they construct this map is a source of some disagreement, as the chapters in this volume attest.

The available cues for target localization are interaural differences in intensity and arrival time, either of the envelope of the sound or of the phase of each cycle in it. Localization is almost certainly not accomplished by directing a narrow beam of sound in various directions like a flashlight. In vespertilionids, with which much of the behavioral work has been done, the emitted sounds are not beamed very directionally (Hartley and Suthers 1989). Even if they were, information must be obtained about many objects, in many directions, with each echo. Because of the combined directionality of sound emission and hearing, either ear is almost equally sensitive to targets anywhere between straight ahead and 30°–40° to the side of that ear. On the other side, sensitivity decreases by 20–30 dB within 30° (see Simmons et al., Chapter 4, Fig. 4.13). The head and external ears (the latter in bats tend to be extremely large and complicated by the presence of a conspicuous tragus in front of the pinna) create strong sound shadows at ultrasonic frequencies. A target 10°–20° to one side of the axis of a bat's head returns an echo that may be 30–40 dB louder at one ear than at the other. Binaural inhibitory interactions further exaggerate interaural differences within the nervous system.

Vertical localization is as important to bats as horizontal localization. In other mammals it depends, at least in part, on the timing of multiple reflections from the pinna into the ear canal (Middlebrooks 1992). The huge pinna and prominent tragus characteristic of bat ears make them good candidates for performing a similar function. The importance of the pinna and tragus to vertical localization has been demonstrated behaviorally (Lawrence and Simmons 1982). Moreover, long CF/FM bats rapidly flick one pinna forward and the other backward with each pulse emission; perhaps this rapid alternation permits correlation of ear position with echo direction. External ear shape and movements introduce differences in intensity as well as timing of reflections. Until it has been determined whether a bat can locate targets in the vertical dimension monaurally, the possibility of binaural comparisons cannot be excluded. Indeed, because hearing is directional in the vertical as well as horizontal dimension, a comparison of interaural intensity differences at three or more different frequencies in an echo (or the whole FM sweep) could, in principle, pinpoint the direction of a target in both planes (Grinnell and Grinnell 1965; Fuzessery and Pollak 1984).

The other major cue to direction is interaural arrival time (or phase) differences. Echoes arrive slightly earlier at the closer ear. In most bats, these time differences are 50 μsec or less. For directions within 15° of the midline, the interaural time differences would be 10 μsec or less. Because there is a trade-off between intensity and response latency, and a 1-dB intensity difference between the two ears is equivalent to approximately a 15-μsec difference in arrival time (Simmons et al. 1990a), it would seem that interaural intensity differences are much the more prominent cues for signal direction. Moreover, there is an orderly array of neurons in the inferior colliculus that respond preferentially to different interaural intensity differences, corresponding to different angles in the azimuth (see Pollak and Park, Chapter 7, Fig. 7.12; and Wenstrup, Fuzessery, and Pollak 1986; Park and Pollak 1993).

On the other hand, the situation becomes enormously more complicated with additional targets. For example, when two targets are at approximately the same distance on either side of the head (e.g., two wires the bat must avoid), the echoes, one slightly louder than the other, arrive at the two ears displaced by a few microseconds. Because time and intensity are interchangeable once they enter the nervous system, cues that are not subject to these transformations would be preferable. Indeed, Simmons and his colleagues (Simmons et al., Chapter 4) argue that intensity and spectral cues are useless for localization in the usual situation in which there are multiple echoes from different directions overlapping in time as they return to the bat. Because the "integration time" of either ear is 300–400 μsec, that is, for this length of time the ear integrates all the information it receives at a given frequency and assigns it a single amplitude, a simple measurement of interaural intensity difference when there are multiple echoes would result in an intermediate and false target localization. Instead, Simmons et al. argue that correct localization can be done only by using the spectral cues generated at each ear by the interference patterns caused by overlapping echoes from different targets (or glints from parts of targets). Arriving at each ear, the overlapping echoes cause peaks and troughs in the echo intensity at different frequencies, the location of which (spectral cues) is the result of differences in arrival time (phase). The height and depth of the peaks and troughs reflect the relative intensity. The spectral peaks and troughs in the composite echo can in principle be deconvoluted to determine the arrival times of each echo at each ear, which, given the ability to resolve interaural time differences as short as 1 μsec, can then be compared to determine the direction of each echo source. This is not a trivial neural computation, and it remains to be demonstrated that bats actually use these cues for localization.

3.4 Overcoming Clutter

Most remarkably, bats can detect and determine the direction and distance of targets and discriminate between them in the presence of an enormous amount of acoustic clutter, the loud outgoing sounds and echoes of multiple targets at different distances and directions (see Simmons et al., Chapter 4, Figs. 4.8, 4.10). This clutter does interfere with fine discrimination abilities; both forward masking (e.g., by the emitted sound and closer objects) and backward masking (e.g., by branches or a wall behind the target) degrade performance, but much less than might be expected (Simmons et al. 1988; Hartley 1992).

To overcome the deafening effect of their own emitted sounds, bats, like other mammals, reduce the sensitivity of their hearing during the emission by contracting the middle ear muscles. They must, of course, be as sensitive as possible to returning echoes. Henson (1965) showed that FM bats (e.g., the Mexican free-tailed bat, *Tadarida*) begin contracting the middle ear muscles a few milliseconds before pulse emission. Contraction was maximal at approximately the onset of emission, and relaxation began immediately thereafter. Sensitivity at the level of cochlear microphonics was reduced by 20–30 dB during pulse emission and was gradually restored to maximum levels within 5–8 msec after the pulse. This mechanism is especially important for the relatively long search pulses emitted at low repetition rates. As the interval between pulses decreases, however, the contractions become progressively more shallow and sustained.

3.4.1 Automatic Gain Control

Contraction of the middle ear muscles strongly attenuates sensitivity to the emitted signal, weakly attenuates response to echoes from nearby targets, and leaves the auditory system maximally sensitive to echoes from distant objects. At the same time, because of the spatial dispersion of both signal and echo energy, echo strength falls sharply with distance. The result is what is now recognized as a form of automatic gain control (Simmons and Kick 1984). The decrease in echo strength with distance is compensated for by the increase in auditory sensitivity associated with pulse–echo interval. The effective echo strength stays relatively constant as a bat approaches a target. Whether this stabilization of echo loudness independent of distance is of major importance remains to be demonstrated, but this is plausible. It should be noted, however, that bats often decrease the intensity of their emitted pulses as they approach a target.

Moreover, there is another form of suppression of response to the emitted sound. A neural inhibition of responsiveness occurs at the level of the lateral lemniscus (Suga and Shimozawa 1974). Like contractions of the middle ear muscles, this inhibition is triggered by the motor program of pulse production. The time course of the inhibition and recovery from it are unclear, but the mechanism may reduce the perceived intensity of echoes in ways that sharply alter the gain control associated with middle ear muscle function that has been documented with cochlear microphonics.

Bats employing long CF/FM signals have a similar but seemingly more severe problem: they are still emitting a loud sound during most of the time the echo is returning. In these bats, the middle ear muscles contract during most of the period of pulse emission, reducing sensitivity to the emitted sounds (Henson and Henson 1972). Surprisingly, their returning echoes are much less suppressed. The principal reason is the extraordinarily sharp tuning of their auditory system. The emitted CF, in a flying bat, is at a frequency of much reduced sensitivity, while Doppler-shifted echoes fall in a region of high sensitivity, with little inhibition from the lower emitted frequencies (Suga, Neuweiler, and Möller 1976; and see Kössl and Vater,

Chapter 5, Fig. 5.5 for a characteristic single-cell tuning curve illustrating the sharpness of tuning near the CF_2).

3.5 Target Discrimination

Some of the earliest experiments with flying bats established that they are remarkably skilled at discriminating one target from another (Webster 1967). Under natural conditions, bats appear to be able to distinguish insects from other objects and preferred insect prey species from undesirable insects. In the laboratory, a bat trained to catch mealworms propelled into the air can identify and capture the worm despite the presence of decoy disks or spheres (Webster 1967; Simmons and Chen 1989).

In long CF/FM bats, this ability depends at least in part on relative movement of the target and the surroundings. Doppler shifts will distinguish a flying insect from nearby stationary objects. Moreover, the wingbeat frequency and degree of modulation of the CF echo differ for different insects, and the CF signal is long enough to encompass one or more complete insect wingbeat cycles. Given the exquisite ability of long CF/FM bats to discriminate frequencies around their CF, frequency and amplitude modulations can be used to distinguish differences in wingbeat frequency as small as 4%–9% (Schnitzler 1970; Goldman and Henson 1977; Emde and Menne 1989; Roverud, Nitsche, and Neuweiler 1991).

The emitted pulses of most FM bats are too short to detect significant change in echo strength resulting from movement of an insect's wings, and FM signals typically are swept in frequency much too quickly to take advantage of (or be misled by) Doppler-shift information. Yet FM bats can discriminate fluttering targets only slightly less well than long CF bats (Sum and Menne 1988; Roverud, Nitsche, and Neuweiler 1991). What cues do the FM bats use?

The phenomenal ability of bats to resolve "glints" or multiple overlapping wavefronts has already been described. Perhaps their temporal resolution is good enough to get an instantaneous image of an insect's shape based on all the glints returning from its different body parts. Alternatively, bats may use differences between the Doppler-shifted echo from moving wings and the (relatively) unshifted echoes of the body (Sum and Menne 1988). These factors are discussed in detail by Moss and Schnitzler in Chapter 3.

In the context of insect identification and capture, there is a fascinating subplot to the story (see Fenton, Chapter 2 for the more complete version). Many insects have developed simple but effective ears to detect bats and to allow evasion. Arctiid moths go one step further, producing loud ultrasonic clicks of their own when they hear bat cries. These prey-generated sounds cause the bats to veer away, abandoning the chase. As arctiid moths appear to be distasteful to bats in captivity, it is probable that the signal is telling the bat, "You don't want me; I taste bad." Alternatively, and less plausibly, the moth-generated sounds may disorient the bat.

4. Specializations of the Auditory System for Echolocation

The bat auditory nervous system is constructed with the same basic elements as that of other mammals: the same identifiable nuclei, cell types, synaptic connections, and pharmacology. However, echolocation poses special requirements: for extraordinarily precise temporal resolution, for accurate sound localization in three-dimensional space, and for formation of an acoustic image with each echo. The circuits and processes responsible probably represent selection and refinement of general mammalian mechanisms rather than evolution of novel pathways or mechanisms.

Most species of bats have large and complex pinnae, a tragus in front of each air canal, and, in many, elaborate nose leaves and other nasal structures associated with sound emission. The inner ear and auditory nervous system are no less dramatically specialized for the use of sound. At each level of the auditory system there are specializations in structure, organization, and physiological processing that are critical to echolocation. Much of this volume is devoted to reviews of the current understanding of these specializations and the mechanisms of information processing used in echolocation. Most of this work has been done with the FM bats *Eptesicus* and *Myotis* and the CF/FM bats *Rhinolophus* and *Pteronotus parnellii*. Not surprisingly, the auditory nervous system of these two groups differs in fascinating ways, reflecting their different signals and echolocation behavior.

4.1 Cochlea

4.1.1 Morphology

The cochlea, while conforming to the general mammalian model, is clearly specialized for the use of high frequencies and, in CF/FM bats, for hyperacuity around the CF (see Kössl and Vater, Chapter 5). Readers wishing to gain more perspective on specializations of the mammalian cochlea are advised to consult the excellent chapter by Echteler, Fay, and Popper (1994) in a companion volume of this series. The basilar membrane is appropriately narrow, with unusual thickenings especially at the basal end. The inner hair cells (IHCs) entirely overlie the bony spiral lamina. The basilar membrane is longer, relative to body weight, than in other mammals, especially in CF/FM bats. In most FM bats, the basilar membrane increases in width systematically and decreases in thickness along its length in typical mammalian fashion, forming a logarithmic topographic map of frequency. In *Trachops*, a frog-eating bat, and perhaps in others taking advantage of prey-generated sounds, the basilar membrane shows special adaptations near the apical end that extend hearing to comparatively low frequencies (Bruns, Burda, and Ryan 1989).

It is in long CF/FM bats, however, that the cochlear specializations become most dramatic. In both *Rhinolophus* (Bruns 1976a,b) and *Pteronotus parnellii* (Henson 1978), much of the basal half of the basilar membrane is approximately the same width and thickness, appropriate to analyzing a narrow band of frequencies around the CF_2 (see Kössl and Vater, Chapter 5, Fig. 5.13); hence the term, "acoustic fovea." Distal to the expanded CF_2 region, there is an abrupt widening and decrease in thickness associated with analysis of lower frequencies. In *Pteronotus*, the cochlear microphonics (CMs) show a pronounced resonance at the frequency of the Doppler-shifted echo CF (60 kHz) that persists after an acoustic stimulus anywhere near that frequency (Henson, Jenkins, and Henson 1982). Curiously, *Rhinolophus*, with even more dramatic discontinuities in basilar membrane stiffness, does not show such resonance, but somehow this species also achieves sharp frequency selectivity for the dominant CF component. The innervation pattern of the hair cells also conforms to the general mammalian pattern, although the few studies that have been done show some interesting quantitative specializations. In *Rhinolophus*, for example, 10%–20% of the afferent fibers innervate outer hair cells (OHCs), compared with about 5% in other mammals (Bruns and Schmieszek 1980). In bats using CF components, there is dense innervation of the basilar membrane in regions corresponding to the representations of the second and third harmonics, with sparse innervation in between (Zook and Leake 1989). The density of innervation at these sites is close to the maximum observed for other mammals (25–35:1); in the FM bat *Myotis*, interestingly, the density is about twice that high (70:1) (Ramprashad, Money, Landolt and Laufer 1978).

The efferent innervation of the cochlea in bats also shows some interesting specializations. As in other mammals, there are both medial and lateral olivocochlear efferent systems. Unlike other mammals, however, the lateral system contains only an ipsilateral component, not one from the contralateral side (see Kössl and Vater, Chapter 5, Fig. 5.19). *Rhinolophus* is unique among the species studied in having no medial olivocochlear system and lacking efferent synapses on the OHCs (Bishop and Henson 1987). *Pteronotus* is unusual in a different way. Each OHC receives only one efferent ending, and each efferent fiber forms terminals on only a few OHCs, in contrast to most other mammals. Because the overall function of the efferents is so poorly understood, it is not easy to interpret the differences among mammals or between bats and other mammals, but an understanding of their significance in bats may provide critical insights into the role of cochlea efferents in general.

Kössl and Vater (Chapter 5) also discuss the implications of these specializations in cochlear structure and innervation pattern for the important "second filter" mechanism of achieving sharply tuned low-threshold responses. *Pteronotus* exhibits particularly prominent otoacoustic emissions (OAEs) at approximately the frequency of its second harmonic CF (60–62 kHz), about 100-fold louder than in other animals. There is no compelling evidence, however, that active movement of OHCs is responsible. On the other hand, active micromechanical processes in the OHCs are likely responsible for the sensitivity peaks at high frequency, with specialization in shape and thickness of the basilar and tectorial membranes shaping the tuning curves.

4.1.2 Physiology

The first physiological study of bat hearing was the recording by Galambos (1942b) of CMs in the little brown bat, *Myotis lucifugus*, which he obtained up to 98 kHz. This finding supported the idea that bats could use frequencies that high, but presented a severe and persistent problem for models of transduction in the cochlea. In fact, most bats utilize frequencies in the 25- to 100-kHz range for echolocation, and some emit and analyze principal components as high as 150 kHz. Although frequencies this high offer advantages for echolocation acuity (and disadvantages for detection of distant targets), they require considerable specialization of the middle and inner ears. In FM bats, recordings of CMs, auditory nerve single units, and the synchronized onset response of the auditory nerve (N_1) all indicate broad sensitivity throughout the range of emitted frequencies. With the exception of cetaceans, this is well beyond the range of high sensitivity in other mammals. Single units show best frequencies (BFs) to more than 100 kHz, and tuning curves similar to those of other mammals, but with $Q_{10\ dB}$ values (the BF divided by the bandwidth of the tuning curve 10 dB above threshold at the BF) up to 30. This tuning is two- to threefold sharper than in most other mammals.

Responses in long CF/FM bats are much more specialized, reflecting the dramatic adaptations for analysis of frequencies around the CF_2 at early stages of sound processing. CMs and N_1 show peak sensitivity slightly above the frequency of the CF second harmonic (CF_2) and sometimes near CF_3 as well (Pollak, Henson, and Novick 1972). On both sides of these maxima, especially the low-frequency side, thresholds rise sharply. Single auditory nerve fibers in this frequency range show $Q_{10\ dB}$ values as high as 400! (see Fig. 1.2; and Kössl and Vater, Chapter 5, Fig. 5.3). Single units tuned to other frequencies are much less numerous than those in the CF region and have $Q_{10\ dB}$ values like those in FM bats. Associated with the steep tuning near the CF_2, especially in *Pteronotus*, is a distinct long-lasting resonance in the CM just above the frequency of the CF (Suga, Simmons, and Jen 1975). Whatever is giving rise to this resonance is presumably also responsible for the sharp tuning of single units and the phenomenally accurate resolution of frequency around the CF_2.

4.2 Brainstem Auditory Nuclei and Specializations in Processing of Information for Echolocation

The mammalian auditory system is a series of parallel frequency-tuned pathways, beginning with the tonotopic distribution of eighth nerve endings along the basilar membrane and maintained as a tonotopic map in (almost) every division of every acoustic nucleus en route to the cortex. In addition, beginning with the input to the cochlea nuclei, the information in each frequency-tuned pathway is duplicated and sent through multiple channels that process it in different ways. Some are specialized for frequency and intensity analysis, others for precise registration of the timing of signals. In bats, this information is integrated at various levels to provide the range of targets, their localization in space, and the dimensions, movement, and surface texture of each target. It seems probable that all this information is eventually combined somehow to produce an acoustic image of the world around the bat.

Figure 2 in Chapter 7 shows diagrammatically the principal nuclei of the auditory system. The earliest recordings were from the inferior colliculi (Grinnell 1963a; Suga 1964), but much has now been done at all neural levels through the cortex, revealing a great deal about mechanisms of information processing for echolocation and the circuitry involved. This work is summarized in chapters by Covey and Casseday (Chapter 6), Pollak and Park (Chapter 7), Wenstrup (Chapter 8), and O'Neill (Chapter 9).

Information from the auditory nerve is sent to three different divisions of the cochlea nucleus (CN), which then project in parallel to different subsets of nuclei in the brainstem, some ipsilaterally, most contralaterally. In many cases, from the superior olivary nuclei centrally, inputs converge from the two ears. These may both be excitatory (EE), but more often one is excitatory, the other inhibitory (EI).

A separation of time and intensity information is seen initially in the medial superior olivary nuclei (MSO) and lateral superior olivary nuclei (LSO). Responses at the MSO tend to be broadly tuned and either sustained or phasic with only one or two spikes to the onset of a sound. In most mammals, the MSO receives approximately equal excitatory input from both cochlea nuclei. The relative timing of these binaural inputs determines the activity of different populations, signaling target azimuth. In contrast, the MSO of bats receives predominantly contralateral input, with analysis of binaural timing apparently shifted to higher neural levels (Covey, Vater, and Casseday 1991). Responses in the LSO, in bats as in other mammals, emphasize frequency and intensity information. Neurons have V-shaped tuning curves much as at the CN, and respond with sustained firing that reflects a balance of excitatory input from the ipsilateral CN and inhibitory input for the contralateral CN.

The nuclei of the lateral lemniscus are complex and much hypertrophied in bats, and clearly are an important part of the temporal analysis pathway. Parts receive binaural input, others are monaural. Of particular interest are neurons in the dorsal nucleus (DNLL) that show strong facilitation to the second of two identical sounds at a specific delay. These "best delay" (BD) neurons are all selective for intervals within the range needed in echolocation (Covey 1993; see also Grinnell 1963b; Suga and Schlegel 1973; Feng, Simmons, and Kick 1978). These neurons probably are responsible for the finding — in the first study of auditory neurophysiology in bats, using the evoked potential (N_4) representing input to the inferior colliculus — that there is strong facilitation of response to the second of a pair of sounds (Grinnell 1963b). The rapidity of full recovery of N_4 (2–3 msec) and exaggeration of recovery at ascending neural levels from auditory nerve to lateral lemniscus is characteristic only of echolocating bats (again, with the exception of cetaceans [Bullock et al. 1968]).

The most prominent of the nuclei of the lateral lemniscus is the columnar division of the ventral nucleus ($VNLL_c$). This nucleus is especially conspicuous in echolocating bats, consisting of several densely packed, highly organized columns of monaurally driven small neurons separated by fiber tracts. These neurons differ from others in the auditory system to this point in that they show broad tuning, presumably by convergence of inputs from cells with widely separated BFs, and they are highly phasic in response, usually firing just one spike at the onset of a sound. The dominant input is from cochlear nucleus axons that form large calyx-like endings on the cell body. Like other cells that have been recorded in the inferior colliculus (Bodenhamer, Pollak, and Marsh 1979), the response latency of these cells is remarkably invariant for any given cell, independent of changes in frequency or intensity (Covey, Vater, and Casseday 1991). Thus they are ideal for detecting wavefront arrival time.

Other parts of the nuclei of the lateral lemniscus encode time in another way, showing sustained, nonadapting responses firing throughout the duration of a signal. These are robust responders that can follow rapid frequency and amplitude modulations and can be viewed as intensity/duration encoders. In Chapter 6, Covey and Casseday help sort out all the complexities of these brainstem nuclei.

4.3 Inferior Colliculus

Information coursing through the brainstem takes many alternative pathways, but all converge at the inferior colliculus. In echolocating bats, the inferior colliculus (IC) is a relatively huge nucleus, readily accessible near the dorsal surface of the skull in most species and thus intensively studied. These studies, and our current understanding of the organization and function of the IC, are summarized in Chapter 7 by Pollak and Park.

The central nucleus of the IC (IC_c) receives input from at least 10 lower auditory nuclei as well as descending input from higher centers. This includes excitatory input from the ipsilateral CN, MSO, intermediate nucleus of the lateral lemniscus (INLL), and contralateral LSO; inhibitory input comes from other parts of the nucleus of the lateral lemniscus and from the ipsilateral LSO. While some neurons receive monaural excitatory input, most are either excited by both ears or excited by inputs from one ear and inhibited by the opposite side. In addition, there are many inhibitory interneurons within the colliculus and large tracts promoting interaction between the colliculi. The result of all this convergence and synaptic interaction is a conspicuous transformation of information.

In the frequency–intensity domain, one of the conspicuous transformations is the sharpening of single-unit tuning curves. Instead of the mostly V-shaped or broad tuning curves of lower centers, a large percentage of collicular neurons have narrower tuning curves with nearly vertical cutoffs on either side of the BF. This has the important consequence that these cells respond at approximately the same time independent of the intensity of the sound, for example, a narrow band within an FM sweep. Moreover, many neurons have "closed" tuning curves, that is, they respond well to faint sounds (e.g., echoes), but are inhibited by high-intensity sounds (e.g., emitted sounds), even at their BFs.

The IC of long CF/FM bats, like all other auditory nuclei, reflects the existence of an acoustic fovea near the bat's CF_2 (Pollak and Bodenhamer 1981). In *Pteronotus*, about a third of the neurons in IC are tuned to 60–62 kHz. As at more peripheral levels, the tuning is exquisitely sharp, with Q_{10} dB values reaching well over 300. Indeed, the tuning of these units is so sharp that many of them are responsive to a Doppler-shifted echo while being almost totally unresponsive to the slightly lower emitted CF_2. These neurons (and their equivalents at the cortex) are also highly sensitive to the slight frequency and amplitude modulations introduced into echoes by the wingbeats of insects (Schuller 1979; Bodenhamer and Pollak 1983). Even frequency modulations as small as 10 Hz on a carrier of 63 kHz evoke clear responses (see Pollak and Park, Chapter 7, Fig. 7.23).

Because it is so large, the slab of IC dealing with CF_2 signals in CF/FM bats has proved useful for discerning the organization of inputs within a single isofrequency contour. Populations of cells receiving monaural input, binaural excitatory input (EE), and binaural excitatory/inhibitory (EI) input are arranged topographically, each receiving a different subset of inputs from brain stem nuclei. Some of the inhibitory inputs are mediated by the neurotransmitter γ-aminobutyric acid (GABA), others by glycine. As in other bats, a large fraction of the EI-driven neurons show inhibition of response at high intensities and even "closed" tuning curves. The EI neurons in *Pteronotus*, and probably in FM bats as well, show a further important form of organization. They are excited via the contralateral ear, inhibited by sounds to the ipsilateral ear, and highly sensitive to interaural intensity differences. Pollak and his colleagues have shown that there is a systematic topographic distribution of the relative effectiveness of inputs from the two sides: at one extreme, the ipsilateral input needed to be 25 dB louder than the contralateral input to effect a 50% reduction in response; at the other extreme, the ipsilateral input caused equivalent inhibition even when it was 15 dB fainter (Wenstrup, Fuzessery, and Pollak 1986). This suggests that the neurons are arranged in an orderly sequence with respect to the angle of incidence of a sound at which ipsilateral inhibition becomes effective.

The IC also marks an important transformation in the processing of temporal information. About one-third of the neurons respond with sustained activity throughout a signal; the rest are strongly phasic. The latter normally respond to an effective stimulus with just one spike. The minimum latency of that response is about 5 msec, but in most cells it is considerably longer. In fact, the response latencies for different IC neurons cover a much wider range than at any earlier nucleus: from 5 to more than 30 msec, compared with a range of 3–6 msec at the lateral lemniscus (Casseday and Covey 1992, Park and Pollak 1993). Response latencies are also topographically ordered. Most of the wide spread of latencies occurs near the dorsal surface of the IC. With increasing depth, the minimum latency and the range of latency both decrease. The mean latency is about 12–17 msec, depending on the intensity and depth of recording.

Unlike many of the neurons of the nucleus of the lateral lemniscus, which show striking specialization for rapid recovery and even facilitation of response to the second of two signals, IC neurons tend to recover relatively slowly. However, the information about echoes must be getting beyond the IC. The explanation is thought to lie in those neurons that respond with latencies so long that echoes from objects up to several meters distance return before the response to the initial sound. In these cases, if the excitatory input from an echo can reach a population of target cells at the same time as the delayed response from these long-delay IC neurons, the coincidence might be strongly excitatory (Park and Pollak 1993; Sullivan 1982b; Jen and Schlegel 1982; Dear, Simmons, and Fritz 1993; and see Pollak and Park, Chapter 7, Fig. 7.46). The specific subpopulation of long-delay IC neurons that provides an input coincident with that from a particular echo potentially allows the nervous system to determine the echo delay. We shall see later that this is in fact the case in the medial geniculate. Park and Pollak (1993) have shown that the wide range of latencies in these collicular neurons is caused, at least in part, by GABA-mediated inhibitory synapses, because locally applied bicuculline greatly shortens the response latency. It is possible that cortical feedback also contributes to the long latency responses.

4.4 Medial Geniculate Body (MGB)

The medial geniculate body is a composite of several structurally and functionally distinct divisions, each a part of different parallel pathways to the cortex. It is the least thoroughly studied of the principal auditory nuclei in bats. Indeed, it has mainly been studied in the CF/FM bat *Pteronotus* with the motivation of understanding the origin of the beautifully organized combination-sensitive neurons of the auditory cortex (see following).

The MGB in bats receives its input mainly from the central nucleus of the ipsilateral IC (in the cat, there is a strong contralateral input as well). While some parts of the MGB, especially the ventral division, are tonotopically organized with sharply tuned neurons sensitive to sound direction, the main interest has been in the populations of neurons in the dorsal and medial divisions that do not show clear tonotopic organization but are instead sensitive to combinations of inputs at different frequencies. As Wenstrup (Wenstrup, Chapter 8) points out, these neurons, which tend to display long latencies and broad tuning, are the first to show strongly facilitated responses to combinations of first and second or first and third harmonic spectral components. This is a remarkable response specificity. Where most investigators expected to find specialized populations of neurons responding to the same narrow band of frequencies in the emitted sound and echo at intervals down to a fraction of a millisecond (which may be the case in FM bats, but see following), the long CF/FM bats — or *Pteronotus* at least — are doing something much more clever. At their CF_2, which provides critical information for target detection and discrimination, they employ a receiver so sharply tuned that there is little response to the emitted signal, but high sensitivity, from cochlear to cortical levels, to Doppler-shifted echoes. At the MGB (and even more so at the auditory cortex), however, such sharp tuning is not enough. Many of these sharply tuned neurons will not respond to the Doppler-shifted CF_2 component unless they receive input at the CF_1 at the same time. The CF_1 component of the emitted sound is relatively faint, but to the bat emitting it, it is adequate to excite input to the MGB combination-sensitive neurons. The combination of CF_1 and CF_2 elicits response up to 50 fold greater than that to either harmonic alone.

Of the neurons sensitive to CF combinations, about 70% are specific to CF_1/CF_2 and 30% to CF_1/CF_3. Interestingly, some of these appear specialized for echolocation when the bat is at rest. They are sharply tuned close to the resting emitted CF_1, but are excited only when the CF_2 (or CF_3) component is delayed with respect to the CF_1; they are strongly inhibited when the two begin simultaneously. Others are specialized for echolocation in flight. They are excited by CF_1 only when it is shifted downward to compensate for Doppler shift (Olsen and Suga 1991a).

There are also large numbers of neurons selective for different FM–FM combinations, almost equal numbers preferring FM_1–FM_2 and FM_1–FM_3, and even a few selected for FM_1–FM_4. These FM_1–FM_n populations are beautifully adapted for encoding target distance. They respond better when the FM_1 signal is louder than the FM_n (echo) component, and each is maximally facilitated within a narrow range of FM_n delays. The best delays range from 0 to 23 msec in *Pteronotus*, with most between 1 and 10 msec, corresponding to the distances of greatest importance in echolocation (Olsen and Suga 1991b; and see Wenstrup, Chapter 8, Fig. 8.13).

The CF_1/CF_n and FM_1–FM_n combination-sensitive neurons are segregated in the MGB, but a finer degree of organization is not clear. In the cortex, on the other hand, the CF/CF neurons are topographically arranged according to amount of Doppler frequency shift and FM/FM neurons are arranged by best delay (see following). Wenstrup (Chapter 8) discusses in detail the circuitry and pharmacology of the combination-sensitive pathways.

4.5 Auditory Cortex

Studies of the auditory cortex of echolocating bats have yielded the clearest pictures of how these animals extract the necessary information from echoes. Four species have been studied in detail: two FM bats (*Myotis lucifugus* and *Eptesicus fuscus*) and two CF/FM bats (*Pteronotus parnellii* and *Rhinolophus ferrumequinum*). The organizational principles for these two groups are so different that it is not easy to generalize about the bat cortex. O'Neill describes this work in Chapter 9.

4.5.1 Long CF/FM Bats

Of particular importance has been the series of elegant studies by Suga and his colleagues of the cortex of the CF/FM mustached bat, *Pteronotus*. The spectacular organization of the auditory cortex in this species (see O'Neill, Chapter 9, Fig. 9.2) illustrates well the advantage of working with an animal for which one knows the signals used and the information needed.

As in all mammals, a major portion of the *Pteronotus* auditory cortex (the AI equivalent) is arranged in tonotopical order. However, within this sequence, the narrow band of frequencies just above the emitted CF_2 (61–63 kHz) is even more overrepresented than at lower auditory centers. It comes to occupy fully one-third of the tonotopic portion of the cortex. In the large "foveal" portion of the tonotopic map dealing with Doppler-shifted, constant-frequency signals (the DSCF area), neurons are extremely sharply tuned, even at high intensity, and neurons with BFs from 61 to 63 kHz are further organized tonotopically in concentric rings. Those tuned to 61 kHz are at the center with an orderly progression to 63 kHz at the periphery. Superimposed on this concentric high-resolution frequency map is a map of "best amplitudes" (at which the response is maximal), arranged radially in an orderly sequence from about 10 dB to 90 dB. Finally, the populations preferring low-amplitude signals tend to receive EE input from both ears; those preferring louder sounds receive EI input and are better adapted for target localization. It is clear from this organization that CF_2 components of echoes will excite different populations in the DSCF area depending on the amplitude, the relative velocity of bat and target, and the direction of the echo.

Lying outside the main tonotopic area are at least seven secondary fields that are not organized tonotopically. In these secondary fields, responses are fascinatingly adapted for various features of echolocation. Neurons respond well only to combinations of sounds, similar to the combination neurons described for the MGB but more dependent on the combination of inputs. One of these nontonotopic fields is an area selective for combinations of CF_1 with CF_2 or, in an immediately adjacent strip, with CF_3. Within these strips, there is an orderly sequence of preference for small to large Doppler shifts of the CF_2 and CF_3 components.

Just dorsal to this area are a series of cortical strips selective for combinations of FM_1 with FM_2, FM_3, or FM_4 (see O'Neill, Chapter 9, Fig. 9.2). These cells are arranged, from one end to the other, according to the most effective delay of the second signal with respect to FM_1; that is, the cells are "delay tuned" (see O'Neill, Chapter 9, Figs. 9.6, 9.7, 9.10). The best delays range from 0.4 to 18 msec, covering most of the range of echo delays of importance to the bat (Suga, O'Neill, and Manabe 1978). These fields, thus, map echo delay or target distance. Several other small outlying regions of the auditory cortex also code echo delay, mostly over restricted parts of the full range of echo delays. The segregation of function between these various secondary nuclei is not clear in all cases, but the ability of the nervous system to construct computational maps of auditory space is evident and has reoriented much of the research on cortical organization in other animals as well.

The other principal long CF/FM bat, *Rhinolophus*, has been studied less extensively but shows qualitatively similar organization and combination-sensitive stimulus requirements. The organization is not so distinct, however, and facilitation by harmonically related combinations of signals is not as clear-cut. These differences are described in detail in Chapter 9 by O'Neill.

4.5.2 Cortical Organization in FM Bats

The auditory cortex of FM bats also has both tonotopically and nontono-topically organized regions, but they are quite different than in CF/FM bats. The tonotopic map covers the range of frequencies heard, with overrepresentation of the frequencies in the FM sweep, but there is no equivalent of the DSCF region. Most cells are more sensitive to an FM sweep than to any pure tone falling within the sweep, and between 10% and 20% respond only to FM sweeps. A much smaller fraction respond only to CF signals. These CF-specific neurons are segregated at one end of the tonotopic map and are tuned to frequencies at the low end of the FM sweep.

Many of the FM-sensitive cells are delay tuned; some have short best delays and fixed response latencies but are strongly facilitated by echoes in a narrow range of delays, and others have long best delays and response latencies that are time locked to the echo (O'Neill and Suga 1982; Sullivan 1982a, b; Dear et al. 1993). Moreover, in contrast to the requirement for different harmonic combinations seen in CF/FM bats, facilitation of response in echo-sensitive units requires instead a pair of FM sounds differing in amplitude (Sullivan 1982a). In *Eptesicus*, best delays extend to more than 30 msec, with cells preferring long delays typically found in the nontonotopic areas while short delays are preferred in the tonotopic region. How delay tuning can be achieved has already been discussed in the context of responses in the IC and MGB.

In general, the models depend on coincidence of inputs via two different pathways with different delay lines (or different amounts of inhibition). One has a high-amplitude signal pathway with long delay lines (or prolonged inhibition), the other a low-amplitude signal pathway with relatively short delay lines (or little or no inhibition). Interestingly, the convergent inputs that elicit the strongest response to pairs of FM sounds are not necessarily to the same frequency component of the two FM signals. Usually, in *Myotis*, best facilitation occurred when the pulse was a frequency about 8 kHz higher than the echo (Berkowitz and Suga 1989). Perhaps this is one step in the direction of the harmonic combination specificity of CF/FM bat neurons. Dear, Simmons, and Fritz (1993) postulated that populations of neurons like these in the IC and cortex, which fire with latencies up to 30 msec, allow the nervous system to process information from targets at all distances simultaneously, helping build a true acoustic image with each pulse.

Another important feature of many cortical neurons in FM bats is that they are tuned to two different frequency ranges, typically harmonically related in the relationship 1:3, with the lower frequency near the low end of the FM sweep (Dear et al. 1993). This may be exactly the property needed for analysis of the spectral peaks and notches generated by overlapping echo "glints" that are postulated to produce the information necessary for bats to resolve the dimensions of targets.

Unfortunately, despite beautiful studies of the directional sensitivity of neurons at cortical and lower auditory levels, especially by Pollak and his colleagues, the mechanisms of target localization in three-dimensional space are not clear.

4.6 Summary and Comments

Field observations and behavioral experiments with echolocating bats document remarkable, sometimes seemingly impossible, skills at detecting, localizing, and discriminating the nature of targets by echoes of emitted sounds. Using either CF/FM or FM sounds in ways adapted to different echolocation conditions, and using different mechanisms of information processing, bats behave as if they can construct a full acoustic image of nearby objects in space by the echoes of each pulse. Ingenious behavioral experiments have led to a number of interesting and controversial models of how they might be obtaining the necessary information.

Neurophysiological experiments with both CF/FM-and FM-emitting species have revealed neural organization and functional specializations associated with each echolocation strategy. Research continues to be driven to a large extent by behavioral experiments. The major emphasis at present is on mechanisms of preserving and analyzing microsecond and submicrosecond differences in arrival time of echoes from different targets and glints from a given target. There is much interest in the possibility that small time differences become expanded at the level of the colliculus and higher, but how are these differences encoded to that point, and how might they be magnified?

The auditory nervous system is extremely complex. Often there seems a surfeit of possible pathways or mechanisms to help explain a given behavioral ability. Based on response properties, one can make educated guesses about the function of different cell populations, such as different divisions of the auditory cortex. To test these guesses, however, it is important to attempt to alter behavior in predictable (and reversible) ways by localized injections of pharmacological agonists or blocking agents, much as Suga and his colleagues (Riquimaroux, Gaioni, and Suga 1992) have done to verify the role of the DSCF area for frequency discrimination around CF_2 in the mustached bat.

The ultimate aim of those studying echolocation is to understand how a bat can form the equivalent of a visual image with the multiple echoes of a single emitted sound, and do the same for each successive sound. This requires independently registering the directions (as well as distances) of all objects returning echoes. (There is a real need for behavioral experiments that begin to test the accuracy of information that can be obtained simultaneously about multiple targets.) It is difficult enough to understand the basis for accurate range discrimination and localization of a single target in the vertical and horizontal directions. To be able to do this simultaneously for many different targets at different directions, and at the same time analyze the fine structure of each echo to identify the dimensions and surface features of the reflecting object, and then put together successive images to determine the rate and direction of movement of each target as well—these abilities frankly boggle the mind.

Several aspects of hearing in bats have been largely neglected to date and deserve much more attention. To what extent is echolocation hard wired, for example, and what role does experience play in the perfection of echolocation abilities? How do neurophysiological adaptations develop ontogenetically? The little work that has been done on development, especially in CF/FM bats, shows that there are coordinated changes in the frequencies of maximal hearing sensitivity and of CF vocalization (Grinnell and Hagiwara 1972a; Rübsamen 1987; Rübsamen, Neuweiler, and Marimuthu 1989; Rübsamen and Schäfer 1990a). Is this primarily the result of growth, or can vocalizations be greatly altered to compensate for changes in hearing capabilities? Can neural plasticity, during development or in mature animals, compensate for abnormalities such as experimental removal of a tragus, change in pinna shape, or partial deafening (see, e.g., Rübsamen and Schäfer 1990b)? Also important are the questions: In what ways do the emitted pulse, or the motor commands for pulse emission, play roles in "priming" the auditory system for echo analysis (see Metzner 1993)? How is information obtained in the sensory pathways translated into flight maneuvers? How is spatial memory for a home territory encoded?

There is also a general need for more comparative research. Much of what we understand about echolocation is the result of natural experiments: evolution of different pulse design and adaptation of different species to different habitats and hunting strategies. Only a small number of species have been studied in the lab or in the field. Study of natural echolocation behavior of many more species is sure to reveal new and important adaptations. These will demand neural correlates and eventually help us not only to understand echolocation behavior, but to gain additional valuable insights into hearing mechanisms in man.

Finally, many bats are gregarious, highly social animals, with most interactions occurring in the dark. Almost certainly the auditory brains that have been revealing such remarkable adaptations for analysis of echolocation sounds are also highly adapted for complex acoustic communication (see Fenton 1985 for a good review of the literature). Study of the wide repertoire of communication sounds, their uses, and their analysis is likely to yield rich rewards.

This seems a long list of unknowns. It is, and it is just a beginning. However, no one would dispute that enormous progress has been made in explaining the remarkable behavior skills of echolocating bats at the neurophysiological level, and that bats constitute a striking example of the adaptability of the mammalian brain.

Acknowledgements It is a privilege to have been asked to write an overview chapter on this exciting subject. Most of the data and ideas are, of course, those of my colleagues in the field, but I would like especially to acknowledge the many happy hours spent discussing bats and echolocation with Drs. D.R. Griffin, H.-U. Schnitzler, J. A. Simmons, R. Roverud, N. Suga, G. Pollak, and G. Neuweiler. I also thank Dr. S.A. Kick for valuable help in refining the manuscript.

References

Berkowitz A, Suga N (1989) Neural mechanisms of ranging are different in two species of bats. Hear Res 41:255–264.

Bishop AL, Henson OW Jr (1987) The efferent cochlear projections of the superior olivary complex in the mustached bat. Hear Res 31:175–182.

Bodenhamer RD, Pollak GD (1983) Response characteristics of single units in the inferior colliculus of mustache bats to sinusoidally frequency modulated signals. J Comp Physiol 153:67–79.

Bodenhamer RD, Pollak GD, Marsh D S (1979) Coding of fine frequency information by echoranging neurons in the inferior colliculus of the Mexican free-tailed bat. Brain Res 171:530–535.

Bullock TH, Grinnell AD, Ikezono E, Kameda K, Katsuki Y, Nomoto M, Sato O, Suga N, Yanagisawa K (1968) Electrophysiological studies of central auditory mechanisms in cetaceans. Z Vgl Physiol 59:117–156.

Bruns V (1976a) Peripheral auditory tuning for fine frequency analysis by the CF-FM bat, *Rhinolophus ferrumequinum*. I. Mechanical specializations of the cochlea. J Comp Physiol 106:77–86.

Bruns V (1976b) Peripheral auditory tuning for fine frequency analysis by the CF-FM bat, *Rhinolophus ferrumequinum*. II, Frequency mapping in the cochlea. J Comp Physiol 106:87–97.

Bruns V, Schmieszek E (1980) Cochlear innervation in the greater horseshoe bat: demonstration of an acoustic fovea. Hear Res 3:27–43.

Bruns V, Burda H, Ryan MJ (1989) Ear morphology of the frog-eating bat (Trachops cirrhosus, Family: Phyllostomidae): apparent specializations for low-frequency hearing. J Morphol 199:103–118.

Casseday JH, Covey E (1992) Frequency tuning properties of neurons in the inferior colliculus of an FM bat. J Comp Neurol 319:34–50.

Covey E (1993) Response properties of single units in the dorsal nucleus of the lateral lemniscus and paralemniscal zone of an echolocating bat. J Neurophysiol 69:842–859.

Covey E, Vater M, Casseday JH (1991) Binaural properties of single units in the superior olivary complex of the mustached bat. J Neurophysiol 66:1080–1093.

Dear SP, Simmons JA, Fritz J (1993) A possible neuronal basis for representation of acoustic scenes in auditory cortex of the big brown bat. Nature 364:620–623.

Dear SP, Fritz J, Haresign T, Ferragamo M, Simmons JA (1993) Tonotopic and functional organization in the auditory cortex of the big brown bat, Eptesicus fuscus. J Neurophysiol 70:1988–2009.

Dijkgraaf S (1946) Die Sinneswelt der Fledermäuse. Experientia (Basel) 2:438–448.

Echteler SM, Fay RR, Popper AN (1994) Structure of the mammalian cochlea. In: Fay RR, Popper AN (eds) Comparative Hearing: Mammals. New York, Springer-Verlag, pp. 134–171.

Emde GVD, Menne D (1989) Discrimination of insect wingbeat-frequencies by the bat Rhinolophus ferrumequinum. J Comp Physiol A 164:663–671.

Feng AS, Simmons JA, Kick SA (1978) Echo-detection and target ranging neurons in the auditory system of the bat, Eptesicus fuscus. Science 202:645–648.

Fenton MB (1985) Communication in the Chiroptera. Bloomington: Indiana University Press.

Fuzessery ZM, Pollak GD (1984) Neural mechanisms of sound localization in an echolocating bat. Science 225:725–728.

Galambos R (1942a) The avoidance of obstacles by bats: Spallanzani's ideas (1794) and later theories. Isis 34:132–140.

Galambos R (1942b) Cochlear potentials elicited from bats by supersonic sounds. J Acoust Soc Am 14:41–49.

Goldman LJ, Henson OW (1977) Prey recognition and selection by the constant frequency bat, *Pteronotus p. parnellii*. Behav Ecol Sociobiol 2:411–419.

Griffin DR (1958) Listening in the Dark. New Haven: Yale University Press.

Grinnell AD (1963a) The neurophysiology of audition in bats: intensity and frequency parameters. J Physiol 167:38–66.

Grinnell AD (1963b) The neurophysiology of audition bats: temporal parameters. J Physiol 167:67–96.

Grinnell AD, Grinnell VS (1965) Neural correlates of vertical localization by echolocating bats. J Physiol 181:830–851.

Grinnell AD, Hagiwara S (1972a) Adaptations of the auditory nervous system for echolocation. Studies of New Guinea bats. Z Vgl Physiol 76:41–81.

Grinnell AD, Hagiwara S (1972b) Studies of auditory neurophysiology in nonecholocating bats, and adaptations for echolocation in one genus, *Rousettus*. Z Vgl Physiol 76:82–96.

Hartley DJ (1992) Stabilization of perceived echo amplitudes in echolocating bats. I. Echo detection and automatic gain control in the big brown bat, *Eptesicus fuscus*, and the fishing bat, *Noctilio leporinus*. J Acoust Soc Am 91:1120–1132.

Hartley DJ, Suthers RA (1989) The sound emission pattern of the echolocating bat, *Eptesicus fuscus*. J Acoust Soc Am 85:1348–1351.

Hanson MM (1978) The basilar membrane of the bat *Pteronotus p. parnellii*. Am J Anat 153:143–159.

Henson MM, Jenkins DB, Henson OW Jr (1982) The cells of Boettcher in the bat, *Pteribitus parnellii*. 7:91–103.

Henson OW Jr (1965) The activity and function of the middle ear muscles in echolocating bats. J Physiol Lond 180:871–887.

Henson OW Jr, Henson MM (1972) Middle ear muscle contractions and their relations to pulse and echo-evoked potentials in the bat, *Chilonycteris parnellii*. In: AIBS-NATO Symposium on Animal Orientation Navigation, Wallops Station, VA, pp. 355–363. Galler, S.R, Schmidt-Koenig, K., Jacobs, G.J., & Belleville, R.E (eds) Publ: Scientific and Technical Information Office, NASA, Washington, D.C.

Jen PH-S, Schlegel PA (1982) Auditory physiological properties of the neurons in the inferior colliculus of the big brown bat, *Eptesicus fuscus*. J Comp Physiol A 147:351–363.

Jepsen GL (1970) Bat origins and evolution. In: Wimsatt WA (ed) Biology of Bats, Vol. 1. New York: Academic Press, pp. 1–64.

Lawrence BD, Simmons JA (1982) Echolocation in bats: the external ear and perception of the vertical positions of targets. Science 218:481–483.

Long GR, Schnitzler H-U (1975) Behavioral audiograms from the bat, *Rhinolophus ferrumequinum*. J. Comp. Physiol. 100:211–219.

Masters WM. Jacobs SC (1989) Target detection and range resolution by the big brown bat (*Eptesicus fuscus*) using normal and time-reversed model echoes. J Comp Physiol A 166:65–73.

Masters WM, Moffat AJM, Simmons JA (1985) Sonar tracking of horizontally moving targets by the big brown bat, *Eptesicus fuscus*. Science 228:1331–1333.

Menne D, Kaipf, I, Wagner J, Ostwald J, Schnitzler, H-U, (1989) Range estimation by echolocation in the bat *Eptesicus fuscus*: trading of phase versus time cues. J Acoust Soc Am 85:2642–2650.

Metzner W (1993) An audio-vocal interface in echolocating horseshoe bats. J Neurosci 13:1899–1915.

Middlebrooks JC (1992) Narrow-band sound localization related to external ear acoustics. J Acoust Soc Am 92:2607–2624.

Mogdans J, Schnitzler H-U, Ostwald J (1993) Discrimination of 2-wavefront echoes by the big brown bat, *Eptesicus fuscus* — Behavioral experiments and receiver simulations. J Comp Physiol A 172:309-323.

Møhl B (1986) Detection by a pipistrellus bat of normal and reversed replica of its sonar pulses. Acoustica 61:75-82.

Moss CF, Schnitzler H-U (1989) Accuracy of target ranging in echolocating bats: Acoustic information processing. J Comp Physiol A 165:383-393.

Neuweiler G, Bruns V, Schuller G (1980) Ears adapted for the detection of motion, or how echolocating bats have exploited the capacities of the mammalian auditory system. J Acoust Soc Am 68:741-753.

Olsen JF, Suga N (1991a) Combination sensitive neurons in the medial geniculate body of the mustached bat: encoding of relative velocity information. J Neurophysiol 65:1254-1274.

Olsen JF, Suga N (1991b) Combination sensitive neurons in the medial geniculate body of the mustached bat: encoding of target range information. J Neurophysiol 65:1275-1296.

O'Neill WE, Suga N (1982) Encoding of target range and its representation in the auditory cortex of the mustached bat. J Neurosci 2:17-31.

Park TJ, Pollak GD (1993) GABA shapes a topographic organization of response latency in the mustache bat's inferior colliculus. J Neurosci 13:5172-5187.

Pollak GD (1993) Some comments on the proposed perception of phase and nanosecond time disparities by echolocating bats. J Comp Physiol A 172:523-531.

Pollak GD, Bodenhamer RD (1981) Specialized characteristics of single units in the inferior colliculus of mustache bats: frequency representation, tuning and discharge patterns. J Neurophysiol 46:605-620.

Pollak GD, Henson OW Jr, Novick A (1972) Cochlear microphonic audiograms in the 'pure tone' bat, *Chilonycteris parnellii parnellii*. Science 176:66-68.

Ramprashad F, Money KE, Landolt JP, Laufer J (1978) A neuroanatomical study of the little brown bat (*Myotis lucifugus*). J Comp Neurol 178:347-363.

Riquimaroux H, Gaioni SJ, Suga N (1992) Inactivation of the DSCF area of the auditory cortex with muscimol disrupts frequency discrimination in the mustached bat. J Neurophysiol 68:1613-1623.

Roverud RC (1993) Neural computations for sound pattern recognition: evidence for summation of an array of frequency filters in an echolocating bat. J Neurosci 13:2306-2312.

Roverud RC, Grinnell AD (1985a) Discrimination performance and echolocation signal integration requirements for target detection and distance determination in the CF/FM bat, *Noctilio albiventris*. J Comp Physiol 156:447-456.

Roverud RC, Grinnell AD (1985b) Echolocation sound features processed to provide distance information in the CF/FM bat, *Noctilio albiventris*: evidence for a gated time window utilizing both CF and FM components. J Comp Physiol 156:457-469.

Roverud RC, Nitsche V, Neuweiler G (1991) Discrimination of wingbeat motion by bats correlated with echolocation sound pattern. J Comp Physiol A 168:259-263.

Rübsamen R (1987) Ontogenesis of the echolocation system in the rufous horseshoe bat, *Rhinolophus rouxi*, (Audition and vocalization in early postnatal development.) J Comp Physiol A 161:899-913.

Rübsamen R, Schäfer M (1990a) Ontogenesis of auditory fovea representation in the inferior colliculus of the Sri Lakan rufous horseshoe bat, *Rhinolophus rouxi*. J Comp Physiol A 167:757-769.

Rübsamen R, Schäfer M (1990b) Audiovocal interactions during development? Vocalisation in deafened young horseshoe bats vs. audition in vocalisation-impaired bats. J Comp Physiol A 167:771–784.

Rübsamen R, Neuweiler G, Marimuthu G (1989) Ontogenesis of tonotopy in inferior colliculus of a hipposiderid bat reveals postnatal shift in frequency-place code. J Comp Physiol A 165:755–769.

Schuller G (1979) Coding of small sinusoidal frequency and amplitude modulations in the inferior colliculus of the CF-FM bat, *Rhinolophus ferrumequinum*. Exp Brain Res 34:117–132.

Schuller G, Pollak GD (1979) Disproportionate frequency representation in the inferior colliculus of horseshoe bats: evidence for an "acoustic fovea" J Comp Physiol 132:47–54.

Schuller G, Beuter K, Schnitzler H-U (1974) Response to frequency-shifted artificial echoes in the bat, *Rhinolophus ferrumequinum*. J Comp Physiol 89:275–286.

Schnitzler H-U (1970) Comparison of echolocation behavior in *Rhinolophus ferrumequinum* and *Chilonycteris rubiginosa*. Bijdr Dierkd 40:77–80.

Schnitzler H-U, Flieger E (1983) Detection of oscillating target movements by echolocation in the greater horseshoe bat. J Comp Physiol 153:385–391.

Simmons JA (1979) Perception of echo phase in bat sonar. Science 204:1336–1338.

Simmons JA (1987) Acoustic images of target range in bat sonar. Naval Res Rev 39:11–26.

Simmons JA (1993) Evidence for perception of fine echo delay and phase by the FM bat, *Eptesicus fuscus*. J Comp Physiol A 172:533–547.

Simmons JA, Chen L (1989) The acoustic basis for target discrimination by FM echolocating bats. J Acoust Soc Am 86:1333–1350.

Simmons JA, Kick SA (1984) Physiological mechanisms for spatial filtering and image enhancement in the sonar of bats. Annu Rev Physiol 46:599–614.

Simmons JA, Moss CF, Ferragamo M (1990a) Convergence of temporal and spectral information into acoustic images of complex sonar targets perceived by the echolocation bat, *Eptesicus fuscus*. J Comp Physiol A 166:449–470.

Simmons JA, Ferragamo M, Moss CF, Stevenson SB. Altes RA (1990b) Discrimination of jittered sonar echoes by the echolocating bat, *Eptesicus fuscus*: the shape of target images in echolocation. J Comp Physiol A 167:589–616.

Simmons JA, Kick AS, Moffat AJM, Masters WM, Kon D (1988) Clutter interference along the target range axis in the echolocating bat, *Eptesicus fuscus*. J Acoust Soc Am 84:551–559.

Simmons JA, Kick SA, Lawrence BD, Hale C, Bard C, Escudie' B (1983) Acuity of horizontal angle discrimination by the echolocating bat, *Eptesicus fuscus*. J Comp Physiol 153:321–330.

Suga N (1964) Recovery cycles and responses to frequency-modulated tone pulses in auditory neurons of echolocating bats. J Physiol 175:50–80.

Suga N (1973) Feature extraction in the auditory system of bats. In: Møller AR (ed) Academic Press, pp. 675–744. New in auditory neurons of echolocating bats. J Physiol 175:50–80.

Suga N, Schlegel P (1973) Coding and processing in the auditory systems of FM-signal producing bats. J Acoust Soc Am 54:174–190.

Suga N, Shimozawa T (1974) Site of neural attenuation of responses to self-vocalized sounds in echolocating bats. Science 183:1221–1213.

Suga N, Jen PH-S (1977) Further studies on the peripheral auditory system of

'CF-FM' bats specialized for fine frequency analysis of Doppler-shifted echoes. J Exp Biol 69:207–232.

Suga N, Simmons JA, Jen PH-S (1975) Peripheral specialization for fine analysis of Doppler shifted echoes in the auditory system of the CF-FM bat *Pteronotus parnellii*. J Exp Biol 63:161–192.

Suga N, Neuweiler G, Möller J (1976) Peripheral auditory tuning for fine frequency analysis by the CF-FM bat, *Rhinolophus ferrumequinum*. IV. Properties of peripheral auditory neurons. J Comp Physiol 106:111–125.

Suga N, O'Neill WE, Manabe T (1978) Cortical neurons sensitive to combinations of information-bearing elements of biosonar signals in the mustached bat. Science 200:778–781.

Sullivan WE (1982a) Neural representation of target distance in auditory cortex of the echolocating bat *Myotis lucifugus*. J Neurophysiol 48:1011–1032.

Sullivan WE (1982b) Possible neural mechanisms of target distance coding in auditory system of the echolocating bat, Myotis lucifugus. J Neurophysiol 48:1033–1047.

Sum YW, Menne D (1988) Discrimination of fluttering targets by the FM bat *Pipistrellus stenopterus*? J Comp Physiol A 163:349–354.

Webster FA (1967) Performance of echolocating bats in the presence of interference. In: Busnel R-G (ed) Animal Sonar Systems: Biology and Bionics. Jouy-en-Josas-78, France: Laboratoire de Physiologie Acoustique, pp. 673–713.

Webster FA, Brazier O G (1965) Experimental studies on target detection, evaluation, and interception by echolocating bats. TDR No. AMRL-TR-65-172, Aerospace Medical Division USAF Systems Command, Tucson, AZ.

Wenstrup JJ, Fuzessery ZM, Pollak GD (1986) Binaural response organization within a frequency-band representation of the inferior colliculus: implications for sound localization. J Neurosci 6:692–973.

Wenstrup JJ, Fuzessery ZM, Pollak GD (1988) Binaural neurons in the mustache bat's inferior colliculus: responses of 60-kHz EI units to dichotic sound stimulation. J Neurophysiol 60:1369–1383.

Zook JM, Leake PA (1989) Connections and frequency representation in the auditory brainstem of the mustache bat, *Pteronotus parnellii*. J Comp Neurol 290:243–261.

2

Natural History and Biosonar Signals

M. BROCK FENTON

1. Introduction

The diversity of bats is reflected in many aspects of their appearance and behavior. There are approximately 900 species of extant bats with most occurring in the tropics and subtropics. From continental settings to islands, including remote oceanic ones such as Hawaii, bats often are prominent members of the mammal fauna. In almost any country in subsaharan Africa, for example, bats are the most diverse group of mammals, with more species than rodents (Smithers 1983). Whether one considers bats by diet or roosting habits, they present an impressive array of approaches to living. This diversity is clearly reflected in their echolocation behaviors.

Echolocation is an active form of orientation that permits animals to be independent of lighting conditions. It is polyphyletic in vertebrates, having evolved in both birds and mammals (classes Aves and Mammalia, respectively; Fenton 1984). Furthermore, echolocation has appeared more than once in both of these classes, in the avian orders Caprimulgiformes and Apodiformes and in the mammalian orders Insectivora, Cetacea, and Chiroptera (Fenton 1984).

Although echolocation often is assumed for all bats, not all bats echolocate, and not all echolocating bats use the same echolocation signals or use echolocation for the same purpose. Differences in echolocation behavior can raise basic questions about the evolution of bats, manifested by arguments about their classification. Traditionally, bats are placed in the mammalian order Chiroptera with living species arranged in two suborders, the Megachiroptera and the Microchiroptera (Table 2.1). While some families of bats include living and fossil representatives (Megachiroptera or Microchiroptera), one is known only as fossils (Table 2.1). The Eocene and Oligocene fossils are typically different genera or species from living taxa.

TABLE 2.1. The classification, fossil, worldwide, and dietary diversity of bats.

Suborder/superfamily, family	First fossils — Geological age	First fossils — Years ($\times 10^6$)	Worldwide distribution	Variation in diet	Number of living species
Palaeochiropterygoidea					
Palaeochiropterygidae (the "old bats")	Eocene	50	Europe and North America	Insects	0
MEGACHIROPTERA					
Pteropodidae (the Old World fruit bats; flying foxes)	Oligocene	30	European fossils; today, Old World tropics	Fruit, nectar, and pollen	150
MICROCHIROPTERA					
Emballonuroidea					
Rhinopomatidae (the rat-tailed bats)	Not known	–	North Africa to Southern Asia and Borneo	Insects	2
Craseonycteridae(hog-nosed bats)	Not known	–	Thailand	Insects	1
Emballonuridae (sheath-tailed bats)	Eocene	50	Tropics	Insects	44
Rhinolophoidea					
Megadermatidae (false vampire bats)	Eocene	50	European fossils; today, Old World tropics	Animals, from insects to vertebrates	5
Nycteridae (slit-faced bats)	Not known	–	Africa to Java and Sumatra	Animals, from insects to vertebrates	13
Rhinolophidae (horseshoe bats)	Eocene	50	Old World	Insects	69
Hipposideridae (Old World leaf-nose bats	Eocene	50	Old World tropics	Insects	56
Phyllostomoidea					
Noctilionidae (bulldog bats)	Not known	–	New World tropics	Insects and fish	2
Mormoopidae (mustached bats)	Not known	–	New World tropics	Insects	8
Mystacinidae (short-tailed bats)	Not known	–	New Zealand	Insects, fruit, nectar, carrion	1
Phyllostomidae (New World leaf-nose bats)	mid-Miocene	22	New World tropics	Insects, fruit and pollen, vertebrate blood	123

	Geological age of first fossils	Millions of years	Geographic distribution	Diet	Number of species
Vespertilionoidea					
Natalidae (funnel-eared bats)	Not known	—	New World tropics	Insects	4
Furipteridae (thumbless bats)	Not known	—	New World tropics	Insects	2
Thyropteridae (New World sucker-footed bats)	Not known	—	New World tropics	Insects	2
Vespertilionidae (plain-nosed bats)	Eocene	50	Worldwide	Insects, fish, and other vertebrates	283
Myzopodidae (Old World sucker-footed bats)	Not known	—	Madagascar	Insects	1
Molossidae (freetailed bats)	Not known	—	Tropical	Insects	82

The family names of bats are shown bold print. The geological name of the time of the first fossils in the family are shown with the approximate are in millions of years.

1.1 Phylogeny, Evolution, and Classification

Classifying bats in one order, the Chiroptera, implies that they are monophyletic, with all living and fossil bats sharing an immediate common ancestor. This position infers that the Megachiroptera and the Microchiroptera are more closely related to one another than either is to any other group of mammals. The alternative view, that bats are diphyletic, has been proposed by various workers during the past 100 years with the specific suggestion that the Megachiroptera and the order Primates share an immediate common ancestor, while the Microchiroptera are more closely related to mammals in the order Insectivora, the shrews and their allies (e.g., Pettigrew et al. 1989).

In one sense, the question of bat phylogeny revolves around the relative importance of similarities versus differences. Although the Megachiroptera share many characteristics with the Microchiroptera, the two groups also differ substantially. Although the wings of bats are generally similar (Thewissen and Babcock 1991), the Megachiroptera and Microchiroptera have quite different cervical vertebrae, with those of megachiropterans resembling the typical mammalian condition. The Microchiroptera, however, have extremely flexible necks and their cervical vertebrae differ strikingly from those of other mammals (Fenton and Crerar 1984). Similarly, the visual pathways of megachiropteran bats are more like those of primates, while those of microchiropterans resemble other mammals (Pettigrew 1991). Several recent papers have explored the similarities and differences between the two suborders in search of the "correct" interpretation (e.g., Pettigrew et al. 1989; Baker, Novacek, and Simmons 1991; Pettigrew 1991; Simmons, Novacek, and Baker 1991) and two schools of thought remain with no clear prevailing view or agreement about "truth" (e.g., Rayner 1991; Thewissen and Babcock 1991, 1992). A recent study of the structure of the epsilon-globin gene supports the view that bats are monophyletic (Bailey, Slightom, and Goodman 1992).

The study of echolocation is affected by the debate about the phylogeny of bats because most of the 150 or so species of Megachiroptera do not echolocate (Hill and Smith 1984; Fenton 1992). Furthermore, the echolocating megachiropteran, *Rousettus aegyptiacus* (the Appendix provides more information about each species of bat mentioned in this chapter), uses a completely different approach to signal production than do species in the Microchiroptera. While *R. aegyptiacus* makes echolocation sounds by clicking its tongue, the microchiropterans studied to date all use sounds produced in the larynx (vocalizations) as echolocation signals (Suthers 1988). The monophyletic position poses two alternatives: (1) that echolocation is an ancestral trait (e.g., Fenton 1974) that has been lost in the Megachiroptera and reacquired in one species; or (2) that echolocation is not an ancestral trait, but has evolved independently in both suborders. The diphyletic position necessitates the independent evolution of echolocation in the two major groups of bats.

The fossil record does not help to resolve the question of bat phylogeny. The first fossil bats occur in Eocene strata, a time when at least eight families were represented. Although there are no fossils of animals that appear to be intermediate between bats and something else, most mammalogists suspect that bats developed from a small, arboreal, insectivorous mammal with long arms and fingers. Flight is presumed to have evolved when the proto-bat glided from one tree to another. The "top-down" theory proposes that the ancestor of bats captured insects along the trunks and branches of trees, working its way up to the top before gliding down to begin again on another tree (Norberg 1990). In a variation on the top-down theory, Jepsen (1970) proposed that the ancestor of bats used membranes between the fingers and body as insect nets so the incipient wing membranes simultaneously may have served at least two functions. The diversity of bats in the Eocene (see Table 2.1) makes it easy to presume that the group has a long history, perhaps originating in the early Palaeocene or late Cretaceous.

In spite of their early appearance and diversity in the fossil record, there are relatively few fossil bats. One currently accepted phylogeny of the Chiroptera (Fenton 1992) reflects the presumed relationships among the living families and superfamilies. Together with the data in Table 2.1, the phylogeny (Fig. 2.1) reflects the diversification of living bats in terms of both diet and geographic distribution. Although the general relationships among the families and superfamilies appear clear, as noted there is still no general agreement about whether the Megachiroptera are the closest living relatives of the Microchiroptera.

There are other questions about relationships among the Chiroptera. In Figure 2.1, the New Zealand Mystacinidae are grouped with New World bats that constitute the superfamily Phyllostomoidea, reflecting a number of morphological and molecular traits (Pierson et al. 1986). Traditionally, the Mystacinidae have been placed in the Vespertilionoidea (e.g., Koopman and Jones 1970). In this way, the arrangement of living Microchiroptera in different families changes with the appearance of new evidence or the reanalysis of older data.

Other points of contention in bat classification concern the numbers of families, with some biologists arguing for fewer families and others for more. In the past, the three living species of vampire bats have been treated as a distinct family, the Desmodontidae, largely because of their specializations for feeding on blood. In their basic structure, however, including skeletal features and sperm morphology, genetics, and biochemical features, the vampires closely resemble bats in the family Phyllostomidae, and most biologists now place them there as a subfamily (Desmodontinae; Koopman 1988). The opposite trend involves some Old World bats. While some workers combine the Hipposideridae and the Rhinolophidae into one family (Rhinolophidae; e.g., Vaughan 1986), others stress the differences between these bats and treat them as distinct families (e.g., Fenton 1992). As recently as 1974 a new family of living bats was described (Hill 1974), suggesting that we may not yet appreciate the full diversity of bats.

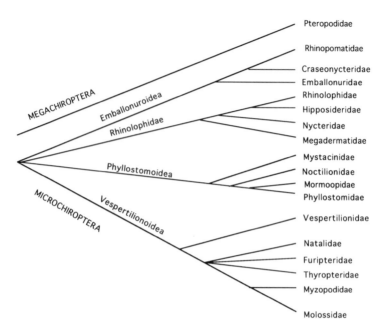

FIGURE 2.1. In this phylogeny of the families of living bats, the suborders Megachiroptera and the Microchiroptera do not share a common ancestor. Furthermore, the Mystacinidae are included among the superfamily Phyllostomoidea rather than with the Vespertilionoidea. The arrangement may or may not reflect reality (after Fenton 1992), and recent molecular evidence (Bailey, Slightom, and Goodman 1992) suggests that both suborders share an immediate common ancestor.

At another level, changes in the classification of bats affect the names by which they are known in the literature. For the student of echolocation it is vital to realize that *Chilonycteris parnellii* is now called *Pteronotus parnellii*, following Smith's (1972) revision of the family Mormoopidae. In some cases the morphological diversity of bats may have been underestimated by earlier classifications. For example, the tribe Plecotini constitutes a distinct group of vespertilionid bats, characterized by enormous ears and, presumably, distinctive foraging behavior. Although there have been various assessments of the systematic status of the bats in this group, the most recent study suggests that the North American forms are distinct from those in Eurasia (Tumlison and Douglas 1992) and should be placed in separate genera (*Corynorhinus*, and *Idionycteris* versus *Plecotus*). This situation also may be reflected by differences in echolocation and foraging behavior (e.g., Simmons and O'Farrell 1977; Anderson and Racey 1991).

1.2 The Diversity of Bats

There are some obvious generalizations about megachiropteran and microchiropteran bats (Hill and Smith 1984; Fenton 1992). Megachiropteran bats occur in the Old World tropics and feed mainly on fruit or nectar and pollen. Their teeth are highly specialized for these diets, and they depend upon flight only to get from one place to another. Their pectoral girdles are relatively unspecialized. The available data suggest that the echolocating species of Megachiroptera uses this form of orientation to find its way in the darkness of its roosts, which are usually located in dark hollows, often in caves.

Microchiropteran bats, by comparison, occur everywhere there are bats. They fill different trophic roles, feeding on a variety of animals, fruit, nectar and pollen, and blood. With the exception of blood feeders, the teeth of Microchiroptera are relatively unspecialized. Many species, however, use flight when feeding and their pectoral girdles are often highly specialized, reflecting aerial agility and maneuverability. Many microchiropterans use echolocation to detect and evaluate targets they treat as prey (see Schnitzler, Simmons, and O'Neill, Chapters 3, 4, and 9, respectively). In various aspects of the brain, reproductive systems, cervical vertebrae, and other features, the Megachiroptera differ from the Microchiroptera, but in general appearance, particularly in wing structure, they are very similar (Hill and Smith 1984; Thewissen and Babcock 1992).

Bats range in size from adult body masses of 2 g to 1500 g, with most species weighing less than 50 g. Length of forearm, a general indication of body size in bats, ranges from about 25 mm to about 230 mm, corresponding to wingspans of about 15 cm and 200 cm, respectively. While most species of bats (see Table 2.1) feed mainly on insects, others eat animals ranging from frogs to fish, from scorpions to millipedes, and from other bats to birds. Larger species tend to take larger animal prey, including small vertebrates, in their diets (Norberg and Fenton 1988). In the Megachiroptera (Family Pteropodidae), most species are thought to feed mainly on fruit, but a few are specialized for nectar and pollen (Hill and Smith 1984; Fenton 1992). This dependence on plant material as food is paralleled in the Neotropics by various subfamilies within the microchiropteran Family Phyllostomidae, the New World leaf-nosed bats. Other phyllostomids feed on animals, and three species, the vampires, depend entirely on the blood of other vertebrates. Fish-eating is known from microchiropteran bats in at least four families (Noctilionidae, Megadermatidae, Nycteridae, and Vespertilionidae), but details of fishing behavior have been studied only in the Noctilionidae (Suthers 1965; Hill and Smith 1984; Fenton 1992).

While some species of bats, perhaps most of them, live relatively solitary lives and roost in foliage, others congregate in huge numbers in roosts such as caves and some artificial structures. A few species live in small social units, often harems (e.g., Bradbury 1977), which appear to be well organized and involve a high level of individual recognition of group members. This organization is most evident in common vampire bats (*Desmodus rotundus*), among whom individuals will regurgitate blood for roost mates that are unsuccessful foragers (Wilkinson 1985). Little is known about the social lives of most of the 900 or so species of bats, but the available data suggest diversity, generally reflecting that shown in dietary habits and morphological structures (Hill and Smith 1984; Fenton 1992).

1.3 Ontogeny

In bats studied to date, as in many other mammals, hearing develops after birth even though neonates begin to vocalize almost immediately (Brown 1976; Brown and Grinnell 1980; Rubsamen, Neuweiler, and Marimuthu 1989; Rubsamen and Schafer 1990a). In some species, echolocation calls appear in the first 10 days or so, coinciding with the development of hearing (Brown 1976). This pattern prevails both in bats such as *Antrozous pallidus*, with typically mammalian patterns of hearing sensitivity (audiograms; Brown 1976), and in species such as *Hipposideros speoris* and *Rhinolophus rouxi*, which have specialized patterns of hearing coinciding with a strikingly different approach to echolocation (see Section 2.2; and Simmons; Kössl and Vater; Casseday and Covey; Pollak and Parks; Wenstrup; and O'Neill and Winer, Chapters 4–10, respectively) apparently specialized for detecting fluttering targets (Rubsamen, Neuweiler, and Marimuthu 1989; Rubsamen and Schafer 1990a). In *Rhinolphus rouxi* (Rubsamen, Neuweiler and Marimuthu 1989) there are developmental shifts in the cochlear frequency-place code associated with specialized hearing, and in *Hipposideros speoris* the development of hearing specialization cannot be explained by a single developmental parameter (Rubsamen, Neuweiler, and Marimuthu 1989). By deafening young animals, Rubsamen and Schafer (1990b) demonstrated that in *Rhinolophus rouxi* echolocation pulse production is under auditory feedback control. The emergence and development of the specialized hearing curve (the so-called auditory fovea) is an innate process that occurs through shifts in frequency tuning.

The development of vocalizations and hearing suggests that in a variety of bats (e.g., *Antrozous pallidus* [Brown 1976], *Myotis lucifugus* [Buchler 1980], *Rhinolophus rouxi* [Rubsamen, Neuweiler, and Marimuthu 1989], and *Hipposideros speoris* [Rubsamen and Schafer 1990a]), young animals can echolocate at least a week to 10 days before they start to fly. The clumsiness of bats' first flights could reflect the mechanical challenges of learning to fly and the difficulties of using echolocation during flight (e.g., Buchler 1980).

2. Biosonar Signals of Bats

Abbreviations are commonly used when discussing bats and their approaches to echolocation, and the literature abounds with terms such as CF (constant-frequency) and FM (frequency-modulated), describing different bat signals (e.g., Busnel and Fish 1980; Hill and Smith 1984; Nachtigall and Moore 1988). In the same volumes, and in other literature, these modifiers often are extended to the bats themselves, giving us CF bats, CF-FM bats, and FM bats (Fig. 2.2). While FM clearly describes a signal (or component) showing frequency modulation over time, CF is more ambiguous because it is used for signals with bandwidths of less than 1 kHz as well as signals with bandwidths of several kilohertz. As noted elsewhere (e.g., Neuweiler and Fenton 1988), it would be more prudent to recognize that the shorthand describes the signals and their components, but not necessarily the bats or their use of them.

This point is particularly important in the context of a bat's ability to use echolocation to detect a fluttering target, usually an insect flapping its wings. Traditionally, flutter detection has been associated with bats that produce CF signals most of the time, "high duty cycle" species (Fig. 2.3) that actually produce signals at least 50% and usually 80% or more of the time they are echolocating (e.g., Neuweiler and Fenton 1988). The high duty cycle bats are horseshoe bats (Rhinolophidae), Old World leaf-nosed bats (Hipposideridae), and the mormoopid *Pteronotus parnellii*. Other microchiropterans are "low duty cycle" species that produce calls less than 20% of the time they are echolocating.

Flutter-detection behavior is not limited to high duty cycle bats using narrowband signals. While Roverud, Nitsche, and Neuweiler (1991) showed that two CF species were better at flutter detection than one low duty cycle FM species (*Rhinolophus rouxi* and *Hipposideros lankadiva*, versus *Eptesicus fuscus*), Summe and Menne (1988) found that the FM bat *Pipistrellus stenopterus* used its echolocation to detect fluttering targets. Furthermore, Casseday and Covey (1991) found "filter" neurons in the inferior colliculus of *Eptsicus fuscus* that were as sharply tuned as those in some high duty cycle CF bats (see Covey and Casseday, Chapter 6).

The overlap between species and intraspecific flexibility in echolocation calls parallels the situation in wing morphology (e.g., Norberg and Rayner 1987), communities (e.g., Aldridge and Rautenbach 1987), or general morphology (e.g., Findley and Wilson 1982), and together these suggest broad overlap between species.

The following sections document the ways that bats vary different aspects of their echolocation signals, demonstrating the danger of generalizing broadly about different approaches to echolocation. Evidence supports two basic approaches, one based on signal production and the other on duty cycle. Beyond these fundamental differences in signal production, abbreviations used to describe signals and their components are most valuable when the broader picture of variation is clearly presented.

The Chiroptera studied to date appear to depend mainly on two kinds of signals for echolocation, the tongue clicks of the echolocating Megachiroptera (the pteropodid *Rousettus aegyptiacus*) and the vocalizations of the

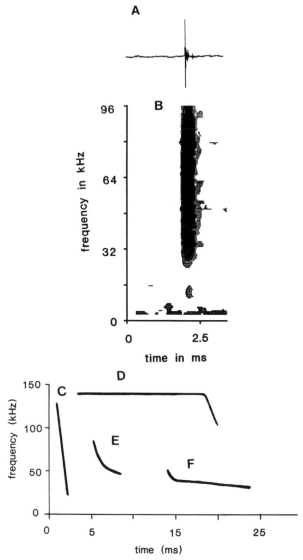

FIGURE 2.2A–F. While the pteropodid *Rousettus aegyptiacus* produces tongue clicks as echolocation calls (A, time-amplitude plot; B, sonogram), the microchiropteran bats use tonal calls that show structured change in frequency over time (C–F). Search-phase bat echolocation calls, include those that would be considered frequency modulated (FM; C, E, F) and constant frequency (CF) ending in an FM sweep (D), reflect low and high duty cycles, respectively. The short, broadband call (C) is typically of low to medium intensity and common among species of *Myotis* that take prey from surfaces (= glean; e.g., *Myotis auriculus* [Fenton and Bell 1979]; *Myotis emarginatus* [Schumm, Krull, and Neuweiler 1991]). The long call dominated by a narrowband (CF) component (D)

FIGURE 2.3A,B. Two time-amplitude plots compare a sequence of echolocation calls produced at low duty cycle (A: 13.6%, *Lasiurus borealis*) and high duty cycle (B: 56.8%; *Rhinolophus landeri*).

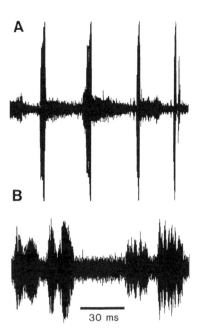

Microchiroptera (Suthers 1988). Echolocating Microchiroptera use a variety of call designs (see Fig. 2.2) to collect information about their surroundings, coinciding with different approaches to echolocation. Bats also produce other vocalizations, apparently for communication with other bats. Many putative social communication calls are much longer in duration than echolocation calls (Fenton 1985), and this, combined with the speed of sound in air, atmospheric attenuation of high-frequency sound, and the flight speeds of bats, probably precludes the use of longer social calls in echolocation. Quite simply, long-duration signals from a flying bat could mask their echoes rebounding from nearby targets. A possible exception is provided by the long, low-frequency calls given by Hawaiian hoary bats (*Lasiurus cinereus*) during intraspecific chases at feeding grounds (Belwood and Fullard 1984). A comparison of some features of bat echolocation calls is presented in Table 2.2.

←

FIGURE 2.2A–F. (*continued*) is typical of high duty cycle bats, rhinolophids, hipposiderids, and the mormoopid *Pteronotus parnellii*, with different species showing different CF frequencies (e.g., Heller and von Helverson 1989). One call (E) combines narrowband and broadband components and is typical of many species that hunt airborne targets in along the edges of woodlands, for example, while the longer call dominated by a narrowband component (F) is commonly associated with species that hunt in more open settings.

TABLE 2.2. The sizes and echolocation features of extant bats.

Suborder/superfamily	Size FA (nm)	Echolocate (yes/no)	Signal type	Duty cycle %	Intensity	Duration (msec)	Bandwidth
MEGACHIROPTERA							
Pteropodidae	40–230	No	–	–	–	–	–
Rousettus aegyptiacus		Yes	Click	≤20	Low	1–2	Broad
MICROCHIROPTERA							
Emballonuroidea							
Rhinopomatidae	55–70	Yes	Vocal	≤20	High	10	Narrow
Craseonycteridae	22–26	Yes	Vocal	≤20	High	5	Broad
Emballonuridae	32–80	Yes	Vocal	≤20	High	10–20	Narrow
Rhinolophoidea							
Megadermatidae	50–115	Yes	Vocal	≤20	Low	1	Broad
Nycteridae	36–60	Yes	Vocal	≤20	Low	1	Broad
Rhinolophidae	30–75	Yes	Vocal	≥80	High	>20	Narrow
Hipposideridae	30–110	Yes	Vocal	≥80	High	10	Narrow
Phyllostomoidea							
Noctilionidae	70–92	Yes	Vocal	≤20	High	10	Broad
Mormoopidae	35–65	Yes	Vocal	≤20	High	5–10	Broad
Pteronotus parnellii		Yes	Vocal	≥80	High	10	Narrow
Mystacinidae	44–48						
Phyllostomidae	30–105	Yes	Vocal	≤20	Low	1–4	Broad
Vespertilionoidea							
Natalidae	27–41	Yes	Vocal	≤20	Low	1–5	Broad
Furipteridae	30–40	Yes	Vocal	≤20	?	?	?
Thyropteridae	27–38	Yes	Vocal	≤20	?	?	?
Vespertilionidae	22–75	Yes	Vocal	≤20	High	5	Broad
		Yes	Vocal	≤20	High	10–20	Narrow
		Yes	Vocal	≤20	Medium	1–5	Broad
Myzopodidae	44–48	(Yes)	(Vocal)	(≤20)	?	?	?
Molossidae	27–85	Yes	Vocal	≤20	High	10–20	Broad

Size is shown as forearm (FA) length in millimeters; signal type is either tongue click (click) or vocalization (vocal).

2.1 Intensity

It has been known for some time that the echolocation calls of bats vary in intensity (measured 10 cm in front of the bat) from low (<60 dB SPL [sound pressure level]) to high (>110 dB SPL) (Griffin 1958). Among echolocating bats there is an obvious gradient in call intensity that includes many species producing calls of intermediate strength. The situations in which bats operate appear to influence call intensity. For example, bats searching for airborne targets produce intense echolocation calls, while those searching for prey on surfaces depend more on quieter calls. Call intensity is consistent in some families; for example, the Phyllostomidae use low-intensity calls and the Molossidae high-intensity calls (Table 2.2). In other families, such as the Vespertilionidae, some species produce high-intensity calls and others low-intensity ones. At least one species, the vespertilionid *Myotis emarginatus*, adjusts the intensity of its call according to the situation in which it is hunting (Schuum, Krull, and Neuweiler 1991).

Differences in call intensity pose important challenges to people studying the echolocation of bats. Species using high-intensity calls are easily monitored with microphones (bat detectors) sensitive to the frequencies in the calls, while those using low-intensity calls are virtually undetectable even by very sensitive bat detectors (e.g., Fenton and Bell 1981; Fenton et al. 1992).

2.2 Durations, Duty Cycles, and Pulse Repetition Rates

Two fundamentally different approaches to microchiropteran echolocation are reflected in the duty cycles of their calls, or the percentage of time that signals are being produced. As noted earlier, most echolocating bats have duty cycles of less than 20% (Table 2.2; Fig. 2.3), and these species appear unable to tolerate overlap between pulse and echo (Schnitzler 1987). In the other approach, species in the families Hipposideridae and Rhinolophidae, and the mormoopid *Pteronotus parnellii* (Table 2.2), duty cycles regularly exceed 80% and these species can tolerate overlap between pulses and their echoes. The auditory systems of high duty cycle bats are highly specialized for exploiting Doppler-shifted echoes generated by fluttering targets (see Moss and Schnitzler, Simmons et al., Kössl and Vater, Covey and Casseday, Pollak and Park, Wenstrup, and O'Neill, Chapters 3–9, respectively).

Low duty cycle bats adjust the durations of individual calls according to the situations in which they are operating. During an attack on an airborne target, species such as *Lasiurus cinereus* shorten their calls from 20 msec to about 1 msec, as they detect, approach, and close in on a target (Barclay 1986; Obrist 1989). Shorter echolocation calls are essential in the final stages of the attack to ensure that the outgoing pulse does not mask the fainter returning echo. Other species, for example, *Myotis daubentoni*

(Kalko and Schnitzler 1989), use shorter echolocation calls (≤ 5 msec) when searching for targets and calls of 1 msec or less when closing with prey.

High and low duty cycle bats that hunt airborne targets increase their pulse repetition rates as individual calls shorten through an attack sequence. When searching for targets, low duty cycle bats produce calls at intervals ranging from 50 to 300 ms, while during attacks (feeding buzzes; Griffin, Webster, and Michael 1960), intercall intervals are closer to 5 msec. Even during feeding buzzes, the duty cycles of these bats rarely exceed 20% (Obrist 1989).

When low duty cycle bats take prey from surfaces (glean) they usually produce short (< 1-msec) echolocation calls (see Table 2.2) during their final approaches, but show no evidence of high pulse repetition rates (feeding buzzes) during actual attacks on targets (e.g., *Myotis auriculus* [Fenton and Bell 1979]; *Trachops cirrhosus* [Barclay et al. 1981]; *Nycteris grandis* and *Nycteris thebaica* [Fenton, Gaudet, and Leonard 1983]; *Myotis evotis* [Faure and Barclay 1992]). At least two species that take prey from surfaces also pursue airborne targets, situations in which they produce feeding buzzes (*Myotis emarginatus* [Schumm, Krull, and Neuweiler 1991]; *Antrozous pallidus* [Krull 1992]). This pattern of behavior probably also occurs in other gleaners.

The high duty cycle bats (rhinolophids, hipposiderids, and *P. parnellii*) produce calls that vary in duration from around 10 msec in hipposiderids to more than 50 msec in rhinolophids. During attacks on airborne targets, these bats shorten the durations of individual calls while maintaining high duty cycles (e.g., Griffin and Simmons 1974; Neuweiler et al. 1987).

2.3 Patterns of Frequency Change over Time

In these patterns, there are two basic kinds of echolocation pulses. The first are short, broadband clicks typical of *Rousettus aegyptiacus* (see Fig. 2.2b) and other echolocators such as birds, shrews, and cetaceans. The second are those of the Microchiroptera, tonal sounds showing structured changes in frequency over time (see Fig. 2.2). There is no evidence that broadband clicks offer any special advantage or disadvantage in echolocation relative to broadband tonal sounds (Buchler and Mitz 1980). It is possible that tonal echolocation sounds originated as communication signals (Fenton 1984).

An array of sonograms of bat echolocation calls (see Fig. 2.2) illustrates variations in patterns of frequency change over time as well as different patterns of bandwidth and duration. Short, broadband calls typically show sharp changes in frequency over time, while longer calls of narrower bandwidth have more gradual changes. While hunting, some species using short calls always show rapid changes in frequency over time (e.g., *Myotis daubentoni*, as noted previously), but others such as *Lasiurus cinereus* use rapid rates of frequency change mainly in feeding buzzes.

2.4 Variations in Echolocation Calls

The various combinations of intensity, duration, duty cycle, and patterns of frequency change over time (see Table 2.2) illustrate the diversity of echo-location call design in bats. Some workers have proposed using variation in echolocation calls to address questions about the evolution of bats (e.g., Simmons and Stein 1980). The recurring use of short, broadband signals or long, narrowband signals by species in several families (Table 2.2), how-ever, weakens the argument that echolocation call design reflects evolution and phylogeny.

Flexiblity is particularly evident in the case of *Myotis emarginatus* (Schumm, Krull, and Neuweiler 1991), which dramatically alters the intensity, durations and patterns of frequency change over time in its calls according to the situation in which it is hunting. When searching for prey on surfaces, these bats use short broadband signals of medium intensity but when the targets are airborne, they use long signals of high intensity that are dominated by narrowband components. There are hints of comparable behavior in other species (e.g., *Megaderma spasma* [Tyrrell 1988]; *Myotis septentrionalis* [Faure, Fullard, and Dawson 1993, and unpublished obser-vations]). Together, the data on flexibility in call design suggest a strong environmental component, with bats adjusting their call design to provide best operation in the specific settings where they are hunting. If this is true, then individuals hunting in the open may be expected to use different call design and pulse repetition rates (narrowband and low pulse repetition rates) than when they are hunting in cluttered situations (broadband and higher pulse repetition rates).

There is also a geographical component to variation in call design as illustrated by the search-phase echolocation calls of *Eptesicus fuscus*. In eastern North America, for example, around Toronto (Canada), *E. fuscus* typically produces five 10-msec-long broadband echolocation calls with en-ergy distributed across a broad bandwidth (Brigham, Cebek, and Hickey 1989). In south central British Columbia (Canada), however, this species produces ten 15-msec-long calls dominated by narrow bandwith components (Obrist 1989). Such variation may reflect more situation-specific behavior; the differences between hunting in open versus in cluttered habitats.

An initial indication of intraspecific variation in echolocation calls was provided by observations of the frequencies dominating the high duty cycle vocalizations of rhinolophid and hipposiderid bats (summarized in Novick 1977; see also Heller and von Helverson 1989). The frequencies of the narrowband (CF) calls of some rhinolphids and hipposiderids varied geo-graphically (e.g., Fenton 1986) as do those in lower duty cycle, broadband calls (e.g., Thomas, Bell, and Fenton 1987; Brigham, Cebek, and Hickey 1989).

More recently, Cebek (1992) has found statistical evidence supporting the preliminary observations of Brigham, Cebek, and Hickey (1989) that *Eptesicus fuscus* uses colony-specific echolocation calls. In both studies,

52 M. Brock Fenton

statistical analysis of different features of echolocation calls (frequency with maximum energy, lowest frequency) allowed a computer to correctly assign bats to some colonies on the basis of features of echolocation calls. J.S. Wilkinson (personal communication, February 1993) has similar data for *Nycticeius humeralis*. In *E. fuscus*, colony-specific echolocation calls do not coincide with genetic differences (Cebek 1992) and may reflect the communicative nature of echolocation calls (see Section 6.1).

Other studies have produced evidence of intraspecific variation in echolocation calls according to the setting in which bats were operating. In a rhinopomatid (Habersetzer 1981), some emballonurids (Barclay 1983), and at least one molossid (Zbinden 1989), bats changed the frequencies of their echolocation calls when flying with other (conspecific) bats. In a more detailed, longer-term study of marked individuals, Obrist (1989) observed that four species of vespertilionids might change the frequencies dominating their calls, call strengths, and pulse repetition rates when flying with other bats. Variations in calls in these settings could reflect jamming, avoidance behavior, or some communicative role for echolocation calls.

Although it is clear that bats vary their echolocation calls, these vocalizations often are significantly less variable in some parameters than their social calls (Fenton, 1994).

2.5 Species-Specific Echolocation Calls

In spite of a growing appreciation of the variations in the echolocation calls of microchiropteran bats just noted (e.g., Thomas, Bell, and Fenton 1987), many species produce quite distinctive echolocation calls (e.g., Ahlén 1981; Fenton and Bell 1981; Thomas and West 1989; Fullard et al. 1991). It is clear that in any location the value of identifying bat species by their echolocation calls will depend upon the local bat fauna and on the measurement instrumentation. For example, at many locations in southern Ontario (Canada), the echolocation calls of some of the eight local species of bats are readily distinguished from one another using a narrowband bat detector (Fig. 2.4). At 20 kHz, the long, chirplike signals of *Lasiurus cinereus* are distinctive, while similar sounds at 40 kHz are usually the calls of *Lasiurus borealis*. Tuned to 40 kHz, however, the same bat detector will pick up echolocation calls of *Eptesicus fuscus*, *Lasionycteris noctivagans*, *Pipistrellus subflavus*, and three species of *Myotis*, complicating the situation. As noted in the original papers about using echolocation calls to identify bats, however, monitoring the calls of individual bats marked with active tags can maximize the chances of correctly identifying the signaller (Bell 1980; Fenton and Bell 1981).

2.6 The Operational Range of Echolocation

Echolocation in air is a short range operation (Griffin 1958). In 1982, Kick reported the results of behavioral experiments which demonstrated that an

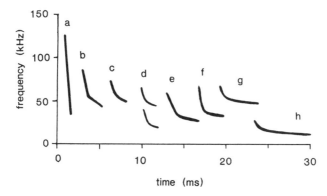

FIGURE 2.4. Eight species of bats that occur in southeastern Ontario produce different search-phase echolocation calls. Shown here are the calls of *Myotis septentrionalis* (*a*), *Myotis lucifugus* (*b*), *Myotis leibii* (*c*), *Pipistrellus subflavus* (*d*), *Eptesicus fuscus* (*e*), *Lasionycteris noctivagans* (*f*), *Lasiurus borealis* (*g*), and *Lasiurus cinereus* (*h*). The calls of *Pipistrellus subflavus* show a distinct second harmonic that is conspicuous when heard over a narrowband bat detector (strongest signals at 20 and 40 kHz). Most of these data are from Fenton and Bell (1979, 1981), Barclay (1986), and Obrist (1989), those for *P. subflavus* from MacDonald et al. (1994).

echolocating *Eptesicus fuscus* first detected a 19-mm-diameter sphere at a range of 5 m, the equivalent of 40 nosetip-to-tailtip lengths. These data confirmed that echolocation in air is a short-range operation, an impression remaining from observation of the distances at which bats reacted to targets and obstacles (e.g., Griffin 1958).

Interpulse intervals probably give an impression of the operational ranges of echolocating bats using low duty cycle calls. These bats are presumed to be intolerant of overlap between the outgoing pulse and the returning echo (Schnitzler 1987; see also Schnitzler, Chapter 3, and Simmons, Chapter 4). When interpulse intervals are considered with respect to the speed of sound in air, the low duty cycle echolocating bats could have operational ranges from 2.4 m to just over 62 m, corresponding to interpulse intervals of 20 msec and 365 msec for *Nycteris grandis* and *Euderma maculatum*, respectively [Fenton 1990]). Interpulse intervals, however, may provide no indication of the range at which bats are actively searching for prey.

3. Echolocation and the Hunting Strategies of Animal-Eating Bats

Microchiropteran bats may or may not use echolocation to detect, track, and assess targets. Species hunting airborne prey, usually insects, appear to depend more on echolocation than do species taking prey from surfaces such as the ground or foliage. Airborne targets are tracked by bats using low duty cycle, broadband or narrowband signals as well as high duty cycle calls dominated by narrowband components (see Table 2.2). While short, broadband, frequency-modulated sweeps provide bats with detailed information about target detail (e.g., Schmidt 1988), particularly at short range, other call designs may serve different functions. A convincing demonstration of this is Roverud's (1987) work showing that the narrowband component in the echolocation calls of *Noctilio albiventris* primes the auditory system for processing echoes by opening a window for analysis. In behavioral experiments, Roverud (1987) could effectively jam these bats' echolocation systems by presenting them with appropriate narrowband pulses of sound. Roverud (1987) suggested similar processes in *Rhinolophus ferrumequinum*.

There remains no clear indication of the function of the broadband FM sweeps that dominate the terminal parts of the echolocation calls of rhinolophid and hipposiderid bats and *Pteronotus parnellii* (e.g., Fig. 2.2b). Evidence that these call components provide details about the target is provided by some studies but not others. Support for this function comes from observations of *Hipposideros caffer* approaching (to within about 15 cm) but not attacking the fluttering target presented by a small electric motor with rotating flaps of masking tape (Bell and Fenton 1984). *Pteronotus parnellii*, however, attacks and grabs a similar fluttering target (Goldman and Henson 1977). While *H. caffer* clearly made an abort decision before contacting the target, *P. parnellii* only rejected the target after contact, suggesting differences in echolocation behavior.

3.1 Airborne Targets

Echolocating bats may hunt airborne targets while in flight or from a perch. While there have been many studies of the echolocation signals and behavior of bats that hunt in flight, less is known about the ones hunting from perches. Furthermore, the approach to foraging may change over time, with young animals hunting from perches and adults from flight (Buchler 1980). Several species switch from one approach to the other (e.g., Vaughan 1976; Fenton and Rautenbach 1986; Neuweiler et al. 1987), particularly when food is scarce (Fenton et al. 1990). Cost of hunting is one important difference between the two strategies, because flying bats consume energy at 7 to 25 times the rate of sitting bats (Thomas 1987). The two strategies also can result in bats encountering suitable prey at different rates.

As noted earlier, bats change the design and pattern of their echolocation calls as they detect, follow, and close with an airborne target, culminating in a feeding buzz (Fig. 2.5). Traditionally, the different stages in the process have been identified, from "search" through "approach" and "terminal" (e.g., Griffin, Webster, and Michael 1960; Barclay 1986; Kalko and Schnitzler 1989), reflecting times when bats are collecting different kinds of

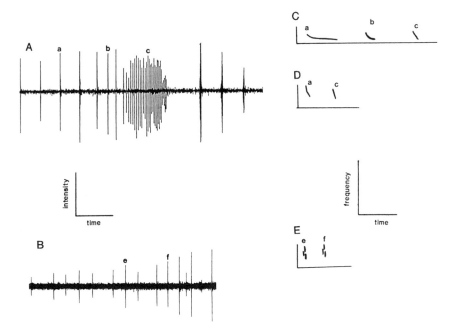

FIGURE 2.5A–E. Bats attacking airborne targets typically produce feeding buzzes (A) as they close with their prey (500-msec-long sequence), while species taking prey from surfaces (B) do not (500-msec-long sequence). While some species dramatically change the durations and patterns of frequency change over time of their calls over a feeding buzz (e.g., *Lasiurus cinereus*; C), others show less striking changes (e.g., *Myotis lucifugus*; D). Species such as *Macrotus waterhousii* show virtually no change in call durations or patterns of frequency change over time during an approach to a target on the ground (E). Calls indentified by lowercase letters in the attack sequences (A, B) correspond to sonograms with the same letters (C, D, E). *Lasiurus cinereus* calls are 15 msec long (*a*), 5 msec long (*b*), and 3 msec long (*c*); *Myotis lucifugus* calls are 3 msec (*a*) and 1 msec (*c*) long; and *Macrotus waterhousii* calls are 0.75 msec long (*e, f*).

information (detection versus tracking and evaluation). Kick and Simmons (1984) proposed "track" as a fourth stage between search–approach and terminal, a position not widely accepted by others (e.g., Kalko and Schnitzler 1989). The situation is complicated because different species show different patterns of behavior. For example, low duty cycle bats using long narrowband calls when searching for targets dramatically change both pulse repetition rates and call design during an attack sequence. However, low duty cycle bats, using shorter broadband signals and higher pulse repetition rates, make more gradual changes in call design and pulse repetition rate during an attack sequence (see Fig. 2.5). In some species, therefore, there are marked changes in call features and call rates over an attack sequence, while in others the changes in call features are less striking.

Apparent changes in echolocation calls over the course of an attack sequence may reflect differences in a bat's position relative to a microphone rather than actual changes in the calls. Photographs of flying *Myotis lucifugus* and *Lasiurus borealis* attacking targets (Griffin, Webster, and Michael 1960) demonstrated the drastic changes in body posture and head position associated with bats detecting and responding to airborne targets. We still lack details about the changes in head position made by foraging bats as they search for airborne targets, although Masters, Moffat, and Simmons (1985) documented changes in head position for *Eptesicus fuscus* using echolocation to track a target while sitting on a platform.

Flying bats actually catch airborne targets in different ways, either in the interfemoral membrane or on parts of the wing (Hill and Smith 1984). Work with captive *Rhinolophus ferrumequinum* revealed that some individuals invariably caught flour-coated mealworms in the wing pocket near their wrists (Trappe and Schnitzler 1982). The available data, much of it anecdotal and descriptive, leaves a strong impression that hunting bats show striking behavioral flexibility, particularly during the final pursuit. For example, *Lasiurus borealis* and *Lasiurus cinereus* dive to within 1 m of the ground in pursuit of a moth, and once a *L. cinereus* made an Immelman turn with a radius of about 1.5 m as it turned back to chase a stone that had been tossed into the air (Fenton, unpublished data). The bat began its feeding buzz within a second of making the turn and contacted the stone, which it then dropped. In other situations, high duty cycle bats have been observed taking prey from surfaces such as the ground, trunks of trees, or foliage (e.g., Bell and Fenton 1984; Link, Marimuthu, and Neuweiler 1986).

It is obvious that bats chasing airborne targets use echolocation to detect and track prey (Campbell and Suthers 1988). Bat attacks on stones and other small objects thrown into the air (e.g., *Lasiurus borealis* and *Lasiurus cinereus* [Hickey and Fenton 1990; Acharya and Fenton 1992]; *Myotis lucifugus* and *Myotis yumanensis* [R.M.R. Barclay and R.M. Brigham, personal communication, October 1992]) cast some doubt on how much prey evaluation is done by echolocation. Temporal and spectral cues are vital to these operations and may play important roles in each stage of the process. A large body of evidence demonstrates the value of temporal information to bats, partly through illustrations of their ability to finely measure time (e.g., Simmons, Saillant, and Dear 1992). Other work highlights the importance of spectral cues to echolocating bats (e.g., Schmidt 1988). More recently, descriptions of the interactions between attacking bats and their arctiid moth prey (Section 6.2.2) have shown that echolocating *Lasiurus borealis* may continue to use temporal and spectral information after they enter the feeding buzz stage of their attacks (Acharya and Fenton 1992).

Narrowband echolocation calls are presumed to provide a greater effective range than broadband calls because more energy is focused in a smaller range of frequencies. Signal theory suggests, however, that calls of this design provide fewer details about targets than signals of broader band-width (Simmons and Stein 1980). As noted previously, in low duty cycle bats interpulse interval may provide an indication of functional range, and concentrating of energy means that long, narrowband calls should con-tribute to greater operational range. Bats with low pulse repetition rates usually use long, narrowband calls, making it difficult to separate cause from effect in this situation.

Signal theory proposes that broadband calls provide better resolution of target detail (Simmons and Stein 1980), but atmospheric attenuation means that calls with a preponderance of higher frequency sounds have shorter functional ranges (Griffin 1971). Bats using short, broadband calls at low duty cycle typically use relatively high pulse repetition rates, presumably reflecting shorter operational ranges.

3.2 Long-Range and Short-Range Bats

Two distinct patterns of hunting behavior associated with narrowband and broadband calls are exemplified by *Lasiurus cinereus* and *Lasionycteris noctivagans* (Barclay 1986). Low pulse repetition rates and long narrow-band echolocation calls are typical of *Lasiurus cinereus* searching for prey. These bats typically make one attack per pass through a concentration of insects (for example, around a light) and appear to fix on targets at ranges of at least 5–10 m. Higher pulse repetition rates and shorter broadband echolocation calls are typical of *Lasionycteris noctivagans* searching for prey. These bats often make several attacks as they move through a concentration of insects and appear to react to targets at ranges of 5 m or less. There is a wealth of comparable data from other species of bats that have been studied in the field (reviewed in Neuweiler and Fenton 1988). This fundamental distinction has been recognized for some time, and Brosset (1966) spoke of insectivorous bats that react to targets at close range (hunting "aux bout de leur nez") and distinguished them from others that worked at longer distances.

A 1-cm-diameter stone tossed into the air where bats are hunting can provide a graphic demonstration of these two basic approaches to echolo-cation and foraging. A bat such as *L. cinereus,* using long, narrowband echolocation calls, will attack an airborne stone and actually capture it, while another like *L. noctivagans* will turn toward the stone, but turn away without making physical contact with it. Geographical variation in *Epte-sicus fuscus* behavior, however, emphasizes that there is a continuum between these two extremes. As noted in Section 2.4, this bat uses shorter broadband echolocation calls in eastern North America and longer narrow-band calls in the west. The difference in call structure appears to coincide with differences in behavior, supporting the long- versus short-range distinction noted earlier (Obrist 1989). A more striking difference is provided by the variation in echolocation calls and feeding behavior of *Myotis emarginatus* (Schumm, Krull, and Neuweiler 1991).

With further evidence of the behavioral flexibility associated with echolocation and foraging, it will becokme increasingly difficult to label most bats by their echolocation calls.

3.3 Nonairborne Targets

Depending upon the setting, a flying insect is a hard target on a soft background, meaning that it offers an echolocating bat a different perceptual challenge than an insect sitting on a surface, which is a hard target on a hard background. While some bats can use echolocation to find hard targets on hard backgrounds (Suthers 1965; Fiedler 1979; Bell 1982), and some bats use the same signals in either situation (Neuweiler 1989), more and more studies show that echolocating bats use prey-generated sounds or vision to find hard targets on hard backgrounds (e.g., Barclay et al. 1981; Fenton, Gaudet, and Leonard 1983; Ryan and Tuttle 1987; Faure and Barclay 1992). As noted in Section 2.2, low duty cycle bats attacking targets on surfaces usually do not produce feeding buzzes. High duty cycle, flutter-detecting bats readily use echolocation to find fluttering targets on surfaces (e.g., *Hipposideros caffer* [Bell and Fenton 1984]).

As in the situation with airborne targets, some bats hunt nonairborne targets from flight (e.g., *Trachops cirrhosus* [Tuttle and Ryan 1981]; *Myotis myotis* [Audet 1990]; *Antrozous pallidus* [Krull 1992]) and others hunt from perches (e.g., *Cardioderma cor* [Vaughan 1976]; *Nycteris grandis, N. thebgaica* [Fenton, Gaudet, and Leonard 1983; Aldridge et al. 1990]; *Cardioderma cor* [Ryan and Tuttle 1987]). At least some species alternate (e.g., *Nycteris* species [Aldridge et al. 1990; Fenton et al. 1990]) between the two foraging modes, perhaps because of the costs of flight (Thomas 1987).

Bats locate nonairborne targets using a variet of cues. When the target is moving, some bats clearly use echolocation to detect and track prey. *Noctilio leporinus*, for example, uses echolocation to detect and track prey moving across the surface of water (Campbell and Suthers 1988). In other situations, bats use other cues. Sounds of movement may be the most common, whether they are the footfalls of mice (*Megaderma lyra* [Fiedler 1979]), the rustling sounds of insects (*Nycteris grandis* or *Nycteris thebaica* [Fenton, Gaudet, and Leonard 1983]; *Macrotus californicus* [Bell 1985]; *Cardioderma cor* [Ryan and Tuttle 1987]; *Plecotus auritus* [Anderson and Racey 1991]), or the fluttering of wings (*Antrozous pallidus* [Bell 1982]; *Myotis evotis* [Faure and Barclay 1992]). Frog calls (*Trachops cirrhosus* [Tuttle and Ryan 1981]) and the calls of some orthopterans (phyllostomine bats [Tuttle, Ryan, and Belwood 1985; Belwood and Morris 1987]) also are important cues for some species, and at least one, *Macrotus californicus*, sometimes hunts by vision (Bell 1985).

Just how much these bats depend on echolocation to find and assess their prey remains unknown. Many of them that obviously take a lot of their prey from surfaces have large, conspicuous ears that effectively amplify the low-frequency sounds associated with movement (Obrist, Fenton, Eger, and Schlegel 1993). The echolocation calls of these species tend to be short (≤ 1 msec), broadband, and low in intensity. This combination of features means that the calls are barely detectable even by very sensitive bat detectors. It is difficult, therefore, to monitor the production of these echolocation calls, a factor contributing to our ignorance about the role echolocation plays in the lives of these bats. Some *Megaderma lyra* use echolocation to find frogs whose heads protrude from the water, but relatively few individual *Megaderma lyra* exhibit this behavior (G. Marimuthu, personal communication). When taking prey from surfaces, several bats are known to grab prey in their mouths (e.g., *Myotis auriculus* [Fenton and Bell 1979]; *Trachops cirrhosus* [Tuttle and Ryan 1981]; *Megaderma lyra* [Fiedler 1979]), while others use their wings to envelop their prey (*Nycteris grandis* and *Nycteris thebaica* [Fenton, Gaudet, and Leonard 1983]). When taking prey from surfaces, bats may use echolocation to assess the background from which they are taking their victims.

4. Echolocation and the Foraging of Bats That Eat Plant Material and Blood

The available evidence, which is not extensive, suggests that *Rousettus aegyptiacus* uses echolocation to gain access to dark roosting sites such as caves. We still have no clear indication of the role that echolocation plays in the lives of the phyllostomid bats that feed on fruit, nectar and pollen, or blood. These bats produce short (≤ 1 msec), broadband, low-intensity echolocation calls (e.g., *Carollia perspicillata* [Hartley and Suthers 1987]), but appear to rely heavily on other cues such as vision and olfaction for orientation and for finding and evaluating food. In the blood-feeding *Desmodus rotundus*, the inferior colliculus contains neurons tuned to the sounds associated with the heavy breathing of sleeping mammals (Schmidt et al. 1991); there are no clear indications of the use these bats make of echolocation. Obviously any of these species could use echolocation to gain access to dark roosts such as caves or tree hollows, but many of them roost in foliage, situations where echolocation should make little difference to roosting (e.g., stenodermine phyllostomids [Morrison 1980]).

Echolocation behavior may reflect a bat's dependence on different food material. Concentrations of protein, for example, are often very low in fruit and nectar, making it an important factor for bats that eat plant material (Thomas 1984). Howell (1974) found that phyllostomids more specialized for flower feeding obtained their protein from pollen, while less specialized taxa supplemented their diets with insects. She related the differences in echolocation acuity to the move from insects to pollen as a source of protein, presumably reflecting a decreased role for echolocation in finding and assessing food.

5. Echolocation and Foraging Ecology

It is well known, through studies of foraging behavior, that echolocation directly affects the ecology of animal-eating bats (e.g., Aldridge and Rautenbach 1987; Norberg and Rayner 1987). Lack of information about the role that echolocation plays in the lives of bats eating fruit, nectar and pollen, or blood precludes drawing conclusions about how echolocation affects their ecology.

By affecting a bat's ability to detect potential obstacles and food in its surroundings, echolocation influences patterns of habitat use. Bats using broadband signals at relatively high pulse repetition rates appear to collect detailed information about their surroundings, permitting them to work in more closed habitats. Bats using broadband signals at higher pulse repetition rates are better suited to dealing with clutter (echoes returning from objects other than the target of interest) than species using narrowband calls at lower pulse repetition rates (Simmons and Stein 1980). For working in thick vegetation such as the closed crowns of trees and shrubs, short, broadband calls and high pulse repetition rates are essential. Many bats working in these settings, however, depend on other cues (e.g., sounds of prey or sight) to detect and locate their prey. Intraspecific variability in echolocation behavior complicates defending broad generalizations about the association between echolocation call design and habitat use. Even so, many bat communities are composed of very different species. For example, *Lasiurus cinereus* uses long narrowband calls and low pulse repetition rates while *Myotis septentrionalis* usually relies on short broadband calls and high pulse repetition rates. Both species eat insects, but while *Lasiurus cinereus* seems to depend entirely on echolocation to find its prey, *Myotis septentrionalis* often uses prey-generated sounds to find its victims (see Fig. 2.4). Although different species like these are sympatric, they may forage in different settings (see also Neuweiler 1989).

Among high duty cycle echolocators in the Rhinolophidae and Hipposideridae, sympatric species use echolocation calls dominated by different narrowband frequencies (Heller and von Helverson 1989). The frequencies are more different than expected by chance, and there also is a relationship between call frequency and bat size, with smaller bats using higher-frequency calls (Heller and von Helverson 1989). The ecological consequences of these differences remain to be determined. Many smaller hipposiderid and rhinolophid species, however, have echolocation calls with most energy at frequencies beyond the range at which moth ears are most sensitive (Fenton and Fullard 1979; see Section 6.2.2). Heller and von

Helverson (1989) also noted that differences in echolocation call frequencies could be associated with communication (see Section 6).

The various approaches to foraging and echolocating outlined here coincide with striking differences in the auditory systems of bats. In the best illustration of this to date, Neuweiler, Singh, and Sripathi (1984) documented differences in hearing, echolocation calls and foraging behavior in a community of sympatric bats in southern India (see Chapters 3, 4, and 5). Hearing and echolocation is one axis along which bat communities may be organized (Neuweiler 1989).

Different foraging strategies and differences in echolocation behavior can influence the prey available to animal-eating bats. By taking nonairborne prey, bats immediately have access to larger animals, including both insects and arthropods, and small vertebrates (Norberg and Fenton 1988). Hunting nonairborne prey may also provide access to more prey because there often are more insects on the ground than in the air (Rautenbach, Kemp, and Scholtz 1988). The energetic implications of differences in prey availability were demonstrated by Barclay (1991) working on the eastern slopes of the Rocky Mountains. He found that *Myotis evotis*, which commonly takes nonairborne prey (Faure, Fullard, and Barclay 1990; Faure and Barclay 1992), bears young in his study area while two sympatric species (*Myotis lucifugus* and *Myotis volans*) that eat airborne targets do not.

Access to prey also is influenced by actual feeding behavior. Comparing the numbers of different-sized noctuid and sphingid moths in the diets of *Lasiurus cinereus* and *Nycteris grandis* (Fig. 2.6) illustrates this point. Both species weigh about 30 g, but while *L. cinereus* hunts and eats on the wing, taking only airborne targets (e.g., Barclay 1986; Hickey 1993), *N. grandis* sometimes hunts on the wing but other times from a perch, taking a mixture of airborne and nonairborne targets (Fenton et al. 1990). Furthermore, *N. grandis* appears to consistently perch while eating, perhaps enhancing its capacity to handle large prey. The differences in behavior coincide with striking differences in prey consumed (insects alone versus a varied diet of insects, other arthropods, and small vertebrates [Hickey 1993; Fenton et al. 1990]), even when the comparison is restricted to two families of moths.

6. Communication Role of Biosonar Signals

The signals one bat produces to collect information about its surroundings also are available to other animals, making information leakage a reality of echolocation. While many intraspecific interactions mediated by echolocation calls clearly meet most definitions of "communication" (e.g., Slater 1983), other interactions do not. The magnitude of information leakage is determined partly by the strength of the echolocation signals and partly by the hearing properties of the eavesdroppers. Using bat detectors to monitor the activity and behavior of echolocating bats (see Section 2) takes

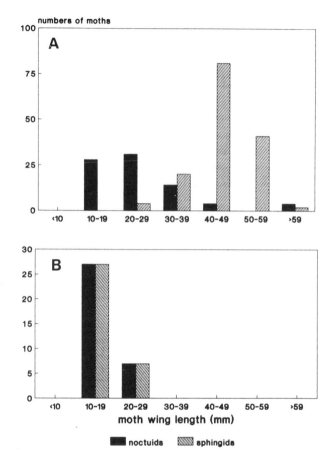

FIGURE 2.6A,B. A comparison of the numbers of noctuid (*black bars*) and sphingid (*shaded bars*) moths of different sizes taken by *Nycteris grandis* (A) and *Lasiurus cinereus* (B). Both bats weigh about 30 g as adults, but while *Nycteris grandis* takes prey from the ground and in flight, *Lasiurus cinereus* attacks only airborne prey. Differences in foraging strategy affect the prey available to a hunting bat. The data for *N. grandis* are from Fenton et al. (1993), and those for *L. cinereus* are from Hickey (1993).

advantage of information leakage. While the calls of some bats are readily detectable over dozens of meters, others are picked up only when the bats fly within 1 or 2 m of the microphone, reflecting differences in call intensity (see Section 2.1).

6.1 Intraspecific Interactions

It is evident that some bats respond to the echolocation calls of conspecifics, but the nature of the response varies by species. Möhres (1967) observed that some captive *Rhinolophus ferrumequinum* appeared to use the echo-location calls of conspecifics to find the locations of preferred roostmates. Using playback presentations, Barclay (1982) demonstrated that free-flying *Myotis lucifugus* flew toward speakers presenting the echolocation calls of conspecifics or sounds of similar bandwidth and duration. These bats responded significantly more to presentations of echolocation calls than to control signals generated by playing the recorded calls backward. These bats are quite gregarious at roost sites and in feeding areas, places where Barclay (1982) found a positive response. He proposed that in either situation the individuals whose calls attracted conspecifics suffered no penalty.

A contrasting picture emerged from similar experiments with *Euderma maculatum*, a species that forages alone over swamps and in open wood-land. These bats showed two patterns of response to playback presentations (Leonard and Fenton 1984). While they sometimes flew directly at the speaker, other times they flew away from it, supporting the suggestion that echolocation calls serve to space individuals in foraging areas.

The sight of one bat chasing another often has been interpreted as evidence of territorial behavior (e.g., Rydell 1986), especially when the interactions are accompanied by vocalizations other than echolocation calls (e.g., Belwood and Fullard 1984; Racey and Swift 1985). But things are not always what they seem. Griffin (1958) proposed that in *Lasiurus borealis* chases occurred when one or more bats used the feeding buzzes of another to cue on an available (vulnerable?) prey. Hickey and Fenton (1990) showed that intraspecific chases in *L. borealis* did not correspond to exclusive use of rich patches of food and that their incidence was not related to prey abundance. Balcombe and Fenton (1988) used playback presentations to demonstrate that in this species feeding buzzes were cues that precipitated chases. The data from individually marked bats and from playback experiments (Balcombe and Fenton 1988; Hickey and Fenton 1990) supported Griffin's original (1958) interpretation. The chases in *L. borealis* did not reflect territorial behavior, but rather the presence of the medium-sized moths that form the bulk of the bats' diet at the study site in southwestern Ontario.

A variety of evidence, including analysis of signals and the results of playback presentations, suggests that echolocation calls mediate some aspects of interactions between mother bats and their young. Thomson, Barclay, and Fenton (1985) showed that infant *Myotis lucifugus* were more attentive to the echolocation calls of their mothers than to those of other females, and Balcombe and McCracken (1992) found evidence that in *Tadarida brasiliensis* echolocation calls were one factor in reunions between mothers and their own offspring.

While echolocation calls by themselves can mediate communication, Suthers (1965) demonstrated that when two *Noctilio leporinus* were on a collision course, one or both bats would alter their echolocation calls by dropping the terminal FM sweep by an additional octave. These "honks" apparently alerted the bats to a possible collision and were associated with one or both animals changing their course. Similar "honks" have been reported from a variety of other bats, including several species of *Myotis* (Fenton and Bell 1979).

6.2 Interspecific Interactions

There are two obvious interspecific audiences of bats: other species of bats, and other kinds of animals. These usually involve predator–prey interactions where bats may be the predators or the prey.

6.2.1 Bat–Bat and Bat–Bird Interactions

Just as chases between bats may not represent territorial behavior, chases involving bats and other animals may reflect other phenomena. Shields and Bildstein (1979) proposed that chases of common nighthawks (*Chordeiles minor*: Aves) by big brown bats (*Eptesicus fuscus*) reflected efforts by the bats to exclude the birds from a rich food source. Alternatively, the chases could reflect missed communication signals because the birds would not have heard the ultrasonic honks of bats on collision courses with them. In this situation, chases reflect the hazard of missed warning signals.

Interspecific communication among bats can involve echolocation calls. Both Barclay (1982) and Balcombe and Fenton (1988) demonstrated that responses to echolocation calls often involved species other than the one whose calls were presented in playback situations. Furthermore, the statistical association that Bell (1980) found between some species responding to rich patches of prey could also result from a communicative role of echolocation calls, perhaps the specific availability of prey as indicated by feeding buzzes.

In parts of South America, Africa, Asia, and Australia, some bats regularly or occasionally eat other bats (Norberg and Fenton 1988). These predators are found in three families: the Phyllostomidae (*Vampyrum spectrum, Chrotopterus auritus*), the Nycteridae (*Nycteris grandis*), and the Megadermatidae (*Cardioderma cor, Megaderma lyra, Macroderma gigas*). Although it is tempting to think that bat-eating bats depend on other species' echolocation calls to locate and identify their prey, this hypothesis remains unproven. Fenton, Gaudet, and Leonard (1983) found that captive *Nycteris grandis* sometimes responded to the vocalizations of other bats, as well as to their wing flutterings. Vaughan's (1976) observations of *Cardioderma cor* attacking, killing and eating some *Pipistrellus* species could have involved echolocation calls as a cue.

Other predators of bats, such as owls or raptors such as bat hawks (*Macheirhamphus alcinus* [Fenton 1992]), are unlikely candidates as eavesdroppers on bat echolocation calls because birds cannot hear ultrasonic

(>20 kHz) sound (Welty 1975). Predators such as small mammals with keen ultrasonic hearing (Bench, Pye, and Pye 1975) are more obvious candidates for using echolocation calls to find and identify bat as prey.

6.2.2 Bat–Insect

The story of moths with ears that detect the echolocation calls of hunting bats (Roeder 1967) is certainly one of the "eye-opening" discoveries in animal behavior (Griffin 1976). We know that insects such as moths, lacewings, and various orthopterans have bat-detecting ears (for reviews, see Fullard 1987; Surlykke 1988), and recently mantids have been added to the list (Yager, May, and Fenton 1990). Insect bat detectors may be located on wings (lacewings), face, thorax or abdomen (moths), or mid venter on the thorax (mantids). While insect ears sensitive to echolocation calls tend to occur in pairs, providing information about direction, mantids have only one ear and show no directionality in their response to the calls of an approaching bat (Yager, May, and Fenton 1990).

The actual response of a moth, lacewing, cricket, or mantid to an approaching bat appears to be mediated by the strength of the bat's echolocation calls, although some species appear to react to changes in pulse repetition rates. The echolocation calls of a distant bat, perceived as quiet calls, produce negative phonotactic behavior as the insect flies away from the bat. A close bat, perceived as loud calls, may cause erratic behavior ranging from diving to the ground to complex spiral and zigzag patterns. Some species of tiger moths (Arctiidae) produce clicks in response to loud bat calls, and these often affect the bats' attacks (Fullard 1987; Surlykke 1988).

Bat-detecting ears in insects range from simple structures with one sensory neuron to more complex ones involving more neurons. Moth ears are not equally sensitive to all frequencies in the bandwidths used by echolocating bats, and the frequencies of their echolocation calls make some bats much less conspicuous to moths than others (Fullard 1987; Surlykke 1988). Intensity also affects the conspicuousness of bat calls to moths and other insects (Faure, Fullard, and Barclay 1990). Although it appears that the "average" moth with ears has 40% less chance of being taken by a bat than a deaf moth (Roeder and Treat 1960), ear structure and sensitivity together combine with characteristics of bat echolocation calls to influence this situation. Fenton and Fullard (1979) demonstrated the impact of bat call frequency and intensity on the conspicuousness of bats to moths. They found that an "average" bat, one using high-intensity echolocation calls with energy between 35 and 60 kHz, was detected by the "average" moth at about 40 m. With the operational range of bat echolocation restricted to a few meters (see Section 2.6), the potential advantage to the moth is considerable.

Moths, including species with bat-detecting ears, form a large part of the diets of some echolocating bats. Sometimes this can be explained by the inability of the moths to hear the bats' echolocation calls. Thus, the small African hipposiderid *Cloeotis percivali* eats many moths (Whitaker and Black 1976), and most of the energy in its echolocation calls is at about 212 kHz (Fenton and Bell 1981), well above the frequencies at which sympatric moths can hear (Fenton and Fullard 1979). In the same way, the North American vespertilionid *Euderma maculatum* sometimes eats moths (Ross 1967), and its echolocation calls range in frequency from 9 to 15 kHz, with most energy at 10–12 kHz (Leonard and Fenton 1984; Obrist 1989), well below the lower frequency threshold of most sympartric moths (Fullard, Fenton, and Fulonger 1983).

Many species of moths with bat-detecting ears, however, are taken by bats using intense echolocation calls with most energy in the range of the moths' best hearing. In the laboratory, L.A. Miller (personal communication) found that some individual *Pipistrellus pipistrellus* learned to thwart the auditory-based defensive maneuvers of lacewings. In the field, Hickey (1993) and Hickey and Fenton (1990) have observed that several *Lasiurus cinereus* and *Lasiurus borealis* often forage together in places where flying moths are abundant. L. Acharya (personal communication) has observed that in these situations moths with ears usually react to attacking bats by flight maneuvers, but the individual that successfully evades one bat may quickly (within a second) be caught by another. This means that although individual *L. cinereus* and *L. borealis* typically succeed in only 40%–50% of their attacks on moths (Hickey 1993; Hickey and Fenton 1990), the vulnerability of the moths is complicated by the presence of several bats. In the final analysis, the hearing-based defenses of moths, like other defensive systems (Edmunds 1974), do not provide absolute immunity from predators.

Dunning and Roeder (1965) noted that the clicks of some arctiid moths affected the behavior of *Myotis lucifugus* trained to take mealworms tossed into the air. They found that if arctiid clicks were presented just as the bats closed with their targets, they veered away, aborting the attack. Three hypotheses have been used to explain the bats' behavior: (1) moth clicks could interfere with or "jam" the bats' echolocation, (2) moth clicks could startle the bats, and (3) moth clicks may alert the bats to bad-tasting arctiids. A number of studies have presented different sets of data on this topic (e.g., Dunning 1968; Fullard, Fenton, and Simmons 1979; Surlykke and Miller 1985; Stoneman and Fenton 1988; Bates and Fenton 1990; Miller 1991; Acharya and Fenton 1992; Dunning et al. 1992). Some laboratory work contradicts the first hypothesis, that moth clicks jam bat echolocation (e.g., Surlykke and Miller 1985; Stoneman and Fenton 1988), while other work supports it (Miller 1991). Some captive animals clearly reject arctiids, and in the field, free-flying bats take arctiids less often than expected by their general incidence in moth populations (Dunning et al. 1992).

Presentations of muted free-flying arctiids (*Hypoprepia fucosa*) to hunting *L. borealis* showed that the moths use their clicks to warn bats of their bad taste (Acharya and Fenton 1992). Free-flying *L. borealis* approached but did not attack intact *H. fucosa*, but attacked, caught, and then rejected these moths that had been muted.

Arctiid moth clicks affect the behavior of bats. Bates and Fenton (1990) used captive *Eptesicus fuscus* to demonstrate that a bat's prior experience determined its response to moth clicks. Bats that have learned to associate moth clicks with bad taste avoid prey that click, while naive bats ignore clicks after an initial period of being startled. Field studies reveal that the clicks of some arctiids act as aposematic signals for bats that feed heavily on moths, namely *L. borealis* and *L. cinereus* (Acharya and Fenton 1992). Further work will certainly reveal variations on these themes, including the likelihood of Batesian mimics depending on acoustic displays (cf. Dunning 1968). It remains to be seen if the sounds of other insects also influence the prey taken by bats (e.g., peacock butterfly clicks [Møhl and Miller 1976]; defense stridulations [Masters 1979]).

7. Biosonar and the Lives of Bats

Although we do not always know just what role echolocation plays in the lives of bats, especially in species producing weak calls, several lines of evidence suggest that this mode of orientation has profoundly affected bats. Echolocation can be expensive, particularly when the cost is measured in information leakage (see Section 6, earlier in this chapter). By making themselves more conspicuous, echolocating bats can affect their chances of avoiding or attracting conspecifics, catching prey, or being taken by predators.

Other costs of echolocation could be in the actual production and processing of sounds. Speakman and Racey (1991), however, used two species of echolocating bats (*Pipistrellus pipistrellus* and *Plecotus auritus*) to demonstrate that the costs of echolocation were not additive to those of flight. They concluded that echolocation was relatively inexpensive for animals already expending so much energy in flying. The situation could be different for bats that hunt from perches.

The short effective range of bat echolocation is probably one factor contributing to bats the small size of (Barclay and Brigham 1991). Being restricted to collecting information from just in front of it may be an acceptable limitation for a small maneuverable animal, but not a large unmaneuverable one. In this context, it is not surprising that faster-flying bats have longer interpulse intervals (= effective range; see Sections 2.2, 2.6) than slow-flying species. Although the fossil record reveals that a much larger vampire bat (30% bigger than the living species, which weigh about 40 g) once lived in South America (Ray, Linares, and Morgan 1988), there are no other records of "giant" bats. It is evident that being small is a predominant feature of most insectivorous bats (Fig. 2.7).

The facial features of bats (Fig. 2.8), including their ears, provide one of the most graphic demonstrations of the impact of echolocation on the physiognomy of bats. Obrist et al. (1993) demonstrated three general trends in bats' ears that are associated with different orientation behaviors. Ear (pinna) size was directly correlated with echolocation call frequency in bats that used high duty cycles (rhinolophids, hipposiderids, and the mormoopid *Pteronotus parnellii*). In low duty cycle species, the relationship between pinna size and frequency varied from being relatively strong in species using narrowband signals to weak in species using broadband calls. Bats that listen to sounds coming from their prey have large ears that amplify low-frequency sounds. These bats occur in several evolutionary lines; the Nycteridae and Megadermatidae, the phyllostomine Phyllostomidae, and some vespertilionids.

Leaflike structures are conspicuous on the faces of bats in the families Hipposideridae, Rhinolophidae, Megadermatidae, and Phyllostomidae, while less marked leaflike structures occur in the Rhinopomatidae, some Vespertilionidae, and Mormoopidae (Fenton 1992). In the Nycteridae, a conspicuous slit dominates the animals' faces. Hartley and Suthers (1987) demonstrated how noseleaf position in the phyllostomid *Carollia perspicillata* affects the pattern of sound radiation from the bat's face and thus its acoustic perception of its surroundings. Still and motion photographs of the

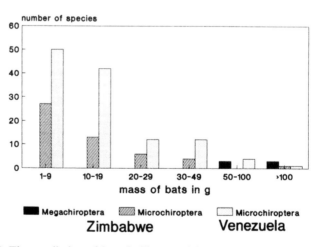

FIGURE 2.7. The small size of bats is illustrated here as the number of species of different adult body mass from Zimbabwe and Venezuela. The data for Zimbabwe come from Smithers (1983) and those for Venezuela from Linares (1986). *Black bars*, Megachiroptera, *shaded bars*, Microchiroptera, Zimbabwe; *clear bars*, Microchiroptera, Venezuela.

A

B

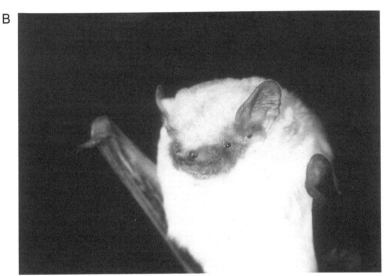

FIGURE 2.8A-H. The faces and ears of bats of bats appear to reflect their echocation behavior. While some species producing low duty cycle, broadband, frequency-modulated (FM) calls have plain faces (e.g., A: *Balantiopteryx plicata*, Emballonuridae; B: *Chalinolobus variegatus*, Vespertilionidae), species using high duty cycle, narrowband calls tend to have facial ornaments (e.g., C: *Rhinolophus hildebrandti*, Rhinolophidae; D: *Hipposideros caffer*, Hipposideridae). Other species with facial ornaments produce low duty cycle, low-intensity, broadband echolocation calls (E: *Nycteris grandis*, Nycteridae; F: *Megaderma lyra*, Megadermatidae; G: *Chrotopterus auritus*; H: *Tonatia brasiliensis*, Phyllostomidae).

C

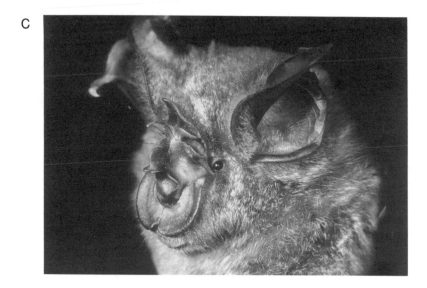

D

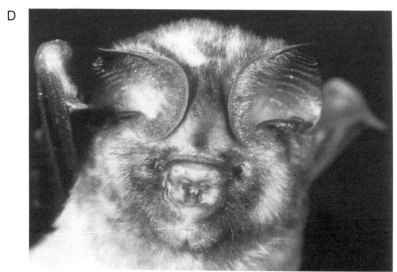

FIGURE 2.8 (*continued*)

phyllostomid *Trachops cirrhosus* approaching frogs suggest changes in noseleaf position in these situations (Tuttle and Ryan 1981), indicating that the noseleaf may serve a dynamic role in orientation.

As noted in Section 4, our lack of knowledge about the role that echolocation plays in the lives of the low-intensity bats precludes strong statements about the relationship between facial features and echolocation. In Hipposideridae and Rhinolophidae, independent ear movements and the presence of a noseleaf coincide with a high duty cycle approach to

FIGURE 2.8 (*continued*)

echolocation, presumably affecting the bats' sonar prowess (Pye and Roberts 1970). The temptation to associate facial features and noseleafs is tempered by two facts: first, the absence of such structures in the mormoopid *Pteronotus parnellii*, another high duty cycle echolocator, and second, their presence in nycterids, megadermatids, and phyllostomids, which use low duty cycle echolocation calls that are broadband and low intensity. The situation is ripe for further investigation.

8. Summary

Echolocation typifies species in the suborder Microchiroptera but is exceptional among Megachiroptera. *Rousettus aegyptiacus*, the echolocating megachiropteran, produces short, broadband pulses of sound by clicking its tongue, while the echolocation calls of microchiropterans are vocalizations, sounds produced in the larynx. The echolocation calls of microchiropteran bats vary considerably in different acoustic parameters such as intensity,

F

FIGURE 2.8 (*continued*)

duration, bandwidth, duty cycles, and patterns of frequency change over time. Each species appears to produce a distinctive echolocation call. Species using echolocation to detect, track, and assess airborne targets generally use low duty cycle ($\leq 20\%$) calls and marked increases in pulse repetition rate during attacks on targets. At some stages in the attack process, these bats use short (≤ 1-msec) echolocation calls dominated by frequency-modulated (FM) components. Species in the families Rhinolophidae and Hipposideridae and the mormoopid *Pteronotus parnellii* use high duty cycle ($\geq 80\%$) calls dominated by narrow bandwidth components. These bats, known as constant-frequency (CF) species, appear specialized for flutter detection.

Among the Microchiroptera, many species hunt nonairborne targets they take from the ground, foliage, or other surfaces. These species, from the families Nycteridae, Megadermatidae, Phyllostomidae, and Vespertilionidae, use short (≤ 1-msec), broadband echolocation calls of low intensity but often depend on prey-generated sound cues or vision to find their targets. Bats in the family Phyllostomidae that feed on blood, fruit, nectar, and pollen also produce short, low-intensity, broadband echolocation calls but

G

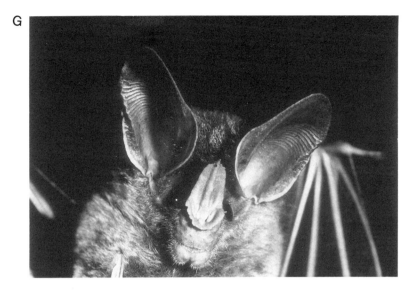

H

FIGURE 2.8 (*continued*)

as yet the role that echolocation plays in their lives is not clear. The echolocation calls of many bats serve a communication function, and playback presentations elicit various responses to the calls of conspecifics. Echolocation calls also mediate interspecific interactions in which bats may be prey or predators. The bat-detecting ears of many insects are among the most notable of interspecific interactions mediated by echolocation calls. Echolocation has profoundly affected the lives of microchiropteran bats, from their body size to the appearances of their faces.

Acknowledgments. I thank Lalita Acharya, Doris Audet, Robert M. R. Barclay, James H. Fullard, David Johnston, Jennifer Long, Cathy Merriman, Martin Obrist, David Pearl, Stuart Perlmeter, and Daphne Syme for reading earlier drafts of this manuscript and making helpful suggestions for improving it. I am particularly grateful to J. E. Cebek, P. A. Faure, J. H. Fullard, M. B. C. Hickey, G. Marimuthu, and G. S. Wilkinson for sharing some of their unpublished results with me and to R. A. Suthers for providing the images of Rousettus sounds. My research on bats has been supported by operating and equipment grants from the Natural Sciences and Engineering Research Council of Canada.

References

Acharya L, Fenton MB (1992) Echolocation behaviour of vespertilionid bats (*Lasiurus cinereus* and *Lasiurus borealis*) attacking airborne targets, including arctiid moths. Can J Zool 70:1292–1298.

Ahlén I (1981) Field identification of bats and survey methods based on sounds. Myotis 18–19:128–136.

Aldridge HDJN, Rautenbach IL (1987) Morphology, echolocation and resource partitioning in insectivorous bats. J Anim Ecol 56:763–778.

Aldridge HDJN, Obrist M, Merriam H G, Fenton M B (1990) Social calls and habitat use: roosting and foraging by the African bat *Nycteris thebaica*. J Mammal 71:242–246.

Anderson MB, Racey PA (1991) Feeding behaviour of captive brown long-eared bats, *Plecotus auritus*. Anim Behav 42:489–493.

Arroyo-Cabrales J, Jones JK Jr (1988) *Balantiopteryx plicata*. Mammal Species 301:1–4.

Avery M I (1991) Pipistrelle. In: Corbet G B, Harris S (eds) The Handbook of British Mammals, 3d Ed. Oxford: Blackwell Scientific, pp. 124–128.

Audet D (1990) Foraging behavior and habitat use by a gleaning bat, *Myotis myotis*. J Mammal 71:420–427.

Audet D, Krull D, Marimuthu G, Sumithran S, Singh JB (1991) Foraging behavior of the Indian false vampire bat, *Megaderma lyra* (Chiroptera: Megadermatidae). Biotropica 23:63–67.

Bailey WJ, Slightom JL, Goodman M (1992) Rejection of the "flying primate" hypothesis by phylogenetic evidence from the epsilon-globin gene. Science 256:86–89.

Balcombe JP, Fenton MB (1988) Eavesdropping by bats: the influence of echolocation call design and foraging strategy. Ethology 79:158–166.

Balcombe JP, McCracken GF (1992) Vocal recognition in Mexican free-tailed bats: do pups recognize their mothers? Anim Behav 43:79–88.

Baker RJ, Novacek MJ, Simmons NB (1991) On the monophyly of bats. Syst Zool 40:216–231.

Barclay RMR (1982) Interindividual use of echolocation calls: eavesdropping by bats. Behav Ecol Sociobiol 10:271–275.

Barclay RMR (1983) Echolocation calls of emballonurid bats from Panama. J Comp Physiol 151:515–520.

Barclay RMR (1986) The echolocation calls of hoary (*Lasiurus cinereus*) and silver-haired (*Lasionycteris noctivagans*) bats as adaptations for long- versus short-range strategies and the consequences for prey selection. Can J Zool 64:2700–2705.

Barclay RMR (1991) Population structure of temperate zone insectivorous bats in relation to foraging behaviour and energy demand. J Anim Ecol 60:165–178.

Barclay RMR, Brigham RM (1991) Prey detection, dietary niche breadth, and body size in bats: why are aerial insectivorous bats so small? Am Nat 137:693–703.

Barclay RMR, Fenton MB, Tuttle MD, Ryan MJ (1981) Echolocation calls produced by *Trachops cirrhosus* (Chiroptera: Phyllostomatidae) while hunting for frogs. Can J Zool 59:750–753.

Bates DL, Fenton MB (1990) Aposematism or startle? Predators learn their responses to prey. Can J Zool 68:49–52.

Bell GP (1980) Habitat use and responses to patches of prey by desert insectivorous bats. Can J Zool 58:1876–1883.

Bell GP (1982) Behavioral and ecological aspects of gleaning by a desert insectivorous bat, *Antrozous pallidus* (Chiroptera: Vespertilionidae). Behav Ecol Sociobiol 10:217–223.

Bell GP (1985) The sensory basis of prey location by the California leaf-nosed bat *Macrotus californicus* (Chiroptera: Phyllostomidae). Behav Ecol Sociobiol 16:343–347.

Bell GP, Fenton MB (1984) The use of Doppler-shifted echoes as a flutter detection and clutter rejection system: the echolocation and feeding behavior of *Hipposideros ruber* (Chiroptera: Hipposideridae). Behav Ecol Sociobiol 15:109–114.

Belwood JJ, Fullard JH (1984) Echolocation and foraging behaviour in the Hawaiian hoary bat, *Lasiurus cinereus semotus*. Can J Zool 62:21130–2120.

Belwood JJ, Morris GK (1987) Bat predation and its influence on calling behavior in neotropical katydids. Science 238:64–67.

Bench RJ, Pye A, Pye JD (eds) (1975) Sound Reception in Mammals (Symposium No. 37). London: Zoological Society of London.

Bradbury JW (1977) Social organization and communication. In: Wimsatt WA (ed) Biology of Bats, Vol. 3. New York: Academic, pp. 2–72.

Brigham RM, Cebek JE, Hickey MBC (1989) Intraspecific variation in the echolocation calls of two species of insectivorous bats. J Mammal 70:426–428.

Brosset A (1966) La biologie des chiroptères. Paris: Masson et Cie.

Brown P (1976) Vocal communication in the pallid bat, *Antrozous pallidus*. Z Tierpsychol 41:34–54.

Brown PE, Grinnell AD (1980) Echolocation ontogeny in bats. In: Busnel R-G, Fish J (eds) Animal Sonar Systems. New York: Plenum, pp. 355–380.

Brown PE, Brown TW, Grinnell AD (1983) Echolocation development and vocal communication in the lesser bulldog bat, *Noctilio albiventris*. Behav Ecol Sociobiol 13:287–298.

Buchler ER (1980) The development of flight, foraging, and echolocation in the little brown bat (*Myotis lucifugus*). Behav Ecol Sociobiol 6:211–218.

Buchler ER, Mitz AR (1980) Similarities in design features of orientation sounds used by simpler, nonaquatic echolocators. In: Busnel R-G, Fish JF (eds) Animal Sonar Systems. (Nato Advanced Study Institutes, Vol. A28.) New York: Plenum, pp. 871–874.

Busnel R-G, Fish JF (eds) (1980). Animal Sonar Systems. (Nato Advanced Study Institutes, Series A28.) New York: Plenum.

Campbell KA, Suthers RA (1988) Predictive tracking of horizontally moving targets by the fishing bat, *Noctilio leporinus*. In: Nachtigall PE, Moore PWB (eds)

Animal Sonar: Processes and Performance, Vol. A156. New York: Plenum, pp. 501–506.

Casseday J H, Covey E (1991) Frequency tuning properties of neurons in the inferior colliculus of an F M bat. J Comp Neurobiol 319:34–50.

Cebek J E (1992) Social and genetic correlates of female philopatry in the temperate zone bat, *Eptesicus fuscus*. Ph.D. Dissertation, Department of Biology, York University, North York, Ontario, Canada.

Dunning DC (1968) Warning sounds of moths. Z Tierpyschol 25:129–138.

Dunning DC, Roeder KD (1965) Moth sounds and the insect-catching behavior of bats. Science 147:173–174.

Dunning DC, Acharya L, Merriman C, Dal Ferro L (1992) Interactions between bats and arctiid moths. Can J Zool 70:2218–2223.

Edmunds M (1974) Defence in Animals. London: Longman.

Faure PA, Barclay RMR (1992) The sensory basis of prey detection by the long-eared bat, *Myotis evotis*, and the consequences for prey selection. Anim Behav 44:31–39.

Faure PA, Fullard JH, Barclay RMR (1990) The response of tympanate moths to the echolocation calls of a substrate gleaning bat, *Myotis evotis*. J Comp Physiol A 166:843–849.

Faure PA, Fullard JH, Dawson JW (1993) The gleaning attacks of northern long-eared bats, *Myotis septentrionalis*, are relatively inaudible to moths. J Exp Biol. 178:173–189.

Fenton MB (1974) The role of echolocation in the evolution of bats. Am Nat 108:386–388.

Fenton MB (1984) Echolocation: implications for the ecology and evolution of bats. Rev Biol 59:33–53.

Fenton MB (1985) Communication in the Chiroptera. Bloomington: Indiana University Press.

Fenton MB (1986) *Hipposideros caffer* (Chiroptera: Hipposideridae) in Zimbabwe: morphology and echolocation calls. J Zool (Lond) 210:347–353.

Fenton MB (1990) The foraging behaviour and ecology of animal-eating bats. Can J Zool 68:411–422.

Fenton MB (1992) Bats. New York: Facts on File.

Fenton MB (1994) Assessing signal variability and reliability: 'to thine ownself be true'. Anim Behav 47:757–764.

Fenton MB, Bell GP (1979) Echolocation and feeding behaviour in four species of *Myotis* (Chiroptera: Vespertilionidae). Can J Zool 57:1271–1277.

Fenton MB, Bell GP (1981) Recognition of species of insectivorous bats by their echolocation calls. J Mammal 62:233–243.

Fenton MB, Crerar LM (1984) Cervical vertebrae in relation to roosting posture in bats. J Mammal 65:395–403.

Fenton MB, Fullard JH (1979) The influence of moth hearing on bat echolocation strategies. J Comp Physiol A 132:77–86.

Fenton MB, Rautenbach IL (1986) A comparison of the roosting and foraging behaviour of three species of African insectivorous bats. Can J Zool 64:2860–2867.

Fenton MB, Gaudet CL, Leonard ML (1983) Feeding behaviour of the bats *Nycteris grandis* and *Nycteris thebaica* (Nycteridae) in captivity. J Zool (Lond) 200:347–354.

Fenton MB, Swanepoel CM, Brigham RM, Cebek JE, Hickey MBC (1990) Foraging behavior and prey selection by large slit-faced bats (*Nycteris grandis*). Biotropica 22:2–8.

Fenton MB, Rautenbach IL, Chipese D, Cumming MB, Musgrave MK, Taylor JS, Volpers T (1993) Variation in foraging behaviour, habitat use and diet of large slit-faced bats (*Nycteris grandis*). Z Saeugetierk 58:65–74.

Fenton MB, Acharya L, Audet D, Hickey MBC, Merriman C, Obrist MK, Syme DM, Adkins B (1992) Phyllostomid bats (Chiroptera: Phyllostomidae) as indicators of habitat disruption in the Neotropics. Biotropica 24:440–446.

Fiedler J (1979) Prey catching with and without echolocation in the Indian false vampire bat (*Megaderma lyra*). Behav Ecol Sociobiol 6:155–160.

Findley JS, Wilson DE (1982) Ecological significance of chiropteran morphology. In: Kunz TH (ed) Ecology of Bats. New York: Plenum, pp. 243–260.

Flemming TH (1988) The short-tailed fruit bat. Chicago: University of Chicago Press.

Fujita MS, Kunz TH (1984) *Pipistrellus subflavus*. Mammal Species 228:1–8.

Fullard JH (1987) Sensory ecology and neuroethology of moths and bats: interactions in a global perspective. In: Fenton MB, Racey PA, Rayner JMV (eds) Recent Advances in the Study of Bats. Cambridge: Cambridge University Press, pp. 244–272.

Fullard JH, Fenton MB, Fulonger CL (1983) Sensory relationships of moths and bats sampled from two nearctic sites. Can J Zool 61:1752–1757.

Fullard JH, Fenton MB, Simmons JA (1979) Jamming bat echolocation: the clicks of arctiid moths. Can J Zool 57:647–649.

Fullard JH, Koehler C, Surlykke A, McKenzie NL (1991) Echolocation ecology and flight morphology of insectivorous bats (Chiroptera) from south-western Australia. Aust J Zool 39:427–438.

Goldman LJ, Henson OW Jr (1977) Prey recognition and selection by constant frequency bat, *Pteronotus parnellii parnellii*. Behav Ecol Sociobiol 2:411–420.

Greenhall AM, Schmitd U (eds) (1988) Natural History of Vampire Bats. Boca Raton: CRC Press.

Griffin DR (1958) Listening in the Dark. New Haven: Yale University Press.

Griffin DR (1971) The importance of atmospheric attenuation for the echolocation of bats (Chiroptera). Anim Behav 19:55–61.

Griffin DR (1976) The Question of Animal Awareness. New York: Rockefeller University Press.

Griffin DR, Simmons JA (1974) Echolocation of insects by horseshoe bats. Nature 250:730–731.

Griffin DR, Webster FA, Michael CR (1960) The echolocation of flying insects by bats. Anim Behav 8:151–154.

Guppy A, Coles RB (1988) Acoustical and neural aspects of hearing in the Australian gleaning bats, *Macroderma gigas* and *Nyctophilus gouldii*. J Comp Physiol A 162:653–668.

Habersetzer J (1981) Adaptive echolocation sounds in the bat *Rhinopoma hardwickei*: a field study. J Comp Physiol A 144:559–566.

Hartley D J, Suthers R A (1987) The sound emission pattern and the acoustical role of the noseleaf in the echolocating bat, *Carollia perspicillata*. J Acoust Soc Am 82:1892–1900.

Heller K-G, von Helverson O (1989) Resource partitioning of sonar frequency bands in rhinolophoid bats. Oecologia 80:178–186.

Herd RM (1983) *Pteronotus parnellii*. Mammal Species 209:1–5.

Herd RM, Fenton MB (1983) An electrophoretic, morphological, and ecological investigation of a putative hybrid zone between *Myotis lucifugus* and *Myotis yumanensis* (Chiroptera: Vespertilionidae). Can J Zool 61:2029-2050.

Hickey MBC (1993) Thermoregulatory and foraging behaviour of hoary bats, *Lasiurus cinereus*. Ph.D. Dissertation, Department of Biology, York University, North York, Ontario, Canada.

Hickey MBC, Fenton MB (1990) Foraging by red bats (*Lasiurus borealis*): do intraspecific chases mean territoriality? Can J Zool 68:2477-2482.

Hill JE (1974) A new family, genus and species of bat (Mammalia, Chiroptera) from Thailand. Bull Br Mus (Nat Hist) Zool 27:303-336.

Hill JE, Smith JD (1984) Bats: A Natural History. London: British Museum of Natural History.

Howell DJ (1974) Acoustic behavior and feeding in glossophagine bats. J Mammal 55:293-308.

Hudson WS, Wilson DE (1986) Macroderma gigas. Mammalian Species 260:1-4.

Jepsen GL (1970) Bat origins and evolution. In: Wimsatt WA (ed) Biology of Bats, Vol. 1. New York: Academic, pp. 1-64.

Kalko EKV, Schnitzler H-U (1989) The echolocation and hunting behavior of Daubenton's bat, *Myotis daubentoni*. Behav Ecol Sociobiol 24:225-238.

Kick SA (1982) Target-detection by the echolocating bat, *Eptesicus fuscus*. J Comp Physiol A145:431-435.

Kick SA, Simmons JA (1984) Automatic gain control in the bat's sonar receiver and the neuroethology of echolocation. J Neurosci 4:2725-2737.

Koopman KF (1988) Systematics and distribution. In: Greenhall AM, Schmidt U (eds) Natural History of Vampire Bats. Boca Raton: CRC Press, pp. 7-19.

Koopman KF, Jones JK Jr (1970) Classification of bats. In: Slaughter BH, Walton DW (eds) About Bats. Dallas: Southern Methodist University Press, pp. 29-50.

Krull D (1992) Jagdverhalten und Echoortung bei *Antrozous pallidus* (Chiroptera: Vespertilionidae). Ph.D. Dissertation, Fakultat fur Biologie der Ludwig-Maximilians Universitat Munchen, Germany.

Kurta A, Baker RH (1990) *Eptesicus fuscus*. Mammal Species 356:1-10.

Leonard ML, Fenton MB (1984) Echolocation calls of *Euderma maculatum* (Chiroptera: Vespertilionidae): use in orientation and communication. J Mammal 65:122-126.

Linarés O (1986) Murcielagos de Venezuela. Caracas: Cuadernos Lagoven.

Link A, Marimuthu G, Neuweiler G (1986) Movement as a specific stimulus for prey catching behavior in rhinolophid and hipposiderid bats. J Comp Physiol 159:403-414.

MacDonald K, Matsui E, Stevens R, Fenton MB (1994) Echolocation calls of *Pipistrellus subflavus* (Chiroptera: Vespertilionidae), the eastern pipistrelle. J Mammal 75:462-465.

Masters MW (1979) Insect disturbance stridulation: its defensive role. Behav Ecol Sociobiol 5:187-200.

Masters MW, Moffat AJM, Simmons JA (1985) Sonar tracking of horizontally moving targets by the big brown bat *Eptesicus fuscus*. Science 228:2332.

Medellin RA (1989) *Chrotopterus auritus*. Mammal Species 343:1-5.

Medellin RA, Artia HT (1989) *Tonatia evotis* and *Tonatia sylvicola*. Mammal Species 334:1-5.

Miller LH (1991) Arctiid moth clicks can degrade the accuracy of range difference discrimination in echolocating big brown bats, *Eptesicus fuscus*. J Comp Physiol A 199:571–579.

Møhl B, Miller LA (1976) Ultrasonic clicks produced by the peacock butterfly: a possible bat repellant. J Exp Biol 64:639–644.

Möhres FP (1967) Communicative characters of sonar signals in bats. In: Busnel R-G (ed) Animal Sonar Systems, Vol. 2. (NATO Advanced Study Institutes.) New York: Plenum, pp. 939–945.

Morrison DW (1980) Foraging and day-roosting dynamics of canopy fruit bats in Panama. J Mammal 61:20–29.

Nachtigall PE, Moore PWB (eds) (1988) Animal sonar: processes and performance. (NATO Advanced Study Institutes 156.) New York: Plenum.

Navarro L-D, Wilson D E (1982) *Vampyrum spectrum*. Mammal Species 184:1–4.

Neuweiler G (1989) Foraging ecology and audition in bats. Trends Ecol Evol 4:160–166.

Neuweiler G, Fenton MB (1988) Behavior and foraging ecology of echolocating bats. In: Nachtigall PE, Moore PWB (eds) Animal Sonar Systems: Processes and performance. (NATO Advanced Study Institutes 156.) New York: Plenum, pp. 535–550.

Neuweiler G, Singh S, Sripathi K (1984) Audiograms of a south Indian bat community. J Comp Physiol A154:133–142.

Neuweiler G, Metzner W, Heilman U, Rubsamen R, Eckrich M, Costa HH (1987) Foraging behavior and echolocation in the rufus horseshoe bat, *Rhinolophus rouxi*. Behav Ecol Sociobiol 20:53–67.

Norberg UM (1990) Vertebrate flight. Zoophysiology 27:1–291.

Norberg UM, Fenton MB (1988) Carnivorous bats? Biol J Linn Soc 33:383–394.

Norberg UM, Rayner JMV (1987) Ecological morphology and flight in bats (Mammalia, Chiroptera): wing adaptations, flight performance, foraging strategy and echolocation. Philos Trans R Soc Lond B Biol Sci 316:335–427.

Novick A (1977) Acoustic orientation. In: Wimsatt WA (ed) Biology of Bats, Vol. 3. New York: Academic, pp. 73–289.

Obrist KM (1989) Individuelle Variabilitat der Echoortung: Vergleichende Freilanduntersuchungen an Vier Vespertilioniiniden Fledermausarten Kanadas. Ph.D. Dissertation, Faculty of Science, Ludwig-Maximilians-Universitat Munchen, Germany.

Obrist M, Aldridge HDJN, Fenton MB (1989) Roosting and echolocation behavior of the African bat, *Chalinolobus variegatus*. J Mammal 70:828–833.

Obrist KM, Fenton MB, Eger JL, Schlegel P (1993). What ears do for bats: a comparative study of pinna sound pressure transformation in Chiroptera. J Exp Biol 180:119–152.

Pettigrew JD (1991) Wings or brain? convergent evolution in the origin of bats. Syst Zool 40:199–216.

Pettigrew JD, Jamieson BGM, Robson SK, Hall LS, McNally KI, Cooper HM (1989) Phylogenetic relations between microbats, megabats and primates (Mammalia: Chiroptera and Primates). Philos Trans R Soc Lond B Biol Sci 325:489–559.

Pierson ED, Sarich VM, Lowenstein JM, Daniel MJ, Rainey WE (1986) A molecular link between the bats of New Zealand and South America. Nature 323:60–63.

Pye JD, Roberts LH (1970) Ear movements in a hipposiderid bat. Nature 225:285–286.

Racey PA, Swift SM (1985) Feeding ecology of *Pipistrellus pipistrellus* (Chiroptera: Vespertilionidae) during pregnancy and lactation. 1. Foraging behaviour. J Anim Ecol 54:205–215.

Ransome RD (1991) Greater horseshoe bat. In: Corbet GB, Harris S (eds) The Handbook of British Mammals, 3d Ed. Oxford: Blackwell Scientific, pp. 88–94.

Rautenbach IL, Kemp A C, Scholtz C H (1988) Fluctuations in availability of arthropods correlated with microchiropteran and avian predator activities. Koedoe 31:77–90.

Ray CE, Linares OJ, Morgan GS (1988) Paleontology. In: Greenhall AM, Schmidt U (eds) Natural History of Vampire Bats. Boca Raton: CRC Press, pp. 19–30.

Rayner JMV (1991) Complexity and a coupled system: flight, echolocation and evolution in bats. In: Schmidt-Kittler N, Vogel K (eds) Constructional Morphology and Evolution. Berlin: Springer-Verlag, pp. 173–191.

Roeder KD (1967) Nerve cells and insect behavior, revised edition. Cambridge: Harvard University Press.

Roeder KD, Treat AE (1960) The acoustic detection of bats by moths. In: XI International Entomological Congress, Vienna.

Ross A (1967) Ecological aspects of the food habits of insectivorous bats. Proc West Found Vertebr Zool 1:205–263.

Roverud R C (1987) The processing of echolocation sound elements in bats: a behavioural approach. In: Fenton MB, Racey P A, Rayner JMV (eds) Recent Advances in the Study of Bats. Cambridge: Cambridge University Press, pp. 152–170.

Roverud RC, Nitsche V, Neuweiler G (1991) Discrimination of wingbeat motion by bats, correlated with echolocation sound pattern. J Comp Physiol A168:259–263.

Rubsamen R, Schafer M (1990a) Ontogenesis of auditory fovea representation in the inferior colliculus of the Sri Lankan rufous horseshoe bat, *Rhinolophus rouxi*. J Comp Physiol A167:757–769.

Rubsamen R, Schafer M (1990b) Audiovocal interactions during development? Vocalisation in deafened young horseshoe bats vs. audition in vocalisation-impaired bats. J Comp Physiol A167:771–784.

Rubsamen R, Neuweiler G, Marimuthu G (1989) Ontogenesis of tonotopy in inferior colliculus of a hipposiderid bat reveals postnatal shift in frequency-place code. J Comp Physiol A165:755–769.

Ryan MJ, Tuttle MD (1987) The role of prey-generated sounds, vision and echolocation in prey location by the African bat, *Cardioderma cor* (Megadermatidae). J Comp Physiol A161:59–66.

Rydell J (1986) Feeding territoriality in female northern bats, *Eptesicus nilssoni*. Ethology 72:329–337.

Schmidt S (1988) Evidence for a spectral basis of texture perception in bat sonar. Nature 331:617–619.

Schmidt U, Schlegel P, Schweizer H, Neuweiler G (1991) Audition in vampire bats, *Desmodus rotundus*. J Comp Physiol A168:45–51.

Schnitzler H-U (1987) Echoes of fluttering insects: information for echolocating bats. In: Fenton MB, Racey PA, Rayner JMV (eds) Recent Advances in the Study of Bats. Cambridge: Cambridge University Press, pp. 226–243.

Schumm A, Krull D, Neuweiler G (1991) Echolocation in the notch-eared bat, *Myotis emarginatus*. Behav Ecol Sociobiol 28:255–261.

Shields WM, Bildstein KL (1979) Birds versus bats: behavioral interactions at a localized food source. Ecology 60:468–474.

Simmons JA, O'Farrell MJ (1977) Echolocation in the long-eared bat, *Plecotus phyllotus*. J Comp Physiol A122:201–214.

Simmons JA, Stein RA (1980) Acoustic imaging in bat sonar: echolocation signals and the evolution of echolocation. J Comp. Physiol A135:61–84.

Simmons JA, Saillant PA, Dear SP (1992) Through a bat's ear. IEEE Spectrum, 29 March, pp. 46–48.

Simmons NB, Novacek M J, Baker R J (1991) Approaches, methods and the future of the chiropteran monophyly controversy: a reply to J.D. Pettigrew. Syst Zool 40:239–243.

Slater PJB (1983) The study of communication. In: Halliday TR, Slater PJB (eds) Animal Behaviour 2. Communication. New York: Freeman, pp. 9–81.

Smith JD (1972) Systematics of the chiropteran family Mormoopidae. Univ Kansas Mus Nat Hist Misc Publ 56:1–132.

Smithers RHN (1983) Mammals of the Southern African Subregion. Pretoria: University of Pretoria Press.

Speakman JR, Racey PA (1991) No cost of echolocation for bats in flight. Nature 350:421–423.

Stoneman MG, Fenton MB (1988) Disrupting foraging bats: the clicks of arctiid moths. In: Nachtigall P E, Moore PWB (eds) Animal Sonar: Processes and performance. (NATO Advanced Study Institutes 156.) New York: Plenum, pp. 635–638.

Summe YW, Menne D (1988) Discrimination of fluttering targets by the FM bat *Pipistrellus stenopterus*. J Comp Physiol A163:349–354.

Surylkke AM (1988) Interactions between echolocating bats and their prey. In: Nachtigall PE, Moore PWB (eds) Animal Sonar: Processes and Performance. (NATO Advanced Study Institutes 156.) New York: Plenum, pp. 635–638.

Surlykke AM, Miller LH (1985) The influence of arctiid clicks on bat echolocation: jamming or warning? J Comp Physiol A156:831–843.

Suthers RA (1965) Acoustic orientation by fish-catching bats. J Exp Zool 158:253–258.

Suthers RA (1988) The production of echolocation signals by bats and birds. In: Nachtigall P E, Moore PWB (eds) Animal Sonar Systems: Processes and Performance. (NATO Advanced Study Institutes 156.) New York: Plenum, pp. 23–46.

Thewissen JGM, Babcock SK (1991) Distinctive cranial and cervical innervation of wing muscles: new evidence for the monophyly of bats. Science 251:934–936.

Thewissen JGM, Babcock S K (1992) The origin of flight in bats. Bioscience 42:340–345.

Thomas DW (1984) Fruit intake and energy budgets of frugivorous bats. Physiol Zool 57:457–462.

Thomas DW, West SD (1989) Sampling methods for bats. Gen. Tech. Rep. PNW-GTR-243. Portland, Oregon: U.S. Dept. of Agriculture, Forest Service, Pacific NW Research Station.

Thomas DW, Bell GP, Fenton MB (1987) Variation in echolocation call frequencies in North American vespertilionid bats: a cautionary note. J Mammal 68:842–847.

Thomas SP (1987) The physiology of flight. In: Fenton MB, Racey PA, Rayner J M V (eds) Recent Advances in the study of bats. Cambridge: Cambridge University Press, pp. 75–99.

Thomson CE, Barclay RMR, Fenton MB (1985) The role of infant isolation calls in mother-infant reunions in the little brown bat (*Myotis lucifugus*). Can J Zool 63:1982–1988.

Trappe M, Schnitzler H-U (1982) Doppler shift compensation in insect-catching horseshoe bats. Naturwissenschaften 69:193–194.

Tumlison R, Douglas M E (1992) Parsimony analysis and the phylogeny of the plecotine bats. J Mammal 73:276–285.

Tuttle MD, Ryan MJ (1981) Bat predation and the evolution of frog vocalizations in the neotropics. Science 214:677–678.

Tuttle MD, Ryan MJ, Belwood JJ (1985) Acoustic resource partitioning by two species of phyllostomid bats (*Trachops cirrhosus* and *Tonatia sylvicola*). Anim Behav 33:1369–1371.

Tyrrell K (1988) The use of prey-generated sounds in flycatcher-style foraging by *Megaderma spasma*. Bat Res News 29:51.

Vaughan TA (1976) Nocturnal behavior of the African false vampire bat (*Cardioderma cor*). J Mammal 57:227–248.

Vaughan TA (1977) Foraging behaviour of the giant leaf-nosed bat (*Hipposideros commersoni*). J East Afr Wildl 15:237–249.

Vaughan TA (1986) Mammalogy, 3d Ed. Philadelphia: Saunders.

Warner RM (1982) *Myotis auriculus*. Mammal Species 191:1–3.

Welty JC (1975) The Life of Birds, 2d Ed. Philadelphia: Saunders.

Whitaker JO Jr, Black H L (1976) Food habits of cave bats from Zambia. J Mammal 57:199–204.

Wilkinson JS (1985) The social organization of the common vampire bat. I. Pattern of cause and association. Behav Ecol Sociobiol 17:111–121.

Wilkinson JS (1992) Information transfer at evening bat colonies. Anim Behav 44:501–518.

Yager DD, May ML, Fenton MB (1990) Ultrasound-triggered, flight-gated evasive maneuvers in the praying mantis *Parasphendale agrionina*. 1. Free flight. J Exp Biol 152:17–39.

Zbinden K (1989) Field observations on the flexibility of the acoustic behaviour of the European bat *Nyctalus noctula* (Schreber, 1774). Rev Suisse Zool 96:335–343.

Appendix

The scientific and common names of bats referred to in the text are shown with forearm lengths (in millimeters) and family name given in parenthesis, followed by information about distribution, diet, and foraging behavior. The asterisk (*) identifies species illustrated in Figure 2.8.

Antrozous pallidus, the pallid bat (53–60; Vespertilionidae), is widespread in western North America. It often takes arthropod prey from surfaces, having located it by passive cues (Bell 1982). *Antrozous* also use echolocation to locate flying targets (Krull 1992).

Balantiopteryx plicata, the least sac-winged Bat (38–44; Emballonuridae), is an insectivorous species of Central America (Arroyo-Cabrales and Jones 1988).

Cardioderma cor, the heart-nosed bat (54–59; Megadermatidae), occursin east Africa and takes animal prey, from insects to bats (Vaughan 1976). This species often catches its food on the ground having located it by passive cues (Ryan and Tuttle 1987).

Carollia perspicillata, the short-tailed fruit bat (40–45; Phyllostomidae), is common and widespread in the neotropics. This bat eats a variety of species of fruit and, occasionally, insects (Fleming 1988).

Chalinolobus variegatus, the butterfly bat (41–45; Vespertilionidae), is a widespread insectivorous species in the African savannah (Smithers 1983). Its echolocation calls have been described by Obrist et al. (1989).

Chrotopterus auritus, the woolly false vampire bat (78–87; Phyllostomidae), is a large "carnivorous" bat that eats large insects, lizards, and small bats. It is widespread in South and Central America, but few details are known about its biology and behavior (Medellin 1989).

Cloeotis percivali, the short-eared trident bat (30–35; Hipposideridae), is a small insectivorous species from southeastern Africa. Known to feed heavily on moths (Whitaker and Black 1976), its echolocation calls are dominated by constant frequency components of more than 200 kHz (Fenton and Bell 1981).

Desmodus rotundus, the common vampire Bat (52–63; Phyllostomidae), is the most abundant and widespread of the blood-feeding vampire bats. Like the others, it occurs in South and Central America. It is more common and better known than the other vampire bats (Greenhall and Schmidt 1988; Schmidt et al. 1991).

Eptesicus fuscus, the big brown bat (41–52; Vespertilionidae), a widespread insectivorous species of North America, also occurs in the West Indies and in the northern part of South America. This has been "the bat" of many echolocation studies (e.g., Kick 1982; Masters, Moffat, and Simmons 1985; Miller 1991), and its biology is well known (Kurta and Baker 1990).

Euderma maculatum, the spotted bat (48–51; Vespertilionidae), is a spectacular black-and-white insectivorous bat of western North America. The bat has enormous ears and uses echolocation calls that are clearly audible to most human observers (Leonard and Fenton 1984).

Hipposideros caffer, Sundevall's leaf-nosed bat (40–48; Hipposideridae), is widespread in savannah regions of subsaharan Africa. It commonly roosts in large numbers in caves and old mines and often is mistaken for the similar *Hipposideros ruber* (e.g., Bell and Fenton 1984; Fenton 1986), which is more common in rain forest situations (Smithers 1983). Moths are among the insects commonly taken by this species.

Hipposideros lankadiva, Kelaart's leaf-nosed bat (48–55; Hipposideridae), occurs in Sri Lanka and southern India, and the ontogeny of its echolocation has been reported by Rubsmen and colleagues (Rubsamen and Schafer 1990a, b; Rubsamen, Neuweiler, and Marimuthu, 1989).

Hipposideros speoris, Schneider's round-leaf bat (48–52; Hipposideridae), occurs in peninsular India and Sri Lanka. Insectivorous, its foraging behavior and echolocation have been studied in southern India (Neuweiler 1989; Neuweiler, Singh, and Sripathi 1984).

Lasionycteris noctivagans, the silver-haired bat (36–45; Vespertilionidae), is widespread in North America. This migratory species feeds on airborne insects and roosts in hollows and crevices around trees (Barclay 1986).

Lasiurus borealis, the red bat (36–42; Vespertilionidae), is widespread in North and Central America. This migratory species roosts in foliage and feeds on airborne prey (Hickey and Fenton 1990).

Lasiurus cinereus, the hoary bat (54–58; Vespertilionidae), occurs from Hawaii to the Galapagos and throughout much of North, Central, and South America. Another migratory species, *Lasiurus cinereus*, roosts in foliage and feeds on airborne insects (Barclay 1986).

Macroderma gigas, the ghost bat (105–115; Megadermatidae), occurs over much of northern Australia. This large bat feeds on prey ranging from large insects to vertebrates such as birds and mice, which it locates mainly by listening to their sounds of movement (Hudson and Wilson 1986).

Macrotus californicus, the California leaf-nosed bat (45–58; Phyllostomidae), occurs in the southwestern United State and adjacent Mexico. It takes prey from surfaces, having located them by prey-generated sounds or acute vision (Bell 1985).

*Megaderma lyra, the Indian false vampire bat (65–70; Megadermatidae), is a widespread species extensively studied in the laboratory (Fiedler 1979; Schmidt 1988; Neuweiler 1989) and field (Audet et al. 1991). This bat takes a range of prey from large arthropods to frogs, mice, and other bats and tends to rely on prey-generated sounds to locate its targets.

Megaderma spasma, the lesser false vampire bat (54–61; Megadermatidae), occurs in southeast Asia and feeds mainly on medium-sized to large insects. In captivity it takes airborne prey and prey sitting on surfaces (Tyrrell 1988).

Myotis auriculus, the Mexican long-eared myotis (37–40; Vespertilionidae), occurs in the American Southwest (Warner 1982) and takes prey from surfaces (Fenton and Bell 1979). In its behavior, *M. auriculus* resembles *Myotis emarginatus*, *Myotis evotis*, and *Myotis septentrionlis*.

Myotis daubentoni, Daubenton's bat (35–39; Vespertilionidae), appears to be the Eurasian counterpart of *Myotis lucifugus*, a small mouse-eared bat that hunts airborne targets or grabs insects from the surface of the water (Kalko and Schnitzler 1989).

Myotis emarginatus, the notch-eared Bat (37–43; Vespertilionidae), of Europe, is a versatile species sometimes taking prey from surfaces, sometimes in the air. It produces a range of echolocation calls (Schumm, Krull, and Neuweiler 1991).

Myotis evotis, the long-eared bat (36–41; Vespertilionidae), is a large-eared *Myotis* of western North America that commonly takes prey from surfaces, having located it by prey-generated sounds (Faure and Barclay 1992; Faure, Fullard, and Barclay 1990).

Myotis lucifugus, the little brown bat (34–40; Vespertilionidae), is a widespread bat of North America that commonly roosts in buildings and hibernates in caves and old mines. This species hunts airborne targets but also takes prey from the surface of water (Fenton and Bell 1979; Buchler 1980; Thomson, Barclay, and Fenton 1985).

Myotis septentrionalis, the northern long-eared bat (35–40; Vespertilionidae), is widespread in North America. Like other large-eared *Myotis*, this species commonly takes prey from surfaces (Fenton 1992).

Myotis volans, the long-legged myotis (36–44; Vespertilionidae), occurs in western United States and Canada where it feeds on airborne targets hunted in the open. For a *Myotis*, this species uses a relatively narrowband echolocation call (Fenton and Bell 1979).

Myotis yumanensis, the Yuma myotis (33–37; Vespertilionidae), occurs in western North America from British Columbia to Mexico. The bat hunts airborne targets and feeds over water (Herd and Fenton 1982).

Noctilio albiventris, the lesser bulldog bat (60–68; Noctilionidae), occurs in the Neotropics where it commonly feeds over water. Its echolocation has been studied in the laboratory (Roverud 1987) and in the field (Brown, Brown, and Grinnell 1983).

Noctilio leporinus, the greater bulldog bat (81–88; Noctilionidae), is another neotropical species that feeds on fish and other animals taken from the surface of water or from land. Its echolocation has been studied in a variety of situations (Suthers 1965; Campbell and Suthers 1988).

*Nycteris grandis, the large slit-faced bat (57–66; Nycteridae), occurs mainly in central Africa and uses low-frequency sounds of movement to locate prey ranging from insects to frogs, bats, birds, and fish (Fenton et al. 1990).

Nycteris thebaica, the Egyptian slit-faced Bat (42–52; Nycteridae), is widespread in Africa from Egypt to South Africa. This species feeds on insects ranging from orthoperans to moths, often taking prey from surfaces (Fenton, Gaudet, and Leonard 1983; Aldridge et al. 1990).

Nycticeius humeralis, the evening bat (34–39; Vespertilionidae), is common in the east-central United States where it roosts in buildings and takes airborne prey (Wilkinson 1992).

Pipistrellus pipistrellus, the pipistrelle (28–35; Vespertilionidae), is one of the more common Eurasian bats that has been well studied in many parts of Europe (e.g., Racey and Swift 1985; Avery 1991).

Pipistrellus stenopterus, the narrow-winged pipistrelle (38–42; Vespertilionidae), is known from Malaysia and Sarawak. Its echolocation behavior has been studied by Summe and Menne (1988).

Pipistrellus subflavus, the eastern pipistrelle (32–36; Vespertilionidae), is a common and widespread species of eastern North America. It appears to feed on airborne insects, but there are few data about its echolocation behavior (Fujita and Kunz 1984).

Pteronotus parnellii, Parnell's mustached bat (50–60; Mormoopidae), is widespread neotropical species noted for its specializations for detecting fluttering targets. As the New World high duty cycle bat, itts echolocation behavior has been widely studied (e.g., Goldman and Henson 1977), but there is less information about its behavior in the wild (Herd 1983).

Rhinolophus ferrumequinum, the greater horseshoe bat (50–59; Rhinolophidae), was once common and widespread in Europe and Asia but has disappeared from large parts of its original range. The topic of many studies of echolocation (e.g., Griffin and Simmons 1974; Trappe and Schnitzler 1982), there is a growing body of information about its behavior in the field (e.g., Ransome 1991).

*Rhinolophus hildebrdandti, Hildebrandt's horseshoe bat (62–66; Rhinolophidae), is the largest of the horseshoe bats and alternates between hunting from a perch and from continuous flight. This insectivorous species is widespread in eastern Africa, where it often forages in riverine forest (Fenton and Rautenbach 1986).

Rhinolophus landeri, Lander's horseshoe bat (42–45; Rhinolophidae), occurs widely in subsaharan Africa and commonly roosts in caves or other hollows (Smithers 1983). This insectivorous species produces echolocation calls dominated by 120 kHz CF components (Fenton and Bell 1981).

Rhinolophus rouxi, the rufous horseshoe bat (45–50; Rhinolophidae), occurs from Sri Lanka to Vietnam. The foraging behavior and echolocation have been studied in Sri Lanka (Neuweiler et al. 1987; Rubsamen and Schafer 1990a).

Rousettus aegyptiaca, the Egyptian fruit bat (90–105; Pteropodidae), is the species of Pteropodidae known to echolocate. This species is commonly exhibited in zoos and has been widely studied in the field and in captivity (Smithers 1983).

Tadarida brasiliensis, the Mexican free-tailed bat (38–45; Molossidae), roosts in huge numbers in some caves. It is widespread from the southern United States into Central and South America, and there are excellent studies of mother–young interactions in this species (e.g., Balcombe and McCracken 1992).

*Tonatia brasiliensis, the pygmy round-eared bat (32–36; Phyllostomidae), is a little-known insectivorous species of South and Central America (Medellin and Arita 1989).

Trachops cirrhosus, the fringe-lipped bat (58–64; Phyllostomidae), is best known for its use of male songs to locate and identify the frogs on which it feeds (Tuttle and Ryan 1981). This bat also eats other prey and although it is not well known, occurs widely in the neotropics (Tuttle, Ryan, and Belwood 1985).

Vampyrum spectrum, Linnaeus' false vampire bat (105–115; Phyllostomidae), is the largest of the New World bats and occurs widely in the Neotropics. It takes prey from birds to bats and other small vertebrates, but has been little studied in the wild (Navarro and Wilson 1982).

3

Behavioral Studies of Auditory Information Processing

CYNTHIA F. MOSS AND HANS-ULRICH SCHNITZLER

1. Introduction

Echolocating bats are nocturnal animals that rely largely on auditory information to orient in the environment and intercept prey. Bats produce high-frequency vocal signals and perceive their surroundings by listening to the features of the echoes reflecting off targets in the path of the sound beam (Griffin 1958). Computations performed on these echoes by the auditory system allow the bat to extract fine spatial information about its world through acoustic channels. This chapter attempts a comprehensive review and synthesis of data on auditory information processing and perception by sonar in echolocating bats.

The information available to a bat's acoustic imaging system is constrained by the characteristics of its species-specific sonar emissions. Species diversity in echolocation signal design therefore gives rise to differences in auditory information processing in bats (see Fenton, Chapter 2). This chapter includes a discussion of psychophysical studies of auditory information processing in species that use both constant-frequency (CF) and frequency-modulated (FM) signals. In this chapter, research on target echo detection, range estimation, horizontal and vertical localization, and movement discrimination are examined. Although some early behavioral work in the field is covered, the emphasis of this chapter is on research conducted since 1979. For a detailed review of echolocation behavior in bats before this period, see Schnitzler and Henson (1980).

2. Detection

Sonar sounds used by bats may exceed 100 dB sound pressure level (SPL) at a distance of 10 cm (Griffin 1958), and yet spherical spreading losses and excess attenuation (Lawrence and Simmons 1982a; Hartley 1989) result in echoes from small targets (e.g., insects) that may be 90 dB weaker than the transmitted signal at a distance of 1.6 m (Kober and Schnitzler 1990). Given the bat's success at capturing small prey, we must infer that it can detect very weak echoes. Indeed, a bat can only process acoustic information about the spatial dimensions of a sonar target if the signal level exceeds the animal's threshold for detection.

Target detection occurs when a bat accurately determines that an echo arriving at its ears has reflected from an object in the path of its sonar transmission. There are many parameters that influence the detection process, including the structure of the emitted signal, the characteristics of the auditory receiver, the reflective properties and the distance of the target, and the acoustic environment (see Møhl 1988). Here, each of these parameters is considered, with particular emphasis on the characteristics of the auditory receiver and the acoustic environment.

Early studies of echo detection relied on methods of obstacle avoidance and behavioral responses to airborne targets to estimate the operating range of bat sonar. A most impressive result of wire avoidance studies is that many species of bat can detect wires with diameters much less than the wavelength of their sonar sounds (summarized in Schnitzler and Henson 1980). From field studies of bats chasing airborne targets, Griffin (1953) found that the big brown bat, *Eptesicus fuscus*, first reacts to a 1-cm-diameter target at a distance of 200 cm, and in the laboratory Webster and Brazier (1965) reported that the little brown bat, *Myotis lucifugus*, reacts to a 2.1-mm-target at a distance of 60 cm and a 4.2-mm-diameter sphere at about 120–135 cm. A basic premise of this work is that the bat's first behavioral reaction to a target serves as a reliable indicator of its detection of that target. Using estimates of the level of the bat's sonar emissions, its detection range, and the reflective properties of the target, echo detection threshold estimates were 23–28 dB SPL for *Myotis lucifugus* (Griffin 1958), 25–34 dB SPL for the greater horseshoe bat *Rhinolophus ferrumequinum* (Sokolov 1972), and 9.2–21.6 dB SPL for *Myotis oxygnathus* (Airapetianz and Konstantinov 1974), all performing in wire avoidance tasks. Threshold estimates were 17 dB SPL for *Eptescius fuscus* reacting to a sphere in the field (Griffin 1958) and 15–30 dB SPL for *Myotis lucifugus* hunting fruitflies in the laboratory (Griffin, Webster, and Michael 1960). This early work and more recent field studies on the behavioral responses to airborne targets are discussed elsewhere in this volume (see Fenton, Chapter 2; Simmons et al., Chapter 4).

Changes in the bat's sonar signals as it approaches obstacles or attempts to intercept prey serve to describe the animal's vocal-motor control of acoustic information guiding spatial perception by sonar. In fact, the very features of the sonar sounds used during different phases of insect pursuit suggest the acoustic information a bat requires at different distances from its prey (Griffin 1958). However, the bat's motor response to an obstacle or target may be delayed with respect to absolute detection, and estimates of the sound level of echolocation cries produced by flying bats can be distorted by the directional properties of the transmitter and microphone. Together, these factors limit the information one can gather on target detection and the operating range of sonar in flying bats. While laboratory studies do not draw upon the richness of the bat's behavior observed under natural conditions, they present the opportunity to carefully control and measure the bat's acoustic environment. Indeed, psychophysical experiments on target detection complement data gathered in the field, and results from these different methods can be assembled to develop a unified description of target detection in bat sonar.

Behavioral studies of the bat sonar receiver have employed diverse methods. Traditional psychoacoustic measures of passive hearing with pure tone stimuli have generated data on absolute sensitivity and frequency selectivity. Experiments requiring the bat to actively interrogate its environment using sonar have yielded complementary data on the animal's detection of targets by echolocation. Collectively these studies show that the bat sonar receiver is well suited for the detection of species-specific echolocation sounds.

2.1 Absolute Sensitivity

The most extensive data on receiver characteristics important for signal detection in echolocating bats are found in behavioral audiograms. Behavioral audiograms obtained in different species show that echolocating bats hear over a broad frequency range, often spanning several octaves. Minimum thresholds are comparable to other mammalian species (see Fay 1988), typically in the range of −5 to +10 dB SPL. Figure 3.1 displays behavioral audiograms for the following species: the big brown bat, *Eptesicus fuscus* (Dalland 1965; Poussin and Simmons 1982), the little brown bat, *Myotis lucifugus* (Dalland 1965), the horsehoe bat, *Rhinolophus ferrumequinum* (Long and Schnitzler 1975), the fish-catching bat, *Noctilio leporinus* (Wenstrup 1984), the Indian false vampire bat, *Megaderma lyra* (Schmidt, Türke, and Vogler 1983), and the Egyptian tomb bat, *Rousettus aegyptiacus* (Suthers and Summers 1980).

The audiogram for *Rhinolophus ferrumequinum* was measured using classical conditioning of the heart rate response, and all other audiograms were obtained using operant conditioning procedures. Above the abscissa of each audiogram in Figure 3.1 is a horizontal bar to indicate the frequency band of the harmonic components of the echolocation sounds produced by that species. The thickness of each bar denotes the relative strength of the individual harmonic components. For all microchiropteran species, the frequency range of the bat's echolocation sound corresponds closely to a frequency region of maximum auditory sensitivity. *Rhinolophus ferrumequinum* shows the lowest threshold to a narrow frequency band, flanked by frequency regions of relative insensitivity. This narrow frequency region of high auditory sensitivity corresponds closely to the constant frequency

90   Cynthia F. Moss and Hans-Ulrich Schnitzler

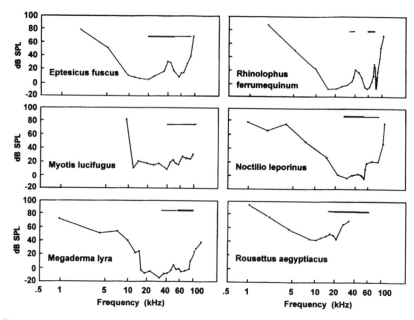

FIGURE 3.1. Behavioral audiograms of six different bat species: *Eptesicus fuscus* (Dalland 1965; Poussin and Simmons 1982), *Myotis lucifugus* (Dalland 1965), *Rhinolophus ferrumequinum* (Long and Schnitzler 1975), *Megaderma lyra* (Schmidt, Türke, and Vogler 1983), *Noctilio leporinus* (Wenstrup 1984), and *Rousettus aegyptiacus* (Suthers and Summers 1980). The horizontal bars above each audiogram show the frequency band of the echolocation sounds of each species. SPL, sound pressure level.

component of its echolocation signal. The species that shows the least sensitivity to pure tones across its audible range is *Rousettus aegyptiacus*, a megachiropteran fruit-eating bat that uses clicks for echolocation (Suthers and Summers 1980).

While most echolocation signals produced by bats are in the range of 25 to 100 kHz, many of the behavioral audiograms show thresholds well below 50 dB SPL at 10 kHz; in fact, the audible range of hearing for *Eptesicus fuscus* and *Megaderma lyra* extends down to about 1 kHz (Poussin and Simmons 1982; Schmidt, Turke, and Volger 1983). Hearing at frequencies below those used for echolocation may play an important role in social communication and detection of prey and predators through passive listening.

2.2 Frequency Selectivity

While it is of interest to establish the bat's absolute sensitivity to sounds in quiet, animals that rely on echolocation must detect signals against background noise. Wind, noise associated with flight, the communication sounds of other nocturnal animals, and jamming signals from other bats and possibly insects may all create acoustic interference for echo detection. For this reason, masked auditory thresholds provide a useful description of an animal's resistance to noise. If a masking stimulus is broadband white noise, only a portion of the noise band actually contributes to the masking of a pure tone stimulus. This was originally demonstrated by Fletcher (1940), who introduced the concept of the critical band, the frequency region about a pure tone that is effective in masking that tone. A large critical band indicates that the noise must be summed over a wide frequency band to mask the signal and therefore indicates relatively poor frequency resolution of the auditory system. By contrast, a small critical band indicates relatively high frequency resolution. Thus, the critical band indexes the animal's resistance to noise as a function of signal frequency (see Long 1994).

Fletcher included in the concept of the critical band a hypothesis proposing that the power of the noise integrated over the critical band equals the power of the pure tone signal at threshold. This implies that a critical band can be determined indirectly by measuring the detection of a pure tone against broadband masking noise, rather than directly by measuring the threshold against a variety of noise bandwidths. If one knows the level of the tone at threshold and the spectrum level of the noise, the ratio of the two provides the necessary information to determine the critical bandwidth based on Fletcher's assumptions. This ratio has been termed the critical ratio (Zwicker, Flottorp, and Stevens 1957).

Critical bands and critical ratios generally show parallel and systematic increases with signal frequency in vertebrates; however, there are note-worthy exceptions, among them the horseshoe bat, *Rhinolophus ferrume-quinum*. This bat shows a sharp decline in critical ratio (i.e., a marked increase in frequency resolution) at 83 kHz, relative to neighboring frequencies (Long 1977). This specialization for frequency resolution at 83 kHz parallels that observed for absolute hearing sensitivity in this bat (see Fig. 3.1) and reflects specialization in the horseshoe bat's cochlea (e.g., Bruns 1980; Neuweiler, Bruns, and Schuller 1980). Critical ratios have also been estimated in the echolocating megachiropteran bat *Rousettus* (Suthers and Summers 1980). The shape of the critical ratio function in this species differs from most other mammals, showing an unusual elevation of critical ratios at frequencies below 8 kHz.

Critical bands have been measured for the false vampire bat, *Megaderma lyra*, a species that uses a broadband FM signal for echolocation (Schmidt 1993). As in other mammals, the width of the auditory filters in *Megaderma* increases with center frequency. However, unlike the auditory filters measured for nonecholocating mammals, the slopes are steeper and the filter shape around the center masker frequency remained symmetrical at high masker levels. Schmidt (1993) proposed that this departure from the standard mammalian critical band filter may reflect adaptations for the perception of broadband stimuli in noisy and echo-cluttered environments.

2.3 Directionality

Directional sensitivity of the bat's hearing also plays a role in echo detection. Behavioral measurements of directional sensitivity as a function of signal frequency have been conducted only in the CF bat *Rhinolophus ferrumequinum* (Grinnell and Schnitzler 1977). This species exhibits a ventral sidelobe of relative insensitivity, yet when these behavioral measurements are combined with acoustic measurements of sonar emission directionality (Schnitzler and Grinnell 1977), the complete echolocation system of *Rhinolophus* shows radially symmetrical sensitivity, with a −6 dB bandwidth of about 15°. The directionality of emission patterns has been studied in detail in other species, including the FM bat *Eptesicus fuscus* (Hartley and Suthers 1989), the grey bat *Myotis grisescens* (Shimozawa et al. 1974), and the leaf-nosed bat *Carollia perspicillata* (Hartley and Suthers 1987); however, behavioral sensitivity measures of directionality in these animals are currently lacking.

2.4 Psychophysical Studies of Echo Detection

In the past decade, psychophysical studies on echo detection in bats have been carried out in laboratory settings where the distance between the bat and the target has been controlled. It is important to note that estimates of echo detection threshold depend on several parameters: the features of the bat's sonar signals, the conditions of the environment, and the characteristics of the bat's sonar receiver.

Sonar signal structure in detection experiments is under the bat's active control and can be adapted to improve target detection under different conditions. Increasing the duration and level of the sonar signal and decreasing the bandwidth all serve to improve conditions for echo detection. It is therefore essential that signal structure be continuously monitored in a detection experiment.

The acoustic environment influences the extent to which masking of the target echo occurs, and there are several potential sources of masking: the bat's own sonar transmission, clutter echoes surrounding the target of interest, noise generated in the environment, and noise internal to the bat's sonar receiver. With the exception of internal noise, all other sources of noise can be monitored and controlled in behavioral studies of detection.

Also of importance to measures of echo detection is the sensitivity of the bat's auditory receiver following each sonar transmission. The bat's hearing sensitivity is reduced during and after signal emission by contractions of the middle ear muscles (Henson 1965; Suga and Jen 1975) and by neural suppression in the central auditory system (Suga and Schlegel 1972; Suga and Shimozawa 1974). There is a gradual recovery of hearing sensitivity following each sonar emission, and therefore changes in echo detection depend on the delay (distance) of the target echo.

2.4.1 Detection of Target Echoes in Quiet

Estimates of echo detection thresholds obtained from psychophysical studies are summarized in Table 3.1. The first behavioral laboratory study to examine detection of targets at controlled distances was conducted by Kick (1982), who trained the FM bat *Eptesicus fuscus* in a two-alternative forced-choice (2-AFC) procedure to respond to the presence of a ball (to the left or right of its observing position) by crawling toward it on a Y-shaped platform. In a 2-AFC procedure, the subject is required to make a response in every two-choice trial, and its performance level should directly reflect the relative difficulty of the task. In this particular experiment the bat's performance was nearly 100% correct when the target was positioned close to the animal and the target echoes were strong. As the target was placed further from the bat, the echoes were weaker, and the bat's performance level dropped off. When the bat was no longer responding to the presence of the target, its mean performance was close to that expected by chance (50% correct). In this study, as well as most others implementing a 2-AFC procedure, the criterion for threshold was the SPL producing 75% correct responses, that is, halfway between 100% correct and chance performance.

A schematic of the experimental conditions in Kick's study is presented in Figure 3.2B. The solid circle represents the target, which was presented either to the bat's left or right on a given trial. The bottom trace in this figure illustrates the bat's emission followed by a target echo; the time separation between these sounds depended on the distance between the bat and the target (see Section 3, Ranging). This particular study minimized echoes from other objects in the room, and hence no clutter echo is shown. The bat's task was to crawl down one arm of the Y-shaped platform, toward the side on which the target was presented. The distance of this target was increased over days of testing, resulting in successively weaker echoes returning to the bat's ears at increasing delay times. Detection of two sizes of spheres was studied (4.8 mm and 19.1 mm in diameter), and the detection range for the two spheres was 2.9 and 5.1 m, respectively. As noted previously, the maximum effective range of bat sonar was previously inferred from changes in the animal's echolocation behavior in obstacle avoidance and target interception tasks, and these data represent the bat's *distance of reaction*. Kick's data represent the bat's *distance of detection*, and they are the first to demonstrate the operation of the bat's sonar receiver for distances beyond 3 m. The threshold for echo detection in Kick's experiment was estimated at 30 and 60 kHz to be $+5$ and -12 dB SPL, respectively, for the 4.8-mm sphere and -2 dB and -21 dB SPL, respectively, for the 19.1-mm sphere. Because the echo level is stronger at 30 kHz than at 60 kHz, Kick suggested that the bat uses this component to detect the presence of the target. Thus, she concluded that the echo detection threshold of *Eptesicus* is approximately 0 dB SPL (see data plotted for the two sizes of spheres in Fig. 3.3A).

TABLE 3.1 Summary of behavioral data on detection by sonar in echolocating bats.

Species	Procedure	Target distance (cm delay/msec)	Speaker angular position	Speaker position relative to target echo	Threshold echo sound level	Author(s) (year)
Eptesicus fuscus	2-AFC simultaneous method of limits	To 510 cm (29.6 msec) real targets	40°	n/a	0 dB SPL	Kick (1982)
Eptesicus fuscus	2-AFC simultaneous method of limits	17 cm/1.0 msec; 110 cm/6.4 msec; sonar sound playback	40°	25 cm/1.45 msec; 2.1 m/12.2 msec after target echo	36 dB SPL; 8 dB SPL	Kick and Simmons (1984)
Eptesicus fuscus	Yes/no sequential staircase	46 cm/2.7 msec; sonar sound playback	Single channel	68 cm/3.9 msec after target echo	37–38 dB SPL	Møhl and Surlykke (1989)
Eptesicus fuscus	2-AFC simultaneous method of limits	80 cm/4.6 msec; canned signal playback	40°	15 cm/0.87 msec before target echo	26 dB SPL	Master and Jacobs (1989)
Eptesicus fuscus and Noctilio leporinus	Yes/no sequential staircase	100 cm/5.8 msec; sonar sound playback	Single channel	100 cm/5.8 msec after target echo	40 dB SPL; 40 dB SPL	Hartley (1992a)
Eptesicus fuscus	2-AFC simultaneous method of limits	54 cm/3.1 msec; 40 cm/2.3 msec; 80 cm/4.6 msec; 160 cm/9.3 msec; all with sonar sound playback	40° 40°	100 cm/5.8 msec after target echo; 12 cm/0.7 msec before target echo;	33 dB SPL; 21 dB SPL; 11 dB SPL; 20 dB SPL	Simmons et al (1988)
Eptesicus fuscus	2-AFC simultaneous method of limits	40 cm/2.3 msec; 80 cm/4.6 msec; sonar sound playback	40°	12 cm/0.7 msec before target echo	23 dB SPL; 12 dB SPL	Simmons, Moffat, and Masters (1992)
Eptesicus fuscus	2-AFC simultaneous method of limits	57 cm/3.3 msec; phantom target sonar sound playback	40°	20 cm/1.16 msec before target echo	55–59 dB SPL	Moss and Simmons (1993)
Pipistrellus pipistrellus	Yes/no sequential staircase	138 cm/8.0 msec; triggered canned signal playback	Single channel	85 cm/4.9 msec before target echo	35 dB peSPL	Møhl (1986)
Eptesicus serotinus	Yes/no sequential staircase	55 cm/3.2 msec; sonar sound playback	Single channel	88 cm/5.1 msec after target echo	40–48 dB peSPL	Troest and Møhl (1986)

2-AFC, two-alternative forced-choice detection experiments; peSPL, peak equivalent sound pressure level SPL, sound pressure level; n/a, not available. level.

DETECTION

A. yes-no real target	B. 2-AFC real target	C. yes-no phantom target	D. yes-no phantom target	E. 2-AFC phantom target	F. 2-AFC phantom target
•	•	O m • s •	s • O m •	O m • • s • •	s • • O m • •
decision platform	decision platform	decision platform	decision platform	decision platform	decision platform
bat echo ◆——◆	bat echo ◆——◆	bat clutter echo ◆—◆——◆	bat clutter echo clutter ◆—◆—◆—◆	bat clutter echo ◆——◆——◆	bat clutter echo clutter ◆—◆—◆—◆

RANGING

G. yes-no real target	H. 2-AFC real targets	I. yes-no phantom target	J. yes-no phantom target	K. 2-AFC phantom target	L. 2-AFC phantom target
• ↕	• •	O ↕ m • s •	s • O ↕ m •	O m • • s • •	s • O m • •
decision platform	decision platform	decision platform	decision platform	decision platform	decision platform
bat echo ◆——◆	bat echo ◆——◆	bat clutter echo ◆—◆——◆	bat clutter echo clutter ◆—◆—◆—◆	bat clutter echo ◆——◆——◆	bat clutter echo clutter ◆—◆—◆—◆

FIGURE 3.2A–L. Schematic of behavioral experiments for target detection (above, A–F) and target range difference discrimination (below, G–L). For all the experiments, the bat was trained to echolocate from the base of a decision platform and to indicate a response by crawling (or flying) to the left or right of its start position (see text). *Filled circles* represent real targets; *open ovals* represent phantom targets; *m*, microphone, illustrated with concentric rings; *s*, speaker, illustrated with concentric rings. Below each schematic is an illustration of the temporal sequence of sounds the bat hears under given experimental conditions: *bat*, bat's echolocation cry; *clutter*, echoes from speaker and microphone; *echo*, target echo or playback signal. *Detection*: In the yes/no detection experiments, the bat was rewarded for reporting whether the target was present or absent. It crawled down one arm of the platform to indicate "yes" (target present) and down the other arm of the platform to indicate "no" (target absent). In the two-alternative forced-choice (2-AFC) detection experiments, the bat was rewarded for reporting whether the target was presented to the left or right of its observing position. It learned to crawl down the arm of the platform on which it detected the target. *Ranging*: In the yes/no ranging experiments, the bat was rewarded for reporting whether the target was presented at a fixed distance (d_1) or at a greater distance (d_2). It crawled down one arm of the platform to indicate "yes" (d_1 presented) and down the other arm of the platform to indicate "no" (d_1 not presented). The *arrows* shown by the targets in G and I illustrate the change in target distance from trial to trial. In the 2-AFC ranging experiments, the bat was rewarded for crawling toward the closer target, which was presented to the its left or right following a pseudorandom schedule.

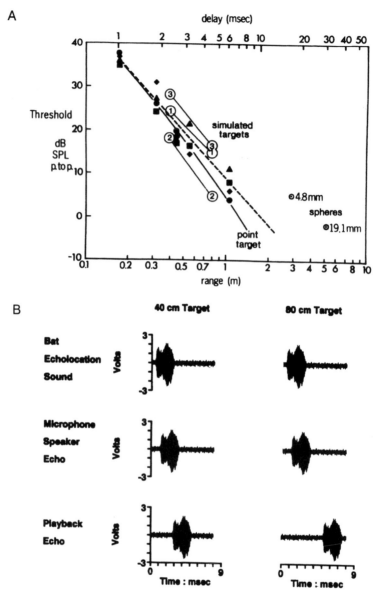

FIGURE 3.3A,B. (A). Echo detection thresholds for *Eptesicus fuscus* as a function of target distance or echo delay. Thresholds were all based on 75% correct detection performance. *Open circles* (4.8- and 19.1-mm spheres) show data taken from Kick (1982; real targets, see Fig. 3.2B); *solid symbols (triangles, squares, diamonds, circles)* represent data for individual animals taken from Kick and Simmons (1984; phantom target echoes; see Fig. 3.2F) *Open circles* marked *1, 2, 3,* are data points from individual animals taken from Simmons, Moffat, and Masters. (1992; phantom target echoes; see Fig. 3.2E). *Dashed line* shows the regression line fit to the solid data points; the *solid line* shows the amplitude of

Subsequent psychophysical studies of echo detection in bats have often electronically simulated echoes from targets. In phantom target experiments, the bat is trained to rest on a platform and emit sonar cries. Broadcast back to the bat through loudspeakers are either delayed playback replicas of its sonar emissions or triggered playbacks of a digitally stored sonar signal. In both instances, the bat is trained to respond to the presence or absence of a sonar signal playback that is delayed to represent an echo from a target positioned at ranges typically from 40 to 120 cm (see Fig. 2C–F).

No psychophysical data are available for bats that use CF FM sounds (e.g., *Rhinolophus ferrumequinum*, or the mustached bat, *Pteronotus parnelli*); however, *Noctilio leporinus* uses short quasi-CF signals (i.e., very shallow FM) in its repertoire (see Wenstrup 1984; Wenstrup and Suthers 1984). The most extensive work on detection in bats has employed the FM bat *Eptesicus fuscus*. Perhaps the most striking comparison shown in Table 3.1 is the difference in echo detection threshold estimates for *Eptesicus* reported by Kick (1982) using real targets and others using target simulation methods. In fact, the detection threshold reported by Kick ranges from 8 to 59 dB lower than estimates obtained with echo playback systems.

Why are Kick's detection threshold estimates so much lower than those obtained using virtual targets? And why are the detection threshold estimates obtained using virtual targets so variable? Most important is the influence of clutter echoes in virtual target experiments. Unavoidable in playback experiments are echoes from objects in the apparatus, such as microphones and speakers, that are used to generate virtual target echoes (in Fig. 3.2, compare A and B with C–F). These clutter echoes may serve as either forward or backward maskers, depending on their temporal relation to the virtual target echoes. For example, clutter echoes from the microphone-speaker pair shown in Figures 3.2C and 3.2E may serve as forward maskers of the phantom target echoes, while clutter echoes from the speakers shown in Figures 3.2D and 3.2F may serve as backward maskers (see also Fig. 3.3B).

2.4.2 Clutter Interference

Several experiments have shown that clutter echoes interfere with the detection of playback echoes, even if these sounds are temporally isolated

---

FIGURE 3.3 (*continued*) echoes from an ideal point target at different distances. (B) Schematic of the temporal relationship between sounds produced by bats in a detection experiment and the resulting clutter echoes and phantom target playback echoes for each sonar emission, based on the 40-cm and 80-cm target echoes studied by Simmons, Moffat, and Masters (1992; Fig. 3.2E). Note that the conditions for forward masking differ between the 40- and 80-cm target distances.

from other signals. For example, Troest and Møhl (1986) showed that the detection threshold of *Eptesicus serotinus* for a phantom target in a single-channel yes/no experiment dropped by 20 dB when the speaker diameter in the playback apparatus was reduced from 75 to 15 mm. Simmons et al. (1988) reported that the zone of clutter interference depends on the absolute range of the target. For a target distance of 54 cm, surrounding objects produce a threshold elevation if separated from the target by less than 15 cm. Hartley (1992a) also demonstrated the effect of clutter on target detection in *Eptesicus fuscus* and *Noctilio leporinus* by placing rings around the loudspeaker to increase the clutter echoes by measured amounts. These experiments showed backward masking of virtual target echoes by clutter. There are also data to suggest that forward masking by clutter can influence echo detection thresholds as well (e.g., Simmons et al. 1988; Moss and Simmons 1993). Thus, the large differences between estimates of echo detection thresholds in bats may be largely the result of the masking effects of clutter echoes. In addition, changes in the bat's sonar emission level across trials and days, directly influencing the level of sonar playback echoes, contribute error to estimates of echo detection thresholds, which in many studies are based on very few measures of the SPL of the bat's sonar transmissions.

2.4.3 Detection of Target Echoes in Noise

In addition to the forward and backward masking effects of clutter, the presence of simultaneous and interrupted noise can disturb target echo detection by sonar in bats. This finding has behavioral relevance to echolocation performance of bats foraging under natural conditions where environmental noise may influence target detection. Roverud and Grinnell (1985a) demonstrated this phenomenon in the CF-FM bat *Noctilio albiventris* under conditions of continuous and pulsed noise (schematic of behavioral testing shown in Fig. 3.2A). Animals were unable to detect the presence of a 19-mm sphere at a distance of 35 cm when white noise (20–80 kHz) was broadcast continuously at levels above 54 dB SPL. When the white noise was pulsed, target detection performance remained above chance levels for noise duty cycles below 80%. While the bats performing in this task increased the repetition rate of their sonar sounds in the presence of noise, there was no evidence of an increase in sound duration. No information was provided in this report on the estimated echo levels required for detection in quiet or the signal-to-noise ratio (S:N) that interferes with echo detection performance. These critical aspects of target detection and masking were carefully considered by Troest and Møhl (1986), who used a single-channel target simulator to measure simultaneous masking in *Eptesicus serotinus* (a schematic of the apparatus is shown in Fig. 3.2D). In this study, the bat's detection threshold in quiet was 40–48 dB peSPL (peak equivalent sound pressure level; see Stapells, Picton, and

Smith 1982) and elevated by 7–8 dB in the presence of continuous broadband noise. Under simultaneous masking conditions, the ratio of signal energy to noise spectrum level ranged from 36 to 49 dB. The loudspeaker used in these masking experiments was positioned 88 cm from the bat and had a 75-mm diameter, producing a strong clutter echo that limited the bat's detection performance (see earlier).

2.4.4 Aural Integration Time

Also using a single-channel target simulator, Møhl and Surlykke (1989) compared simultaneous and backward masking in *Eptesicus fuscus*, but in this experiment used a smaller diameter speaker (15 cm in diameter at a distance of 68 cm) to reduce the effect of clutter. In this experiment, they found that echo detection in quiet was about 37.5 dB SPL. When continuous broadband noise was presented, the detection thresholds were elevated by about 15 dB relative to those measured in quiet, and the signal energy to noise spectrum level ratio was about 35 dB. The effect of masking was relatively stable over masker durations from 2 to 37.5 msec, and backward masking was evident only for echo–noise delays as long as 2 msec. The duration of the bats' sounds was 1.7–2.4 msec, suggesting that masking only occurred for noise pulses that overlapped the playback echoes. The results of this experiment suggest that aural integration time in the bat is tied to the duration of the bat's sonar cries. This estimate of integration time is an order of magnitude lower than that reported in the human psychophysical literature (e.g., de Boer 1985) and suggests an auditory specialization for echolocation in bats.

2.4.5 Echo Gain Control

As a bat flies toward a stationary target, the echoes returning to its ears increase in amplitude, because the distance over which spherical spreading losses occur is decreased. For each twofold reduction in target range, there is a 12-dB increase in echo amplitude. This increase in echo amplitude could produce a systematic change in the discharge latency of auditory nerve fibers in response to echoes, because neural response latency decreases with increasing stimulus level (see Simmons, Moss, and Ferragamo 1990). In turn, a decrease in response latency could disturb the bat's accurate estimate of echo delay, its perceptual cue for target distance (e.g., Simmons 1973; Simmons et al. 1990; see also Section 3, Ranging). Thus, the increase in echo sound level that occurs when a bat approaches a target could potentially distort its perception of target range.

Kick and Simmons (1984) proposed that there is little or no change in the sensation level of echoes over a target distance of about 1.5 m, largely because of middle ear muscle modulation of hearing sensitivity. When a bat emits sonar pulses, the middle ear muscles contract, at least in part to protect the inner ear from damage caused by the intense sound levels of its vocalizations. These middle ear muscle contractions serve to reduce hearing sensitivity, producing a maximum rise in threshold around the time of vocal emission (Henson 1965; Suga and Jen 1975). As the middle ear muscles relax during the several milliseconds following each sonar emission, hearing sensitivity increases. In addition, there is evidence for central attenuation of neural responses to emitted sounds, which could affect sensitivity to echoes occurring within a restricted time window (Suga and Schlegel 1972). Thus, the bat should be more sensitive to echoes returning from more distant targets at longer delays than to closer targets at shorter delays. Kick and Simmons (1984) suggested that the net effect is an automatic gain control mechanism whereby the bat's sensitivity to echoes increases by about 11–12 dB for each doubling of target range, effectively offsetting the effect of spherical spreading losses on signal detectability over a distance of 1.5 m.

The gain control hypothesis developed from data on the hearing sensitivity of *Eptesicus fuscus* to phantom target echoes at simulated distances ranging from 17 to 110 cm. Kick and Simmons (1984) trained bats in a 2-AFC procedure to echolocate from the base of a Y-shaped platform and to indicate the presence of a target echo to the left or right of its observing position. The bat's echolocation sounds were picked up by two condenser microphones placed beyond the arms of the platform approximately 10 cm from the observing position (schematic shown in Fig. 3.2F). These sounds were played back to the bat on either the left or right side through an electrostatic speaker whose distance determined the total delay (or perceived distance) of the phantom target echo. The total delay of the phantom target echoes was set by the travel time of the bat's sound to the microphone and that from the speaker back to the bat. The sound level of the playback echo was determined by the bat's own emission level, and echo level could be modified by attenuating the signal output to the speaker. The peak-to-peak SPL producing 75% correct detection performance was estimated from recordings of sounds produced by bats during the task. A measure of 105–110 dB was taken and used to assign thresholds for each delay or simulated range studied. The results of this study show that the threshold for detecting a phantom target echo decreased by about 11 dB for each doubling of target distance, a change that would approximately cancel the spherical spreading losses in echo level of approximately 12 dB for each doubling of distance (data plotted as solid symbols in Fig. 3.3A).

As described previously, the detection of a target echo is strongly influenced by the presence of clutter echoes from objects present in the behavioral testing setup. Using phantom targets necessitates the presence of a microphone and a loudspeaker whose echoes can serve as both forward and backward maskers, depending on their spatial relationship with respect to the playback echo. In addition, the sonar transmission itself can function as a forward masker if the echo delay is short with respect to the duration of the bat's echolocation sound. In the Kick and Simmons study, a phantom target echo was presented at a shorter distance, or delay, than the clutter echo returning from the loudspeaker (see Fig. 3.2F). As the phantom target was presented at shorter distances, the speaker was moved closer to the bat, thus reducing the time interval between the playback echo and the clutter echo. The time between the bat's sonar transmission and echo also shortened, increasing the effectiveness of the bat's own echolocation sound as a forward masker. Moreover, as the speaker–bat distance decreased, the amplitude of the speaker echo increased. Both the reduction in temporal separation between the target and clutter echoes and the increase in clutter echo level could serve to increase the magnitude of backward masking effects on target detection at shorter distances. Without measures of the contribution of clutter to the echo detection thresholds in this study, the change in threshold with target distance is difficult to evaluate.

In an attempt to minimize backward masking effects in estimating the slope relating hearing sensitivity and target distance, Simmons, Moffat, and Masters (1992) used a modified experimental setup. The speakers in this second 2-AFC experiment remained fixed at a distance of 22 cm from the bat, and a variable electronic delay of sound playbacks simulated target echoes at two different distances, 40 and 80 cm from the position of the bat (see Fig. 3.2E). The data from this study also showed that the bat's detection threshold decreased by about 11 dB for a doubling of target range (see open symbols numbered 1, 2, and 3 in Fig. 3.3A). However, this experiment was not free of masking effects either. In this study, the playback echo delays were 2.32 and 4.64 msec for the 40- and 80-cm target distances, respectively. The microphone and speaker pair, positioned at a distance of 22 cm, produced clutter echoes at a delay of 1.27 msec. Echolocation sound durations recorded in this study were 2–4 msec, and thus there was considerable overlap between echoes from the speaker-microphone pairs and the phantom target echoes, particularly for the 40-cm target (see Fig. 3.3B). In addition, each of the bat's sonar transmissions may have interfered with detection of target echoes, especially at the lesser of the two distances. Therefore, in this experiment, forward masking may have influenced detection thresholds, and presumably such masking was stronger for the 40-cm target than that for the 80-cm target. Here again, the slope relating detection threshold to target distance presumably includes some contribution of masking effects.

Hartley (1992a), having noted the potential contribution of backward masking to the data reported by Kick and Simmons (1984), used a different experimental paradigm to study gain control in both *Eptesicus fuscus* and *Noctilio leporinus*. With a one-channel phantom target playback apparatus, Hartley trained bats to report which of two successive echoes sounded louder. He presented pairs of sounds to the bats, simulating echoes from targets at distances ranging from 0.28 to 11.48 m; the second echo playback in the pair arrived at approximately double the delay of the first. The speaker-microphone pair that picked up and played back the bat's sounds was positioned approximately 10 cm from the animal's head and produced a clutter echo of approximately 60 dB SPL. The echo playback level of the pairs of sounds was modified on line by attenuators adjusted to present suprathreshold echoes in the range of 70–80 dB SPL.

Hartley's data indicate that bats perceive two echoes to be equally loud when their delay (target distance) differs by a factor of two and their SPL differs by about 6.7 dB for *Eptesicus fuscus* and 7.2 dB for *Noctilio leporinus*. From these results, Hartley inferred that the slope of the echo gain control function is about 6–7 dB per halving of target distance, rather than the 11–12 dB reported by Simmons and colleagues. Furthermore, Hartley measured the sonar emission levels of *Eptesicus* tracking an approaching target (maximum range 1 m) and found that the bat reduced pulse intensity by 6 dB per halving of target range (Hartley 1992b). This distance-dependent intensity compensation of sonar emissions, together with the automatic gain control, would serve to stabilize the sensation level of echoes over a target range of about 1–2 m. Thus, two very different approaches to the question of echo gain control in bats generated conflicting results, but congruent conclusions.

How do we reconcile the results and interpretation of these studies? First, it may be that the slope of the gain control function is level dependent: it may be steeper at echo levels close to threshold for detection (Simmons, Moffat, and Masters 1992) than at suprathreshold levels used for loudness discrimination (Hartley 1992a). A complete test of this idea would require estimating the slope of the gain control function at several different echo levels relative to threshold. There is also evidence to suggest that the level of a bat's sonar emissions changes with target size and distance (Reetz and Schnitzler 1992; Hartley 1992a,b), a factor that was not considered in using a fixed estimate of sonar emission level to determine detection thresholds for targets at different distances (Kick and Simmons 1984; Simmons, Moffat, and Masters 1992). Using measures of bat sonar transmission levels for each of the target distances tested might change estimates of echo detection thresholds and possibly alter the slope of the gain control function.

As noted, it is also possible that measures of the gain control slope based on echo detection thresholds are influenced by forward masking by the sonar transmission and by clutter echoes from microphones and speakers, but the magnitude of this effect was not estimated in the study by Simmons, Moffat, and Masters (1992). Measures of the signal to clutter ratios in detection experiments (see for example, Møhl 1986; Troest and Møhl 1986; Masters and Jacobs 1989; Hartley 1992a) at different distances could help to clarify the contribution of backward and forward masking to estimates of echo gain control.

2.5 Summary

In conclusion, estimates of echo detection thresholds in bats yield data differing by more than 50 dB in the same species. Echo detection in bats reflects the combined properties of the signal, environment, target, and receiver, all of which varied across the studies reviewed here. Moreover, the level of the sonar transmissions used by the bats may have differed across individual animals, conditions, and experimental trials within a single study (Hartley 1992a,b; Reetz and Schnitzler 1992; Denzinger and Schnitzler 1994), contributing error to estimates of echo detection calculated from a few measures of sonar transmission level. Indeed, the critical parameters determining the limits of echo detection were not measured in many of the studies cited. For future work in this area, we would like to emphasize the importance of careful control and measurement of these parameters.

3. Ranging

Insectivorous bats require an accuracy of about 1–3 cm to successfully intercept prey (Webster and Griffin 1962; Trappe 1982); however, some behavioral studies of the bat's target ranging performance in the laboratory demonstrate distance discrimination accuracy several orders of magnitude higher. Psychophysical estimates of ranging accuracy depend on the bat's perceptual task, the presentation of the stimuli, and the behavioral methods employed. Most important is the perceptual task, and in this section the discussion is organized around three types of tasks employed in the study of target ranging in bats: range difference discrimination, range jitter discrimination, and range resolution/two-wavefront discrimination.

3.1 Range Difference Discrimination

Echolocating bats use the time delay between sonar emission and returning echo to determine the distance to a target (Hartridge 1945). This was first demonstrated experimentally by Simmons (1973), who trained bats in a 2-AFC procedure to discriminate the distance between two targets positioned at different ranges. He first estimated the bat's distance discrimination threshold using two spheres, positioned to the left and right of the animal's observing position. The bat was trained to scan the two targets from the base of a Y-shaped platform and crawl down one arm of the platform toward the closer of the targets (schematic of the apparatus is shown in Fig. 3.2H). The closer target's left or right position was randomized from trial to trial, and the difference in distance between the two was gradually reduced until the bat's performance fell to chance (see Fig. 3.4A). Simmons then replaced the spheres with playback replicas of the bat's sonar sounds whose delays were adjusted to simulate target echoes at different ranges (schematic shown in Fig. 3.2L). (The two-way travel time of the bat's sonar emission and the returning echo results in an echo delay of about 58 μsec for each centimeter of target distance.) Simmons found that the performance curves and range difference thresholds using real targets and virtual targets were similar, suggesting that the bat experienced the playback

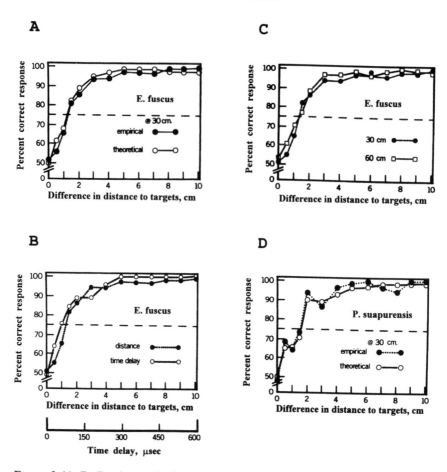

FIGURE 3.4A–D. Psychometric functions on range difference discrimination for *Eptesicus fuscus* (A–C) and *Pteronotus suapurensis* (D) using a two-alternative forced choice (2-AFC) procedure. (A) Empirical and theoretical curves for *Eptesicus* at 30 cm. (B) Discrimination performance using real and phantom targets is compared for *Eptesicus* at 30 cm. (C) Behavioral performance curves for *Eptesicus* using real targets at 30 and 60 cm. (D) Empirical and theoretical curves for *Pteronotus* using real targets at a distance of 30 cm. Theoretical curves are based on the envelope of the autocorrelation function of the species' sonar transmissions, corrected for head movements in the two-choice task, (From Simmons 1973.)

echoes at different delays as targets at different ranges (Fig. 3.4B). The behavioral comparison in performance using real and phantom targets was conducted only for the species *Eptesicus fuscus*.

Simmons (1973) reported that *Eptesicus* can discriminate the range difference of two targets separated by about 12 mm (70-μsec echo–delay difference), and that this threshold is relatively stable over test distances between 30 and 240 cm (e.g., see the performance data at 30 and 60 cm shown in Fig. 3.4C). He also measured range difference thresholds using real targets in the horseshoe bat *Rhinolophus ferrumequinum*, the naked-backed bat *Pteronotus suapurensis*, and the spear-nosed bat *Phyllostomus hastus*. Both *Pteronotus suapurensis* (see Fig. 3.4D) and *Phyllostomus hastus* were tested with targets at different distances, and, like *Eptesicus*, showed that range difference threshold is relatively independent of absolute distance. The stability of range difference thresholds over a distance of 30 to 240 cm has important implications for evaluating receiver models, and this is discussed in Section 4.

During the past 20 years, range difference (or echo delay difference) thresholds have been measured in a variety of bat species. These data are summarized in Table 3.2. While estimates of echo detection thresholds in bats differ widely (see Table 3.1), behavioral measures of range difference thresholds are remarkably similar, even across different experimental methods and species. (For schematic diagrams of different methods used to study range difference discrimination, see Fig. 3.2G–L.) This finding suggests that range difference discrimination may be less sensitive to the effects of clutter and therefore more stable under varying experimental conditions. Most species studied show range difference thresholds between about 8 and 15 mm, corresponding to echo-delay differences of 46 and 87 μsec. The two striking exceptions are *Rhinolophus ferrumequinum* (Simmons 1973; Airapetianz and Konstatinov 1974), tested in a 2-AFC simultaneous discrimination task (see Fig. 3.2H), and the lesser bulldog bat, *Noctilio albiventris* (Roverud and Grinnell 1985a), tested in a single-target yes/no sequential discrimination (see Fig. 3.2G), both of which show range difference discrimination thresholds of about 30–41mm (174–238 μsec of echo–delay difference).

The higher discrimination thresholds obtained for *Rhinolophus ferrumequinum* may directly reflect limitations of the information carried in its sonar signals about target range. It is widely accepted that the CF-FM bat uses the FM component of its echolocation sound for target distance estimation, and the bandwidth of the FM signal influences the bat's ranging performance (e.g., Simmons 1973; Schnitzler and Henson 1980). Surlykke (1992) reported data consistent with this view. She demonstrated that the range difference threshold in *Eptesicus fuscus* was elevated by a factor of 2–3 when it listened to 1-msec triggered-playback echoes of its FM sounds that were either 40 kHz lowpass or 40 kHz highpass filtered. Summarized in Table 3.2 is the approximate bandwidth of the FM components used by different species of bats performing in ranging tasks. The bandwidth of the FM component of *Rhinolophus*'s sound is narrower than that of other species showing lower range difference discrimination thresholds. The role of bandwidth on echo ranging performance is discussed further in Section 4, Receiver Models.

The elevated threshold of *Noctilio albiventris* in the single-target sequen-

TABLE 3.2 Summary of behavioral data on range difference discrimination in bats.

Species	Procedure	Target Distance (cm delay/msec)	Target Angular Separation	Signal Bandwidth (strongest harmonic, kHz)	Threshold (mm/delay/μsec)	Author(s) (year)
Eptesicus fuscus	2-AFC simultaneous method of limits	30 cm/1.74 msec; 60 cm/3.48 msec; 240cm/13.9 msec; real targets	40°	25–35	12 mm/70 μsec 13 mm/75 μsec 14 mm/81 μsec	Simmons (1973)
Eptesicus fuscus	2-AFC simultaneous method of limits	30 cm/1.74 msec; sonar sound playback	40°	25–35	10 mm/58 μsec	Simmons (1973)
Eptesicus fuscus	2-AFC simultaneous method of limits	80 cm/4.6 msec; canned signal playback	40°	25–35	7 mm/41 μsec– 17 mm/99 μsec	Masters and Jacobs (1989)
Eptesicus fuscus	Yes/no sequential staircase	54–74 cm, 3.1–4.3 msec; canned signal playback	Single channel	25–35	6 mm/35 μsec– 11 mm/64 μsec	Miller (1991)
Eptesicus fuscus	Yes/no sequential staircase	40–46 cm, 2.32–2.67 msec; canned signal playback	Single channel	25–35	10 mm/58 μsec– 20 mm/116 μsec	Surlykke (1992)
Eptesicus fuscus	2-AFC simultaneous method of limits	30 cm/1.74 msec; sonar sound playback	40°	25– 35	13.8 mm/80 μsec	Denzinger and Schnitzler (1994)
Phyllostomus hastus	2-AFC simultaneous method of limits	60 cm/3.48 msec; 120 cm/6.9 msec; real targets	40°	35	12 mm/70 μsec 12 mm/70 μsec	Simmons (1973)
Pteronotus suapurensis	2-AFC simultaneous method of limits	30 cm/1.74 msec; 60 cm/3.48 msec; real targets	40°	22.5	15 mm/87 μsec 17 mm/99 μsec	Simmons (1973)
Rhinolophus ferrumequinum	2-AFC simultaneous methods of limits	30 cm/1.74 msec; real targets	40°	15	30 mm/174 μsec	Simmons (1973)
Rhinolophus ferrumequinum	2-AFC simultaneous method of limits	100 cm/5.8 msec; real targets	13°	15	41 mm/238 μsec	Airapetianz and Konstantinov (1974)

Species	Method	Distance/duration	Angle		Resolution	Reference
Myotis oxygnathus	2-AFC simultaneous method of limits	100 cm/5.8 msec; real targets	13°	85	8 mm/46 μsec– 12 mm/70 μsec	Airapetianz and Konstantinov (1974)
Pipistrellus pipistrellus	2-AFC simultaneous method of limits	24 cm/1.39 msec; real targets	75°	65	15 mm/87 μsec	Surlykke and Miller (1985)
Noctilio albiventris	2-AFC simultaneous method of limits	35 cm/2.03 msec; real targets	40°	20	13 mm/75 μsec	Roverud and Grinnell (1985a)
Noctilio albiventris	Yes/no sequential method of limits	35 cm/2.03 msec; real targets	Single target	20	30mm/174 μsec	Roverud and Grinnell (1985a)

tial discrimination task compared to that obtained in the 2-AFC simultaneous discrimination are the only data showing a large difference in the bat's ranging performance with changes in the behavioral paradigm. In the yes/no sequential task, the bat was trained to recognize a sphere at 35 cm and respond differentially between this target range and a more distant one (see Fig. 3.2G). Thus, the bat learned to compare a target's distance with a stored representation of 35-cm range, and it is perhaps not surprising that its performance under these conditions was poorer than in the simultaneous discrimination task. However, other researchers have used the single-target sequential discrimination method and reported range difference thresholds of about 6–15 mm in *Eptesicus fuscus* (e.g., Miller 1991; Surlykke 1992), which compare well with estimates obtained using a modified method of limits and a 2-AFC procedure in the same species (e.g., Simmons 1973; Masters and Jacobs 1989; Denzinger and Schnitzer 1994). The congruence of the threshold estimates using different methods with *Eptesicus fuscus* prompt us to further evaluate the conflicting range difference data in *Noctilio albiventris*.

The experimental conditions that apparently differentiate the single-target sequential discrimination study by Roverud and Grinnell (1985a) and others cited (Miller 1991; Surlykke 1992) include both the use of real targets and the psychophysical procedure. It may be that the real targets impose difficulties for the bat discriminating range differences from successive stimulus presentations, because any variation in the animal's start position on the platform will influence its estimate of target distance. In a single-channel phantom target ranging task, however, the bat's start position on the platform would not influence its estimate of target distance if it learns to use the echo from the speaker or microphone as a fixed reference from which to measure target echo delay (Miller 1991; Surlykke 1992), and this might explain the lower threshold estimates using virtual targets in single-channel range discrimination tasks.

The psychophysical procedure employed by Roverud and Grinnell (1985a) might have also contributed to a higher threshold estimate in the yes/no task. Roverud and Grinnell used a modified method of limits, in which all of the trials on a given test day contained the sphere at a range of 35 cm and one alternative at a greater distance; on succeeding test days, the range difference between the two stimuli was decreased. By contrast, others employing a single-target sequential presentation of targets used a staircase method, in which the range difference between the stimuli was decreased by a fixed step size for each correct response and increased for an incorrect response. The staircase method allows the bat many more trials close to threshold, and changes in the range of the alternative (more distant stimulus) perhaps provide the bat with information that facilitates its discrimination performance. This factor may be of importance in the design of future behavioral studies on echolocation in bats. The modified method of limits has been widely used in 2-AFC simultaneous experiments, and it appears to yield data that are reliable across species and tasks. By contrast, the bat may show its best performance in single-target sequential discrimination tasks when presented with many comparison stimuli around some reference, as in a staircase procedure.

3.1.1 Range Difference Discrimination in the Presence of Interfering Signals

When echolocating bats forage, sounds from the environment can disrupt ranging accuracy, interfering with estimates of target echo arrival time. This environmental noise might take the form of sonar cries from neighboring bats, echoes from nearby vegetation, or ultrasound clicks from insects. The influence of interference signals on range estimation in bats has been investigated in several species. Roverud and Grinnell (1985a) reported for *Noctilio albiventris* that range difference discrimination of two spheres separated by 5 cm deteriorated in the presence of white noise, presumably because of decreased S:N. With continuous white noise at 93.5 dB SPL, the bat's discrimination performance fell to chance. When white noise at this high level was presented in pulses, the bat's distance discrimination performance depended on the range separation of the targets and the duty cycle of the noise pulses. To reliably discriminate range differences of 5–10 cm, the bat required a minimum silent interval of 300 msec. Shorter intervals disrupted range discrimination, which may be explained by backward or forward masking of the echoes by the noise pulses.

In another set of experiments, *Noctilio*'s discrimination of two spheres separated by 5 cm deteriorated in the presence of high-amplitude interference signals that contained both CF and FM components similar to those found in species-specific sonar sounds (Roverud and Grinnell 1985b). As with white noise, the magnitude of this effect depended on the intensity and duty cycle of the artificial sonar pulses. Isolated CF and FM interference signals presented randomly with respect to the target echo did not disrupt range discrimination of 5 cm. However, FM signals that were time locked to the target echo interfered with distance discrimination if they occurred within 8–27 msec after the onset of the FM component of the bat's own pulse and within a correspondingly shorter interval of the FM component of the target echo. Roverud and Grinnell (1985b) interpreted these results to demonstrate that the CF component of the echo activates a gating mechanism that sets up a time window for processing distance information contained in the FM sweep. The width of the postulated time window depends, however, on the SPL of the interfering signal, suggesting that this hypothesis may require modification.

Roverud (1989a) conducted a set of distance discrimination experiments with the roufous horseshoe bat, *Rhinolophus rouxi*, that were similar to those of Roverud and Grinnell (1985b). *Rhinolophus rouxi*'s discrimination of two spheres positioned at 35 and 43 cm was disrupted by free-running CF-FM pulses with no regular temporal relationship to the target echoes when the interference signals were presented at frequencies near the first or second harmonic of this species' echolocation sounds (Roverud 1989a). When a free-running, 2-msec CF sound preceded an FM sound by about 10–65 msec, range discrimination performance also deteriorated. FM pulses alone only disrupted the bat's ranging behavior when triggered by the bat's own sonar emissions and played back after a delay of approximately 45–65 msec. The triggered playback FM sounds arrived close in time to the FM echoes from the spheres. Roverud (1989b) proposed a gating mechanism for the processing of distance information in *Rhinolophus* similar to that in *Noctilio*, and suggested that the temporal requirements of this gating mechanism depend on the duration of the CF component of the individual species' echolocation signals. For *Rhinolophus rouxi*, he did not test whether the width of the time window depends on the SPL of the interfering signal, leaving open the interpretation of this result.

Surlykke and Miller (1985) measured range difference thresholds in *Pipistrellus pipistrellus* in the presence of broadband clicks that occurred randomly with respect to echoes from real targets. The purpose of this study was to test the hypothesis that clicks, such as those produced by arctiid moths in response to bat echolocation sounds (see Fullard and Fenton 1977; Miller 1983), might interfere with the bat's ability to estimate distance. The clicks were playbacks of sounds produced by noxious arctiid moths (the ruby tiger, *Phragmatobia fuliginosa*) recorded in response to computer-generated bat sonar cries. These clicks were broadcast from electrostatic speakers built into two balls used for the range difference discrimination experiment. Using a 2-AFC procedure, Surlykke and Miller (1985) reported that *Pipistrellus pipistrellus* discriminated a range difference of 1.5 cm in a baseline experiment and of 1.0 cm in the presence of clicks. Thus, in this experiment, the bat's range difference discrimination threshold was not disturbed by random presentation of ultrasound clicks.

When the presentation of interfering clicks is synchronized to target echoes, a different pattern of results is observed. Miller (1991) studied range difference discrimination of phantom targets in *Eptesicus fuscus* in experiments in which arctiid moth click trains were time locked to the playback echoes. In this study, he found that the bat's range difference threshold increased by a factor of almost 40 when the interfering click train began about 760 μsec before the phantom target echo.

A general conclusion that can be drawn from the experiments on range difference discrimination in the presence of interfering signals is that the bat's performance is disturbed largely when such signals coincide close in time with the arrival of the target echo. Future experiments should be directed at better understanding the mechanisms involved in this process.

3.2 Range Jitter Discrimination

Range difference thresholds in bats are no greater than 40 mm for any species studied, which is close to the accuracy estimated for bats catching insects (Webster and Griffin 1962). Using other measures of ranging performance, however, the psychophysical and insect capture data diverge. In particular, behavioral data using a range jitter discrimination task developed by Simmons (1979) suggests that the bat can detect changes in target distance many orders of magnitude smaller than appear biologically relevant or neurophysiologically possible (but see Simmons et al. in press).

In Simmons's (1979) first range jitter discrimination task, the bat was trained to rest on the base of a Y-shaped platform and emit echolocation sounds into two microphones, positioned at the end of each arm of the platform. The bat's sounds were picked up by the microphones, delayed, and played back to the bat through two loudspeakers positioned 40° apart to the left and right of the animal's observing position (Fig. 3.5A). A voltage comparator determined the microphone receiving the stronger signal (the side toward which the bat's head was aimed) and activated the loudspeaker on that side. Thus, as the bat scanned its frontal field, it received delayed replicas of its sonar transmissions, delivered sequentially from the loudspeakers. Through one loudspeaker, the bat's sounds were returned at a fixed delay, simulating echoes from a stationary target at a distance of about 50 cm. Through the other loudspeaker, the bat's sounds were returned at a delay that alternated between two time values, simulating echoes from a jittering target, also at a distance of about 50 cm (see Fig. 3.5B). The delay of the playback echo was selected to minimize temporal overlap between the bat's outgoing sonar transmissions and clutter echoes from speakers and microphones with the phantom target echoes.

The side on which the jittering target appeared was randomized from trial to trial, and the bat's task was to crawl down the arm of the platform toward the jittering target. For each of the bat's sonar transmissions, it received only one playback echo, and the level, duration, bandwidth, and sweep rate were all controlled by the bat. Simmons found that the bats performing in this task could discriminate jitter in echo delay of 1–2 μsec (based on a 75% correct criterion taken from Fig. 1; Simmons 1979). This astonishing result stimulated great controversy, prompting further investigation (e.g., Menne et al. 1989; Moss and Schnitzler 1989) using digital playback systems that would not produce spectral artifacts associated with delay that are inherent to the analog system used in Simmons's 1979 study.

Range jitter discrimination performance has since been studied in *Eptesicus fuscus* by several investigators, using both single-channel yes/no (Moss and Schnitzler 1989) and two-channel 2-AFC (Menne et al. 1989; Simmons et al. 1990; Moss and Simmons 1993) psychophysical procedures. The echo jitter values tested in these studies were typically between 0.4 and

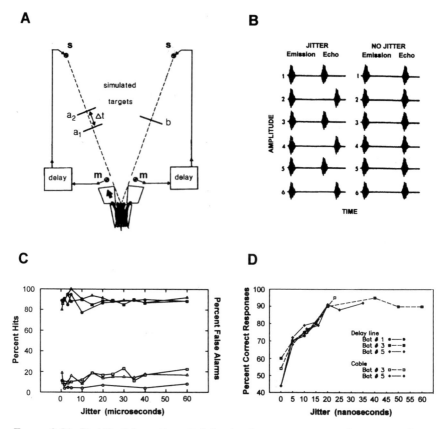

FIGURE 3.5A–D. (A) Schematic of behavioral test apparatus for range jitter discrimination (from Simmons 1989). (B) Schematic of echo time waveforms under conditions with and without jitter in playback delay. For each sonar emission, the bat receives a single playback echo (adapted from Moss and Schnitzler 1989). (C) Jitter discrimination performance of three individual bats for jitter values between 0.4 and 60 μsec (adapted from Moss and Schnitzler 1989). In this experiment, the bat performed in a yes/no discrimination task; *solid symbols* plot percentage hits and *open symbols* plot percentage false alarms. (D) Jitter discrimination performance for three individual bats tested in a 2-AFC task with temporal jitter between 0 and 60 nsec. Bat 3 (*squares*) and bat 5 (*diamonds*) were tested with both analog delay lines (*solid symbols*) and coaxial cables differing in length (*open symbols*). Bat 1 was tested using only the analog delay line (*solid circle*) (from Simmons, et al. 1990).

100 μsec (range jitter, 0.07–17 mm). In addition, Simmons et al. (1990) tested sensitivity to jitter in echo delay in the nanosecond range.

Using a playback apparatus that could digitally generate changes in echo playback delay as small as 0.4 μsec, Moss and Schnitzler (1989; see Fig. 3.5C) and Menne et al. (1989) found that *Eptesicus fuscus* showed stable discrimination performance (>85% correct) down to the smallest jitter

3. Behavioral Studies of Hearing in Bats   113

values that could be tested, which were smaller than those reported by Simmons (1979). Moss and Schnitzler examined the role of echo bandwidth in range jitter discrimination and found that *Eptesicus's* performance was not disturbed when the playback echoes were low-pass filtered at cutoff frequencies as low as 40 kHz (filtering out the upper echo frequencies of the first harmonic and all energy in the higher harmonics of the bat's sonar signals). The effect of low-pass filtering in this study was, however, inconclusive, because performance could not be assessed at jitter values around threshold for the unfiltered echo condition. High-pass filtering of playback echoes at 40 kHz disrupted the bat's behavior, and the animals in the study refused to perform the jitter discrimination task under these echo conditions, suggesting that the lower frequencies (22–40 kHz) of *Eptesicus's* sonar sound are important for carrying information about target range. This frequency band also contains the most energy in this species' echolocation sound.

Menne et al. (1989) were the first to demonstrate that *Eptesicus fuscus* is sensitive to the phase of echoes relative to emissions, an idea first proposed by Simmons (1979). Bats performed in a baseline jitter discrimination experiment and in an experiment in which the phase of the jittering echo alternated between $\pm 45°$ (90° phase change between echoes of the pair). In the baseline experiment, sensitivity to echo jitter was studied for values between 0.4 and 120 μsec, echo–delay alternations that the bats consistently discriminated at levels well above chance. In the phase jitter experiment, these time delay values were combined with 90° phase alternations between echoes of the jittering target, and again the bat successfully discriminated all jitter/phase-shift pairs. They also found that the bat was sensitive to a phase alternation alone (jitter = 0 μsec), a result which has since been replicated (Simmons et al. 1990; Moss and Simmons 1993). The study by Menne et al. also bears importance to the assessment of receiver models (see section 4 below).

While Moss and Schnitzler (1989) and Menne et al. (1989) presented data consistent with Simmons's original (1979) report that *Eptesicus* is sensitive to changes in echo delay of about 1 μsec, some have proposed that the bat does not use the timing of echo arrival to perform the jitter discrimination task. For example, Pollak (1993) has proposed that the bat may use a spectral cue for echo jitter discrimination. He argues that clutter echoes from the loudspeakers and microphones of the target simulator might overlap the playback replicas of the bat's sonar emissions, resulting in spectral interference patterns whose features would vary with changes in echo delay (or jitter). Thus, the jittering target would have a spectral color that alternates between echo playbacks and the nonjittering target would have a spectral color that remains constant. Overlap between playback echoes and clutter echoes occurs when the bat's sounds exceed a particular duration, which would depend on the spatial arrangement of the loudspeakers and microphones relative to the bat. The spectral information arising from interference between playback and clutter echoes that might potentially characterize jitter in echo delay of only a few microseconds is, however, well outside the range of hearing in the bat, and theoretical arguments suggest the use of spectral cues in jitter discrimination to be highly improbable (Menne et al. 1989; Moss and Schnitzler 1989). In addition, Simmons et al. (1990) presented experimental data to support the view that bats use echo delay alone to discriminate target jitter and that spectral artifacts do not contribute to the bat's performance, at least in the microsecond range.

Simmons et al. (1990) exploited the phenomenon of amplitude-latency trading to determine whether echo delay is the bat's perceptual cue in performing the jitter discrimination task. The logic behind this study is as follows: If the bat's cue for jitter discrimination is the relative timing of the playback replicas of its sonar emissions, its performance should be susceptible to amplitude-induced latency shifts in the auditory system. For stimulus levels approximately 15 dB above threshold, brainstem auditory evoked responses in *Eptesicus fuscus* show a latency change of -13 to -18 μsec for each decibel increase in the amplitude of a simulated biosonar sound (Burkard and Moss 1994; Simmons, Moss, and Ferragamo 1990) (Fig. 3.6A). Thus, one would predict that an amplitude change in the playback echoes in the jitter experiment would influence the bat's estimate of echo arrival time and disturb its perception of target distance. This is in fact what Simmons et al. (1990) reported. In a jitter discrimination study, they conducted two control experiments in which the amplitude of the echoes of the jittering target pair differed by 1 dB. In one experiment, the amplitude of the echo with the longer delay was increased by 1 dB relative to the echo with the shorter delay. In another experiment, the amplitude of the echo with the shorter delay was decreased by 1 dB relative to the echo with the longer delay (see Fig. 3.6B). For each playback echo, the absolute amplitude was controlled by the bat: The stronger its sonar emissions, the stronger the playback echoes.

Variation in sonar emissions and playback echoes presumably occurs and could disturb the bat's performance in the jitter discrimination task. The bat, however, appears able to overcome this potential difficulty, perhaps by

---→

FIGURE 3.6A–C. (A) Evoked potentials recorded from the inferior colliculus of the bat (*inset*). Stimuli were FM sweeps, ranging in level between 34 and 79 dB SPL (peak to peak). Latency of N_1 response as a function of stimulus level relative to threshold. At about 15 dB SL, the slope of the regression line fit to the data is about -13 μsec/dB (from Simmons, Moss, and Ferragamo 1990). (B) Schematic of jitter apparatus when one of the echoes from the jitter–echo pair is adjusted in amplitude by 1 dB (a_1, -1 dB; a_2, $+1$ dB) to produce an amplitude-induced latency shift. (C) Percentage errors in jitter discrimination of a single bat under conditions when the amplitude of $a_1 = a_2$ (0 dB), when a_1 is reduced by 1 dB, and when a_2 is increased by 1 dB (from Simmons et al. 1990).

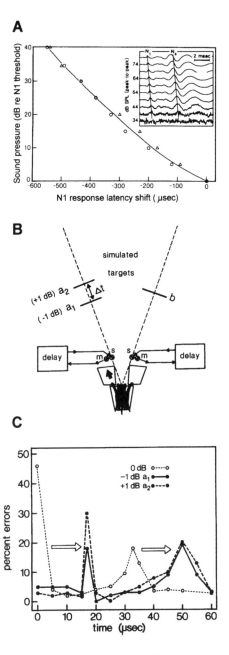

FIGURE 3.6 Caption on facing page.

actively stabilizing its sonar emissions over short durations as it probes the sonar targets. In this control study, Simmons et al. measured the bat's jitter performance using a digital playback system for delay changes of 5–60 μsec and found that the bat's performance dropped when the echoes of the jittering target differed by 1 dB and the echoes jittered in delay by about 17 μsec (Fig. 3.6C) These results suggest that amplitude-induced latency shifts can cancel the bat's perception of target jitter under experimentally controlled conditions. Spectral cues would not be affected by the amplitude manipulation in this experiment, indicating that the bat is using echo delay alone to discriminate target jitter in the microsecond range.

Simmons et al. (1990) also measured behavioral performance of jitter discrimination using time changes in playback echoes much smaller than those used by Moss and Schnitzler (1989) and Menne et al. (1989). In this study, they presented to the bat jitter values in the nanosecond range using two different methods, analog delay lines and coaxial cables differing in length to produce desired amounts of electronic delay. Transfer functions revealed no systematic delay-dependent spectral variations in the playback echoes, and they suggested that the only cue available to the bat for discriminating between jittering and nonjittering targets was echo arrival time. Using either the analog delay line or the cables of varying lengths, they found that the threshold for jitter discrimination in *Eptesicus* was about 10 nsec (range jitter < 0.0017 mm; see Fig. 3.5D). This report has been met with great skepticism (see, e.g., Pollak 1993), and indeed it is difficult to imagine how the nervous system might encode such small differences in echo arrival time, especially considering movements of the bat associated with its respiration and vocalization that must occur between echoes. The control experiment described earlier showing that amplitude-induced latency shifts can affect the bat's perception of target distance was only conducted with jitter in echo delay in the microsecond range. The technical difficulty of studying this phenomenon in the nanosecond range precluded the use of this control at around threshold, and confirmation of the 10-nsec jitter discrimination threshold estimates requires further study using a digital system. Moreover, the amplitude-latency trading data have implications for evaluating the threshold estimate of 10 nsec, and this is discussed further in Section 4, Receiver Models in Bat Sonar.

3.3 Range Resolution and the Discrimination of Two-Wavefront Targets

In the range difference and range jitter experiments just described here, the bat's task was to discriminate differences in the arrival time of echoes from simple targets with single reflecting surfaces. Thus, for each sonar emission produced by the bat, it received back a single echo, the delay of which conveyed information that guided its discrimination behavior. Such experiments are thus designed to measure the bat's accuracy of range estimation, but do not speak to the range resolution of its sonar receiver. Many natural targets of interest to the bat contain multiple reflecting surfaces whose separation may convey information about depth structure or range profile. Perception of the spatial separation of multiple reflecting surfaces might draw upon the range resolution of the bat's sonar receiver.

Most research on range resolution by sonar in bats has employed target stimuli containing (or simulating) two closely spaced reflecting surfaces (or wavefronts). In these experiments, the time delay corresponding to the range separation of the targets is small with respect to the duration of the bat's echolocation sounds. Thus, echoes from the two surfaces are largely overlapping. This results in a spectral interference pattern, with frequency regions of cancellation and reinforcement that correspond directly to the time separation between the sounds. For example, playback replicas of two FM sonar sounds separated by 100 μsec (1.74 cm) result in spectral notches at 10-kHz intervals, starting at 25 kHz in the echolocation signal. As the delay separation between overlapping echoes increases, the interval between spectral notches decreases (Fig. 3.7).

Three general questions have guided research on range resolution by echolocation: What is the minimum range separation between two closely spaced reflecting surfaces that the bat can resolve? What acoustic information in the echo does the bat use to perform in a range resolution task? How are targets with closely spaced reflecting surfaces represented in the bat sonar receiver? Each of these questions is considered next. Here, the discussion is organized around studies employing two different perceptual tasks, one in which the bat discriminates between two-wavefront targets that differ in the range/delay of their component echoes and another in which the bat discriminates between a one-wavefront target and a two-wavefront target that varies in the range/delay separation of its component echoes.

3.3.1 Discrimination Between Two-Wavefront Targets Differing in the Range/Delay Separation of Component Echoes

Two range resolution studies have been conducted on FM bats using real targets (Simmons et al. 1974; Habersetzer and Vogler 1983), and both show that the bat can discriminate range differences of less than 1 mm. The stimuli in these experiments consisted of Plexiglas plates with holes drilled to different depths. Simmons et al. (1974) trained *Eptesicus fuscus* in a 2-AFC procedure to discriminate between Plexiglas plates with holes drilled at a depth of 8 mm and a smaller depth, ranging between 6.5 and 7.6 mm, and reported that bats can discriminate differences as small as 0.6–0.9 mm in the depth of holes. The intensity of the echoes reflecting from the plates differed by less than 0.5 dB across all the pairs of stimuli presented to the bat, while the spectra of the echoes differed systematically with plate hole

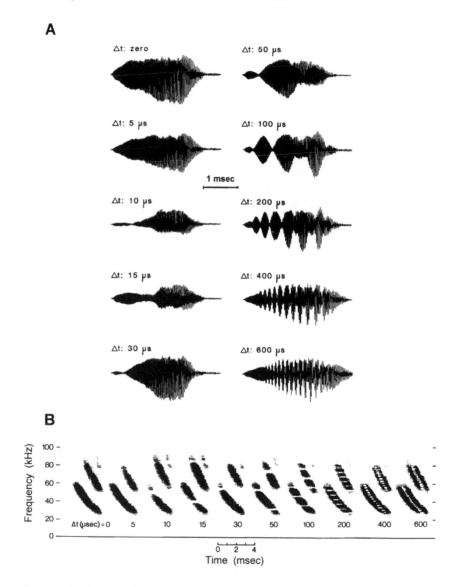

FIGURE 3.7A,B. (A) Time waveforms of two overlapping FM sounds (two-glint echoes) separated by delays ranging between 0 and 600 μsec. The signals are based on the sonar sound used by *Eptesicus fuscus* (from Simmons 1989). (B) Spectrograms of the same signals displayed in (A) (from Simmons 1992).

depth. The largest differences in the spectra appeared between 30 and 60 kHz, with notches occurring at lower frequencies for deeper holes. Simmons et al. suggested that *Eptesicus* can discriminate targets on the basis of information contained in the echo spectra. They also noted that the same signal-processing operation may be common to ranging (echo delay) and range resolution (echo spectrum) within the bat's auditory system, just as time-domain and frequency-domain displays of waveforms are mutually related through the Fourier transform. This might be interpreted to suggest that the bat experiences the overlapping echoes from the closely spaced surfaces as echoes arriving at different delays.

In the Habersetzer and Vogler study, *Myotis myotis* performed the discrimination task using two different reference targets, one with an 8-mm hole depth (as in Simmons et al. 1974) and the other with a 4-mm hole depth. They found that the bat's discrimination threshold was about 1 mm for the 8-mm reference target and about 0.8 mm for the 4-mm reference target. These data are in general agreement with those reported by Simmons et al. (1974) for *Eptesicus fuscus*.

Questions about the information the bat uses to discriminate echoes from closely spaced surfaces and the representation of such complex targets has been the subject of more recent research using target simulation methods. Target simulation methods permit control of the acoustic information arriving at the bat's ears and eliminate the resonance phenomena associated with real targets, such as the plates drilled with holes.

Schmidt (1988, 1992) studied the performance of *Megaderma lyra* trained to discriminate between virtual targets containing two replicas of the bat's sonar sounds at different delays (Fig. 3.8A). The delays were selected to simulate overlapping echoes from two closely spaced planar surfaces with different spatial offsets. In most of her experiments, the bat was rewarded in a 2-AFC task for selecting a reference delay offset of 7.77μsec, which simulated a distance difference between reflecting surfaces of 1.3 mm. Schmidt found that the smallest difference in temporal offset the bat could reliably discriminate was about 1 μsec (spatial separation of 0.17 mm), but discrimination performance was not a monotonic function of the difference in echo delay separation of the targets (see Fig. 3.8B). The reference target contained its first spectral notch at 64.4 kHz, and other echo delay offsets producing notches at this frequency (e.g., 23.3 μsec) caused a small drop in discrimination performance. From these data Schmidt hypothesized that the bat uses echo spectral information to discriminate target surface structure. She also used modeling to evaluate this hypothesis, which is discussed in Section 4 of this chapter.

Mogdans, Schnitzler, and Ostwald (1993) conducted a study on two-wavefront discrimination by *Eptesicus fuscus*, also using a 2-AFC behavioral paradigm. In one experiment, *Eptesicus* was first rewarded for selecting a two-wavefront target with an internal delay of 12 μsec, and its discrimination performance was tested against two-wavefront targets with internal delays of 28, 32, 36, 40, and 44 μsec. In a subsequent experiment, *Eptesicus* was rewarded for selecting a two-wavefront target with an internal delay of 36 μsec against others with internal delays of 28, 32, 36, 40, and 44 μsec. The bat successfully discriminated two-wavefront echoes with delay separations of either 12 or 36 μsec from those with delay separations of 28, 32, 40, or 44 μsec; however, discrimination between two-wavefront echoes with delay separations of 12 and 36 μsec fell to chance. This result

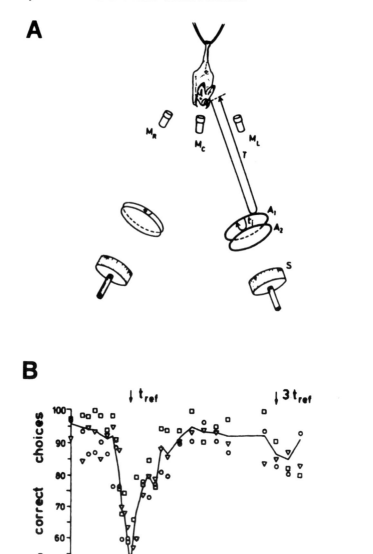

FIGURE 3.8A,B. (A) Schematic of behavioral setup for two-wavefront discrimination experiments with *Megaderma lyra*. M_R, right microphone; M_L, left microphone; M_C, center microphone. A_1 and A_2 represent the individual echo components of the two-wavefront target; S, playback speaker. (B) Behavioral data from three individual bats (*squares, circles, triangles*) trained to discriminate a two-wavefront target (t_{ref}) with an internal echo delay of 7.77 μsec from other two-

can be explained by a prominent spectral notch at about 42 kHz, which was common to the two-wavefront echoes with internal delays of 12 and 36 μsec (Fig. 3.9A). The authors noted, however, that these data do not establish that the bat uses a spectral representation of the signal to perform the task, because a temporal representation can also account for the data. Indeed, the energy density spectrum and cross-correlation function of a signal are mathematically related by the Fourier transform, an issue that is discussed further later in this chapter.

3.3.2 Discrimination Between One-Wavefront and Two-Wavefront Phantom Target Echoes

In another set of experiments, Mogdans, Schnitzler, and Ostwald (1993) trained *Eptesicus fuscus* to discriminate between a one-echo wavefront and a two-echo wavefront whose components differed in delay separation (see also Mogdans and Schnitzler 1990). The amplitude of the one-wavefront echo in this study was twice that of the individual components of the two-wavefront echo, and therefore the total energy of the one-wavefront echo was equal to that of the two-wavefront echo with an internal delay of 0 μsec. When the internal delay of the two-wavefront echo changed, so did its total energy and spectrum.

Mogdans, Schnitzler, and Ostwald (1993) reported that *Eptesicus* successfully discriminated between the targets, but only for particular temporal offsets between the echoes of the two-wavefront targets (see Fig. 3.9B). While the total energy of the two-wavefront target differed from that of the one-wavefront target, and the magnitude of this difference depended on the delay separation between the echoes of the two-wavefront target, the bat's pattern of performance indicates that it did *not* use total echo energy in the task. Instead, the data suggest that the spectrum of the playback echoes determined the bat's discrimination performance. When the spectral dissimilarity between the one- and two-wavefront targets was large, discrimination performance was well above chance. As mentioned earlier, these data do not demonstrate that a spectral model alone accounts for the bat's performance, because a time-domain representation can be derived from a frequency-domain representation (illustrated in Fig. 3.9B; see also Section 4, Receiver Models in Bat Sonar).

The basic pattern of results obtained from a one-wavefront versus

---

←

FIGURE 3.8. (*continued*) wavefront targets differing in internal delay. Percent correct responses plotted as a function of the temporal (*t*, μsec) and spatial (*s*, mm) offset of the two echoes of the test signal. Along the bottom abscissa is the frequency location of the first spectral notch of the two-wavefront test echo. The *solid line* shows the linear interpolation of the mean performance of three animals. (From Schmidt 1992.)

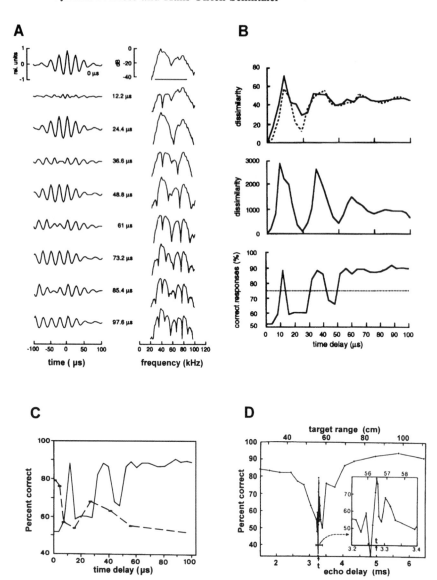

FIGURE 3.9A–D. (A) Cross-correlation functions of a representative echolocation sound used by *Eptesicus fuscus* and two-wavefront echoes at different internal time delays, ranging from 0 to 97.6 μsec (*left*). Amplitude spectra of the two-wavefront echoes at the same internal time delays (*right*). (B) Top panel: comparison of spectral dissimilarity between one-wavefront and two-wavefront echoes differing in internal delay (calculated after Schmidt 1988). Spectral similarity was based on the total frequency range of the sound (*solid line*) and the first harmonic (*stippled line*). Middle panel: dissimilarity of the cross-correlation function (CCF) between a sonar sound and a one-wavefront echo and the CCF between a sound and a two-wavefront echo differing in internal delay. Bottom panel: mean performance

two-wavefront discrimination experiment can be strongly influenced by the initial task conditions. For example, in another experiment *Eptesicus fuscus* was trained in a 2-AFC procedure to detect a test echo at a fixed delay in the presence of clutter (Simmons et al. 1989). The clutter targets were single echoes, presented to the left and right of the bat's observing position. On one side, a clutter echo appeared alone, and on the other side, a test echo and a clutter echo appeared together where the test and clutter echoes were overlapping (i.e., one-wavefront versus two-wavefront). In this experiment, the amplitude of the individual echoes of the two-wavefront target (test plus clutter echoes) was the same as that of the one-wavefront target (clutter echo alone). Thus, for a 0-μsec internal delay of a two-wavefront target (no offset between test and clutter echoes), its energy was twice that of the one-wavefront target. For delay separations between the test echo and clutter echo of less than 100 μsec, the bat's performance fluctuated; however, the shape of the behavior curve differed markedly from that obtained by Mogdans, Schnitzler, and Ostwald (1993; see Fig. 3.9C).

In evaluating the data on two-wavefront discrimination, Mogdans, Schnitzler, and Ostwald (1993) also considered the representation of the complex target in the bat's sonar receiver. Based on the bat's near-chance performance for particular delay offsets, they conclude that it does not perceive the two-wavefront targets as containing distinct surfaces (or glints) along the range axis. This view contrasts with that presented by Simmons, Moss, and Ferragamo (1990), who proposed that a two-wavefront target is displayed in the bat's sonar receiver as discrete glints along the range axis.

In the study by Simmons, Moss, and Ferragamo (1990), the bat was trained to discriminate between a complex target containing two-echo wavefronts, each at 15 dB above threshold and offset by 100 μsec, and a simple target containing a one-echo wavefront. The delay (range) and amplitude of the complex target was fixed, while the delay and the

---

←

FIGURE 3.9   (*continued*) of six bats discriminating between a one-wavefront echo and two-wavefront echoes differing in internal delay. Note that the behavioral performance curve resembles the dissimilarity curves shown in top and middle panels of B (from Mogdans, Schnitzler, and Ostwald 1993). (C) Comparison of *Eptesicus*'s performance on a two-wavefront discrimination task (*solid line*; replotted from Fig. 3.9B, bottom panel) and on a detection task in which a clutter echo was presented at different delays with respect to the target echo (*stippled line*; replotted from Simmons et al. 1989, with an expanded time scale; see also Fig. 3.9D). Both curves plot behavioral performance as a function of the temporal separation between two echoes. Note that the shape of the two performance curves differs dramatically, reflecting differences in the test procedures in the two experiments (see text). (D) Detection of a test target echo at 3.27 msec in the presence of clutter (two-wavefront echo) as a function of the temporal separation between the test and clutter echoes (taken from Simmons et al. 1989). Mean performance of five individual bats. The inset shows a magnified segment of the curve between 3.2 and 3.4 msec.

amplitude of the simple target changed over the course of the experiment. The delay of the simple target varied in 25-µsec steps from 150 µsec before to 175 µsec after the first glint of the complex–echo pair; the amplitude of the simple target varied in 3-dB steps, from 6 dB below to 9 dB above the individual components of the complex target. In this experiment, the bat showed an increase in errors when the simple target was presented at the same delay and amplitude as the first or second echoes of the complex target, and they suggested that this rise in errors can serve as an index of the bat's perceived distance of the simple target with respect to the individual components of the complex target. From a series of experiments in which bats discriminated simple and complex target echoes, they concluded that the bat sonar receiver converts the spectral information contained in a two-wavefront echo (see Fig. 3.7) into a temporal representation, encoding the underlying delay or range separation of the component target echoes (see Simmons et al., this volume, Chapter 4).

In conclusion, research on range resolution suggests that the bat, at least in early stages of information processing, may use the spectrum of echoes to discriminate targets with two closely spaced reflecting surfaces; however, it is important to note that frequency-and time-domain representations of the echoes are mathematically equivalent. We would also like to emphasize that the shape of the bat's performance curve appears strongly dependent on the initial training procedures used in behavioral experiments, and for this reason it is often difficult to make clear comparisons across seemingly similar studies.

3.4 Summary

The behavioral studies summarized in this section contribute to our understanding of the processes important to the perceptual dimension of target range in echolocating bats and suggest that this perceptual dimension is not a single entity. Data on range difference discrimination, range jitter discrimination, and range resolution each appear to tap different perceptual processes in the bat sonar receiver. Range difference data describe the bat's accuracy at estimating the distance to a target and show that all bat species can discriminate distance with an accuracy of less than about 40 mm, and some species down to about 6–12 mm. Range jitter discrimination data reflect the bat's sensitivity to changes in target range, or perhaps to apparent target motion along the range axis. Data from range jitter experiments show that bats are sensitive to changes in echo delay/range of less than 0.5 µsec/0.1 mm. Studies on range resolution examine the bat's perception of target depth structure, and range resolution data show that bats are sensitive to very small spatial-temporal offsets of two-wavefront stimuli. FM bats can discriminate the separation of two closely spaced reflecting surfaces of about 1 mm or the difference in delay offset of two overlapping playback echoes of about 1 µsec. To discriminate a one-wavefront target from a two-wavefront target, the delay offset between the component echoes of the two-wavefront target must be about 12 μsec. The representation of complex targets in the sonar receiver of the bat is not well understood, and further research on the representation of depth-structured targets is particularly important for building our knowledge of spatial information processing by sonar.

4. Receiver Models in Bat Sonar

Since the modern-day discovery of echolocation in bats, engineers and biologists alike have explored the similarities between animal and man-made sonar/radar systems (e.g., Griffin 1944; McCue 1966). The comparison most commonly discussed is that between the bat and an ideal or matched filter receiver. An ideal receiver cross-correlates the outgoing sonar transmission and the incoming echo and reads a time-domain representation of the signal. If the echo reflects from a point target, the autocorrelation function of the sonar signal is a good approximation to the cross-correlation function of the sonar transmission and echo.

An ideal receiver may be either coherent or semicoherent, referring to the use of phase information in the signals. An ideal coherent receiver uses all the information contained in the cross-correlation of the sonar emission and echo to estimate target range. By contrast, an ideal semicoherent receiver does not make use of the phase information and uses the envelope of the correlation function for range estimation. The accuracy of time estimation depends on the form of the correlation function (which is determined by the signal type used) and internal and external noise. Wideband signals produce distinct central peaks with small sidelobes in the correlation function, whereas narrowband signals produce a less distinct central peak and prominent sidelobes (Simmons and Stein 1980).

The six panels in Figure 3.10 show different representations of the FM echolocation sound of *Eptesicus*. A time waveform representation of the sound is shown in Fig. 3.10A, the corresponding spectrogram in Fig 3.10B, and the amplitude spectrum in Fig. 3.10C. Figure 3.10D displays the autocorrelation function of this sound, which is also shown with an expanded time scale in Fig. 3.10E. These panels illustrate that the central peak of the autocorrelation function of a broadband signal is a good time marker. In Fig. 3.10F the bat's FM sound is cross-correlated with a time-reversed echo, and the prominent central peak (Figs. 3.10D and 3.10E) is absent.

4.1 Detection

A matched filter receiver concentrates the signal energy within a small time interval. This compression not only allows for an optimal estimate of echo

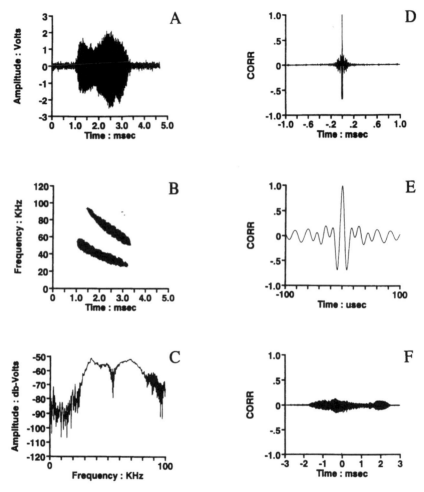

FIGURE 3.10A–F. Echolocation sound of *Eptesicus fuscus*. (A) Time waveform. (B) Spectrogram. (C) Spectrum. (D) Autocorrelation function. (E) Autocorrelation function on expanded time scale. (F) Cross-correlation function of signal and a time-reversed replica.

arrival time but also for optimal detection performance, because of an effective increase in S:N (see Skolnik 1980). If a bat uses a matched filter receiver for detection, the threshold should be very low and depend on the similarity between the echo and the receiver's signal template (Møhl 1988).

The very low threshold for echo detection calculated from Kick's (1982) data on *Eptesicus fuscus* can be interpreted as evidence in favor of coherent reception by bats (Møhl 1988). In another study, Moss and Simmons (1993) showed that the detection performance of *Eptesicus* depended on changes in the arrival time and the phase of phantom target playback signals. Bats were trained in a 2-AFC procedure to detect playback echoes that jittered in delay and in phase. The bat's detection performance for echoes that jittered in delay was improved compared to baseline (no jitter); however, for echo jitter of about 30 μsec, the bat's detection threshold was unchanged from baseline. When the phase of one echo from the jittered echo pair was inverted, detection performance increased compared to baseline, except for jitter values around 15 and 45 μsec. It appears that the temporal jitter in echo arrival time enhances detection performance, similar to a flickering target's enhancement of visual detection (Kulikowski and Tolhurst 1973). One possible interpretation of these data is that perceptual ambiguity of target jitter would arise from the sidelobes of the cross-correlation function of the bat's signal and returning echo at 30 μsec for phase-normal echoes, and at 15 and 45 μsec for phase-inverted echoes. The results of this study do not, however, demonstrate cross-correlation processing.

Arguments against the hypothesis of a matched filter for detection come from experiments using time-reversed playback signals and masking of echoes by white noise. In *Pipistrellus pipistrellus* (Møhl 1986) and *Eptesicus fuscus* (Masters and Jacobs 1989), species using broadband FM sounds that sweep in frequency from high to low, echo detection thresholds for time-reversed FM sweeps were unchanged from baseline measures with natural echolocation signals. (For a comparison of the cross-correlation functions of the baseline and time-reversed signals, see Fig. 3.10F.) Furthermore, Troest and Møhl (1986) reported that echo detection thresholds in noise were much higher than would be predicted of an ideal receiver. Together, these studies do not support the hypothesis that the bat uses a matched filter receiver for echo detection.

4.2 Ranging

Considerable debate has surrounded the question of whether the echolocating bat performs as an ideal receiver for range estimation (e.g., Simmons 1979; Schnitzler and Henson 1980; Schnitzler, Menne, and Hackbarth 1985; Hackbarth 1986; Menne and Hackbarth 1986; Simmons 1993). An ideal receiver achieves high ranging accuracy by cross-correlating the sonar emission and returning echo and using the time of the cross-correlation function peak as its best estimate of echo delay (target range). Here we consider the results of experiments on target ranging in bats and discuss their implications for assessing sonar receiver models.

4.2.1 Range Difference Discrimination

In 1973, Simmons reported a relationship between the width of the envelope of the autocorrelation function of the bat's echolocation sound and its ranging performance. Simmons compared the bat's range discrimination performance and that predicted from the envelope of the signal correlation function for four species, *Eptesicus fuscus, Phyllostomus hastus, Ptero-*

notus suapurensis, and *Rhinolophus ferrumequinum,* and found a corre-spondence between the empirical and predicted performance curves for each of these species. The width of the correlation functions for each species depends on the bandwidth of the animal's sonar sounds. Simmons reported that species using broadband signals with narrow correlation function peaks (sharper registration along the delay axis) showed lower range difference discrimination thresholds than species using narrowband signals with broad correlation function peaks. He used the general correspondence between empirical performance curves for the four species and theoretical curves derived from the envelope of the autocorrelation functions to argue that the bat operates as an ideal semicoherent (phase-insensitive) receiver (see, for example, empirical and theoretical curves shown in Figs. 3.4A and 3.4D). However, Schnitzler, Menne, and Hackbarth (1985) pointed out that the shape of the behavioral performance curve does not provide direct evidence for cross-correlation processing.

Masters and Jacobs (1989) conducted a range difference discrimination study with *Eptesicus fuscus* using normal and time-reversed FM playback echoes. In this experiment, they found that the bat's range difference discrimination threshold for time-reversed echoes was about 18 fold higher than that measured in baseline experiments. While this result can be explained by other receiver models (e.g., Menne 1988), it is consistent with a matched filter receiver for range estimation.

Other experimental results are not consistent with the hypothesis that the bat performs as an ideal receiver for the estimation of target distance. For example, Simmons (1973) reported that the bat's range discrimination performance was relatively stable at several test distances (30–240 cm; see Fig. 3.4C), a finding that is not consistent with the predictions of matched filtering. In this experiment, the S:N presumably decreased with increasing range, and one would predict a rise in threshold with test distance; however, because the S:N was not measured at these different test distances, this issue is difficult to evaluate.

4.2.2 Range Jitter Discrimination

Using the bat's performance in the range jitter discrimination task, Sim-mons (1979) proposed that the bat operates as an ideal coherent (phase-sensitive) receiver. This idea is based on the shape of the bat's performance curve (i.e., percent correct performance as a function of echo jitter) in this task rather than the absolute threshold for range jitter discrimination. In particular, the bat showed a rise in errors for a jitter in echo delay of about 30 μsec relative to smaller and larger jitter delay values (Fig. 3.11). Simmons reasoned that an animal performing the neural equivalent of matched filtering would experience perceptual ambiguity for the presence of jitter at 30 μsec, because it would confuse the central peak and the first sidelobe of the correlation function. Using a computer model to simulate a fully

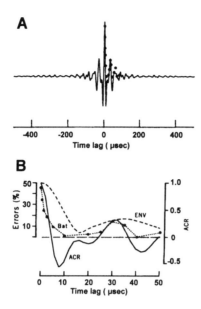

FIGURE 3.11A,B. Jitter discrimination performance compared with the autocorrelation function of the sonar sound used by *Eptesicus fuscus* (from Simmons 1979). (A) Percentage error data points plotted along the autocorrelation function of *Eptesicus*'s echolocation signal for time lag ± 400 μsec. (B) Percentage errors plotted for jitter discrimination performance for temporal jitter between 0 and 50 μsec. The *solid line* plots the autocorrelation function (ACR) of *Eptesicus*'s sonar sound; the *dashed line* shows the envelope (ENV) of the ACR.

coherent cross-correlation receiver, Menne and Hackbarth (1986) found no sidelobe (increase in errors) at 30 μsec in the predicted performance curves. A sidelobe in the performance curve, however, can be demonstrated using other receiver models, for example, a bank of filters with envelope processing (Hackbarth 1986).

In a later experiment, Simmons et al. (1990) showed that *Eptesicus* can discriminate a 180° phase alternation in playback echoes. The results of this study also show a drop in the bat's jitter discrimination performance for a combination of echo phase reversal and temporal jitter of 15 or 45 μsec. These findings are consistent with those predicted of an ideal coherent receiver (Altes 1981), and a computational model predicts this pattern of performance using cross-correlation processing (Saillant et al. 1993).

Additional evidence for an ideal coherent receiver in bat sonar comes from Simmons et al. (1990); they reported that *Eptesicus fuscus* can discriminate jitter in echo delay of 10 nsec with a S:N of about 30 dB, a value close to that predicted in a computer simulation (Menne and Hackbarth 1986). This corresponds to a change in target distance of about 0.0017 mm. Although appropriate calibration of the equipment was made and there were no obvious stimulus artifacts the bat might have used to perform in the task, this measurement demands replication using a digital system capable of producing changes in the delay of playback echoes in the nanosecond range.

The results of some jitter discrimination data are inconsistent with the predictions of a matched filter receiver. For example, Menne et al. (1989) studied the bat's performance with both time and phase jitter. They found that *Eptesicus fuscus* could discriminate changes in echo delay of 0.4 μsec (the smallest value tested) and 90° phase alternation with no added temporal jitter. While these data indicate that the bat is sensitive to very small changes in echo delay and phase, they do not support the hypothesis that the bat performs as an ideal receiver. For a coherent matched-filter receiver, an echo–phase alternation of 90° should cancel the bat's perception of jitter when combined with a 7-μsec alternation in echo delay. Thus, one would predict a drop in discrimination performance for this particular combination of phase and delay jitter. The closest jitter value they tested was 8 μsec, and for this value, the performance was not disturbed. While it is possible that a change in performance was not observed because 7 μsec was not tested, data from Simmons (1979; Simmons et al. 1990) suggest that errors associated with the sidelobe of the correlation function occur over a rather large time window. These data therefore suggest that the bat does not experience the perceptual ambiguity predicted for a matched filter receiver presented with a particular combination of phase and delay jitter.

The report by Simmons et al. (1990) that *Eptesicus* can discriminate jitter in echo delay of 10 nsec is not consistent with the idea that the bat's autocorrelation function can be used to predict its ranging performance (e.g., Simmons 1979; Simmons 1987; Simmons and Stein 1980), for this threshold jitter value is a factor of 10 smaller than the width of the central peak of the correlation function (see Fig. 3.10E). Furthermore, if the bat's range jitter discrimination threshold is 10 nsec, why would temporal ambiguity associated with the side peak of the correlation function contribute to the bat's performance over a time window that is several microseconds wide? And finally, it is difficult to imagine that the bat's head movements between sonar emissions and small changes in the sound level of returning echoes do not introduce variability in estimates of echo arrival time (see Fig. 3.6) that is greater than a few nanoseconds. Clearly more work is required in this area to resolve these issues.

4.2.3 Range Resolution

Receiver models have also been tested using behavioral data from two-wavefront discrimination tasks. For example, Schmidt (1988, 1992), proposed that a spectral correlation model best accounts for the performance of *Megderma lyra* in a task in which the bat is trained to discriminate between two-wavefront targets differing in internal delay. She contrasts the spectral correlation model with a highly simplistic temporal model, which does not yield predictions consistent with the empirical performance curves.

Mogdans, Schnitzler, and Ostwald (1993) used data from an experiment in which *Eptesicus fuscus* was trained to discriminate a one-wavefront target from a two-wavefront target to demonstrate that both spectral correlation and cross-correlation models predict the bat's range resolution performance (see Fig. 3.9B). They noted that the time-domain and frequency-domain models are mathematically related through the Fourier transform, and that it is not possible to differentiate between the two models using the bat's performance in the behavioral task.

4.3 Summary

The behavioral tests of receiver models in bat sonar indicate that detection and ranging are different processes. Many questions still surround the view that the bat performs as an ideal coherent receiver for the estimation of target range, and future research can directly address many of these questions through careful experimental design and stimulus control.

5. Horizontal and Vertical Sound Localization

To successfully maneuver in the environment and intercept prey, echolocating bats must continuously determine the azimuth, elevation, and range of objects in space. Bats use the time delay between sonar emission and returning echo to estimate target distance, and spatial perception along the range axis was discussed in detail in sections of this chapter. Here, we focus on behavioral studies of target localization in the horizontal and vertical planes.

5.1 Horizontal Localization

Mammals compare the arrival time, level, and spectrum of a sound at their two ears to determine azimuthal position (Yost and Gourevitch 1987; Wightman and Kistler 1993; Brown 1994). For a bat, the sound source is a target reflecting an echo of its sonar emission, and the directionality of its sonar emission will impart additional information about the spatial location of a target (e.g., Schnitzler and Grinnell 1977; Hartley and Suthers 1987, 1989).

The maximum arrival time difference occurs when a sound source is located at 80°–90° with respect to the midline of the head, and for bats with short interaural distances this time difference is small. For example, in *Eptesicus fuscus* whose head diameter is less than a centimeter, the maximum interaural time difference is approximately 40–50 μsec. Maximum interaural level differences measured with pure tones for the same species are about 25–30 dB (Jen and Chen 1988). Interaural spectral differences vary continuously with azimuth because of complex interactions between sound reflections off the pinna and tragus (see Shimozawa et al. 1974).

The threshold for horizontal sound localization has been estimated for the FM bat *Eptesicus fuscus*. Using a 2-AFC operant task, Simmons et al. (1983) found that the bat's threshold for discriminating the angular separation of thin rods was 1.5°. This angular separation creates interaural level differences of 0.3 to 1.5 dB, depending on the absolute direction of the targets with respect to the head aim of the bat. The interaural time difference associated with a 1.5° angular separation is about 1 μsec, a time separation that Simmons et al. (1983) suggested may be discriminated by *Eptesicus fuscus*.

5.2 Vertical Sound Localization

The cues associated with vertical sound localization in bats differ across species. Some species make large pinna movements that create interaural intensity differences, which can be used for vertical sound localization. For example, *Rhinolophus ferrumequinum* moves its pinna in coordination with its sonar emissions. Its highly mobile pinnae scan the sound field, moving in opposite directions. As one ear moves forward, the other ear moves backward, resulting in large differences in the echo levels at the two ears (Schneider and Möhres 1960).

Mogdans, Ostwald, and Schnitzler (1988) demonstrated the importance of ear movements for vertical localization in *Rhinolophus ferrumequinum*. Using a wire-avoidance task, they found that surgical immobilization of the pinnae selectively disrupted performance in vertical localization. In the baseline (preoperative) condition, bats successfully flew through vertical and horizontal arrays of wires separated by 15 cm. After the bat's pinnae were immobilized, it continued to successfully fly through the vertical array of wires (requiring horizontal localization) but showed many more collisions with the horizontal array of wires (requiring vertical localization). The bat's vertical localization was not, however, entirely impaired, suggesting that changes in the bat's head position during the task may have also provided information guiding flight through the horizontal wire array.

In other species, pinna movements are small by comparison with *Rhinolophus*, and the cues for vertical sound localization originate largely from the acoustic properties of the external ear. When sound enters the ear, it reflects off the pinna and tragus, creating spectral interference patterns from overlapping echoes. The interference patterns depend on the delay path between the reflecting surfaces of the external ear, which changes systematically with the vertical location of a sound source. In *Eptesicus fuscus*, sounds entering the ear produce primary and secondary reflections, separated by 45–60 μsec (Lawrence and Simmons 1982b). These reflections presumably represent echoes from the pinna and tragus, which mix to create notches in the echo spectrum that can be used by the bat to encode sound location in the vertical plane.

A psychophysical study of vertical sound localization has been conducted in *Eptesicus fuscus* using a 2-AFC procedure (Lawrence and Simmons 1982b). In this experiment, the bat was trained to discriminate between pairs of horizontal rods separated by different vertical angles. Threshold estimates for angular discrimination in the vertical plane was about 3° and depended on the intact tragus. When the bat's tragus was deflected, the bat's vertical discrimination performance deteriorated.

5.3 Summary

There are psychophysical data on horizontal and vertical sound localization thresholds for one species of FM bat, *Eptesicus fuscus*, but there are no comparable threshold data from short CF-FM or long CF-FM bats. In future work, it would be of interest to compare the sound localization capabilities of bats using different types of echolocation signals.

6. Fluttering Target Detection and Classification

Fluttering insects continuously change their reflecting surfaces, thus producing modulations in the echoes from the sonar signals of bats. When emitting short FM signals (below 5 msec), bats such as *Myotis lucifugus* and *Eptesicus fuscus* may use amplitude variations across echoes to recognize flying insects (Griffin 1958; Roeder 1962; Kober and Schnitzler 1990). Such FM sounds are too short to carry echo information during an insect's complete wingbeat cycle (Moss and Zagaeski 1994). By contrast, the rather long CF-FM signals (10–100 msec) that are characteristic of rhinolophids, hipposiderids, noctilionids, and of the mormoopid bat *Pteronotus parnellii* can carry much more acoustic information about a fluttering insect, often over more than one wingbeat cycle, thus increasing the probability of detecting fluttering insects (e.g., Goldman and Henson 1977).

6.1 Echoes from Fluttering Insects

Fluttering insects mounted in the acoustic beam of a CF signal reflect sounds that are rhythmically modulated in amplitude and frequency and in synchrony with the insect's wingbeat. These amplitude and frequency modulations not only reveal that there is a fluttering insect, they also contain species-specific information, such as wingbeat rate and movement patterns, that may be used by the bat to classify its prey. Furthermore, the acoustic parameters carried by fluttering insects may be used to encode the target's angular orientation with respect to the incident sound (Schnitzler and Henson 1980; Schnitzler and Ostwald 1983; Schnitzler et al. 1983; Schnitzler 1987; Ostwald, Schnitzler, and Schuller 1988; Kober and Schnitzler 1990).

The most prominent features of echoes from fluttering insects are very short and strong amplitude peaks or "acoustical glints," which are produced when the wings are perpendicular to the impinging sound waves. The wings act as "acoustic mirrors," causing amplitude peaks as much as 20 dB above the echo from the insect's body when positioned to create maximum echo reflection. The moving wings also produce spectral broadening in the echoes of CF signals from Doppler shifts. These spectral changes are largest when the wing velocity is highest, and their characteristics depend on the direction of the wing movement at the instant of glint production. For example, when the wings move down and backward, sound waves from ahead (0°) produce a negative Doppler shift and sound waves from the rear (180°) produce a positive Doppler shift (Fig. 3.12A). All other aspect angles also produce orientation-specific frequency deviations. Information about an insect species may be contained in the glint rate (encoding wingbeat rate), in the magnitude of the Doppler shift (encoding the size of the wings and the velocity of their motion), and in the temporal and spectral fine structure of the echo, related to other movements and morphological features of the insect (Kober and Schnitzler 1990).

6.2 Fluttering Target Detection and Discrimination by Bats Using CF Signals

Many species of bats that use CF-FM echolocation sounds are particularly well suited to make use of echo information from fluttering targets, as these bats compensate for Doppler shifts in echoes introduced by their own flight speed (Schnitzler 1968). For example, *Rhinolophus ferrumequinum* lowers the frequency of sonar emissions in an amount proportional to its flight velocity, receiving echoes from stationary objects at about 83 kHz, a frequency to which the bat shows high sensitivity (see Fig. 3.1). Thus, Doppler shift compensation behavior enables the bat to decouple echo Doppler shifts created by its own flight velocity from those created by moving targets in the environment (e.g., fluttering insect wings) and allows the bat to detect small shifts in echo frequency produced by the fluttering wings of insects.

Several observations indicate that bats with CF-FM signals use echo information produced by target flutter to recognize insect echoes even in dense clutter. In a laboratory study, *Rhinolophus ferrumequinum* caught fluttering moths not only when they were flying in the middle of the room but also on the walls or on the ground. Stationary moths were ignored (Schnitzler and Henson 1980; Trappe 1982). *Pteronotus parnellii* (Goldmann and Henson 1977) and *Hipposideros ruber* (Bell and Fenton 1984), and *Rhinolophus rouxi*, *Hipposideros bicolor*, and *Hipposideros speoris*

A

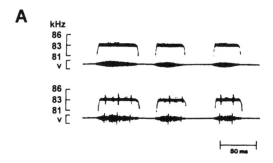

B

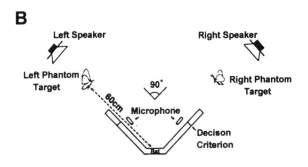

C

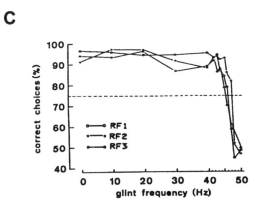

FIGURE 3.12A–C. (A) Sonar sounds used by *Rhinolophus ferrumequinum* (*top*) and playback echoes containing glints (spectral broadening and amplitude peaks) produced by fluttering insect wings (*bottom*). (B) Schematic of behavioral apparatus used to study fluttering target discrimination in *Rhinolophus ferrumequinum*. The bat was trained to make echolocation sounds into two microphones, positioned to the left and right of its start position. The signals picked up by the microphones were modified online and played back through two speakers, positioned behind the microphones. The playback sounds simulated echoes returning from fluttering insects (see echoes shown in A). (C) Fluttering target discrimination performance of three animals. Bats discriminated between an insect beating its wings at 50 Hz and at a slower, variable rate. Discrimination threshold (75% correct criterion) is about 4 Hz (adapted from von der Emde and Menne 1989).

(Link, Marimuthu, and Neuweiler 1986), also pursued fluttering insects whether flying or sitting and ignored the prey when they did not move their wings. Furthermore, hunting behavior in these bats was aborted when the insects stopped fluttering. A rotating propeller that produces an echo glint pattern similar to that of a fluttering insect (von der Emde and Schnitzler 1986) has been shown to attract *Pteronotus parnellii* (Goldmann and Henson 1977; Schnitzler et al. unpublished data) and *Rhinolophus ferrumequinum* (Schnitzler, unpublished data) in the laboratory. Such a fluttering target apparatus has also attracted the insect-hunting bats *Hipposideros ruber* (Bell and Fenton 1984) and *Rhinolophus ferrumequinum* and *Pteronotus parnellii* (Schnitzler et al. unpublished data), in the field. Together, these data suggest that bats using CF-FM signals respond selectively to echo changes produced by target movement and may use acoustic glint information to recognize insect echoes even in dense background clutter. Another type of echo modulation is used by the greater bulldog bat *Noctilio leporinus*, a species that pursues only jumping fish. Interference between echoes from the body of the fish and the water disturbances it creates by jumping produce a spectral pattern that the bat presumably uses to identify its prey (Schnitzler et al. 1994).

Laboratory studies also show that bats with long CF-FM signals are able to detect sinusoidal frequency modulations in their echoes. *Rhinolophus ferrumequinum* is much more sensitive than hipposiderid bats. In an experiment in which the bats had to discriminate an oscillating target from a similar motionless one, *Rhinolophus ferrumequinum* required a minimum oscillation frequency of 40 Hz (typical wingbeat frequency in many moths) at the threshold oscillation amplitude of 0.24 mm (Schnitzler and Flieger 1983). In *Hipposideros lankadiva*, the threshold amplitude of the oscillating target was about five fold higher than in *Rhinolophus ferrumequinum*, and *Hipposideros speoris* never learned the task (von der Emde and Schnitzler 1986). However, both hipposiderids reacted strongly to a glint-producing moving propeller.

The species differences in discrimination performance appear to correspond to differences in signal duration and duty cycle. *Rhinolophus ferrumequinum* produces very long signals (50–60 msec), whereas *Hippodideros lankadiva* produces signals that last about 10 msec and *Hippodideros speoris* signals of about 5 msec. While there are differences in the signal duty cycles among these species, it is interesting to note that they all increased sound duration or repetition rate when exposed to a fluttering target. These changes in the temporal features of the bat's sounds may indicate the bat's active effort to obtain acoustic information about target movement (Trappe and Schnitzler 1982; Schnitzler and Flieger 1983; von der Emde and Schnitzler 1986).

As acoustic information contained in echoes from fluttering targets differs across insect species, it has been asked whether bats can use echo flutter to classify their prey (Schnitzler and Ostwald 1983). In the laboratory, *Rhinolophus ferrumequinum* often selected one species of insect over another, presumably by using wingbeat rate as a cue (Trappe 1982). Studies of the food habits of this species also suggest selective prey acquisition (Ransome and Stebbings, personal communication). By contrast, there is no evidence for prey selection in *Hipposideros ruber* (Bell and Fenton 1984). A reason for this species difference could be the much shorter signal duration used by hipposiderid bats, thus limiting the acoustic information about wingbeat signature in insect echoes.

The role of wingbeat rate as a possible cue for target characterization in bats with CF-FM signals has been studied in *Rhinolophus ferrumequinum* (von der Emde and Menne 1989), in *Rhinolophus rouxi* and *Hipposideros lankadiva* (Roverud, Nitsche, and Neuweiler 1991), and in *Pteronotus parnellii* (Schnitzler and Kaipf 1992). All experiments showed that bats are able to discriminate different wingbeat rates and that the performance of a given species is correlated with the duty cycle of its signals. The lowest flutter discrimination threshold was reported for the horseshoe bat, the species using the highest duty cycle. These bats were able to sense differences in wingbeat rate of 4%–9% (see Fig. 3.12C), whereas *Pteronotus parnellii* needed a difference of 20% and *Hipposideros lankadiva* a difference of 15%.

Many insects have rather similar wingbeat rates so that this cue alone would not be very suitable for the classification of prey. In discrimination experiments in which fluttering insects were simulated as phantom targets, *Rhinolophus ferrumequinum* and *Pteronotus parnelli* were able to discriminate different insect species exhibiting the same wingbeat rate (von der Emde and Schnitzler 1990; Schnitzler and Kaipf 1992). In a generalization experiment, bats were initially trained to select phantom target echoes of a particular insect species viewed from the side and later tested for recognition of the same species at novel aspect angles. The bat's mean performance for recognizing insects from new perspectives was mostly greater than 80%. Because the angular orientation of the prey strongly affects the modulation pattern in the echo (Kober and Schnitzler 1990), this kind of discrimination reveals astonishing cognitive abilities. This finding suggests that bats may develop a three-dimensional representation of a fluttering insect from acoustic information contained in echoes from a single view. The echo of a fluttering insect ensonified from a single orientation may thus be sufficient to construct a complete three-dimensional representation of the moving prey, and bats may use this representation for target classification (von der Emde and Schnitzler 1990).

The ability to use flutter information is not only helpful for target classification; it also allows CF-FM bats to hunt in areas where insect echoes are masked by strong background clutter (Neuweiler et al. 1988). Scattered field observations on feeding behavior in these bats show that they hunt for insects close to or in bushes and trees, along walls, and near the ground. Some species also hunt for insects from perches like a flycatcher (reviewed in Ostwald, Schnitzler, and Schuller 1988).

6.3 Fluttering Target Discrimination in Bats Using FM Signals

Fluttering target information may also be important to bats that hunt out in the open and produce rather long signals (up to 20 msec) of nearly constant frequency when searching for prey. As these bats never produce such signals in the laboratory, it is not possible to determine whether they use glint frequency in quasi-CF echoes to classify prey. However, it is evident that a glint in the echo of search pulses increases the likelihood of target detection and also perhaps indicates the presence of a fluttering insect (Moss and Zagaeski 1994).

In the laboratory, these bats emit short wideband FM sounds, and fluttering target discrimination has been studied with two species using these signals. In experiments in which a rotating propeller simulated a fluttering insect, *Pipistrellus stenopterus* (Sum and Menne 1988) and *Eptesicus fuscus* (Roverud, Nitsche, and Neuweiler 1991; Moss et al. 1992) were able to discriminate between different simulated wingbeat frequencies; however, the fluttering target discrimination threshold was higher in these FM bats than reported for *Rhinolophus* using long CF-FM signals.

What acoustic information does an FM bat use to discriminate between targets fluttering at different rates? In the behavioral discrimination experiments, the duration of the bat's echolocation sounds was too short and the duty cycle too low for glints in echoes to carry information about wingbeat period. It may be that the interference pattern between the Doppler-shifted echo from the moving blades of the propeller and the unmodulated echo from stationary parts of the wingbeat simulator contained the information that was used by the bats (Sum and Menne 1988). It is not clear whether this kind of information allows FM bats in the field to identify fluttering insects.

6.4 Summary

Flutter information is used by different species of bats in different ways. Bats with very long CF-FM signals (more than 20 msec), such as all rhinolophids, the mormoopid bat *Pteronotus parnellii*, and some hipposiderids (e.g., *Hipposideros lankadiva*), not only use flutter information to identify insect echoes and to discriminate them from unmodulated background clutter but may also use echo information from fluttering insects for target classification. In bats that use shorter CF-FM signals, such as most of the hipposiderids and the two species of noctilionids, fluttering target discrimination is not as well developed. This also appears to be the case for bats that hunt insects in the open, using quasi-CF pulses when searching for insects. For bats that use sounds of short duration and low duty cycle, echoes from fluttering insects may be most important in signaling the presence of prey and less useful for carrying detailed information about the wingbeat rate of the prey. In future research it would be of interest to examine the perceptual representations of fluttering insects in bat species using different echolocation signal design.

7. Conclusion

The computations performed by the auditory system of the bat support both passive listening and active echolocation processes. In studies of passive listening, absolute sensitivity data suggest that the bat's auditory system is well adapted to detect species-specific echolocation and communication signals. Basic psychoacoustic data on other aspects of hearing in bats are limited, but what exist suggests that the bat's auditory system may incorporate some specializations into the general mammalian plan.

The bulk of research on auditory information processing in bats has focused on the active echolocation system, requiring the animal to probe its environment by emitting sonar signals and listening to echoes from objects in the path of the sound beam. This chapter surveys research on the bat's active biosonar system for the detection, localization, and identification of targets. In each of these areas, we have examined data that reveal the limits of the bat's echolocation system for extracting information about the environment through acoustic channels. Moreover, when possible we have also considered data that speak to the representation of sonar targets in the bat's auditory system. In attempting to synthesize data from different experiments, we continuously encountered variations in the experimental design, the bat's task, and the acoustic environment that make cross-study comparisons difficult. In future studies on echolocation performance in bats, we propose that researchers pay careful attention to the design of each experiment and the details of the acoustic environment. Moreover, because bats have active control over the echo information returning to their ears, it is essential to frequently monitor the duration, bandwidth, and level of the sounds the bat produces under experimental conditions. We believe that future work in this field will yield exciting and important insights in comparative psychoacoustics.

Acknowledgments Preparation of this chapter was supported by a National Science Foundation Young Investigator Award to CFM and by the Deutsche Forschungsgemeinschaft SFB 307. We thank Richard Fay, Marc Hauser, Arthur Popper, and Doreen Valentine for comments on the manuscript and Ingrid Kaipf and Anne Grossetête for assistance with some of the figures.

References

Altes RA (1981) Echo phase perception in bat sonar? J Acoust Soc Am 69:505–509.

Airapetianz ESH, Konstantinov A I (1974) Echolocation in nature. Leningrad: Nauka. (English translation, Joint Publications Research Service, no. 63328, 1000 North Glebe Road, Arlington, VA 22201.)

Bell GP, Fenton MB (1984) The use of Doppler-shifted echoes as a clutter rejection system: the echolocation and feeding behavior of *Hipposideros ruber* (Chiroptera: Hipposideridae). Behav Ecol Sociobiol 15:109–114.

Brown CH (1994) Sound localization. In: Popper AN, Fay RR (eds) Comparative Hearing: Mammals. New York: Springer-Verlag, pp. 57–96.

Bruns V (1980) Structural adaptation in the cochlea of the horseshoe bat for the analysis of long CF-FM echolocating signals In: Busnel RG, Fish J F (eds) Animal Sonar Systems. New York: Plenum, pp. 867–869.

Burkard R, Moss C F (1994) The brainstem auditory evoked response (BAER) in the big brown bat (*Eptesicus fuscus*) to clicks and frequency modulated sweeps. J Acoust Soc Am 96:801–810.

Dalland JI (1965) Hearing sensitivity in bats. Science 150:1185–1186.

de Boer E (1985) Auditory time constants: A paradox? In: Michelsen A (ed) Time Resolution in Auditory Systems. Berlin: Springer, pp. 141–158.

Denzinger A, Schnitzler H-U (1994) Echo SPL influences the ranging performance of the big brown bat, *Eptesicus fuscus*. J Comp Physiol *175*:563–571.

Fay RR (1988) Hearing in Vertebrates. A Psychophysics Databook. Winnetka, IL: Hill-Fay.

Fletcher H (1940) Auditory patterns. Rev Mod Phys 12:47–65.

Fullard JH, Fenton M B (1977) Acoustic and behavioural analyses of the sounds produced by some species of Nearctic Arctiidae (Lepidoptera) Can J Zool 55:1213–1224.

Goldman LJ, Henson OW (1977) Prey recognition and selection by the constant frequency bat, *Pteronotus p. parnellii*. Behav Ecol Sociobiol 2:411–419.

Griffin D (1944) Echolocation by blind men, bats and radar. Science 100:589–590.

Griffin D (1953) Bat sounds under natural conditions, with evidence for the echolocation of insect prey. J Exp Zool 123:435–466.

Griffin D (1958) Listening in the Dark. New Haven: Yale University Press. (Reprinted by Cornell University Press, Ithaca, NY 1986.)

Griffin D, Webster FA, Michael CR (1960) The echolocation of flying insects by bats. Anim Behav 8:141–154.

Grinnell AD, Schnitzler H-U (1977) Directional sensitivity of echolocation in the horseshoe bat, *Rhinolophus ferrumequinum*. II. Behavioral directionality of hearing. J Comp Physiol A 116:63–76.

Habersetzer J, Vogler B (1983) Discrimination of surface-structured targets by the echolocating bat, *Myotis myotis*, during flight. J Comp Physiol A 152:275–282.

Hackbarth H (1986) Phase evaluation in hypothetical receivers simulating ranging in bats. Biol Cybern 54:281–287.

Hartley DJ (1989) The effect of atmospheric sound absorption on signal bandwidth and energy and some consequences for bat echolocation. J Acoust Soc Am 8:1338–1347.

Hartley DJ (1992a) Stabilization of perceived echo amplitudes in echolocating bats. I. Echo detection and automatic gain control in the big brown bat, *Eptesicus fuscus*, and the fishing bat, *Noctilio leporinus*. J Acoust Soc Am 91:1120–1132.

Hartley DJ (1992b) Stabilization of perceived echo amplitudes in echolocating bats. II. The acoustic behaviour of the big brown bat, *Eptesicus fuscus*, when tracking moving prey. J Acoust Soc Am 91:1133–1149.

Hartley DJ, Suthers RA (1987) The sound emission pattern and the acoustical role of the noseleaf in the echolocating bat, *Carollia perspicillata*. J Acoust Soc Am 82:1892–1900.

Hartley DJ, Suthers RA (1989) The sound emission pattern of the echolocating bat, *Eptesicus fuscus*. J Acoust Soc Am 85:1348–1351.

Hartridge H (1945) Acoustical control in the flight of bats. Nature 156:490–494; 692–693.

Henson OW Jr (1965) The activity and function of the middle ear muscles in echolocating bats. J Physiol (Lond) 180:871–887.

Jen PHS, Chen D (1988) Directionality of sound pressure transformation at the pinna of echolocating bats. Hear Res 34:101–118.

Kick SA (1982) Target detection by the echolocating bat, *Eptesicus fuscus*. J Comp Physiol A 145: 431–435.

Kick SA, Simmons JA (1984) Automatic gain control in the bat's sonar receiver and the neuroethology of echolocation. J Neurosci 4:2725–2737.

Kober R, Schnitzler H-U (1990) Information in sonar echoes of fluttering insects available for echolocating bats. J Acoust Soc Am 87:882–896.

Kulikowski JJ, Tolhurst DJ (1973) Psychophysical evidence for sustained and transient detectors in human vision. J Physiol (Lond) 232:149–162.

Lawrence BD, Simmons JA (1982a) Measurements of atmospheric attenuation at ultrasonic frequencies and the significance for echolocation by bats. J Acoust Soc Am 71:585–590.

Lawrence BD, Simmons JA (1982b) Echolocation in bats: the external ear and perception of the vertical positions of targets. Science 218:481–483.

Link A, Marimuthu G, Neuweiler G (1986) Movement as a specific stimulus for prey-catching behaviour in rhinolophid and hipposiderid bats. J Comp Physiol A 159:403–413.

Long G (1977) Masked auditory thresholds from the bat, *Rhinolophus ferrumequinum*. J Comp Physiol A 116:247–255.

Long G (1994) Psychoacoustics. In: Fay RR, Popper AN (eds) Comparative Hearing: Mammals. New York: Springer-Verlag pp 18–56.

Long G, Schnitzler H-U (1975) Behavioral audiograms form the bat *Rhinolophus ferrumequinum*. J Comp Physiol A 100:211–220.

Masters WM, Jacobs S C (1989) Target detection and range resolution by the big brown bat (*Eptesicus fuscus*) using normal and time-reversed model echoes. J Comp Physiol A 166:65–73.

McCue JJG (1966) Aural pulse compression in bats and humans. J Acoust Soc Am 40:545–548.

Menne D (1988) A matched filter bank for time delay estimation in bats. In: Nachtigall PE, Moore PWB (eds) Animal Sonar Processes and Performance. New York: Plenum, pp. 835–842.

Menne D, Hackbarth H (1986) Accuracy of distance measurement in the bat *Eptesicus fuscus*: theoretical aspects and computer simulations. J Acoust Soc Am 79:386–397.

Menne D, Kaipf I, Wagner I, Ostwald J, Schnitzler H-U (1989) Range estimation by echolocation in the bat *Eptesicus fuscus*: trading of phase versus time cues. J Acoust Soc Am 85:2642–2650.

Miller LA (1983) How insects detect and avoid bats. In: Huber F, Markl H (eds) Neuroethology and Behavioral Physiology. Berlin: Springer-Verlag, pp. 251–266.

Miller LA (1991) Arctiid moth clicks can degrade the accuracy of range difference discrimination in echolocating big brown bats, *Eptesicus fuscus*. J Comp Physiol A 168:571–579.

Mogdans J, Schnitzler H-U (1990) Range resolution and the possible use of spectral information in the echolocating bat, *Eptesicus fuscus*. J Acoust Soc Am 88:754–757.

Mogdans J, Schnitzler H-U, Ostwald J (1993) Discrimination of 2-wavefront echoes by the big brown bat, *Eptesicus fuscus*: behavioral experiments and receiver simulations. J Comp Physiol A 172:309–323.

Mogdans J, Ostwald J, and Schnitzler H-U (1988) The role of pinna movement for the localization of vertical and horizontal wire obstacles in the greater horseshoe bat, *Rhinolophus ferrumequinum*. J Acoust Soc Am 84:1676–1679.

Møhl B (1986) Detection by a pipistrellus bat of normal and reversed replica of its sonar pulses. Acustica 61:75–82.

Møhl B (1988) Target detection by echolocating bats. In: Nachtigall PE, Moore PWB (eds) Animal Sonar Processes and Performance. New York: Plenum Press, pp. 435–450.

Møhl B, Surlykke A (1989) Detection of sonar signals in the presence of pulses of masking noise by the echolocating bat, *Eptesicus fuscus*. J Comp Physiol A 165:119–124.

Moss CF, Schnitzler H-U (1989) Accuracy of target ranging in echolocating bats: acoustic information processing. J Comp Physiol A 165:383–393.

Moss CF, Simmons JA (1993) Acoustic image representation of a point target in the bat *Eptesicus fuscus:* evidence for sensitivity to echo phase in bat sonar. J Acoust Soc Am 93:1553–1562.

Moss CF, Zagaeski M (1994) Acoustic information available to bats using frequency-modulated echolocation sounds for the perception of insect prey. J Acoust Soc Am 95:2745–2756.

Moss CF, Gounden C, Booms J, Roach J (1992) Discrimination of target movement by the FM bat, *Eptesicus fuscus*. Abstracts of the 15th Midwinter Research Meeting of the Society for Research in Otolaryngology, p. 142.

Neuweiler G, Bruns V, Schuller G (1980) Ears adapted for the detection of motion, or how echolocating bats have exploited the capacities of the mammalian auditory system. J Acoust Soc Am 68:741–753.

Neuweiler G, Link A, Marimuthu G, Rübsamen R (1988) Detection of prey in echocluttering environments. In: Nachtigall PE, Moore PWB (eds) Animal Sonar Processes and Performance. New York: Plenum, pp. 613–618.

Ostwald J, Schnitzler H-U, Schuller G (1988) Target discrimination and target classification in echolocating bats. In: Nachtigall PE, Moore PWB (eds) Animal Sonar Processes and Performance. New York: Plenum, pp. 413–434.

Pollak GD (1993) Some comments on the proposed perception of phase and nanosecond time disparities by echolocating bats. J Comp Physiol A 172:523–531.

Poussin C, Simmons JA (1982) Low-frequency hearing sensitivity in the echolocating bat, *Eptesicus fuscus*. J Acoust Soc Am 72:340–342.

Reetz G, Schnitzler H-U (1992) Signal design in the bat *Eptesicus fuscus* when detecting targets differing in range and size. In: Elsner, N and Richter, D.W. (eds.) Stuttgart: Georg Thieme Verlag. Proceedings of the 20th Göttingen Neurobiology Conference, p. 213.

Roeder KD (1962) The behaviour of free-flying moths in the presence of artificial ultrasonic pulses. Anim Behav 10:300–304.

Roverud RC (1989a) Harmonic and frequency structure used for echolocation sound pattern recognition and distance information processing in the rufous horseshoe bat. J Comp Physiol A 166:251–255.

Roverud RC (1989b) A gating mechanism for sound pattern recognition is correlated with the temporal structure of echolocation sounds in the rufous horseshoe bat. J Comp Physiol A 166:243–249.

Roverud RC, Grinnell AD (1985a) Discrimination performance and echolocation signal integration requirements for target detection and distance determination in the CF/FM bat, *Noctilio albiventris*. J Comp Physiol A 156:447–456.

Roverud RC, Grinnell AD (1985b) Echolocation sound features processed to provide distance information in the CF/FM bat, *Noctilio albiventris*: evidence for a gated time window utilizing both CF and FM components. J Comp Physiol A 156:457–469.

Roverud RC, Nitsche V, Neuweiler G (1991) Discrimination of wingbeat motion by bats correlated with echolocation sound pattern. J Comp Physiol A 168:259–263.

Saillant PA, Simmons JA, Dear SP, McMullen T A (1993) A computational model of echo processing and acoustic imaging in frequency-modulated echolocating bats: The spectrogram correlation and transformation receiver. J Acoust Soc Am 94:2691–2712.

Schmidt S (1988) Evidence for a spectral basis of texture perception in bat sonar. Nature 331:617–619.

Schmidt S (1992) Perception of structured phantom targets in the echolocating bat, *Megaderma lyra*. J Acoust Soc Am 91:2203–2223.

Schmidt S (1993) Perspectives and problems of comparative psychoacoustics in echolocating bats. In: Abstracts of the Sixteenth Midwinter Research Meeting of the Association for Research in Otolaryngology, St. Petersburg Beach, FL, p. 145.

Schmidt S, Türke B, Vogler B (1983) Behavioural audiogram from the bat, *Megaderma lyra* (Geoffroy, 1810; Microchiroptera). Myotis 21/22:62–66.

Schneider H, Möhres F P (1960) Die Ohrbewegungen der Hufeisenfledermäse (Chiropetera, Rhinolophidae) und der Mechanismus des Bildhörens. Z Vgl Physiol 44:1–40.

Schnitzler H-U (1968) Die Ultraschall-Ortungslaute der Hufeisen-Fledermaüse (Chiroptera-Rhinolophidae) in verschiedenen Orientierungssituationen. Z Vgl Physiol 57:376–408.

Schnitzler H-U (1987) Echoes of fluttering insects: information for echolocating bats. In: Fenton MB, Racey P, Rayner JMV (eds) Recent Advances in the Study of Bats. Cambridge: Cambridge University Press, pp. 226–243.

Schnitzler H-U, Flieger E (1983) Detection of oscillating target movements by echolocation in the greater horseshoe bat. J Comp Physiol A 153:385–391.

Schnitzler H-U, Grinnell AD (1977) Directional sensitivity of echolocation in the horseshoe bat, *Rhinolophus ferrumequinum*. I. Directionality of sound emission. J Comp Physiol A 116:51–61.

Schnitzler H-U, Henson OW Jr (1980) Performance of airborne animal sonar systems: 1. Microchiroptera. In: Busnel RG, Fish JE (eds) Animal Sonar Systems. New York: Plenum, pp. 109–181.

Schnitzler H-U, Kaipf I (1992) Classification of insects by echolocation in the mustache bat, *Pteronotus parnellii*. Deustche Gesellshaft für Sägertierkunde, Vol. 66. Karlsruhe: Hauptversammlung.

Schnitzler H-U, Ostwald J (1983) Adaptation for the detection of fluttering insects by echolocation in horseshoe bats. In: Ewert JP, Capranica RR, Ingle DJ (eds) Advances in Vertebrate Neuroethology. New York: Plenum, pp. 801–827.

Schnitzler H-U, Menne D, Hackbarth H (1985) Range determination by measuring time delays in echolocating bats In: Michelsen A (ed) Time Resolution in Auditory Systems. Berlin: Springer-Verlag, pp. 180–204.

Schnitzler H-U, Kalko EKV, Kaipf I, Grinnell A (1994) Fishing and echolocation behavior of the greater bulldog bat, *Noctilio leporinus*, in the field. Behav Ecol Sociobiol 35:327–345.

Schnitzler H-U, Menne D, Kober R, Heblich K (1983) The acoustical image of fluttering insects in echolocating bats. In: Huber F, Markel H (eds) Neuroethology and Behavioral Physiology. Heidelberg: Springer-Verlag, pp. 235–249.

Shimozawa T, Suga N, Hendler P, Schuetze S (1974) Directional sensitivity of echolocation system in bats producing frequency modulated signals. J Exp Biol 60:53–69.

Simmons JA (1973) The resolution of target range by echolocating bats. J Acoust Soc Am 54:157–173.

Simmons J A (1979) Perception of echo phase information in bat sonar. Science 204:1336–1338.

Simmons JA (1987) Acoustic images of target range in the sonar of bats. Nav Res Rev 39:11–26.

Simmons JA (1989) A view of the world through the bat's ear: the formation of acoustic images in echolocation. Cognition 33:155–199.

Simmons JA (1992) Time-frequency transforms and images of targets in the sonar of bats. In: Bialek W (ed) Princeton Lectures on Biophysics–1991. Princeton: NEC Research Institute.

Simmons JA (1993) Evidence for perception of fine echo delay and phase by the FM bat, *Eptesicus fuscus*. J Comp Physiol A 172:533–547.

Simmons JA Stein RA (1980) Acoustic imaging in bat sonar: echolocation signals and the evolution of echolocation. J Comp Physiol A 135:61–84.

Simmons JA, Moffat AJM, Masters WM (1992) Sonar gain control and echo detection thresholds in the echolocating bat, *Eptesicus fuscus*. J Acoust Soc Am 91:1150–1163.

Simmons JA, Moss CF, Ferragamo M (1990) Convergence of temporal and spectral information into acoustic images of complex sonar targets perceived by the echolocation bat, *Eptesicus fuscus*. J Comp Physiol A 166:449–470.

Simmons JA, Ferragamo MJ, Dear SP, Haresign T, Fritz J (1995) Auditory computations for acoustic imaging in bat sonar. In: Hawkins H, McMullem T (eds) Springer Handbook of Auditory Research: Auditory Computations. New York: Springer-Verlag, (in press).

Simmons JA, Ferragamo M, Moss CF, Stevenson SB, Altes RA (1990) Discrimination of jittered sonar echoes by the echolocating bat, *Eptesicus fuscus*: the shape of target images in echolocation. J Comp Physiol A 167:589–616.

Simmons JA, Freedman EG, Stevenson SB, Chen L, Wohlgenant TJ (1989) Clutter interference and the integration time of echoes in the echolocating bat, *Eptesicus fuscus*. J Acoust Soc Am 86:1318–1332.

Simmons JA, Kick SA, Moffat AJM, Masters WM, Kon D (1988) Clutter interference along the target range axis in the echolocating bat, *Eptesicus fuscus*. J Acoust Soc Am 84:551–559.

Simmons JA, Kick SA, Lawrence BD, Hale C, Bard C, Escudié B (1983) Acuity of horizontal angle discrimination by the echolocating bat, *Eptesicus fuscus*. J Comp Physiol A 153:321–330.

Simmons JA, Lavender WA, Lavender BA, Doroshow CA, Kiefer SW, Livingston R, Scallet AC, Crowley DE (1974) Target structure and echo spectral discrimination by echolocating bats. Science 186:1130–1132.

Skolnik MI (1980) Introduction to Radar Systems. New York: McGraw-Hill.

Sokolov BV (1972) Interaction of auditory perception and echolocation in bats (Rhinolophidae) during insect catching. Vestn Leningr Univ Ser Biol 27:96–104 (in Russian).

Stapells DR, Picton TW, Smith AD (1982) Normal hearing thresholds for clicks. J Acoust Soc Am 72:74–79.

Suga N, Jen P (1975) Peripheral control of acoustic signals in the auditory system of echolocating bats. J Exp Biol 62:277–311.

Suga N, Schlegel P (1972) Neural attenuation of responses to emitted sounds in echolocating bats. Science 177:82–84.

Suga N, Shimozawa T (1974) Site of neural attenuation of responses to self-vocalized sounds in echolocating bats. Science 183:1211–1213.

Sum YW, Menne D (1988) Discrimination of fluttering targets by the FM-bat *Pipistrellus stenopterus?* J Comp Physiol A 163:349–354.

Surlykke A (1992) Target ranging and the role of time-frequency structure of synthetic echoes in big brown bats, *Eptesicus fuscus*. J Comp Physiol A 170:83–92.

Surlykke A, Miller L A (1985) The influence of arctiid moth clicks on bat echolocation: jamming or warning? J Comp Physiol A 156:831–843.

Suthers RA, Summers C A (1980) Behavioral audiogram and masked thresholds of the megachiropteran echolocating bat, *Rousettus*. J Comp Physiol A 136:227–233.

Trappe M (1992) Verhalten und Echoorting der Grossen Hufeisemase beim Isektenfang. Ph.D. dissertation, University of Marburg.

Trappe M, Schnitzler H-U (1982) Doppler-shift compensation in insect-catching horseshoe bats. Naturwissenschaften 69:193–196.

Troest N, Møhl B (1986) The detection of phantom targets in noise by serotine bats; negative evidence for the coherent receiver. J Comp Physiol A 159:559–567.

von der Emde G, Schnitzler H-U (1986) Fluttering target detection in hipposiderid bats. J Comp Physiol A 14:43–55.

von der Emde G, Menne D (1989) Discrimination of insect wingbeat-frequencies by the bat *Rhinolophus ferrumequinum*. J Comp Physiol A 164:663–671.

von der Emde G, Schnitzler H-U (1990) Classification of insects by echolocating greater horseshoe bats. J Comp Physiol A 167:423–430.

Yost WA, Gourevitch G (eds) (1987) Directional Hearing. Berlin: Springer-Verlag.

Zwicker E, Flottorp G, Stevens SS (1957) Critical bandwidths in loudness summation. J Acoust Soc Am 29:548–557.

4

Auditory Dimensions of Acoustic Images in Echolocation

James A. Simmons, Michael J. Ferragamo, Prestor A. Saillant, Tim Haresign, Janine M. Wotton, Steven P. Dear, and David N. Lee

1. Introduction

1.1 Echolocation and Hearing

Echolocation in bats is one of the most demanding adaptations of hearing to be found in any animal. Transforming the information carried by sounds into perceptual images depicting the location and identity of objects rapidly enough to control the decisions and reactions of a swiftly flying bat is a prodigious task for the auditory system to accomplish. The exaggeration of aspects of auditory function to achieve spatial imaging reflects the vital role of hearing in the lives of bats — for finding prey and perceiving obstacles to flight (Neuweiler 1990). It also highlights the mechanisms behind these functions to make echolocation a useful model for studying how the auditory system processes information and creates auditory perceptions in the most extreme circumstances.

The auditory capacity most at a premium for echolocation with wideband signals (see Fenton, Chapter 2) is the ability to determine the arrival time of sonar echoes, both to perceive the delay of individual echoes and to separately perceive several closely spaced echoes at their slightly different delays. Behavioral evidence accumulated from a variety of recent experiments reveals that bats can determine echo delay with an accuracy and fine resolution in the range of at least 1–5 μsec, and under certain conditions with an accuracy of 10–40 nsec (Menne et al. 1989; Moss and Schnitzler 1989; Moss and Simmons 1993; Simmons 1979, 1989, 1993; Simmons et al. 1990; see Moss and Schnitzler, Chapter 3). However, this degree of temporal acuity has been seen as unattainable from neural responses typically recorded in the bat's auditory system (Pollak et al. 1977; Pollak and Casseday 1989; Schnitzler, Menne, and Hackbarth 1985). Nevertheless, the recent behavioral evidence is so compelling (Simmons 1993) that we have reexamined earlier, widely accepted experimental results and found that they, too, reveal an underlying echo–delay accuracy and resolution in the microsecond range. This chapter summarizes the findings of our reexamination.

1.2 Scope of This Chapter

Our chapter reviews acoustic and behavioral evidence for fine temporal precision and resolving power as the basis for acoustic imaging in echolocation. This evidence is drawn largely from experiments conducted with bats in naturalistic situations that have not previously been considered relevant to determining the detailed content of the bat's sonar images. The conceptual framework for our treatment of echo processing by bats has been introduced elsewhere (Simmons 1989, 1992; Simmons, Moss, and Ferragamo 1990), and the computational validity of this approach has been demonstrated in a quantitative model of echolocation (Saillant et al. 1993). A companion chapter in another forthcoming volume of the *Springer Handbook of Auditory Research* series describes this imaging process and how it is supported by neural responses in the bat's auditory system (see Simmons et al. in press).

In this chapter, we examine the content of the sonar images perceived by bats that broadcast frequency-modulated (FM) sonar sounds. We focus on determining what echolocating bats perceive along the axis of distance, or *target range*, from the arrival time of FM sonar echoes returning to their ears. While this chapter thus initially appears to have limited scope— concentrating on depth perception—our concern for the dimension of target range will expand to encompass perception of other aspects of targets such as shape and direction.

Many of the behavioral experiments considered here were carried out rather early in the history of research on echolocation; indeed, some of them were first used to demonstrate the existence of sonar in bats. While the basic issues they bring forth were recognized quite early from a physiological perspective (Grinnell 1967; Suga 1967), the full implications of their results for understanding the images perceived by bats have only now come to light. Our goal here is to discover what the results of these behavioral experiments reveal about the acoustic dimensionality of the images underlying echolocation by FM bats. A methodologically oriented review of the more recent, controversial experiments on FM echolocation that have drawn attention to the question of temporal acuity in bats is given in another chapter in this volume (see Moss and Schnitzler, Chapter 3). Several other chapters in this volume of the Handbook provide additional background for our chapter.

2. Echolocation by FM Bats

2.1 The Big Brown Bat, Eptesicus fuscus

The experiments discussed in this chapter have been carried out with the North American big brown bat, *Eptesicus fuscus* (family Vespertilionidae) or with several related species of FM bats in this same family. *Eptesicus* is an insectivorous bat that uses echolocation to find its prey (see Fenton, Chapter 2). Figure 4.1 shows the big brown bat in the act of capturing a mealworm suspended on a fine thread. The bat has located and approached the target by sonar and is just about to seize the insect in its tail membrane, which is a mode of capture employed by many species of bats (Webster and Griffin 1962). Details about the feeding habits and general biology of *Eptesicus* have been conveniently summarized by Kurta and Baker (1990).

2.2 Echolocation Signals of Eptesicus

Broadcasting sounds is a means for probing the environment in depth, and bats use FM echolocation sounds for this purpose (Simmons 1973; Simmons and Grinnell 1988). The transmitted signal illuminates or ensonifies objects located in the beam of sound to reflect or scatter back to the bat. Figure 4.2 illustrates spectrograms of echolocation sounds broadcast by the big brown bat, *Eptesicus fuscus*, during interception of prey. General information about the echolocation sounds of different species of bats is given in Chapter 2 of this volume; however, several details about the broadcast waveforms are relevant to our discussion and need to be summarized here. The signals in Figure 4.2 are FM sounds, with two

FIGURE 4.1. A big brown bat, *Eptesicus fuscus*, is about to capture a mealworm hanging on a fine filament. The bat uses its tail membrane to seize the target. (Photograph by S. P. Dear and P. A. Saillant.)

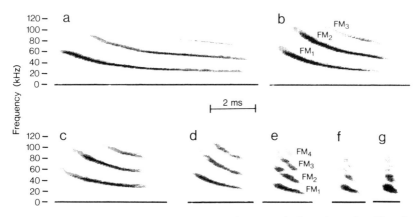

FIGURE 4.2. Spectrograms (a–g) of echolocation sounds broadcast by *Eptesicus* during flight toward prey. These frequency-modulated (FM) signals contain 3–4 harmonics (FM$_1$–FM$_4$) that collectively cover the frequency range of about 15–100 kHz. The bat progressively shortens its signals during the approach, with duration about equal to the two-way travel time of the sound, or echo delay.

prominent harmonics sweeping downward from about 55 kHz to about 20–25 kHz (the first harmonic, labeled FM$_1$ in Fig. 4.2b) and from about 100 kHz to 40–50 kHz (second harmonic, labeled FM$_2$ in Fig. 4.2b). Segments of the third harmonic (FM$_3$; Fig. 4.2a–d), and even the fourth harmonic (FM$_4$; Fig. 4.2e–g) are often present, too, at frequencies sweeping from 90 kHz to about 75 kHz (Griffin 1958; Simmons 1989; see also Fenton, Chapter 2). During the bat's approach to a target, the characteristics of the sounds being broadcast are adjusted from one transmission to the next according to the decreasing distance to the target, indicating that the bat monitors distance throughout its flight (Griffin 1958; Schnitzler and Henson 1980; Simmons 1989).

3. Acoustic Basis for the Bat's Sonar Images

3.1 The Echo Stream as Stimulus

When a bat transmits a sonar sound, the broadcast waveform travels outward into the environment to impinge on objects arrayed at different *distances* and then return as a series of echoes arrayed at different *times*. The image the bat perceives is a reconstruction of information about the *spatial* array of objects from the time series of echoes. Each reflecting point or surface returns a more-or-less complete replica of the incident sound back toward the bat's ears, with modifications caused by propagation through the air and also by the process of reflection itself. Echoes reach the bat's ears later than the transmission, at *delays* determined by target range, at *amplitudes* weaker than the transmission according to target size and distance, and with *spectral characteristics* determined by sound propagation in air, by the target's direction, and by its shape (Griffin 1958, 1967; Novick 1977; Pye 1980; Schnitzler and Henson 1980; Simmons 1989; Simmons and Kick 1984; Kober and Schnitzler 1990; Moss and Zagaeski 1994). The details of these acoustic effects constitute the physical basis for echolocation by defining the stimuli at the bat's ears.

The delay of echoes from targets at different ranges is 5.8 msec/m. An array of objects located at different distances from the bat yields a series of echoes at different arrival times, so that each of the bat's FM sonar transmissions is accompanied by a set of echoes that return within a specified window of time following the broadcast. For *Eptesicus*, with a maximum sonar operating range of about 5 m for insect-sized targets (Kick 1982), this time window is about 30 msec long. Physiological studies confirm that the auditory system of *Eptesicus* processes echo–delay information for arrival times to about 30–35 msec (Dear, Simmons, and Fritz 1993; Dear et al. 1993; see Simmons et al. in press). In the most general sense, the only information the bat has for reconstructing images of targets is the temporal sequence of echoes received by its two ears during this brief interval after each sound is broadcast. Consequently, the arrival time of echoes is the most important dimension for segregating individual reflections within the incoming stream of sound, which, in turn, makes the distance to objects the primitive perceptual quality underlying the images the bat perceives. To grasp the overriding significance of the arrival times of echoes in FM echolocation, the reader needs to know about some peculiarities of echoes formed by small objects in air.

3.2 The Inferential Nature of Echolocation

Echolocation is a classic example of an inverse problem: in sonar, targets are *inferred* from echoes using the properties of propagating sound as rules for reconstructing the locations and identifying features of scattering surfaces and points in space. The requirement that an inferential step formally intervene between the acoustic stimuli and the bat's images gives a sharp focus to our discussion of the perceptual mechanisms embodied in echolocation. For sonar in air (where all objects are significantly higher in acoustic impedance than the medium for propagation), a sonar target is a collection of one or more discrete points in space that reflect or scatter sounds back toward the sound source. In the process of reflection, the object becomes a source of sound itself; active sonar consists of inducing otherwise silent objects into returning echoes. Each reflecting point is called a glint, and in FM echolocation characterizing objects means locating and describing their constituent glints (Simmons 1989).

3.3 The Acoustic Nature of Targets

3.3.1 Targets as Groups of Glints

Contrary to the intuitive view of each target as a discrete object that reflects "an echo" that is straightforwardly separate from the echo reflected by another target, small objects such as flying insects behave as though they consist of two or more discrete reflecting points — that is, glints — which move to different locations within the target at different orientations as the target flutters and moves through the air (Simmons and Chen 1989; Kober and Schnitzler 1990; Moss and Zagaeski 1994). For such objects, there is no "target" apart from its constituent glints.

Figure 4.3 summarizes measurements of the echoes reflected by two types of targets that have been used as stimuli in important experiments with echolocating bats. One object is a mealworm, and the other object is a small plastic disk of roughly the same size as the mealworm (Simmons and Chen 1989). These targets are simplified versions of the more complex animated flying insects that bats encounter; they illustrate the principles of echo

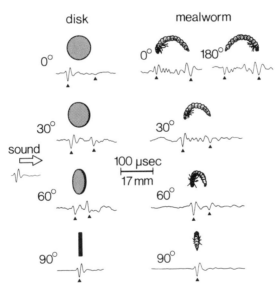

FIGURE 4.3. The acoustic nature of insect-sized targets in air (from Simmons and Chen 1989). A sonar sound (an impulse) is broadcast toward each target at different orientations (0°, 30°, 60°, 90°) in the incident sound field. The echo from the disk shows an impulse for the leading and trailing edges or glints at most orientations, and the echo from the mealworm shows an impulse from the head glint and the tail glint. As a first approximation, each target appears as a two-glint dipole whose most important feature is the separation of the principal glints along the axis of target range, or the time separation of the echo replicas from these two glints.

structure relevant to echolocation. Figure 4.3 shows the echo impulse responses for a mealworm and a disk at orientations of 0°, 30°, 60°, and 90° relative to the axis of the incident sound (an impulse containing the same frequencies as the bat's signals: labeled *sound* in Fig. 4.3). The disk yields a separate echo from its leading and trailing edges for all orientations except those very close to 0° or 90°, with a time separation between these echoes that corresponds to the range separation of the leading and trailing edges. The disk thus appears as an acoustic "dipole" target that reflects a two-glint echo. The target's shape consists chiefly of the momentary separation of its two glints along the range axis, with added secondary effects from more complex aspects of reflectivity (for disks, see Lhémery and Raillon 1994). The mealworm, too, returns two principal echoes at different delays from glints that correspond to its head and tail. At some orientations, weaker secondary echoes also are returned from other points along the mealworm's body, notably the legs and body segments. At most orientations, however, the mealworm is primarily a dipole target reflecting two-glint echoes. Consequently, the mealworm's most prominent shape feature is the separation of its head and tail along the range axis.

3.3.2 Flying Insects as Animated Glints

The echoes recorded from the mealworm and the disk (see Fig. 4.3) reveal a "quantal" character to the sounds reflected by these targets. The acoustic return from each object consists of several complete replicas of the impulse-like incident sonar sound arriving at slightly different delays. The leading and trailing edges of the target deliver the most prominent echo replicas, with finer structure in the target contributing a corresponding finer structure to the echoes between these principal reflections. It is evident that a complete description of these objects in terms of their echoes consists of the strength and temporal spacing of the individual reflected replicas of the incident sound. Echoes from flying insects contain added information, chiefly about the rhythmic motion of glints in relation to the beating of the wings at rates of about 10–100 Hz. Important information about the identity of prey can be obtained from wingbeat rate, which FM bats prove unexpectedly good at discriminating (Sum and Menne 1988; Roverud, Nitsche, and Neuweiler 1991; see Moss and Schnitzler, Chapter 3). Bats progressively increase the repetition rate of their sounds as they approach a target, from a rate of 5–10 Hz when at long range to as much as 150–200 Hz at close range (see Fenton, Chapter 2), and at some point in this progression the rate of emission momentarily matches the insect's wingbeat rate of 10–100 Hz. When this occurs, several successive echoes reveal the insect in a stroboscopically frozen posture, providing an opportunity to judge the target's shape independent of the wingbeat frequency itself (Feng, Condon, and White 1994; Moss and Zagaeski 1994; see Fig. 5D in Kick and Simmons 1984).

3.4 Structure of Echoes from Complex Targets

3.4.1 Overlap of Echoes from Glints

If the bat's sonar sounds were short-duration impulses such as the incident sound in Figure 4.3, the arrival of several echo replicas in a short span of time would be easy to recognize from their separate peaks in the received waveform. Instead, however, the bat's sonar sounds have relatively long durations of several milliseconds (see Fig. 4.2) compared to the short intervals of tens or hundreds of microseconds between echoes from different parts of the same target (see Fig. 4.3). Consequently, the waveform actually returned to the bat from the mealworm or the disk at different orientations is not as easily interpreted as the two primary "head" or "tail" impulses in the representative echoes in Figure 4.3 (Simmons 1989). The echoes reaching the bat's ears from these targets will indeed contain two primary reflections that are separated by a few tens of microseconds, but the individual reflected replicas will correspond to the long-duration FM sounds broadcast by the bat rather than short impulses. The waveform returning to the bat from a real object will contain several echo replicas that mix together to interfere (reinforce and cancel) at different frequencies according to the time separation of the replicas (Simmons et al. 1989; Mogdans and Schnitzler 1990; see Moss and Schnitzler, Chapter 3; also Fig. 21 in Dear et al. 1993).

3.4.2 Information on Time and Frequency Dimensions of Spectrograms

Figure 4.4 illustrates spectrograms of echoes composed of two reflected replicas of an incident *Eptesicus* sonar sound with a duration of 2.5 msec. These spectrograms display the principal information contained in two-glint echoes that have different time separations of the replica reflections from 0 to 600 μsec (for details about these waveforms, see Chapter 2; see also Altes 1980, 1984; Beuter 1980; Mogdans and Schnitzler 1990; Simmons 1992). These delay separations correspond to differences of 0–10 cm in the range spacing of glints, which covers the sizes of airborne objects encountered by bats, most insects having dimensions of a few centimeters at most. The broadcast signal for the echoes in Figure 4.4 contains two prominent harmonics sweeping from 55 to 25 kHz (FM_1) and from 90 to 50 kHz (FM_2), with a third, less prominent harmonic segment from 90 to 75 kHz (FM_3) (see 0-μsec echo at left in Fig. 4.4; this is the same as the incident sound).

In the two-glint echoes in Figure 4.4, the overlapping replicas of this incident sound interfere with each other so that the compound echo as a whole has a complicated spectrogram which is different from the spectrogram of the incident sound itself. In particular, although the two principal FM sweeps are visible in the echo spectrograms, they are streaked by horizontal bands at different spacings along the vertical frequency axis depending on the amount of two-glint echo–delay separation along the

James A. Simmons et al.

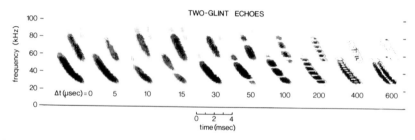

FIGURE 4.4. Spectrograms of two-glint echoes with different time separations of the reflected replicas of the incident sound, which is a 2.5-msec *Eptesicus* sonar signal (Simmons 1992). The incident sound is the same as the 0-μsec spectrogram (*left*). The integration time of these spectrograms is 350 μsec. Echo replicas that arrive separated by more than the integration time (400–600 μsec) are recognizable as separate FM signals, but echoes that arrive closer together (5–200 μsec) merge together; they interfere to form a single spectrogram with horizontal bands or spectral notches at frequencies determined by the reciprocal of the time separation. At short time separations, the only evidence that there are two echo components is this spectral structure.

horizontal time axis. In Figure 4.4, for delay separations of 400–600 μsec, there obviously are two echoes present. Each harmonic appears as a double FM signal with a short interval of time between the signals.

At these large echo–delay separations, the presence of two glints and their approximate spacing in time is immediately evident along the horizontal time axis of the spectrograms. In contrast, for two-glint echo–delay separations of 5, 10, 15, 30, 50, 100, and 200 μsec, the two separate echoes merge to form a single spectrogram. Each two-glint, compound echo contains just one FM sweep for each harmonic rather than a double sweep. Moreover, the appearance of these sweeps has been altered by the fact of overlap to form horizontal bands of interference at specific frequencies. These bands are notches in the spectrum of the echo, and their placement along the vertical frequency axis of the spectrograms is the only clue that the compound echoes really contain two overlapping components. For example, at a delay separation of 100 μsec these notches are spaced at frequency intervals of 10 kHz, while at a delay separation of 50 μsec the notches are 20 kHz apart ($\Delta f = 1 / \Delta t$).

The spectrogram for the two-glint echo with a delay separation of 400 μsec shows characteristics that are intermediate between the fully merged spectrograms of 5 - to 200-μsec echoes and the fully separated spectrograms of 400 - to 600-μsec echoes. At 400 μsec, information about the two glints is seen in both the time and frequency dimensions of the spectrogram, while at 400–600 μsec the information is entirely in the time separation of the sweeps and at 5–200 μsec it is entirely in the frequency spacing of the interference notches. Not only do double sweeps appear in this 400-μsec spectrogram, but a fine pattern of spectral notches also runs vertically along the frequency axis. This dual time–frequency quality of the spectrograms for two-glint echoes at time separations close to the integration time is also present in the bat's images (Simmons 1992), confirming the spectrogram-like time–frequency structure of the bat's auditory representation of echoes (see also Simmons, Moss, and Ferragamo 1990).

4. Reconstructing Target Scenes from Echo Streams

4.1 Implications of Spectrograms for Reconstruction of Glints

The crucial step for producing images in FM sonar is establishing an identity between each reflected version of the transmitted sound (each echo replica at different time separations in Fig. 4.4) and the corresponding glint that brought that echo replica into existence by reflecting or scattering the original broadcast sound back toward the receiver. Here, we consider an individual echo to be the echo replica reflected by a single glint. Because most real objects encountered by bats contain several glints, the target's "echo" in common parlance is really a "packet" of several discrete echo replicas of the incident sound that overlap each other to form a complex waveform (Beuter 1980; Altes 1984; Simmons 1989). The reason for making this distinction is not just that many targets consist of two or more glints.

Bats often encounter several targets together, located at about the same distance but in different directions. In this case, the bat would receive reflections from glints contained in different objects at about the same time. The waveform at the bat's ears still contains two or more overlapping echo replicas, as in Figure 4.4, but the "target" would not be a single two-glint object; it would be several objects. Because what would ordinarily be considered as "the echo" in fact contains several echo replicas from different glints located in different directions, the spatial scene composed of objects cannot be reconstructed from the acoustic stimulus composed of the whole echo packet without first decomposing it into its component echo replicas, one for each glint.

Accordingly, the fundamental role of the bat's auditory system for echolocation must be to recognize echo replicas as discrete components of sounds, even when they arrive so close together that their waveforms merge into a single spectrogram (see Fig. 4.4). Moreover, the fundamental limitation on echolocation performance must be the capacity to resolve two closely spaced replicas as separate reflections with distinct arrival times. However, the presence of two different kinds of information in echo spectrograms about the time separation of glints, separate spectrograms of each echo for long delay differences as opposed to bands of spectral notches for short delay differences (Fig. 4.4), means that inferences about glints must be carried out differently for long delay separations than for short delay separations (Simmons 1989). The location of the transition from the regime of representation that yields two separate spectrograms rather than a single, spectrally notched spectrogram evidently must be a fundamental characteristic of the bat's sonar receiver.

4.2 Auditory Integration Time and Segregation of Echoes

Bats frequently encounter echo replicas that overlap each other on reception and which must be separated to account for the bat's perceptions, that is, for behaviors demonstrably guided by perception of individual glints. The difficulty is, because echo replicas arrive at closely spaced delays, that an interval of time corresponding to the duration of a single sonar sound might have to be subdivided into a number of different estimates of arrival time for multiple, overlapping echo replicas. This might seem impossible; the duration of the sonar sound could be construed as the minimum interval of time for which a single delay value can be specified. This assumes, however, that the bat's ear records each echo in the manner of an oscilloscope, with one sweep of the screen covering the waveform of one transmission or one whole echo. This is equivalent to specifying the minimum integration time of echo reception as equal to the duration of the broadcast signal or one of its echoes (see discussion of integration-time measurements for bats in Chapter 3 by Moss and Schnitzler).

In fact, however, the bat's auditory system filters the wide span of frequencies covered in the FM sweep of each harmonic into numerous, narrower frequency segments (for details about tuning, see Chapter 6 by Covey and Casseday; for estimation of integration time, see Simmons et al. 1989; for discussion of temporal resolution, see Menne 1985; for auditory coding in relation to integration time, see Suga and Schlegel 1973). This auditory representation approximates a spectrogram, but with a frequency axis that is roughly hyperbolic instead of linear (see Simmons et al. in press).

As a result of auditory frequency tuning, an FM sweep that lasts several milliseconds is segmented into shorter pieces, each lasting for a fraction of the duration of the sound as a whole. The minimum integration time for echo reception corresponds to the duration of these segments (Beuter 1980; Altes 1984; Menne 1985; Simmons et al. 1989). This shortening of the effective duration of the FM sweeps is illustrated in Figure 4.4. The integration time of the spectrograms is the time axis width of the dark smear that traces each of the FM sweeps; it was set to be 350 μsec when the spectrograms were made. At a glint separation of 400 or 600 μsec in Figure 4.4, each of the echo replicas shows a distinct FM sweep for each harmonic, with no overlap of the segments at each frequency (even though the whole, unsegmented sweeps do overlap because the duration of the broadcast signal is 2.5 msec while the glint separation is only a fraction of a millisecond).

In contrast, at glint separations of 200 μsec or less, which are shorter than the width of the spectrogram smear, the two echo replicas merge to form a single, horizontally banded spectrogram. The effective duration of the bandpass-filtered segments of the sweep has been estimated from the sharpness of peripheral auditory tuning in *Eptesicus* and *Myotis* to be about 300–400 μsec (Simmons et al. 1989). Behavioral measurement of integration time in *Eptesicus* yields a value of about 350 μsec for double echoes to merge into a single echo (Simmons et al. 1989), which is in close agreement with neural integration time measured from recovery cycles to pairs of sounds or to amplitude-modulated sounds (Grinnell 1963; Suga 1964; Pollak and Casseday 1989; Covey and Casseday 1991; also see Covey and Casseday, Chapter 6).

In physiological terms, two separate sounds will evoke separate volleys of neural discharges as long as they are more than 300–400 μsec apart. Using integration time as a guide, bats should have no difficulty distinguishing two glints that are as close to each other as 5–7 cm in distance because their echo replicas will be 300–400 μsec apart, whereas glints that are closer together in range should be perceived as part of a single object because their echoes merge together into a single spectrogram (see Fig. 4.4). The difficulty with this simple interpretation of integration time is that a variety of behavioral studies unambiguously demonstrate that *Eptesicus* and other echolocating bats have little trouble separately perceiving the locations of glints that are closer together in range than 5–7 cm, which is the limit set by integration time alone.

4.3 Direct Psychophysical Measurements of Two-Glint Resolution

Several different experiments have been carried out specifically to assess the ability of *Eptesicus* to perceive echoes that arrive close together in time. Most of these experiments test the bat's ability to discriminate between two-glint echoes and single-glint echoes without determining whether the bat actually perceives the second glint as such (see Moss and Schnitzler, Chapter 3). However, one group of experiments (Simmons, Moss, and Ferragamo 1990; Simmons 1992, 1993) explicitly evaluates whether the bat can assign a discrete arrival-time value to each of two closely spaced echoes. Figure 4.5 shows the performance of two *Eptesicus* in an experiment that compares two-glint echoes (0-μsec, 10-μsec, 20-μsec, or 30-μsec echo–delay separations, corresponding to glint separations of 0 mm, 1.7 mm, 3.4 mm, or 5.2 mm, respectively) with single-glint echoes at different arrival times (using jittered echoes; see Simmons 1993; Simmons et al. 1990; see also Moss and Schnitzler, Chapter 3). The bats in Figure 4.5

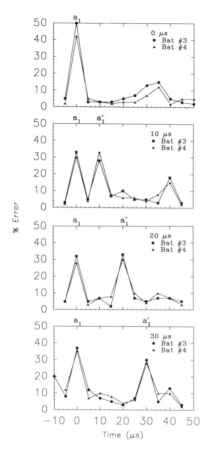

FIGURE 4.5. Performance of *Eptesicus* (bat #3 and #4) at perceiving the arrival times of two closely spaced echoes with delay separations of 0, 10, 20, or 30 μsec (from top to bottom) (Simmons 1993). In each case, the bats make a significant number of errors when the experimental probe echo (a_2) aligns with the delay of either the first glint (a_1) or the second glint (a_1') in the test echo. These error peaks indicate that the corresponding test glint has been encoded in terms of echo arrival time.

make significantly more errors in their discrimination when the arrival time of the single-glint echoes coincides with the arrival time of either the first-glint echo (a_1) or the second-glint echo (a_1'). Thus, the error peak at 0 μsec in Figure 4.5 corresponds to the bat's perception of the leading glint (a_1) at 3.2 msec (see legend), and the error peaks at 10, 20, or 30 μsec correspond to perception of the second glint (a_1') located slightly further away. The bats' errors indicate that both of the two-glint echoes were encoded in terms of their arrival times, even though the echo replicas for the individual glints arrive so close together in time that they merge into a single, spectrally complex spectrogram (see 10-μsec and 30-μsec spectrograms in Fig. 4.4). The results shown in Figure 4.5 demonstrate that the bat can translate the subtle changes in the spectrum of echoes caused by interference within the integration-time window to infer the value of the time separation of the echo replicas reflected by the two glints, so that both the first and the second replicas can have a known arrival time (see Menne 1985 for the concept of resolution and its enhancement; see Saillant et al. 1993 for deconvolution enhancement in echolocation).

The curves in Figure 4.5 show that *Eptesicus* has no difficulty perceiving the second of two glints (both error peaks are the same height) when the overlapping echoes arrive as close together as 10, 20, or 30 μsec. Evidently, the bat's limit of resolution is smaller than 10 μsec, or 1.7 mm in range. Figure 4.6 shows the performance of as many as three bats in a series of two-glint experiments with echo–delay separations of 10 μsec, 5 μsec, 3.8 μsec, 2.6 μsec, and 1.4 μsec (Saillant et al. 1993). The bats all show a pattern of errors similar to Figure 4.5, with a peak at 0 μsec (actual delay is

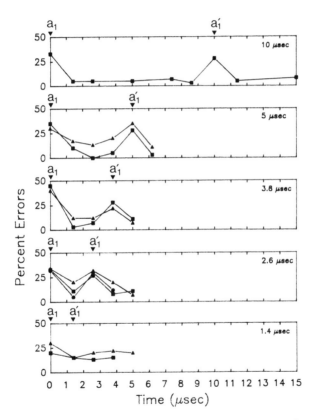

Figure 4.6. Limit of two-glint resolution for *Eptesicus* (three bats; data points are *squares, triangles, circles*). The bats perceive two arrival times for echo–delay separations of 10, 5, 3.8, and 2.6 μsec, (from top to bottom) but they perceive only one arrival time for a delay separation of 1.4 μsec (Saillant et al. 1993). The two-glint echo-resolution threshold is about 2 μsec, which corresponds to a separation in range of about 0.3 mm.

3.2 msec) for the first glint and a peak at 10, 5, 3.8, or 2.6 μsec for the second glint. At a separation of 1.4 μsec in Figure 4.6, however, the bats no longer register an error peak for the second glint, indicating that the second glint's echoes were not assigned a value of arrival time distinguishable from the arrival time of echoes for the first glint. From Figure 4.6, the limit of two-glint resolution in *Eptesicus* must be about 2 μsec (see also Simmons et al. 1989). This is an extraordinarily fine resolving power: first, because it is equivalent to only about 0.3 mm in range, and second, because it is shorter than the periods for any of the frequencies in the bat's sonar signals. *Eptesicus* broadcasts sounds containing frequencies approximately from 20 to 100 kHz (see Fig. 4.2), which have periods from 50 μsec down to 10 μsec. Resolution carried out within the time window of the shortest period in the sounds is within the Rayleigh limit for the signals themselves, which is a significant achievement that amounts to extrapolation of the broadcast spectrum to frequencies higher than those actually contained in the bat's sounds (Saillant et al. 1993).

4.4 Is There Other Evidence for Fine Temporal Acuity in Bats?

The ability of bats to separately perceive the arrival times of two echoes that reach the ears as close together as 2–30 μsec (Figs. 4.5 and 4.6) demonstrates that they are very good at segmenting the stream of incoming echoes following each broadcast into a series of discrete arrival-time values (see also Simmons 1992). The bat's 2-μsec limit of two-glint resolution is, of course, far shorter than the duration of its echolocation sounds, which is several milliseconds (Simmons et al. 1990; Simmons 1993), and it is shorter, too, than the integration-time window of 300–400 μsec. Consequently, the bat must be able to "read" spectrograms to determine time separations from information contained in their frequency axis as well as in their time axis.

The fine temporal acuity shown here, and the dual time-frequency process that must underlie it, have been considered physiologically improbable from the perspective of certain types of single-unit data (e.g., Pollak et al. 1977), leading to rejection of the types of behavioral experiments that generate such results (Pollak 1993). It therefore is important to know whether other types of behavioral experiments, whose results are widely accepted, might in fact conceal independent evidence for fine temporal acuity in echolocation. The following sections review the implications for sonar imaging of several different kinds of experiments that are considered classical in their impact on documenting the existence and natural history of echolocation, but that have not previously been brought to bear on determining the composition of the images themselves.

5. Performance of Echolocation in Natural Tasks

5.1 Interception of Targets in Simple and Complex Situations

5.1.1 Pursuit of Prey

In Figure 4.1, the bat's flight to the target, its detection of the target, and its identification of the object as potential prey are all guided by sonar (Schnitzler and Henson 1980). The behavior of bats during interception is surprisingly stereotyped, with a stable pattern of sonar emissions (see Fig. 4.2) and other responses occurring at regular points along the flight path (see Fenton, Chapter 2; also see Kick and Simmons 1984; Simmons 1989). The bat aims its mouth (the transmitter) and its external ears (the receiving antennas) at the target during the approach and will seize the target in its tail membrane at the moment of capture (Webster and Griffin 1962).

Figure 4.7 gives two examples of the bat's approach to a target in video motion-analysis studies of the interception process. These observations were initiated to evaluate the information the bat uses to control its approach flight (Lee et al. 1992), and they yield a wealth of data to reconstruct the stimuli reaching the bat's ears and the images it perceives each time a sound is broadcast (Saillant et al. 1993). In Figure 4.7, the bat is shown as a series of stick figures that trace its flight to the mealworm over the last 80 cm of the approach. Each stick figure depicts the bat's location and posture in a three-dimensional reconstruction of the bat's flight path taken with two synchronized video cameras at a frame rate of 60 Hz. In Figure 4.7A, the bat flies directly up to a single mealworm (mw) suspended on a fine thread and seizes the mealworm in its tail membrane before eating it. In Figure 4.7B, the bat is given a choice of two mealworms (mw 1, mw 2) hanging on threads; in this example the bat captures the target located on its right slightly farther away (mw 2). It approaches to seize this mealworm from its thread while at the same time withdrawing its right wing to avoid striking the thread holding the other mealworm (mw 1). These studies reveal no important distinctions in the acoustic behavior of bats approaching a single target as opposed to two targets.

5.1.2 Duration of Echolocation Signals for *Eptesicus*

For present purposes, the *duration* of the individual FM sonar sounds is their most critical feature for constraining subsequent processing of echoes to form images of targets. The duration of the signals shown in Figure 4.2 (*a–g*) progressively decreases from 8 msec to less than 0.5 msec as the bat approaches nearer to the target and the delay of echoes becomes shorter. At

A

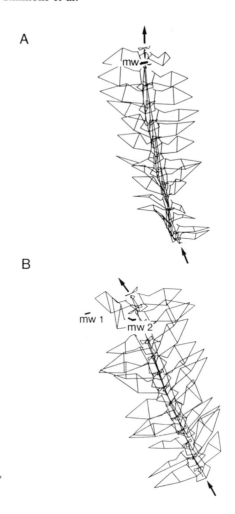

B

FIGURE 4.7A,B. Reconstruction of the flight path and posture of *Eptesicus* during captures of a mealworm hanging on a fine filament (see Fig. 4.1). The stick figures view the bat from the front based on 60-Hz video motion-analysis images as it flies up to the mealworm and seizes it in the tail membrane. In A, the bat captures a single mealworm (mw). In B, the bat captures one of two mealworms (mw 2) while simultaneously withdrawing its right wing to avoid striking the filament holding the other mealworm (mw 1). *Arrows* show direction of flight.

any point in the bat's approach to a target, the duration of the transmitted sound usually is only slightly shorter than the two-way travel time of the echo. Consequently, the pathlength to and from the target is nearly filled up with each sound (Hartley 1992; Saillant et al., in manuscript). Most experiments on echolocation focus on questions about what the bat perceives when it is at distances of about 0.5–1 m from targets, when the sonar transmissions typically are nearly 3–6 msec in duration. With such long durations for the incident sonar sounds, the individual echo replicas from two different parts of the same target (the mealworm's head and tail in Fig. 4.7A) or from different parts of two closely spaced targets (heads and tails of both mealworms in Fig. 4.7B) overlap each other for most of their duration.

5.1.3 Acoustic Stimuli During Capture of Tethered Mealworms

In Figure 4.7A the mealworm returns an echo packet composed of two overlapping echo replicas of roughly equal strength from its head and its tail (see Fig. 4.3). The maximum head-to-tail separation of a mealworm is about 17 mm, so the largest time separation of echo replicas from the head and tail glints will be only about 100 μsec. Thus, as the bat approaches the target in Figure 4.7A, it will receive echoes corresponding to examples of 5- to 100-μsec glint separations in Figure 4.4. In Figure 4.7B, the two mealworms collectively return an echo packet composed of four overlapping echo replicas of roughly equal strength from their heads and tails (see Fig. 4.3). The head-to-tail separation within each mealworm will be less than 17 mm, but the separation of the heads and tails between mealworms depends on the difference in distance from one mealworm to the other as well as on their orientations. In Figure 4.7B, the overall range difference is only about 2 cm, so the head of one mealworm could easily be at the same distance from the bat as the tail of the other.

Because each mealworm really is a dipole (Fig. 4.3), the spacing of the glints frequently will place two parts of different targets in closer range proximity than *two parts of the same target*, so that the objects cannot be identified along the range axis simply from which glints are nearer each other and which glints are farther apart. Consequently, the bat cannot select one of the two targets for capture on the basis of the echo packet alone; all the echo replicas from both targets arrive within a single integration-time window and will merge into a single, spectrally complex packet.

During its approach to the two mealworms in Figure 4.7B, the bat will receive echoes corresponding to the examples of 5-μsec to 100-μsec two-glint echoes (see Fig. 4.4) for *intra*target spacings mixed with examples of 5 μsec to about 300–400 μsec for *inter*target spacings. To perceive the mealworm in Figure 4.7A as an object composed of two principal glints, the bat has to decompose the pair of overlapping echo replicas, typically separated by 5 μsec to no more than 100 μsec, from the head and tail (Simmons and Chen 1989) and assign each echo replica a specific, and different, arrival time. To be sure, the bat does not actually have to perceive that there are two glints in the echo from the mealworm, only locate the mealworm well enough to intercept it.

In stark contrast, however, to perceive the two mealworms in Figure 4.7B as separate objects and guide its flight to intercept just one of them, the bat does have to decompose the series of four overlapping echo replicas from the heads and tails of both mealworms and assign each one a specific arrival time. In this case, perception of the glint structure of the target scene is necessary to locate one target in the presence of the other because the echo packet by itself portrays a nonexistent target in the wrong location. Moreover, the segregation of individual echo replicas in Figure 4.7B has to precede regrouping of replicas into pairs corresponding to the two meal-worms because the bat has to avoid incorrectly associating the head of one mealworm with the tail of the other. Such a misassociation would lead the bat to attack a much lengthened "phantom mealworm" located halfway between the two real mealworms — an error of orientation that is not observed.

5.1.4 Interception of Targets in Clutter

Eptesicus and other FM bats routinely intercept prey in open spaces and clearings, with just the insect near the bat reflecting echoes that arrive as isolated events at the bat's ears (see Fenton, Chapter 2). Other objects such as vegetation or the walls of buildings are located at least a meter or two further away than the insect. However, once an insect is detected, the bats are often capable of following the target into the thick of vegetation, tracking it through an environment of leaves and branches located at similar distances all around (Webster and Brazier 1965; Webster 1967).

Figure 4.8 shows a multiple-exposure photograph of the little brown bat, *Myotis lucifugus* (a close relative of *Eptesicus*), pursuing a moth through branches. The bat is shown in numbered images (#1–8) at strobe-flash intervals of 100 msec, with the corresponding image of the moth connected to each image of the bat by a dashed line (#1–3). In image #1 the bat is about 50 cm from the moth, and by image #3 (200 msec later) it has closed to within 10–15 cm. At this point, the background vegetation has components that rival the moth in being at similarly close range, and the overall strength of the echoes from all the vegetation probably exceeds the strength of the echoes from the moth.

In Figure 4.8, the bat clearly has the moth in its mouth in image #6 and probably is already in contact with the moth in image #5. In spite of the numerous extraneous echoes from many branches that might confuse perception of echoes from the moth, the bat comes out of the clutter with the moth in its mouth (images #7–8). Not only does the bat intercept the moth, but it avoids collisions with the vegetation located all around its flight path. The clutter does not prevent the bat from tracking the moth to a successful capture in much the same manner as would have occurred out in the open, but the clutter probably constrains the bat's flight path while capture takes place (Webster 1967). Figure 4.8 documents the bat's ability to segregate the echo replicas created by the insect from what might seem to be an overwhelming number of interfering replicas from objects located at

FIGURE 4.8. Multiple-exposure photograph of *Myotis lucifugus* pursuing and capturing a moth that flies into vegetation during the bat's interception maneuver (Webster and Brazier 1965). Each numbered image of the bat is connected to the corresponding image of the moth with a *dashed line*.

such a wide range of different distances that there would always be some echo components from the background arriving at about the same time as the echo from the insect.

5.2 Avoidance of Obstacles

5.2.1 Wire-Avoidance Tests

Demonstrations of the ability of FM bats such as *Eptesicus fuscus* or *Myotis lucifugus* to perceive the locations of several targets at once are not confined to photographs or videorecordings of bats intercepting insects in clutter. Echolocation was first shown to exist by flying bats through arrays of obstacles and recording their sounds while they actively maneuvered past the obstacles, and most of the characteristics of echolocation (e.g., use of FM or constant-frequency [CF] sounds by different species; changes in the duration and repetition rate of sonar signals during approach) can be observed in this type of task as well as in pursuit maneuvers (Griffin 1958; Grinnell and Griffin 1958; for obstacle avoidance by *Eptesicus*, see Jen and Kamada 1982; for further reviews, see Novick 1977; Schnitzler and Henson 1980).

Figure 4.9 shows two examples of three-dimensional reconstructions of the flight of *Eptesicus* through an array of vertically stretched wires (0.7 mm diameter) with a spacing of 25 cm between wires (the bat's maximum wingspan is about 30 cm). In each case, the bat is viewed as approaching the observer, with about 80 cm of the bat's flight path shown by the stick figures. In Figure 4.9A, the bat flies up to the middle pair of wires (wires b and c) and passes more or less directly between them, tilting its right wing up and left wing down during the most critical part of its passage between the wires. At each point in its flight the bat is approximately equidistant from the wire on its left (c) and the wire on its right (b). Measurements made from the three-dimensional reconstruction of the flight in Figure 4.9A reveal that the difference in distance from the bat to the left and the right wires is no greater than 2 cm throughout the bat's approach.

In Figure 4.9B, the bat flies almost directly toward one of the wires (wire b) while keeping approximately equidistant from the next two wires (wires a and c). Just before it would have collided with the wire located to its front (b), the bat swerves sharply to its right while sharply tilting its right wing up and its left wing down to avoid striking the wire. The bat slides past the nearest wire (b) with a distance of about 2–3 cm to spare at its closest point. Significantly, the bat remains about the same distance from the wire located on its left (c) and the wire located on its right (a) even while coping with the presence of the wire located straight ahead (b), halfway between the two other wires (a, c). Until the bat swerves in flight, the difference in distance from the bat to these other wires (a and c) is only about 2–3 cm.

5.2.2 Acoustic Stimuli During Obstacle Avoidance

Each of the wires in an obstacle array returns an echo replica to the bat from the point along the wire where the wire is perpendicular to an imaginary line connecting the bat's head and ears to the wire itself. That is, each wire acts as though it consists primarily of a single glint located at this point of perpendicularity. Changes in the bat's height during flight merely move this principal glint along the wire to new locations. Because the wires are stretched to form parallel elements of an array, the array appears to the bat as a series of point reflectors located at distances and directions

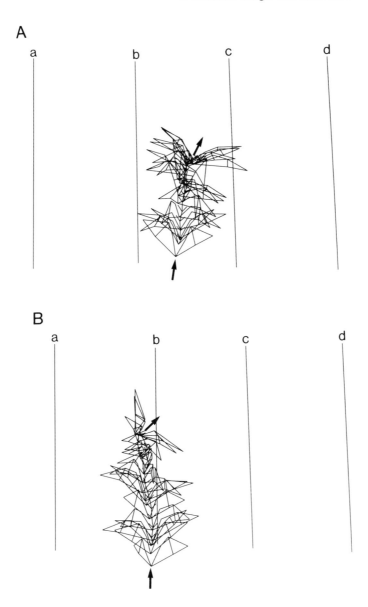

FIGURE 4.9A,B. Reconstruction of the flight path and posture of *Eptesicus* during obstacle-avoidance flights. The bat is viewed from the front as it approaches and avoids hitting the wires *(a-d)*. In A, the bat passes between two adjacent wires; in B, the bat heads straight for one wire but dodges around it at the last moment before colliding. *Arrows* show direction of flight.

corresponding to the positions of the wires in the horizontal plane relative to the bat.

From observations of its flight path during obstacle-avoidance tests, the bat must frequently have selected which wires to pass between by the time it has approached to a distance of 0.5–1 m. The angular separation of the adjacent wires thus is only about 15°–30° when the bat has made at least a preliminary evaluation of their locations. From recordings of the sounds during approach to the wires, the bat progressively shortens its sonar signals and increases their repetition rate, giving further evidence that it is monitoring the location of the obstacles as it flies nearer (see Griffin 1958; Novick 1977; Schnitzler and Henson 1980; for *Eptesicus*, see Jen and Kamada 1982). Crucially, the duration of each signal is far larger than the relatively short interval that separates echoes from the wire on the left and the wire on the right.

The chief factor determining the structure of the stimuli is the distance from the bat to each wire. The bat often makes a nearly perpendicular approach to the plane of the wires (Fig. 4.9A), and their angular separation of only about 15°–30° at distances of 0.5–1 m results in range differences being only a few centimeters. Echoes from two adjacent wires thus will overlap each other because they are at approximately the same distance from the bat, one on the right and one on the left. Consequently, during the bat's approach to the obstacle array, it will receive echoes resembling the examples shown in Figure 4.4. Just which of the two-glint examples actually corresponds to the echo received from the two nearest wires following a given sonar emission depends on how close the bat comes to flying along a path that keeps it equally far from each wire. From one sonar emission to the next, the difference in range from one wire to the other is no greater than several centimeters, so the difference in the arrival time of their echoes is nearly always less than several hundred microseconds and often only a few tens of microseconds.

In contrast to the small delay separation of echoes, however, the durations of the sounds themselves are several milliseconds, so the echoes overlap each other when they reach the bat's ears. Echoes from wires located further away on the left and right arrive later, too, and these will overlap with each other as well as overlap with the echoes from the nearest wires. The delay separation of echoes from the nearest pair of wires and the next nearest pair depends on how far the bat is from the array. For example, in the experiments with *Eptesicus* shown in Figure 4.9, when the bat is 1 m from the array as a whole, the nearest wires on the left and the right are about 101 cm away and the next nearest wires are about 107 cm away, for a difference in range of only 6 cm. Thus, the delay separation of the two sets of echoes is only about 350 μsec, and the delay separation of the echoes from the single wire on the left and right in each set will be smaller still.

5.2.3 Implications of Stimuli for Perception of Wires

As a consequence of the acoustic geometry of the parallel vertically stretched wires, following each broadcast the bat receives a packet of several echo replicas from the left and several echo replicas from the right, with the overlap times as small as a few tens of microseconds for corresponding pairs of wires (the nearest wires, the next nearest wires, etc.) located approximately equidistant from the bat on the left and the right. Because the echoes from each pair of wires will have a single spectrogram with interference notches along the frequency axis (see Fig. 4.4), an inference based on the undifferentiated echo packet from the wire on the left and the wire on the right would indicate the existence of only a single, nonexistent "target" located midway between the two real wires.

The bat cannot locate either of the real wires without decomposing the spectrogram into estimates of the arrival times of both echo replicas. The bat's ability to avoid the obstacles, by dodging between the adjacent wires in Figure 4.9A and by swerving to pass close to the wire located straight ahead in Figure 4.9B while also receiving echoes from the wires located further to the right and left, demonstrates that it can segregate the incoming stream of overlapping echoes which have spectrograms of the type shown in Figure 4.4 into estimates of the arrival time of each echo replica so that a distance and direction can be attributed to each wire.

5.3 Discrimination of Airborne Targets

5.3.1 Capture of Mealworms and Rejection of Spheres and Disks

There is yet another naturalistic task used with echolocating bats that gives evidence for the bat's ability to segment a stream of echoes into closely spaced echo replicas. In a series of very important experiments, flying bats (the FM bats *Myotis lucifugus* and *Eptesicus fuscus*) have been trained to discriminate between targets of different shapes thrown up into the air to tumble along trajectories that the bat can intercept (Griffin, Friend, and Webster 1965; Webster and Brazier 1965). One of the targets is an edible mealworm, which the bat is trained to catch, while the other targets are inedible plastic spheres or disks of different sizes, which the bat learns to reject. Figure 4.10 shows a series of drawings made from a multiple-exposure stroboscopic-flash photograph of a discrimination trial with *Myotis* intercepting a mealworm thrown into the air along with two small disks (Webster and Brazier 1965). The images of the bat (B_3, B_4, B_5), the mealworm (M_3, M_4, M_5), and the two disks (a_3, a_4, a_5; b_3, b_4, b_5) are numbered to correspond to each other at strobe-flash intervals of 100 msec. Each frame in the drawing contains a square marking the location in space that will be occupied by the bat's mouth just before capture; the trajectories of the targets can be judged from this fixed reference point, along with the

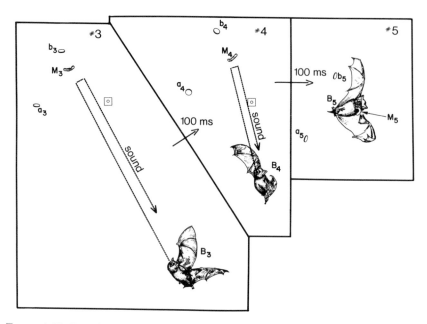

FIGURE 4.10. Drawings made from a multiple-exposure photograph of *Myotis* discriminating an airborne mealworm (M) from two disks (*a, b*). The original drawing (Webster and Brazier 1965) contained six strobe images 100 msec apart; here, the third, fourth, and fifth images have been separated into panels (#3, #4, #5) to show the image of the bat (B_3,B_4,B_5), the mealworm (M_3,M_4,M_5), and each disk (a_3,a_4,a_5; b_3,b_4,b_5) at their corresponding locations. A *small square* containing an *open circle* has been added to show the location that will be occupied by the bat's mouth in image #5. The approximate spatial extent of the bat's sonar sounds has been plotted as a *dashed line (sound)* based on durations of sounds recorded in these experiments.

bat's approach path. In tests of this kind, bats achieve 80%–90% correct captures of the mealworm.

5.3.2 Acoustic Stimuli During Airborne Discrimination Tests

Each frame in Figure 4.10 contains a solid line (labeled sound) going from the bat's mouth to the targets and back again. This line traces the approximate spatial extent of the bat's sonar signals from the average duration of sounds recorded at each distance from the mealworm. The bat's sounds have a duration of several milliseconds in time that spreads their waveforms in space over most of the travel path from the bat out to the target and back again. Note, however, that the mealworm and the disks are located at about the same distance from the bat and are separated by differences in distance that are small compared to the total distance to any one of these targets. Consequently, their echoes are separated by an interval of time smaller than the duration of the incident sonar sounds. The echoes from the mealworm and the disks thus arrive at the bat's ears in the form of one continuous, overlapping sound several milliseconds long. Nevertheless, as in the case of obstacle-avoidance experiments, the bat perceives the locations of several targets at once, but now the bat also perceives enough additional information to identify the mealworm and the disks even though their echoes do overlap.

The bat's sonar sounds return from the mealworm and the disks as discrete echo replicas at time separations for different target orientations given by the impulse reflections in Figure 4.3. The time separation of the echo replicas from different glints within a mealworm or a disk is in the region of 0–100 μsec, with typical separations of only a few tens of microseconds at most orientations. Consequently, the spectrograms of echoes with glint separations of 5–100 μsec in Figure 4.4 are typical of the stimuli returned by a single mealworm or disk. If a mealworm or a disk were presented alone, with no other targets near by, the echoes from the mealworm or the disk should differ enough in their interference spectra that the bat ought to perceive this difference after learning to compare the targets (Simmons and Chen 1989). The bat receives as many as a half-dozen echoes during its approach, while the target rotates as it soars along its trajectory to expose different orientations, and, taken together, these echoes provide enough information for the bat to determine whether the target is a mealworm or a disk on any particular trial (Griffin 1967).

The critical point is that for sequential presentation of an isolated mealworm on one flight and a disk on the next, the bat does not need to perceive the individual glints in the target, just the interference spectrum of the echoes from each target separately. That is, when the targets are considered one at a time, the echo interference spectrum is a plausible basis for the bat's discrimination of mealworms from disks (Griffin 1967; Simmons and Chen 1989). The problem is that the bat can perform the discrimination even when a mealworm and several disks are presented simultaneously, at roughly the same distances, on the same trial, as in Figure 4.10. In many trials a mealworm and one or more disks will follow trajectories that place them at about the same distance from the bat, so that the time separation of echo replicas from different glints between the targets will often be as small as the time separation of echo replicas within each target.

Figure 4.11 shows examples of the bat's approach to a mealworm and a lump of clay hanging on fine threads during six flights in discrimination tests for video motion studies. During its approach, the bat is capable of determining which target is the mealworm and capturing it even when the targets do not turn and tumble, as they do in airborne tests, and when the two targets are always within a few centimeters of the same distance from the bat (see also Fig. 4.7B). These examples serve to confirm that the bat can locate and identify one target from echoes that arrive virtually

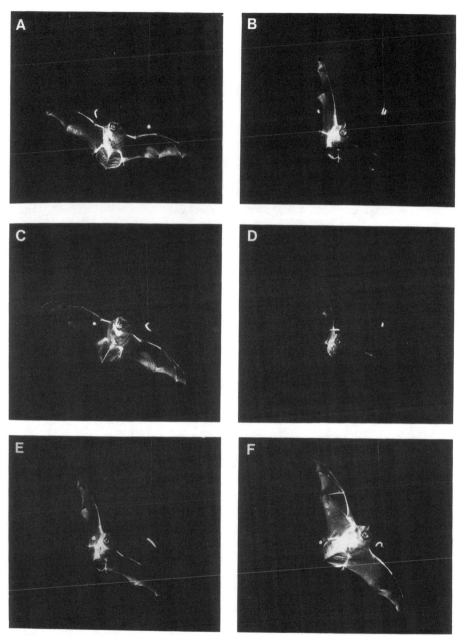

FIGURE 4.11A–F. The approach of *Eptesicus* to a mealworm and a small lump of clay from six flights in which the bat caught the mealworm. In each example, the bat's flight path kept it roughly equidistant from the two targets during the attack. (Photographs by S. P. Dear and P. A. Saillant.)

completely overlapped with echoes from another target located only a short distance away.

Figure 4.12 schematically illustrates the worse-case acoustic situation that often prevails in simultaneous discrimination trials with airborne targets. As the bat approaches a mealworm and a disk, which are thrown upward together and frequently remain close to each other throughout their brief flight, it receives echoes of about the same strength from two primary glints in both targets (a, b in mealworm; a, b in disk). Each target in Figure 4.12 consists of two discrete glints (a, b), each of which reflects a replica of the incident sound, so the whole echo packet contains two components from the mealworm and two from the disk all arriving at only slightly different delays. These four echo replicas overlap each other almost completely because the bat's sonar broadcasts are comparable in duration to the two-way path delay of echoes (sound lines in Fig. 4.10), which amount to several milliseconds, not to the much smaller delay separation between echoes from the two glints in each target (tens of microseconds between a and b in the mealworm or a and b in the disk) or the delay separation

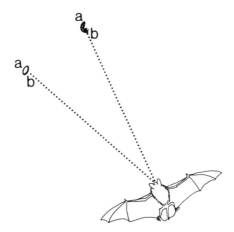

FIGURE 4.12. Diagram of worse-case situation in a bat's approach to a mealworm and a disk. The two targets are equidistant from the bat and are separated by an angle that keeps both targets within the directional beam of the sonar (see Fig. 4.13). The principal echo sources are the leading *(b)* and trailing *(a)* edges of the disk and the head *(b)* and tail *(a)* of the mealworm (see Fig. 4.3). Because the head of the mealworm and the leading edge of the disk are closer together in range than the tail is to the head or the trailing edge is to the leading edge, the bat will receive echo replicas that arrive closer together in time from glints *between* targets than *within* targets. To avoid misperceiving the two nearer glints as being part of the same target, the bat has to segregate all four echo replicas and perceive their locations in space before reassembling them into objects and choosing the one to attack.

between echoes from glints in different targets (tens or hundreds of microseconds between a in the mealworm and b in the disk, for instance).

The spectrograms in Figure 4.4 show examples of two-glint echo packets similar to the overall echo from a mealworm or a disk; the echo packet from a mealworm and a disk together is similar but more complex, with horizontal interference notches at spacings determined by the arrival time differences of all four principal glints from both targets. In situations such as shown in Figure 4.12, the echo packet taken as an undifferentiated whole has an arrival-time corresponding to a nonexistent "target" located about in the center of the volume of space collectively occupied by the four principal glints (a, b) in the mealworm and the disk. Nevertheless, the bat nearly always goes for the mealworm and approaches its location correctly, with no evidence for significant distortion of its perceived location as a consequence of the disk's presence close by. Moreover, the worse-case example in Figure 4.12 is a simplification because bats can successfully discriminate a mealworm from more than one disk presented on each trial (see Fig. 4.10). Figure 4.7B shows what might in fact be the most difficult situation of all — the bat locates and intercepts one mealworm in the presence of a second mealworm at about the same distance. Although each mealworm consists primarily of a head glint and a tail glint whose echoes all overlap when they reach the bat's ears at the about the same time, the bat nevertheless can separate the echo replicas into a pair for one mealworm and a pair for the other. The conclusion seems inescapable that the bat perceives the echo packet in terms of its constituent echo replicas, each with a discrete arrival time at each ear, to determine the location and identity of the mealworm.

6. Inferences About Images from the Bat's Performance in Naturalistic Tasks

6.1 Evidence for Decomposition of Targets into Glints

The performance of bats in several well-known, naturalistic tasks involving echolocation reveals that the bat can separately perceive the locations of multiple objects which are present together in the bat's sonar "field of view." To perceive the locations of different targets, the bat necessarily has to segregate the echoes returning from these objects so that each target can be located from its own specific reflections. Because the echoes from different objects often arrive within 300–400 μsec of each other, they overlap to form echo packets (see Fig. 4.4) that must be "unpacked" to reveal their composition in terms of echo replicas. This step is necessary because the characteristics of the echo packet betray no useful information about the individual objects, but instead about a "phantom" object located approximately at the "center of gravity" for the cluster of glints that collectively make up these objects.

The bat's ability to react to the location of one target when other targets are present at about the same distance means that the echo packets have been broken down into their constituent replicas so that the glints they represent can be separately perceived at their corresponding locations in space. However, information about the presence of different glints is distributed across more than one dimension of the spectrogram representation of echoes (see Fig. 4.4; Simmons 1989, 1992; see Moss and Schnitzler, Chapter 3). Some echo replicas can be segregated by the occurrence of discrete FM sweeps (glint separations greater than 300–400 µsec in Fig. 4.4), but others can only be segregated by recognizing that the notches in the spectrogram signify closely spaced glints (glint spacings of 5–200 µsec in Fig. 4.4). The boundary between separate registration of two echo replicas in time and joint spectral representation of overlapping replicas that are smeared together is fixed by the value of 300–400 µsec for the integration time of echo reception by *Eptesicus*, which also was used for the spectograms in Figure 4.4 (Beuter 1980; Altes 1984; Menne 1985).

If bats can perceive closely spaced glints, as obstacle-avoidance experiments and experiments on interception and discrimination of targets seem to demonstrate, they must be able to use both time axis registration and spectral interference notches to determine the time separation of overlapping echo replicas for reconstructing the spatial separation of the glints themselves in the final images (Simmons 1989). In particular, translation of the spectrum of overlapping echo replicas into arrival time estimates is essential for segregation of echoes from different objects if these objects are located at approximately the same distance from the bat (Saillant et al. 1993).

6.2 Echo–Delay Acuities Estimated from Performance

6.2.1 Minimal Requirements for Single-Target Interception

Most targets of interest to bats are small, with spatial extents (distances between glints) of no more than a few centimeters. Consequently, they reflect several echo replicas that arrive within an interval of no more than 100–200 µsec at the bat's ears. These replicas are blurred together into a single echo packet because the arrival times of the replicas differ by less than the integration time of echo reception, which is 300–400 µsec in *Eptesicus*. In Figure 4.4, the spectrograms for two-glint echoes with time separations of 5 µsec to 100–200 µsec are typical of the echoes encountered from single targets; actual echoes from flying insects have a similar appearance but are rhythmically animated (Kober and Schnitzler 1990; Moss and Zagaeski 1994).

If there is only one object near the bat to reflect echoes, the packet of sound it reflects probably is a good guide to the location of this target's "center of gravity," which is enough to approach the target and seize it in the tail membrane (see Fig. 4.1). With only one target present in the bat's "field of view," relatively uncomplicated echo-processing mechanisms are capable of tracking the target and guiding the bat's flight to a successful interception (Kuc 1994). Except for the fact that bats regulate the duration and repetition rate of their transmissions according to target range, determination of distance is not even formally required to accomplish interception (Kuc 1994), although, if echolocating bats can routinely intercept flying prey without accurate knowledge of target range, they must be "very lucky" (Grinnell 1967). From observations of the bat's reaching response at the moment of capture, an echo–delay accuracy of 50–100 μsec would be adequate to complete most interceptions (see Simmons 1989).

Estimates of the bat's best possible echo–delay acuity from the jitter observed in neural response latencies of single cells is also in the region of 50–100 μsec (Pollak et al. 1977; Bodenhamer and Pollak 1981; Covey and Casseday 1991). Furthermore, the bat does not need to perceive the separate glints in a single target merely to intercept it, so the ability to resolve two echo replicas as distinct could be limited to 300–400 μsec, the integration time of echo reception, without having any particularly deleterious effects on successful captures of isolated targets.

Although the target's spatial extent (separation of its glints) would not be perceived explicitly in spatial terms, the spectrum of the echo packet created by interference between the echo replicas from different parts of the target would be sufficient to distinguish one shape from another in many circumstances, especially for flying insects with different wingbeat rates (Simmons and Chen 1989; Kober and Schnitzler 1990; Neuweiler 1990; Schmidt 1992; Feng, Condon, and White 1994; Moss and Zagaeski 1994). This account of the physiological limitations on the content of the bat's images is widely accepted (Pollak et al. 1977; Pollak and Casseday 1989), but it nevertheless fails to account for obstacle avoidance or for interception in the presence of multiple targets, which are well-founded, classical observations about echolocation.

6.2.2 Minimal Requirements Implicit for Performance in Complex Conditions

To understand echolocation, the bat's performance cannot be characterized from observations made in the simplest cases; it has to be studied in realistically complicated situations with appropriately complex stimuli. Experiments have to be designed to engage whatever sophisticated mechanisms the bat might have for coping with complicated conditions, not just the minimal capabilities required to handle simple conditions. To account for the bat's ability to intercept one of two mealworms (see Fig. 4.7B), to capture an insect in clutter (Fig. 4.8), to avoid striking the adjacent wires in an obstacle-avoidance test (Fig. 4.9AB), to discriminate a mealworm from several disks (Fig. 4.10), or to choose a mealworm from a lump of clay when both are at the same distance (Fig. 4.12), it is not enough for the bat to perceive each packet of sound reaching its ears as an "echo" coming from a single "object." The bat has to segment the incoming stream of echoes into discrete estimates of the arrival time of each reflected replica of the broadcast signal (see Figs. 4.5 and 4.6). Without this explicit step, the bat cannot infer the locations of real objects in complicated environments, just "phantom" objects represented by the echo packets taken as whole units; the real objects are not accessible from the sounds at all except through the intermediary of their glints.

To account for successful obstacle avoidance on flights where the wires immediately on the left and right remain within roughly 5 cm of the same distance from the bat, the bat's ability to resolve two echo replicas as separate must be as good as 300 μsec, which is the lower limit of integration time and a feasible level of performance under the minimal assumptions given here that account for interception of single targets. It is also compatible with a straightforward interpretation of physiological recovery times for responses to the second of two sounds, which can occur for separations as short as 300–500 μsec (Grinnell 1963; Suga 1964; Pollak and Casseday 1989; Covey and Casseday 1991). However, if the bat keeps its flight path within 2 cm of the same distance to the wires, the bat's ability to resolve two echo replicas has to be be as good as 100 μsec, not 300 μsec. This figure is several times better than the minimum separation explainable from integration time alone. Furthermore, video tracking of bats during obstacle flights indicates that avoidance is successful even when the two nearest wires are as little as 1 cm apart in range, so the bat's ability to separate two overlapping echo replicas must be better than 50–60 μsec.

The earliest physiological studies obtained values of about 500 μsec for minimum recovery times in bats, but these experiments were interpreted with the knowledge from behavioral observations such as those described earlier that bats clearly must have true recovery times at least as small as 100 μsec, so a small population of neurons that could respond at time separations as short as 100 μsec was assumed to exist (Grinnell 1963, 1967). Although the fact has not received much attention more recently, the obstacle-avoidance results remain of direct concern for the bat's ability to perceive two echoes as separate: the value of 50–100 μsec for echo–delay resolution (an outside estimate from obstacle flights) is comparable in magnitude to the 50- to 100-μsec accuracy of delay determination for seizing a target in the tail membrane. However, resolution refers to segregation of two overlapping echoes, not accuracy for determining the delay of only one echo at a time (see Menne 1985; Schnitzler, Menne, and Hackbarth 1985). Because each echo replica must be assigned a specific arrival time along a scale of delay to locate its corresponding glint in range, the bat's ability to segregate two replicas only 50–100 μsec apart requires that the accuracy for specifying each of the arrival times by itself must be better than the difference between them (see Menne 1985; Simmons 1989). Figures 4.5 and 4.6 confirm that in well-regulated conditions the bat's ability to resolve two closely spaced echoes extends to separations as small as 2–30 μsec.

To account for interception of one mealworm in the presence of another object at about the same distance places even more severe demands on the bat's perceptual capacities because the glints in one object must be separated from the glints in the other object to locate the desired target, and that cannot occur unless all the glints first are separated out and assigned locations. In Figure 4.7B, the bat flies toward and seizes one mealworm in the presence of a second mealworm located close enough to the same target range that the echo replicas from the glints are completely intermingled; recognition of the individual glints in this case requires the ability to resolve replicas at least as close together as 10–20 μsec, and there are actually four primary glints to be sorted out, not two.

Consequently, the task shown in Figure 4.7B indicates that *Eptesicus* almost certainly can resolve echo replicas as close together as 5–10 μsec while in flight. This outside estimate of minimal echo–delay resolution is roughly 50 times smaller than the integration time of 300–400 μsec. Furthermore, because each echo replica has to be assigned a discrete arrival time, a resolution of 5–10 μsec implies that the accuracy for determining the arrival time of each replica by itself is better than 5–10 μsec as well.

6.2.3 Minimal Requirements Implicit for Target Localization

The estimates of required arrival-time acuity given here treat echo delay as a single imaginary axis from the bat's head to the target and back. The bat indeed broadcasts its sounds from a single site on the head, its open mouth, but it receives echoes at two sites, the left and right ears (see Fig. 4.1). Consequently, there is a separate echo–delay axis for each ear rather than a single axis for the bat's head as a whole. Segregation of echo replicas thus entails first of all separating them by their arrival times in the echo stream at each ear, so the bat really determines the distance to each glint at each ear. Beyond this, for avoiding obstacles or intercepting targets the bat perceives the directions of different objects, even when they are located at about the same distance; otherwise, it could not steer around wires or fly toward the mealworm. In the naturalistic tasks described earlier, there are implications of the stimulus regime for directional segregation and localization of targets as well as segregation along the range dimension.

It is widely assumed that bats use interaural differences in the amplitude and spectrum of echoes to determine the azimuth of a target because the bat's head is so small that interaural arrival-time differences are usually assumed be too small to be detected (Pollak 1988; Pollak and Casseday 1989; see also Schnitzler and Henson 1980; Schnitzler, Menne, and Hackbarth 1985). There are some observations of neural responses sensitive to small binaural time differences in FM bats (Harnischpfeger, Neuweiler, and Schlegel 1985), but these have been dismissed as too few in number (Pollak and Casseday 1989). However, the chief study that did not find such time sensitivity in FM bats (Pollak 1988) also did not use phase-controlled FM stimuli, whereas the study that did find time sensitivity did use phase-controlled stimuli (Harnischpfeger, Neuweiler, and Schlegel 1985), so the assumption that binaural time sensitivity is too poor in bats may be incorrect on physiological grounds.

In *Eptesicus*, interaural time differences amount to about 0.75 μsec per degree of azimuth, which is indeed small (Haresign et al., in manuscript). However, the overlap of echoes from multiple targets in the obstacle-avoidance task and during interception or discrimination of multiple targets renders interaural differences in the amplitude and spectrum of echoes useless for determining the azimuth of targets. This is a consequence of the integration time for echo reception; several echoes that arrive closer together than 300–400 μsec are assigned a single amplitude and a single spectrum over the whole interval, even though there are several glints actually responsible for the sounds. (Integration time *means* only one such estimate for the whole time window; allowing more than one amplitude estimate within this window is equivalent to reducing the length of the integration time itself.)

The single amplitude and single spectrum at each ear cannot be reduced to an estimate of the azimuth of one of the real glints, just a "phantom" glint at some intermediate location derived from binaural amplitude and spectral differences in the echo packets at the two ears. However, each echo packet has a spectrum that reflects primarily the time separation of the echo replicas, not the direction from which they come independent of this separation. Only by determining the arrival time of each replica separately at each ear and applying the time difference between ears to estimating azimuth can the bat perceive the correct location of a real target. Small as they are, interaural time differences are the only reliable cues for direction in obstacle-avoidance and airborne discrimination trials.

If the bat's accuracy for determining target azimuth is roughly 5° in obstacle-avoidance trials and during interception of mealworms, then its interaural echo–delay accuracy must be as good as 4 μsec (from an interaural time-difference ratio of 0.75 μsec/degree). If the bat's azimuth accuracy is 1.5° (a good approximation; see Simmons et al. 1983; Masters, Moffat, and Simmons 1985), then its interaural echo–delay accuracy must be about 1 μsec. These values refer to the separate registration of a single echo replica in both ears to determine the disparity in arrival time; because several echo replicas arrive in an interval of perhaps 100 μsec from different glints, the process of segregating echo replicas at each ear to determine the distance to each glint must really take place with an accuracy of no poorer than 1–4 μsec for all the glints. Otherwise, none of the real objects could be located in azimuth.

6.3 Reexamination of Conventional Localization Cues

The conclusion that FM bats must be able to use interaural arrival times to determine azimuth with an acuity of 1-4 μsec, and therefore that each delay estimate must have an associated accuracy of 1-4 μsec, may seem strong for being based on the performance of flying bats in tasks with continually changing bat–target relationships. In fact, however, during the bat's approach to vertically stretched wires or to airborne mealworms and disks, the critical overlap of echo replicas from several glints nearer each other in range than 5-7 cm is often present throughout the flight. Even though the bat flies progressively nearer the targets, so that the delay of echoes progressively shortens, it flies closer to each target at about the same rate. Consequently, the difference in range is relatively invariant on many trials. The bat's flight path toward several targets at once (see Figs. 4.7–4.12) keeps the separation of echo replicas smaller than the integration time of 300–400 μsec for the duration of entire flights. This hidden constancy in many of the obstacle-avoidance or airborne discrimination tests is documented by video motion-analysis studies, which track the distance from the bat to each part of several targets that are present in front of the bat, but it also can be concluded from descriptions of the original results from even the earliest experiments of these types (e.g., the bat's flight path is just about perpendicular to the plane of the wires [Griffin 1958; Grinnell and Griffin 1958; Jen and Kamada 1982]).

Second, the seemingly improbable conclusion from these naturalistic experiments that bats must have an inherent echo–delay acuity of at least 1-4 μsec becomes more plausible when the requirements for use of the conventional binaural amplitude and spectral cues in these same tasks are closely examined. The ruling constraint is the integration time for echo reception, which ensures that several echoes arriving at each ear within a 300 -to 400-μsec window will be assigned a single amplitude and spectrum at that ear (see Beuter 1980; Altes 1984; Menne 1985). Because all the objects whose echo replicas arrive within this window are lumped together for purposes of amplitude determination across frequencies (see 5-μsec to 200-μsec two-glint echoes in Fig. 4.4), there is no way to determine the binaural difference for each object, just for the echo packet as a whole. But there is no real object at the location represented by the binaural features of the entire packet. Moreover, each of the real objects is only accessible through the echo replicas it contributes to the packet; any information coming from other objects only complicates the problem by distorting the resulting estimate of that object's location from the packet alone. This distortion is nontrivial; in obstacle-avoidance flights, the packet would depict a phantom object located between two real wires, but the bat does not avoid this empty space, it flies through it (see Fig. 4.7A).

To make the binaural amplitude and spectral features of each echo replica available to determine each target's direction separately, the replicas first must be segregated out of the echo packet by their times of arrival. They cannot be separated by their amplitudes or their spectra because these dimensions have been combined across all the different replicas within the integration-time window. The spectrum of the packet can be used to determine the arrival-time separations of the replicas within the packet (Simmons 1989; Simmons et al. 1989; Saillant et al. 1993), but the separate spectra of the replicas cannot be assessed without first segregating the replicas by arrival time, and this requires temporal acuity much finer than the integration time itself.

Even disregarding the requirement that binaural time differences be used, the minimum two-glint resolving power implicit in the airborne interception results is 5–10 μsec. When the bat's ability to perceive the directions of different targets is considered, the minimum temporal resolution needed to segregate echo replicas and assign them their own binaural amplitude and spectral parameters is about as small as the temporal acuity required to locate the objects by interaural arrival-time differences alone. For these reasons, asserting that bats locate targets chiefly from binaural amplitude and spectral cues means that, in obstacle-avoidance and airborne discrimination tests, they must have a temporal resolving power of only a few microseconds to achieve this.

7. Directionality of Echolocation in *Eptesicus*

7.1 Directional Segregation of Targets?

It is sometimes assumed that the sonar of bats is sufficiently directional for echoes from two different objects to be isolated by aiming the sonar broadcasts and the ears toward one target rather than the other. For the directionality of the broadcast to be of practical use in segregating echoes from two different objects, the transmitted beam would have to be narrow enough to ensonify one object strongly while leaving the other object only weakly ensonified. Similarly, for the directionality of hearing to be useful, the receiving beam would have to be narrow enough to select echoes from one direction in preference to echoes from another direction. A commonly used index of target "separation" from directionality is reduction of echo strength from unwanted objects to half the amplitude (-6 dB) or half the energy (-3 dB) of echoes from the target of interest. This is only an arbitrary index of beamwidth, however; in practice, to effectively isolate echoes of one target from echoes of another, the unwanted echoes would have to be reduced sufficiently far that their presence does not impair determination of the delay, amplitude, and spectrum of the desired echoes. A more reasonable index of the minimal necessary separation would be nearer to a tenfold reduction of undesired echo amplitude (-20 dB) or power (-10 dB).

Observations of insect pursuit, obstacle avoidance, or target discrimination reveal that *Eptesicus* and other FM bats can perceive the locations of several objects simultaneously over a region of space that extends roughly from 30°–50° left to 30°–50° right (see Figs. 4.7–4.12). To isolate one object from another, the broadcasting or receiving beam would have to select a narrow window of azimuth from within this broad span of azimuths so that echoes from just the desired object become 10–20 dB stronger than echoes from unwanted objects located in the same overall region of space. In *Eptesicus*, neither the directionality of the sonar transmissions (Hartley and Suthers 1989) nor the directionality of hearing (Haresign et al. in manuscript; Jen and Chen 1988; Wotton 1994). is adequate to explain the bat's ability to perceive each of the adjacent wires in an obstacle-avoidance test or to locate one of two targets at about the same distance.

7.2 Broad Directionality of Transmissions

The open mouth of the bat in Figure 4.1 serves as the bat's broadcasting antenna, and the dimensions of the mouth in relation to the wavelengths of the sonar signals make this a directional antenna whose main axis is pointed to the front (Hartley and Suthers 1989). The equivalent acoustic diameter of the transmitting aperture is estimated to be about 9 mm, which is reasonably close to the actual dimensions of the bat's open mouth. Figure 4.13A illustrates the horizontal directionality of the bat's transmissions at two frequencies, 40 kHz and 60 kHz, that are representative of the sound as a whole (see Fig. 4.2). (Figure 4.13 shows a span of directions from 40° left to 40° right, covering the zone of frontal space within which the bat demonstrably can perceive several different objects at once.) At 40 kHz, the 6-dB width of the beam is ±30°, and at 60 kHz the width of the beam is ±25°. The equivalent 10-dB beamwidths are about ±50°–60°. These are surprisingly wide; in obstacle-avoidance or airborne discrimination flights, virtually all the objects that are present fall well within this subtended span of azimuths until the bat has approached quite close.

For the bat to ensonify one wire in an obstacle-avoidance flight with an incident sound that is as much as 10 dB stronger than the other wire, the bat would have to point its head directly at one wire when it has flown near enough for the other wire to be 50°–60° off to the left or right. For a wire array with a 25-cm distance between adjacent wires (Jen and Kamada 1982; see Fig. 4.7), the bat would have to be nearer than 25–30 cm to achieve 10-dB separation of the wires from the incident sounds alone. However, video studies of obstacle-avoidance flights (e.g., Fig. 4.7) show that the bat usually has settled on its flight path before reaching this distance, when it is as far away as 0.5–1 m and the angular separation of the adjacent wires is only about 15°–30°. At such distances, the largest possible amplitude separation of the broadcast sounds impinging on adjacent wires is only 3–6 dB (Fig. 4.13A).

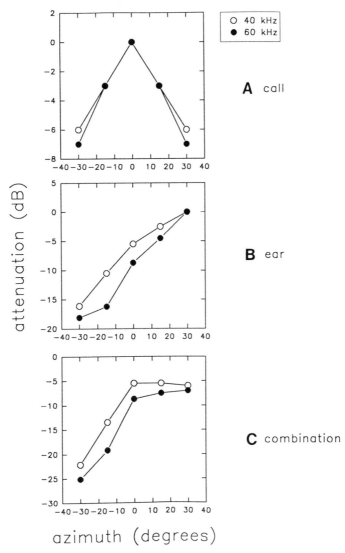

FIGURE 4.13A–C. Directional diagrams for the echolocation of *Eptesicus* (Wotton 1994). (A) The strength of the bat's sonar transmissions at azimuths from left (−) 30° to right (+) 30° at 40 kHz *(open circle)* and 60 kHz *(solid circle)* (Hartley and Suthers 1989); (B) the sensitivity of the right ear to 40 and 60 kHz frequency components in FM sounds from different directions; (C) the combined sensitivity of the sonar system at 40 and 60 kHz to a constant-strength target located in different directions, measured at the right ear. The broadcast is aimed to the front (0°), the right ear is aimed about 30°–40° to the right side at these frequencies, and the sonar system as a whole is omnidirectional in sensitivity to targets on the same side as the ear.

Moreover, the video motion studies confirm that the bat does not make large steering movements of its head to keep just one of the wires straight ahead in the middle of the broadcast beam, which is at 0° in Figure 4.13A. Thus, effective isolation of one wire from another beyond a few decibels clearly does not occur as a consequence of the directionality of the incident sound. During approaches to tethered mealworms (see Fig. 4.7B), the targets which the bat demonstrably can perceive as separate at distances of 0.5–0.75 m are located in a zone of azimuths anywhere from about 12° to 17° wide, so separation of echoes from the broadcast beam cannot exceed 3–4 dB. The bat still appears to have decided which target to capture before it has flown near enough that the unwanted target is as much as 50°−60° off to one side. Consequently, the crude directionality of transmissions is not adequate to segregate targets in well-documented tasks in which perceptual segregation in fact is achieved.

7.3 Broad Directionality of Echo Reception

The bat's ears act as receiving antennas for echoes, and their size and shape in relationship to the wavelengths of the ultrasonic signals make them about as directional for reception of sounds as the mouth is for emitting sounds. However, the bat's ears are not pointed toward the front but off to the side (see Fig. 4.1), and the shape of the receiving beam is not symmetrical (Haresign et al., in manuscript; Jen and Chen 1988; Wotton 1994). Figure 4.13B shows the receiving directionality for the right ear of *Eptesicus* at 40 and 60 kHz in the FM sweep of echolocation signals (Haresign et al., in manuscript; Wotton 1994; see also Jen and Chen 1988 for pure-tone directionality, which exaggerates the sharpness of the receiving beam). At both of these frequencies, which were presented in sweeps rather than as tones, the acoustic axis of the right ear is about 30°–40° off to the right side of the head, so the directional pattern in Figure 4.13B appears as a smooth decline in sensitivity for all directions to the left of this axis (left of +30° in Fig. 4.13B), with a shift to a steeper slope at directions toward the left side of the head (left of 0° in Fig. 4.13B).

Thus, for each ear, reception of echoes is progressively less sensitive at directions located toward the contralateral side of the acoustic axis. If the bat pointed its right-ear axis at the right-hand wire in the obstacle-avoidance array (Fig. 4.7A), the left-hand wire would have to be located 30°–45° contralateral for its echoes to be 10 dB weaker in the same right ear. To achieve this 10-dB separation of echoes in each ear, the bat would have to fly closer than 30–35 cm from the wire array. However, the bat frequently has determined its flight path at greater distances, so the crude directionality of hearing, like the crude directionality of the broadcasts, is apparently not the basis for perception of one target in the presence of others.

7.4 Directionality of the Sonar System

Both the bat's sounds and its hearing are directional, and the two directional patterns add together to determine the strength of echoes actually stimulating the bat from different azimuths. It would be possible for them to reinforce each other so that the directionality of the whole sonar system is sharper than either the broadcast or receiving component. However, the acoustic axis of the transmissions points straight ahead, while the axis of each ear points well off to the side. The combined directional patterns do not reinforce each other; instead, they yield a curious combination of omnidirectional sensitivity to targets on the same side as each ear and directionally graded rejection of echoes from targets on the opposite side.

Figure 4.13C shows this combined directionality for *Eptesicus* at frequencies of 40 and 60 kHz (Wotton 1994). For the right ear, a target located anywhere from straight ahead (0°) to at least 30° off to the right will return echoes of constant amplitude to the ear. A target located anywhere to the left will return echoes at progressively weaker strengths depending on how far to the left it is. For example, at 30° to the left, the echoes will be reduced by about 22–25 dB as a consequence of the combined directionality of transmissions and reception. This pattern of directional sensitivity is mirrored in the left ear. Thus, the bat's sensitivity to targets is omnidirectional but lateralized, with the ear pointing in the target's direction receiving a uniformly strong echo from the same target in many directions and the opposite ear receiving a weaker echo depending on how far in the contralateral direction the target is located.

The directionality of echolocation (see Fig. 4.13C) leads to an unanticipated pattern of stimulation during the bat's approach to obstacles. In Figure 4.7A, the nearest wire on the bat's left (*c*) returns a strong echo to the left ear, while the nearest wire on the right (*b*) returns a strong echo to the right ear. These same wires return substantially weaker echoes to the opposite ears. The arrival times of these echoes depend on the bat's flight path, but the stronger echo from the wire on the same side as the ear will typically be 1–50 μsec earlier than the weaker echo from the wire on the opposite side if the bat's head is exactly equidistant from the two wires. (The echo from the nearer wire, which is on the same side as the ear, arrives first.)

When the bat is at a distance of 0.5–1 m from the wires and the azimuthal separation of the wires is 12°–17°, the opposite ear from each wire receives an echo that is about 13–20 dB weaker than the ear on the same side. Thus, each ear receives a strong echo and a weak echo, at arrival times determined by the distance from each wire to each ear. However, the strong or weak echo in one ear comes from a different wire than the strong or weak echo in the other ear. Furthermore, as a consequence of the bat's nearly perpendicular approach to the wire array, all these echoes will typically fall within the 300 -to 400-μsec integration-time window at each ear. Interaural amplitude and spectral differences by themselves will be useless for localizing the wires because these binaural parameters will be dominated by the stronger echo in each ear, which comes from a different target on the left than on the right.

Because of a combination of the broad directional sensitivity pattern for targets (see Fig. 4.13A) and the nearly symmetric arrangement of the wires (see Fig. 4.7A), the obstacle-avoidance test actually provides a strong rejection of interaural amplitude and spectral hypotheses concerning horizontal localization of targets by *Eptesicus*. The integration time for echo reception so severely constrains this situation that the only way to determine the locations of both wires is to segregate the two echo replicas going to each ear from the spectral notches they create through interference (see Fig. 4.4) and then compare their arrival times between ears. (Behavioral experiments reveal that spectral information can be used to estimate echo–replica separations even when the sources of the overlapping replicas are as far apart as 40° in azimuth [Simmons, Moss, and Ferragamo 1990].) No other method can suppress the dominating effect of the strong echo from the nearest wire in each ear. Instead of being based on crude directionality, the bat's ability to separately perceive each target in an acoustic scene is derived from auditory computations that create images by segregating the glints making up the scene and then comparing the echo streams at the two ears in a glint-by-glint manner.

8. Summary

The bat's acuity for determining the arrival time of echo replicas and its acuity for resolving two overlapping echo replicas as separate are the most fundamental capacities underlying the formation of acoustic images in FM echolocation. In psychophysical experiments, the big brown bat *Eptesicus fuscus* can resolve echoes as separate even when they arrive as close together as 2–30 μsec. The bat's resolving power is independently estimated, from performance in obstacle-avoidance and airborne-interception tests, to be 5–10 μsec. The chief constraint on echo processing is the integration time of echo reception (300–400 μsec in *Eptesicus*), which combines all reflected replicas of the broadcast sound arriving within this window into a single echo "packet" and gives it a single amplitude and spectrum over this window.

Without segregation of closely spaced echo replicas by their arrival times, however, even within the integration-time window, the bat's images will not adequately depict the configuration of vertical wires or the location of one target (e.g., a mealworm) when other targets (e.g., a mealworm or several disks) are present also. Moreover, without very sharp registration of the delay of individual replicas at each ear, two closely spaced replicas cannot be segregated and then assigned individual arrival times, either to depict their distances or to determine their directions from binaural cues. Interaural arrival-time acuity is estimated, from psychophysical experiments, to be about 1 μsec, and independently from obstacle-avoidance and airborne-interception studies to be 1–4 μsec. Even the use of binaural amplitude and spectral cues requires prior temporal processing of echoes with an acuity and resolution of several microseconds to recognize which echoes come from which targets.

Acknowledgments   This survey of evidence concerning the content of the bat's sonar images was supported by ONR Grant No. N00014-89-J-3055, by NIMH Research Scientist Development Award No. MH00521, by NIMH Training Grant No. MH19118, by NSF Grant No. BCS 9216718, by McDonnell-Pew Grant No. T89-01245-023, and by NIH Grant No. DC00511,and by DRF and NATO grants.

References

Altes RA (1980) Detection, estimation, and classification with spectrograms. J Acoust Soc Am 67:1232–1246.

Altes RA (1984) Texture analysis with spectrograms. IEEE Trans Sonics–Ultrasonics SU-31:407–417.

Beuter KJ (1980) A new concept of echo evaluation in the auditory system of bats. In: Busnel RG, Fish JF (eds) Animal Sonar Systems. New York: Plenum, pp. 747–761.

Bodenhamer RD, Pollak GD (1981) Time and frequency domain processing in the inferior colliculus of echolocating bats. Hear Res 5:317–355.

Covey E, Casseday JH (1991) The monaural nuclei of the lateral lemniscus in an echolocating bat: parallel pathways for analyzing temporal features of sound. J Neurosci 11: 3456–3470.

Dear SP, Simmons JA, Fritz J (1993) A possible neuronal basis for representation of acoustic scenes in auditory cortex of the big brown bat. Nature 364:620–623.

Dear SP, Fritz J, Haresign T, Ferragamo M, Simmons JA (1993) Tonotopic and functional organization in the auditory cortex of the big brown bat, *Eptesicus fuscus*. J Neurophysiol 70:1988–2009.

Feng AS, Condon CJ, White KR (1994) Stroboscopic hearing as a mechanism for prey discrimination in FM bats? J Acoust Soc Am 95:2736–2744.

Griffin DR (1958) Listening in the dark. New Haven: Yale University Press. (Reprinted by Cornell University Press, Ithaca, NY, 1986.)

Griffin DR (1967) Discriminative echolocation by bats. In: Busnel RG (ed) Animal Sonar Systems: Biology and Bionics. France: Jouy-en-Josas-78, Laboratoire de Physiologie Acoustique, pp. 273–300.

Griffin DR, Friend JH, Webster FA (1965) Target discrimination by the echolocation of bats. J Exp Zool 158:155–168.

Grinnell AD (1963) The neurophysiology of audition in bats: temporal parameters. J Physiol 167:67–96.

Grinnell AD (1967) Mechanisms of overcoming interference in echolocating animals. In: Busnel R-G (ed) Animal Sonar Systems: Biology and Bionics. France: Jouy-en-Josas-78, Laboratoire de Physiologie Acoustique, pp. 451–481.

Grinnell AD, Griffin DR (1958) The sensitivity of echolocation in bats. Biol Bull 114:10–22.

Harnischpfeger G, Neuweiler G, Schlegel P (1985) Interaural time and intensity coding in the superior olivary complex and inferior colliculus of the echolocating bat, *Molossus ater*. J Neurophysiol 53:89–109.

Hartley DJ (1992) Stabilization of perceived echo amplitudes in echolocating bats: II. The acoustic behavior of the big brown bat, *Eptesicus fuscus*, while tracking moving prey. J Acoust Soc Am 91:1133–1149.

Hartley DJ, Suthers RA (1989) The sound emission pattern of the echolocating bat, *Eptesicus fuscus*. J Acoust Soc Am 85:1348–1351.

Jen PH-S, Chen DM (1988) Directionality of sound pressure transformation at the pinna of echolocating bats. Hear Res 34:101–118

Jen PH-S, Kamada T (1982) Analysis of orientation signals emitted by the CF-FM bat *Pteronotus p. parnellii* and the FM bat *Eptesicus fuscus* during avoidance of moving and stationary obstacles. J Comp Physiol 148:389–398.

Kick SA (1982) Target detection by the echolocating bat, *Eptesicus fuscus*. J Comp Physiol 145:431–435.

Kick SA, Simmons JA (1984) Automatic gain control in the bat's sonar receiver and the neuroethology of echolocation. J Neurosci 4:2725–2737.

Kober R, Schnitzler H-U (1990) Information in sonar echoes of fluttering insects available for echolocating bats. J Acoust Soc Am 87:874–881.

Kuc R (1994) Sensorimotor model of bat echolocation and prey capture. J Acoust Soc Am 96: 1965–1978.

Kurta A, Baker RH (1990) *Eptesicus fuscus*. Mamm Species 356:1–10.

Lee DN, van der Weel FR, Hitchcock T, Matejowsky E, Pettigrew JD (1992) Common principle of guidance by echolocation and vision. J Comp Physiol A 171:563–571.

Lhémery A, Raillon R (1994) Impulse-response method to predict echo responses from targets of complex geometry: II. Computer implementation and experimental validation. J Acoust Soc Am 95:1790–1800.

Masters WM, Moffat AJM, Simmons JA (1985) Sonar tracking of horizontally moving targets by the big brown bat, *Eptesicus fuscus*. Science 228:1331–1333.

Menne D (1985) Theoretical limits of time resolution in narrow band neurons. In: Michelsen A (ed) Time Resolution in Auditory Systems. New York: Springer-Verlag, pp. 96–107.

Menne D, Kaipf I, Wagner I, Ostwald J, Schnitzler HU (1989) Range estimation by echolocation in the bat *Eptesicus fuscus*: trading of phase versus time cues. J Acoust Soc Am 85:2642–2650.

Mogdans J, Schnitzler H-U (1990) Range resolution and the possible use of spectral information in the echolocating bat, *Eptesicus fuscus*. J Acoust Soc Am 88:754–757.

Moss CF, Schnitzler H-U (1989) Accuracy of target ranging in echolocating bats: acoustic information processing. J Comp Physiol A 165:383–393.

Moss CF, Simmons JA (1993) Acoustic image representation of a point target in the bat, *Eptesicus fuscus*: evidence for sensitivity to echo phase in bat sonar. J Acoust Soc Am 93:1553–1562.

Moss CF, Zagaeski M (1994) Acoustic information available to bats using frequency-modulated sounds for the perception of insect prey. J Acoust Soc Am 95:2745–2756.

Neuweiler G (1990) Auditory adaptations for prey capture in echolocating bats. Physiol Rev 70:615–641.

Novick A (1977) Acoustic orientation. In: Wimsatt WA (ed) Biology of Bats, Vol. 3. New York: Academic Press, pp. 73–287.

Pollak GD (1988) Time is traded for intensity in the bat's auditory system. Hear Res 36:107–124.

Pollak GD (1993) Some comments on the proposed perception of phase and nanosecond time disparities by echolocating bats. J Comp Physiol A 172:523–531.

Pollak GD, Casseday JH (1989) The Neural Basis of Echolocation in Bats. New York: Springer-Verlag.

Pollak GD, Marsh DS, Bodenhamer R, Souther A (1977) Characteristics of phasic on neurons in inferior colliculus of unanesthetized bats with observations relating to mechanisms for echo ranging. J Neurophysiol 40:926–942.

Pye JD (1980) Echolocation signals and echoes in air. In: Busnel RG, Fish JF (eds) Animal Sonar Systems. New York: Plenum, pp. 309–353.

Roverud RC, Nitsche V, Neuweiler G (1991) Discrimination of wingbeat motion by bats, correlated with echolocation sound pattern. J Comp Physiol A 168:259–263.

Saillant PA, Simmons JA, Dear SP, McMullen TA (1993) A computational model of echo processing and acoustic imaging in frequency-modulated echolocating bats: the spectrogram correlation and transformation receiver. J Acoust Soc Am 94:2691–2712.

Schmidt S (1992) Perception of structured phantom targets in the echolocating bat, *Megaderma lyra*. J Acoust Soc Am 91:2203–2223.

Schnitzler H-U, Henson OW Jr (1980) Performance of airborne animal sonar systems: I. Microchiroptera. In: Busnel RG, Fish JF (eds) Animal Sonar systems. New York: Plenum, pp. 109–181.

Schnitzler H-U, Menne D, Hackbarth H (1985) Range determination by measuring time delay in echolocating bats. In: Michelsen A (ed) Time Resolution in Auditory Systems. New York: Springer-Verlag, pp. 180–204.

Simmons JA (1973) The resolution of target range by echolocating bats. J Acoust Soc Am 54:157–173.

Simmons JA (1979) Perception of echo phase information in bat sonar. Science 207:1336–1338.

Simmons JA (1989) A view of the world through the bat's ear: the formation of acoustic images in echolocation. Cognition 33:155–199.

Simmons JA (1992) Time-frequency transforms and images of targets in the sonar of bats. In: Bialek W (ed) Princeton Lectures on Biophysics. River Edge, NJ: World Scientific, pp.291–319.

Simmons JA (1993) Evidence for perception of fine echo delay and phase by the FM bat, *Eptesicus fuscus*. J Comp Physiol A 172:533–547.

Simmons JA, Chen L (1989) The acoustic basis for target discrimination by FM echolocating bats. J Acoust Soc Am 86:1333–1350.

Simmons JA, Grinnell AD (1988) The performance of echolocation: the acoustic images perceived by echolocating bats. In: Nachtigall P, Moore PWB (eds) Animal Sonar: Processes and Performance. New York: Plenum, pp. 353–385.

Simmons JA, Kick SA (1984) Physiological mechanisms for spatial filtering and image enhancement in the sonar of bats. Annu Rev Physiol 1984 46:599–614.

Simmons JA, Moss CF, Ferragamo M (1990) Convergence of temporal and spectral information into acoustic images of complex sonar targets perceived by the echolocating bat, *Eptesicus fuscus*. J Comp Physiol A 166:449–470.

Simmons JA, Ferragamo M, Moss CF, Stevenson SB, Altes RA (1990) Discrimination of jittered sonar echoes by the echolocating bat, *Eptesicus fuscus*: the shape of target images in echolocation. J Comp Physiol A 167:589–616.

Simmons JA, Freedman EG, Stevenson SB, Chen L, Wohlgenant TJ (1989) Clutter interference and the integration time of echoes in the echolocating bat, *Eptesicus fuscus*. J Acoust Soc Am 86:1318–1332.

Simmons JA, Kick SA, Lawrence BD, Hale C, Bard C, Escudié B (1983) Acuity of horizontal angle discrimination by the echolocating bat, *Eptesicus fuscus*. J Comp Physiol 153:321–330.

Simmons JA, Saillant PA, Ferragamo MJ, Haresign T, Dear SP, Fritz J, McMullen TA Auditory computations for biosonar target imaging in bats. In: Hawkins HL, McMullen TA, Popper AN, Fay RR (eds) Auditory computation. New York, Springer-Verlag. (in press)

Suga N (1964) Recovery cycles and responses to frequency modulated tone pulses in auditory neurons of echolocating bats. J Physiol 175:50–80.

Suga N (1967) Discussion (of presentation by O. W. Henson, Jr.). In: Busnel R-G (ed) Animal Sonar Systems: Biology and Bionics. Laboratoire de Physiologie Acoustique, France: Jouy-en-Josas-78, pp. 1004–1020.

Suga N, Schlegel P (1973) Coding and processing in the nervous system of FM signal producing bats. J Acoust Soc Am 84:174–190.

Sum YW, Menne D (1988) Discrimination of fluttering targets by the FM-bat *Pipistrellus stenopterus*? J Comp Physiol A 163:349–354.

Webster FA (1967) Performance of echolocating bats in the presence of interference. In: Busnel RG (ed) Animal Sonar Systems: Biology and Bionics. France: Jouy-en-Josas-78, Laboratoire de Physiologie Acoustique, pp. 673–713.

Webster FA, Brazier OG (1965) Experimental studies on target detection, evaluation, and interception by echolocating bats. TDR No. AMRL-TR-65-172, Aerospace Medical Division, USAF Systems Command, Tucson, AZ.

Webster FA, Griffin DR (1962) The role of the flight membrane in insect capture by bats. Anim Behav 10:332–340.

Wotton JM (1994) The basis for vertical sound localization of the FM bat, *Eptesicus fuscus*: acoustical cues and behavioral validation. Ph.D. dissertation, Brown University, Providence, RI.

5

Cochlear Structure and Function in Bats

Manfred Kössl and Marianne Vater

1. Introduction

The mammalian cochlea must extract loudness and frequency information about different and overlapping acoustic events from a single input channel. Unlike the visual system, where input from different spatial sources is separated early and distributed in parallel channels, all input to the cochlea converges in the middle ear to induce movement at the membrane of the oval window. The oval window acts as a point source of mechanical waves dissipating in the fluid-filled spaces of the cochlea. It is now left to the organ of Corti to filter relevant information from the total input. Because this is a difficult task, it is no wonder that the acquisition of highly developed cochleae that are able to analyze low-level signals at many different frequencies is a relatively recent step in animal evolution, one that is confined to higher vertebrates. The high-frequency hearing capabilities of reptiles and birds are restricted to frequencies below about 12 kHz (for review, see Manley 1990). As a specific mammalian adaptation, the middle ear and the cochlea have an extended sensitivity in the high-frequency range. For processing of high frequencies, the cochlea has developed macro- and micromechanical specializations and employs active processes that enhance frequency tuning and sensitivity. The hair cells are clearly distinguished as inner hair cells (IHCs) that are contacted by nearly all afferent nerve fibers to the brain and outer hair cells (OHCs) that function to a lesser degree as receptors but more as effectors. Outer hair cells are thought to be involved in amplifying and filtering the mechanical input in collaboration with additional structures such as the basilar and tectorial membranes (Dallos and Corey 1991). In the cochleae of bats, mechanisms to aquire sensitive and sharp tuning at high frequencies up to 160 kHz are fully exploited. The investigation of these cochleae may give clues to open questions about cochlear function in mammals, in particular about the frequency limitation of active processes, electromechanical feedback, and mechanical tuning mechanisms.

2. Evolutionary Adaptations of the Cochlea of Bats

A significant factor in the evolution of the cochlea in bats has doubtless been the use of echolocation for prey capture. In bats, the cochlea is not only involved in passive listening tasks but is an integral part of active orientation. This functional context is reflected in the sheer size of the cochleae of bats employing different echolocation strategies (Habersetzer and Storch 1992; Fig. 5.1). Most Macrochiroptera do not echolocate and instead rely on their eyes and olfactory system to find edible fruits. Their cochlea, consequently, is smaller relative to the skull size than that of echolocating Microchiroptera. According to the characteristics of their echolocation call, Microchiroptera can be divided into two groups (for a detailed discussion of the evolution of echolocation see Fenton, Chapter 2). The first group uses short broad-band calls that consist of different harmonics of a downward frequency-modulated (FM) component. The second group employs CF-FM calls, composed of a long, constant-frequency (CF) component followed by a short FM sweep, and displays characteristic Doppler-shift compensation behavior (Schnitzler 1968, 1970). CF-FM bats detect small frequency modulations in the CF component of the echo caused by the wingbeat of prey insects, even within a densely cluttered environment such as dense vegetation (see Neuweiler 1990). These capabilities crucially depend on a high cochlear frequency resolution.

Hydromechanical specializations in the cochlea create enhanced tuning and an acoustic fovea, an expanded representation of the biologically most important CF frequency range. As a consequence of the foveal frequency representation, typical CF-FM bats such as Rhinolophidae, the related Hipposideridae, and *Pteronotus parnellii* have a larger cochlea than less specialized FM bats such as Vespertilionidae (Fig. 5.1). Cochlear size depends more on functional requirements related to the echolocation signal than on the taxonomic relationships. This is clearly evident in the genus *Pteronotus*, New World bats of the mormoopid family (see Appendix in Fenton, Chapter 2, this volume). The species *Pteronotus personatus, P. davyi*, and *P. suapurensis*, which use FM calls, have much smaller cochleae than the species *P. parnellii*, whose different subspecies use CF-FM calls. This demonstrates the high degree of adaptability of the cochlear size to different environmental constraints (Habersetzer and Storch 1992). The size of the cochlea of extinct bats from the middle Eocene is at the lower margin of the vespertilionid family, indicating that the larger cochleae associated with CF call components may be a relatively recent development.

3. Physiology of the Cochlea

3.1 Auditory Threshold Curves

The absolute threshold of hearing in bats is obtained most reliably from data on single neurons of the eighth nerve or auditory brain stem and from

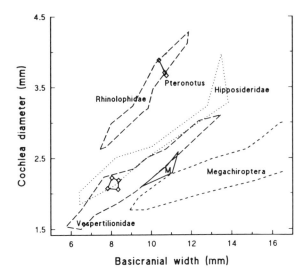

FIGURE 5.1. Relative size of the cochlea in different bat families, plotted as the cochlear diameter versus the basicranial width (after Habersetzer and Storch 1992). The cochleae of (CF-FM) bats (Rhinolophidae, Hipposideridae) are larger than in FM bats (Vespertilionidae), in Macrochiroptera, and in extinct bats from Messel (M). *Open symbols* mark the cochleae of different species of *Pteronotus* (see text).

behavioral audiograms. Cochlear microphonic (CM) potentials depend on the geometry of the cochlea in relation to the recording site and show a frequency-specific bias that makes them less useful for threshold determination. The N_1 evoked potential of the auditory nerve depends on the synchronicity of spike discharge in different nerve fibers. There are mechanical on/off processes in the cochlea of CF-FM bats at certain frequencies (Grinnell 1963, 1973), and the synchronization of onset spikes is influenced by cochlear resonance (Suga, Simmons, and Jen 1975). Therefore the N_1 potential does not unambiguously reflect the threshold of hearing.

Even in nonecholocating mammals, the behavioral threshold can extend up to about 100 kHz (Fig. 5.2A). However, auditory sensitivity steeply deteriorates for frequencies above 60 kHz and in the house mouse is at about 68 dB SPL (sound pressure level) at 100 kHz. In FM bats such as *Carollia perspicillata* or *Megaderma lyra* (Fig. 5.2A,B), the auditory thresholds of neurons measured in the inferior colliculus (Rübsamen, Neuweiler, and Sripathi 1988; Sterbing, Rübsamen, and Schmidt 1990) remain close to 0 dB SPL up to frequencies of about 100 kHz. Both bat species are extremely sensitive to sound frequencies between about 15 and 30 kHz with thresholds below − 10 dB SPL. This frequency range is probably important for passive listening or communication because the range of the broad-band echolocation signals is higher in both species

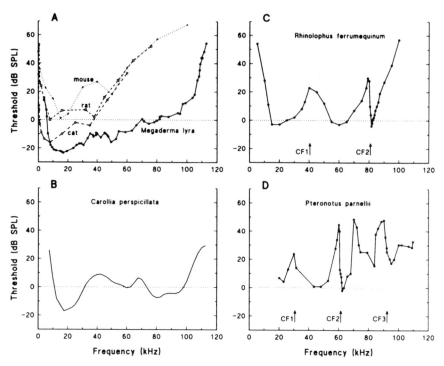

FIGURE 5.2A–D. Auditory threshold curves. (A) Behavioral thresholds in cat (from Neff and Hind 1955), rat (from Kelly and Masterton 1977), and the house mouse (from Ehret 1974) compared to a neuronal audiogram of *Megaderma lyra* (from Rübsamen, Neuweiler, and Sripathi 1988: inferior colliculus). (B) Neuronal audiogram in *Carollia perspicillata* (from Sterbing, Rübsamen, and Schmidt 1990: inferior colliculus). (C) Behavioral audiogram in *Rhinolophs ferrumequinum* (from Long and Schnitzler 1975). (D) Neuronal audiogram in *Pteronotus parnellii* (from Kössl and Vater 1990b: cochlear nucleus). CF_1, CF_2, and CF_3 indicate the first, second, and third harmonic of the constant frequency echolocation signal component. SPL, sound pressure level.

(40–100 kHz in *Megaderma*, 60–110 kHz in *Carollia*). In *Megaderma*, the high sensitivity around 20 kHz is partly caused by an amplifying effect of the large pinnae of the outer ear. The pinna gain is at a maximum of about 17 dB at 20 kHz (Obrist et al. 1993). In *Carollia*, a second minimum in the hearing threshold is evident within the frequency range of the echolocation signal (from 80 to 100 kHz).

In bats that use long CF-FM orientation calls, there are sharp maxima and minima in the threshold curves that are related to the CF frequencies. In *Rhinolophus ferrumequinum* (Fig. 5.2C), a narrow threshold minimum at about 83–85 kHz coincides with the range of the second-harmonic CF component (CF_2; Long and Schnitzler 1975). Between 79 and 82 kHz, there is a pronounced threshold maximum. A second threshold maximum is found close to the first CF component (CF_1) at about 41 kHz. During Doppler-shift compensation behavior, the frequency of the emitted CF_2 component is decreased by about 3 kHz (Schnitzler 1968; Schuller, Beuter, and Schnitzler 1974) and thus shifted into the region of acoustic insensitivity. The frequency of the Doppler-shifted CF_2 echo is now at the threshold minimum. Neuronal audiograms of *Rhinolophus rouxi* obtained in the inferior colliculus and lateral lemniscus (Schuller 1980; Metzner and Radtke-Schuller 1987) show a similar pattern; the CF_2 frequency of about 78 kHz is located at a steep slope between threshold maximum and minimum. In *Pteronotus parnellii*, the neuronal threshold curve shows multiple maxima and minima close to the CF_1, CF_2, and CF_3 frequencies (Fig. 5.2D). Again, the CF_2 frequency is located at a steeply sloping region of the audiogram.

Because the duration of the CF component ranges from 20 to 50 msec in both rhinolophids and *Pteronotus parnellii*, the outgoing call and the returning echo inevitably overlap in time (see Henson et al. 1987). The threshold maximum close to the call CF_2 component and the minimum at the echo frequency can be seen as an adaptation to minimize the cochlear response to the call and to maximize the responses to the echo (Henson et al. 1987). In hipposiderid CF-FM bats, the calls are much shorter (4–7 msec). Consequently, there is less overlap between call and echo, and the variations in the CF_2 frequency emitted by individual bats are larger (*Hipposideros speoris*, 0.5% –2%; *Hipposideros bicolor*, 0.75%) than in *Rhinolophus rouxi* (0.2%) (Schuller 1980; Habersetzer, Schuller, and Neuweiler 1984). Maxima and minima in the CF_2 ranges of the neuronal audiograms of both hipposiderid species are less pronounced than in long CF-FM bats or may even be absent (Schuller 1980; Rübsamen, Neuweiler, and Sripathi 1988). Hipposiderid bats may be seen as intermediate between FM and long CF-FM bats. This is also reflected in the absolute size of their cochleae (see Fig. 5.1).

3.2 Frequency Tuning

Data on the tuning properies of IHCs and OHCs are not yet available for bats. In the guinea pig, the tuning curves of IHCs are matched quite accurately by the activity of single auditory nerve fibers (Evans 1972; Robertson and Manley 1974; Russell and Sellick 1978). Therefore, the tuning measured in single units in the auditory nerve or the cochlear nucleus of different bat species can be taken as a close indicator of cochlear tuning. Figure 5.3 shows neuronal tuning curves from the auditory periphery of two CF-FM bats, *Rhinolophus ferrumequinum* and *Pteronotus parnellii*. The area above each curve corresponds to frequency and level combinations that lead to excitatory responses of the individual neuron. The frequency at which a neuron is most sensitive to pure tone bursts is called the best frequency (BF). The BFs vary between about 10 and 100 kHz in *Rhinolo-*

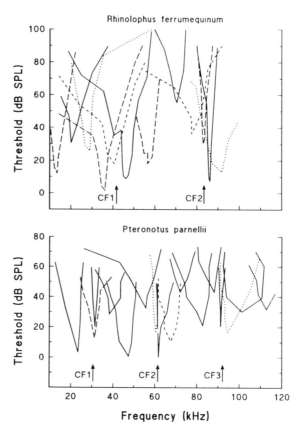

FIGURE 5.3. Tuning curves of single neurons in the cochlear nucleus of *Rhinolophus ferrumequinum* (*top*) (after Suga, Neuweiler, and Möller 1976) and *Pteronotus parnellii* (*bottom*) (after Kössl and Vater 1990b).

phus ferrumequinum and between about 10 and 115 kHz in *Pteronotus parnellii*. In neurons with BFs close to the CF_2 and the CF_3 frequency (only *Pteronotus*), the excitatory areas are narrower and the tips of the tuning curves are much sharper than for other BFs. The sharpness of tuning expressed as Q_{10dB} value (BF divided by the bandwidth of the tuning curve 10 dB above the threshold at the BF) varies from about 5–30 at non-CF frequencies to about 400 around the CF_2 frequency (Fig. 5.4) (Suga, Neuweiler, and Möller 1976). FM bats such as *Myotis lucifugus* have Q_{10dB} values below about 30 over their whole range of hearing (Suga 1964), and are comparable to nonecholocating mammals such as the cat and guinea pig in which maximum Q_{10dB} values of about 12 are found (Evans 1975: auditory nerve; Russell and Sellick 1978: inner hair cells). Hence, in CF-FM bats, cochlear processes that lead to sharp tuning go far beyond the scope seen in other mammals. The enhanced tuning around the CF frequencies guarantees the high-frequency resolution necessary to detect small fre-

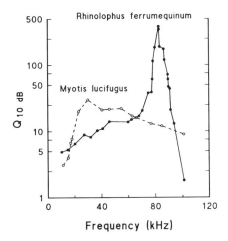

FIGURE 5.4. Q_{10dB} values of single neurons in the cochlear nucleus of *Rhinolophus ferrumequinum* and *Myotis lucifugus* (after Suga 1964; Suga, Neuweiler, and Möller 1976).

quency variations in the CF echoes caused by the wingbeat of prey insects, and is related to hydromechanical specializations in the cochlea (e.g., Neuweiler, Bruns, and Schuller 1980; Neuweiler 1990; see Section 5).

When stimulating CF-FM bats with tone bursts close to the CF ranges, pronounced neuronal on/off responses are observed (Grinnell 1973; Suga, Simmons, and Jen 1975; Suga, Neuweiler, and Möller 1976). The pure tone stimulus is only able to elicit a neuronal response at its onset or after offset (Fig. 5.5; Kössl and Vater 1990b). The on/off responses are not caused by neuronal inhibition and most probably are generated at the level of cochlear mechanics (Grinnell 1973; Suga, Simmons, and Jen 1975). They may be caused by the mechanisms that contribute to enhanced tuning and to the threshold insensitivity slightly below CF_2. Tuning curves of neurons with BFs just above the CF_2 frequency of *Pteronotus* not only show an on/off region on the low-frequency boundary but also show a shallow slope at the high-frequency side (Fig. 5.5). This is in contrast to normal mammalian tuning curves, and may arise from an increased longitudinal coupling of the basilar membrane in a cochlear region located basally to the cochlear place of the CF_2 frequency, where pronounced basilar membrane thickenings are present (see Sections 5.2 and 6.2.2). Similar tuning curves are also found in *Rhinolophus rouxi* (Metzner and Radtke-Schuller 1987).

3.3 Cochlear Emissions

The mammalian cochlea can emit sound at those frequencies at which the hearing sensitivity is high (Kemp 1978; Zwicker and Schloth 1984; review by Probst, Lonsbury-Martin, and Martin 1991). Spontaneous otoacoustic emissions (OAEs) are usually below the threshold of perception, but they can be measured as sharp spectral peaks with sensitive microphones placed in the outer ear. Evoked OAEs (EOAEs) appear as delayed oscillations

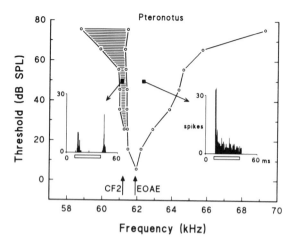

FIGURE 5.5. Tuning curve of a neuron in the cochlear nucleus of *Pteronotus parnellii* (from Kössl and Vater 1990b). Within the shaded area the neuron shows on/off temporal response properties. The arrows indicate the CF_2 frequency and the frequency of the evoked otoacoustic emission (EOAE).

coming from the ear after stimulation with click stimuli. Typically, each ear has multiple emissions at different frequencies. These emissions can also be measured as acoustic interference patterns in the frequency response obtained with pure tone sweeps. In humans, such "stimulus-frequency" OAEs saturate at stimulus levels of about 20–30 dB SPL. Active mechanical processes in the cochlea are thought to be involved in the generation of both spontaneous and evoked emissions.

In *Pteronotus*, each ear emits a single evoked OAE at about 62 kHz (Kössl and Vater 1985a). The stimulus-frequency OAE start to saturate at input levels above about 60 dB SPL (Fig. 5.6A). The maximum level of the evoked OAE is about 70 dB SPL, implying that the evoked emission is about 100 times stronger in this bat species than in other mammals. The evoked OAEs sometimes convert, for periods of a few days, to spontaneous OAEs (SOAEs) with a level up to 40 dB SPL (Fig. 5.6B). The spontaneous OAEs can be suppressed by additional sound stimuli of higher frequency (Fig. 5.6B). Prominent hydromechanical specializations found at and basal to the cochlear representation place of 62 kHz probably play a role in emission generation and enhanced tuning (see Sections 5.2 and 6.2.2). In *Rhinolophus rouxi*, another long CF-FM bat that was examined for otoacoustic emissions, only a few individuals produced weak stimulus-frequency OAEs. These were about 300 Hz above the CF_2 frequency of 78 kHz. Associated changes in the phase of the recorded signal were much broader than in *Pteronotus* (Kössl 1994), indicating a stronger damping of the underlying resonant mechanism (Henson, Schuller, and Vater 1985).

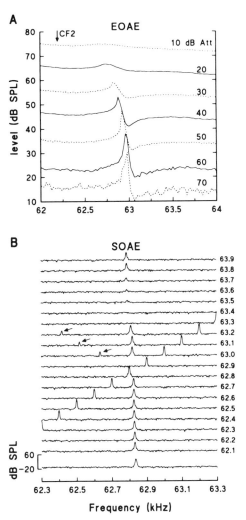

FIGURE 5.6A,B. (A) Evoked otacoustic emission (EOAE) measured with a micro-phone at the tympanum of *Pteronotus parnellii* (after Kössl 1994). A continuous pure tone is swept upward in frequency at different attenuations (indicated at each curve). At about 63 kHz, an outgoing emission interferes with the incoming stimulus, and maxima and minima are evident in the frequency response. (B) spontaneous otoacoustic emission (SOAE) in *Pteronotus* (after Kössl 1994). The lowest trace shows the SOAE without concomitant sound stimulation. The upper traces display the behavior of the SOAE during application of an additional pure tone stimulus of 40 dB SPL. The stimulus was moved across the range of the SOAE; its frequency is indicated to the left of each trace. *Arrows* indicate $2f_1$-f_2 distortion products produced by the SOAE and the stimulus. For stimulus frequencies between 63.3 and 63.7 kHz, the SOAE is suppressed.

3.4 Cochlear Resonance in Pteronotus parnellii

At the frequency of the evoked OAE, a pronounced maximum of the amplitude of cochlear microphonic potentials can be measured, and the CM threshold is at a minimum (Pollak, Henson, and Novick 1972; Henson, Schuller, and Vater 1985; Kössl and Vater 1985a). After stimulation with short tone bursts, the CM potentials show a strong long-lasting ringing that was attributed to a cochlear resonance (Suga, Simmons, and Jen 1975) and also could be measured as a delayed evoked OAE in the outer ear canal (Kössl and Vater 1985a). The cochlear resonance frequency (which equals the frequency of the OAE and the CM maximum) is about 400–900 Hz above the CF_2 frequency of *Pteronotus* (Kössl and Vater 1985a, 1990b; Kössl 1994). Neuronal thresholds in the cochlear nucleus reach minimum values at and slightly above the resonance frequency (Fig. 5.7). Maximum neuronal tuning sharpness, and hence highest Q_{10dB} values of about 400, are found for frequencies about 300 Hz below the resonance frequency (Fig. 5.7), that is, at the steep slope between maximum and minimum thresholds. This indicates that the cochlear mechanism that generates the evoked OAE is involved in creating sharp tuning close to the CF_2 frequency. It has to be pointed out, however, that sharply tuned neurons with Q_{10dB} values up to 300 are also found around 90 kHz, where there is no conspicuous OAE. Enhanced tuning is also found in the 30-kHz range where neuronal Q_{10dB} values can increase to about 70 (Suga and Jen 1977). In both cases, the

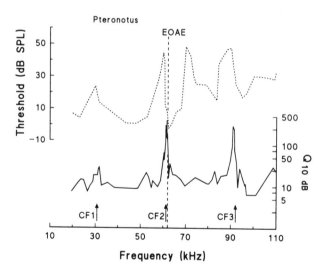

FIGURE 5.7. Lowest threshold *(dotted line)* and Q_{10dB} values *(solid line)* of single neurons in the cochlear nucleus of an individual of *Pteronotus* (from Kössl and Vater 1990b). The three harmonics of the constant-frequency component of the echolocation call are indicated by *arrows*. The frequency of the evoked otoacoustic emission (EOAE) is given by a *dashed line*.

maximum $Q_{10 dB}$ values are located at the slopes between threshold maxima and minima.

The frequency and amplitude of the evoked OAE and corresponding CM potentials are affected by anaesthesia and changes in body temperature (Kössl and Vater 1985a; Henson et al. 1990). The largest variations in the resonance frequency are about 500 Hz, which is less than 1% of the OAE frequency. The magnitude of temperature-induced frequency shifts of the resonance frequency (39 Hz/°C) and of associated neuronal BFs (33 Hz/°C) (Huffman and Henson 1991, 1993a,b) is about 0.06%/°C. The frequency shifts are comparable to changes of spontaneous OAEs in the humans during menstrual or diurnal cycles (Wit 1985, maximally 0.12%/°C; Wilson 1985, 0.1%/°C or less). In *Pteronotus*, the small Q_{10} value (resonance frequency/resonance frequency measured at a 10°C lower temperature, of about 1.006 suggests that the cochlear resonance depends on complex mechanical interactions that are not greatly influenced in frequency by enzymatic or metabolic reactions (Q_{10} range, 2–3).

3.5 Cochlear Distortions

In the past few years, the measurement of acoustic two-tone distortions from the cochlea has gained increasing popularity. During stimulation with two tones of different frequency (f_1, f_2), pronounced cubic and quadratic distortion products can be measured acoustically in the outer ear canal (Probst, Lonsbury-Martin, and Martin 1991). The occurrence of distortion products results from nonlinear cochlear mechanics involved in the ampli-fication of low-level sound stimuli. OHC motility is thought to contribute to the "cochlear amplifier". The distortion product otoacoustic emissions (DPOAEs) are a noninvasive indicator of hearing ability (Brown and Gaskill 1990; Gaskill and Brown 1990) and have the advantage that they appear over the whole range of hearing of an animal, in contrast to spontaneous and evoked OAEs, which are restricted in frequency. The DPOAEs are measured as distinct spectral peaks at both sides of the two primary stimuli (f_1, f_2) at the frequencies $(n + 1)f_1 - nf_2$ and $(n + 1)f_2 - nf_1$. The first lower sideband DPOAE at a frequency of $2f_1$-f_2 is most prominent.

In FM and CF-FM bats, the DPOAEs are measurable from 5–100 kHz up to the frequency limit of the spectrum analyzer used, and they are affected by salicylate (Kössl 1992a,b), which is known to block OHC motility (Dieler, Shehata-Dieler, and Brownell 1991). Fig. 5.8 shows two examples of distortion products from *Pteronotus parnellii* for primary frequencies around 20 and 61 kHz. In both examples the frequency separation between the two stimuli was adjusted so that maximum $2f_1$-f_2 distortion level could be measured by giving an optimum overlap of the two corresponding traveling waves. The optimum frequency separation, Δf, between the two primary stimuli is taken as an indirect measure of the relative mechanical

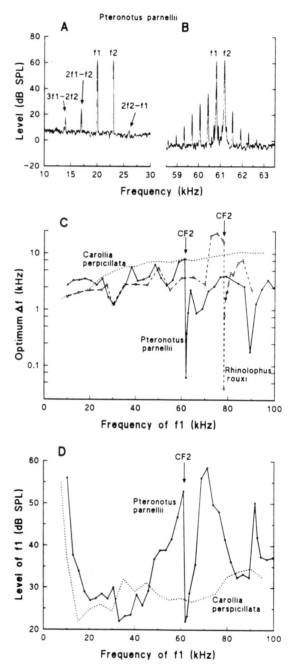

FIGURE 5.8A–D. Acoustic distortion products measured at the tympanum of different bat species. (A,B) The distortion products appear as distinct side peaks both on the low, and high-frequency side of the two stimuli (f_1, f_2; level of about 65 dB SPL). The stimulus frequencies were in the 20-kHz (A) and 60-kHz (B) range of

frequency separation in the cochlea (Kössl 1992b). In FM bats (Fig. 5.8C, *Carollia*), the optimum Δf continuously rises from about 2 to 10 kHz for f_1 frequencies between 10 and 90 kHz. In contrast, in the CF-FM bats *Pteronotus parnellii* and *Rhinolophus rouxi* there are sharp minima of the optimum Δf close to CF_2 and CF_3 (*Pteronotus* only; Fig. 5.8C). Minimum values of about 30–100 Hz are measured when f_1 exactly matched the frequency of the evoked OAE in *Pteronotus* (480 Hz above CF_2) and a frequency range about 300 Hz above CF_2 in *Rhinolophus*. The minima of the optimum Δf correlate well with neuronal tuning measurements (see Fig. 5.7). Additionally, the DPOAEs provide a quite accurate measure of the relative course of the hearing threshold. Figure 5.8D shows $2f_1$-f_2 threshold curves from *Carollia perspicillata* and *Pteronotus parnellii*. In both cases the curves run approximately parallel to the neuronal data (see Fig. 5.2) but are about 15–35 dB higher. In both *Pteronotus parnellii* and *Rhinolophus rouxi*, the CF_2 frequencies of the emitted call coincide with a maximum of the distortion threshold.

The distortion measurements indicate that in long CF-FM bats both the steep variations in threshold and the enhanced frequency tuning are implemented on the level of cochlear mechanics. The mechanical insensitivity at CF_2 may reduce the cochlear response to the emitted signal of the call to guarantee an undisturbed processing of the echoes during Doppler-effect compensation behavior.

4. Cochlear Frequency Maps

Single neuron recordings in the ascending auditory pathway of CF-FM bats show that a large proportion of the cells is tuned to a small frequency band around CF_2 (see Pollak and Park, Chapter 7). To investigate if this overrepresentation of CF_2 is caused by central neuronal connectivity or by cochlear specializations, it was necessary to obtain cochlear frequency maps.

An effective method to explore the frequency representation along the basoapical extension of the basilar membrane (BM) is the HRP frequency mapping technique. The neuronal tracer horseradish peroxidase (HRP) is injected intracellularly in auditory nerve fibers (Liberman 1982) or extra-

---

FIGURE 5.8A–D. (*continued*) *Pteronotus parnellii*. (C) The optimum frequency separation Δf between f_1 and f_2 to elicit maximum levels of the $2f_1$-f_2 distortion is plotted for different f_1 frequencies in *Carollia perspicillata*, *Pteronotus parnellii*, and *Rhinolophus rouxi*. (D) Threshold of the $2f_1$-f_2 distortion in *Carollia perspicillata* and *Pteronotus parnellii*. Plotted is the level of f_1 sufficient to produce a distortion of -10 dB SPL. For each f_1 frequency, the frequency of f_2 was adjusted to match the optimum Δf (adapted from Kössl 1992a,b, 1994).

cellularly close to the synaptic endings of the auditory nerve in the cochlear nucleus (Vater, Feng, and Betz 1985). The frequency tuning of the nerve fiber or the target neuron in the cochlear nucleus is determined before the injection. The tracer is transported toward the spiral ganglion cell body and also labels the afferent terminals at the IHCs. With multiple injections at different BFs, the frequency representation valid for IHCs can be obtained. HRP frequency maps have been published for *Rhinolophus rouxi* (Vater, Feng, and Betz 1985) and *Pteronotus parnellii* (Kössl and Vater 1985b; Zook and Leake 1989). In both CF-FM bat species, a narrow range around the CF_2 frequency is largely expanded in terms of BM length (Fig. 5.9A). The frequency ranges between about 76–83 kHz (*Rhinolophus*) and 59–66 kHz (*Pteronotus*) are expanded to about 30% of the BM length. In both species the maximum frequency expansion is found at CF_2 with about 40 mm BM/octave. In other parts of the cochlea, 2–3 mm of BM length are used for representation of one octave, which is similar to the frequency maps obtained in nonspecialized mammals such as the cat or rat (Fig. 5.9B). A frequency map of *Rhinolophus ferrumequinum*, obtained by analyzing morphological changes in OHCs after loud pure tone exposure (Bruns

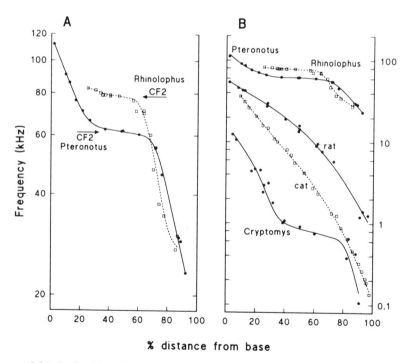

FIGURE 5.9A,B. Cochlear frequency maps of *Rhinolophus rouxi* (from Vater, Feng, and Betz 1985) and *Pteronotus parnellii* (from Kössl and Vater 1985b, 1990a) compared to maps from the rat (from Müller 1991), cat (from Liberman 1982), and mole rat *Cryptomys hottentottus* (from Müller et al. 1992).

1976b), shows a similar expansion of the CF_2 frequency range. However, this map is shifted toward the base (for discussion, see Vater 1988). The only other mammal with an expanded frequency representation of a narrow frequency range found so far is the African mole rat *Cryptomys hotten-tottus* (see Fig. 5.9B). In *Cryptomys*, frequencies between 0.6 and 1 kHz are expanded on about 50% of BM length (Müller et al. 1992), and the respective frequency correlates with a minimum in hearing threshold (Müller and Burda 1989). The sharp threshold minima of *Rhinolophus* and *Pteronotus* are within the range of the expanded cochlear region.

As a possible consequence of cochlear frequency expansion, high-frequency resolution and sharp tuning around the CF_2 frequency could emerge. However, the expanded frequency range extends a few kilohertz both apically and basally of CF_2 and reaches into frequency bands where neuronal tuning is poor. Enhanced cochlear tuning seems to be restricted to a small part of the expanded region. Moreover, in *Pteronotus* there is no conspicuous cochlear frequency expansion around 90 kHz, where also very sharp neuronal tuning and a small optimum frequency separation during distortion measurement are found. This leads to the conclusion that cochlear frequency expansion and enhanced tuning are not necessarily causally linked. Of course, an expanded cochlear region should be well suited to resolve sharp mechanical tuning provided from a different source.

5. Cochlear Anatomy

5.1 General Morphological Features

A schematic illustration of the gross anatomy of the bat cochlea (Fig. 5.10) serves as a reference for further detailed descriptions in later chapters. The basic structural composition of the cochlear duct is identical to the typical mammalian scheme (e.g., Lim 1986), but the relative dimensions of components and the organization of the anchoring system of the BM reflect adaptations for high-frequency hearing (Henson 1970; Firbas 1972; Bruns 1980; Vater, Lenoir, and Pujol 1992).

The receptor cells are organized into one row of IHCs and three rows of OHCs. The dimensions of OHC bodies and stereocilia are small as compared to mammals with good low-frequency hearing (guinea pig, cat, human; see Section 5.4.3.) The IHCs are situated medial to the pillar cells (PC) on the primary osseous spiral lamina (OSL) and therefore are not directly displaced by BM motion (e.g., Bruns 1980). The hair cell stereocilia are in close contact with the tectorial membrane (TM), which varies in shape in different regions of the cochlea (see also Section 5.3). The afferent innervation of hair cells is derived from the peripheral processes of spiral ganglion (SG) cells (see also Section 5.4.5). In addition to the sensory cells, the organ of Corti contains different types of well-developed supporting

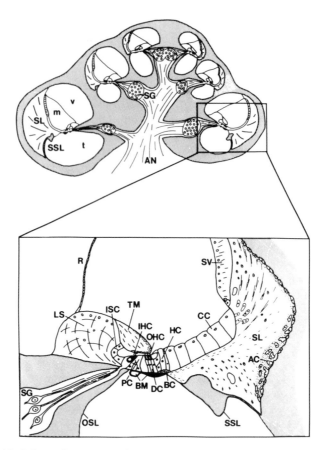

Figure 5.10. Schematic cross sections through the cochlea of *Rhinolophus rouxi* illustrate the composite structures of the organ of Corti in the second half-turn in further detail. Fluid spaces, *v*: scala vestibuli, *m*: scala media, *t*: Scala tympani are marked only in the first basal half-turn. AC, anchoring cells; AN, auditory nerve; BM, basilar membrane; BC, Boettcher cells; CC, Claudius cells; DC, Deiters cells; HC, Hensens cells; IHC, inner hair cell; ISC, inner sulcus cells; LS, spiral limbus; OC, otic capsule; OHC, outer hair cell; OSL, osseous spiral lamina; PC, pillar cells; R, Reissners membrane; SG, spiral ganglion; SL, spiral ligament; SSL, secondary spiral lamina; SV, stria vascularis; TM, tectorial membrane. (See text.)

cells. Sturdy pillar cells (PC) border the tunnel of Corti, and large Deiters cells (DC) provide the structural link between OHCs and the BM. Hensens cells (HC), Böttcher cells (BC), and the Claudius cells (CC) are situated lateral to the sensory epithelium (Henson, Jenkins, and Henson 1982, 1983).

The medial attachment of the BM is formed by the OSL and the lateral attachment by the spiral ligament (SL). Cross sections of the BM reveal two thickened portions (Figs. 5.10 and 5.11): pars arcuata (PA) situated beneath the pillar cells and pars pectinata (PP) lateral to the pillar cells. These thickenings are most prominent in the basal turn but can be found throughout most of the cochlear spiral except for the extreme apex. Such thickenings are also present in the basal turn of nonecholocating mammals with good high-frequency hearing and non-CF-FM bats, but the structure of the PP in the cochlear base of CF-FM bats is qualitatively different (see Section 5.2).

The spiral ligament is considerably enlarged in the basal turns and is composed of a matrix containing a network of extracellular fibers and

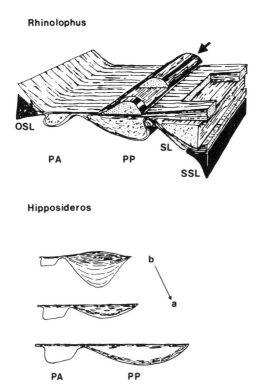

FIGURE 5.11. Structural organization of the basilar membrane in the basal turn of *Rhinolophus ferrumequinum* (*top*; adapted from Bruns 1980) and of *Hipposideros bicolor* (*bottom*; adapted from Dannhof and Bruns 1991). In *Rhinolophus*, the thickening of the scala tympani side of the pars pectinata (PP) contains radially directed fibers whereas the thickening at the scala vestibuli side is composed of longitudinally directed fibers. The thickening of the PP in Hipposideros (shown for different basoapical positions: 0.4 mm, 4 mm, 7 mm) only contain radially organized filaments. OSL, osseous spiral lamina; PA, pars arcuata; SL, spiral ligament; SSL, secondary spiral lamma.

several cell types (see Fig. 5.10). The stress fibers of the ligament and the anchoring cells (AC) along the bony otic capsule (OC) may be involved in creation of radial tension on the BM (Henson, Henson, and Jenkins 1984; Henson and Henson 1988). As a unique feature in *Pteronotus parnellii*, a local enlargement of the spiral ligament is found in the middle of the basal turn at the transition between the sparsely innervated and densely inner-vated cochlea region (Fig. 5.12) (Henson 1978; Kössl and Vater 1985b).

At the scala tympani side of the SL, a bony secondary spiral lamina (SSL) is present throughout most of the cochlea of bats whereas in other mammals it is confined to the basal turns (Firbas 1972). In some CF-FM bats (*Rhinolophus, Hipposideros*), the SSL has a specialized shape (Figs. 5.10, 5.11) that changes systematically along the cochlear spiral (Bruns 1976a; Dannhof and Bruns 1991).

The size of the perilymphatic spaces — scala vestibuli and scala tympani — decreases rather regularly and symmetrically from base to apex in the cochlea of *Rhinolophus* (Bruns 1976a) and *Myotis* (Ramprashad et al. 1979). In the mustached bat (*Pteronotus parnellii*) and the frog-eating bat (*Trachops cirrhosus*), however, deviations from this pattern were noted. In *Pteronotus parnellii*, the scala vestibuli of the basal turn is subdivided into two large chambers by a bony indentation that creates a focal narrowing (Henson, Henson, and Goldman 1977; Kössl and Vater 1985b). The volume of the scala tympani does not change in parallel; rather, it reaches its

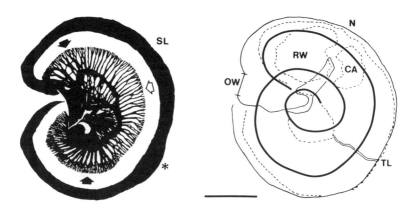

FIGURE 5.12. *Left*: Whole-mount preparations of the basal turn of the cochlea of *Pteronotus* illustrates local changes of nerve fiber densities (*black arrows*, location of maxima in innervation density; *white arrows*, sparsely innervated region) and in size of the spiral ligament (SL; *asterisk*). *Right*: Horizontal projection of the cochlea shows the course of the basilar membrane for all turns (*thick line*) and the size of fluid spaces of the basal turn (scala vestibuli; *thin lines*; scala tympani; *stippled lines*; N, narrowing of scala vestibuli), the location of the oval window (OW), the enlarged size of the round window (RW), and the cochlear aqueduct (CA) and the location of the apical border of the "thick lining" (TL). (Adapted from Henson and Henson 1991; Kössl and Vater 1985b.) Calibration bar, 1 mm.

maximal size beneath the narrowest point of scala vestibuli. The round window and the cochlear aquaeduct are located in this region and are much larger in size than in other species (Henson 1970). A further unique feature of the scala tympani in the basal turn of *Pteronotus* is the presence of a "thick lining" composed of cells associated with thick extracellular osmiophilic substance (Jenkins, Henson, and Henson 1983). The functions of these specializations are unknown. In the frog-eating bat *Trachops cirrhosus*, a species that appears to have exceptionally good low-frequency hearing (Bruns, Burda, and Ryan 1989), the volume of the scala tympani in the apical 50% of the cochlea is remarkably smaller than that of the scala vestibuli. Such asymmetries in scala volume are said to be typical for small mammals with good low-frequency hearing (Müller et al. 1992).

5.2 Basilar Membrane

The BM performs the first important steps in cochlear frequency analysis. Systematic increases in width and decreases in thickness from base to apex produce a stiffness gradient that creates a regular frequency representation along the membrane: high-frequency signals maximally displace basal cochlea regions where stiffness is high, and low frequencies are mapped progressively more apically following the decrease in stiffness (von Békésy 1960).

In general, the BM of Microchiroptera is significantly longer than that of nonecholocating mammals if relative body weight is taken into account (Table 5.1). The longest BMs are found in CF-FM bats. This feature is neither related to a widening of absolute hearing range nor typical for particular taxonomic groups, but results from the presence of an acoustic fovea (see also Sections 2 and 4). Furthermore, the BM of Microchiroptera is narrower and thicker than in most nonecholocating mammals, particularly those with good (cat, guinea pig) or predominantly (mole, rats: *Spalax, Cryptomys*) low-frequency hearing (Table 5.1). The highest values of BM thickenings are found in the basal cochlear regions of CF-FM bats. Although the thickening of the PP in the basal turn only contains radially directed filaments in *Hipposideros* (Dannhof and Bruns 1991) (see Fig. 5.11), the morphology of PP in bats emitting long CF-FM calls differs from the common mammalian pattern. In *Rhinolophus* (Bruns 1980) (see Fig. 5.11) and *Pteronotus* (Vater, unpublished data), the PP is composed of two parts: the part facing the scala tympani contains radially directed filaments, whereas the part facing the scala media is composed of longitudinally directed filaments. The latter feature suggests the presence of a longitudinal mechanical coupling within the membrane and has been suggested to create shallow high-frequency slopes in neurons with BFs slightly above CF_2 (Kössl and Vater 1990b; see also Figs. 5.3 and 5.5). Because the thickenings of PA and PP are not continuous but are linked via a thin segment, it is questionable if they can be viewed as structures simply increasing BM

TABLE 5.1. Basilar membrane (BM) dimensions in various mammalian species.

Species	Body weight	BM length (mm)	BM thickness (μm)		BM width (μm)		Reference
			Base	Apex	Base	Apex	
Rhinolophus ferrumequinum (horseshoe bat)	18 g	16	35	2	90	150	Bruns 1976a
Pteronotus parnellii (mustached bat)	12 g	14.3	22	2	50	110	Henson 1978; Kössl and Vater 1985b
Hipposideros fulvus	10 g	8.8	28	2	60	100	Kraus 1983
Hipposideros speoris	10 g	9.2	23	2	70	90	Dannhof and Bruns 1991
Molossus ater	37 g	14.6	10	1	60	130	Fiedler 1983
Rhinopoma hardwickii		11.2	14	2	70	150	Kraus 1983
Taphozous kachensis		14.4	12	2	60	140	Fiedler 1983
Myotis lucifugus	8 g	6.9	10	2	60	115	Ramprashad et al. 1979
Megaderma lyra (false vampire)	48 g	9.9	14	2	80	140	Fiedler 1983
Trachops cirrhosus (frog-eating bat)		14.9	19	2	60	160	Bruns, Burda, and Ryan 1989
Mus musculus (mouse)	40 g	6.8	15	1	100	170	Ehret and Frankenreiter 1977
Rattus norwegicus (rat)	400 g	10.4	19	8	140	240	Roth and Bruns 1992
Spalax ehrenbergi (mole rat)		13.7			110	200	Bruns et al. 1988
Cryptomys hottentottus (mole rat)	80 g	12.9	5	1.5	130	175	Müller et al. 1992
Cavia porcella (guinea pig)	400 g	18	7.4	1.3	100	245	Fernandez 1952
Felis cattus (cat)	3 kg	23.6	12	5	105	420	Cabuzedo 1978
Homo sapiens (human)	60 kg	32			104	504	Nadol 1988
Bos bovis (cattle)	500 kg	38					von Békésy 1960
Elephas max (elephant)	4000 kg	60					von Békésy 1960

stiffness. Rather, this arrangement might allow for relatively independent motion of inner and outer BM segments (Steele 1976; Ehret and Frankenreiter 1977; Bruns 1980).

The basoapical gradients in BM dimensions of different bat species are illustrated in Figure 5.13 together with measurements obtained in the cat, an auditory generalist with good low- and high-frequency hearing, and the African mole rat *Cryptomys*, specialized on low-frequency hearing. The gradients are clearly species specific, but common features are noted among those species sharing certain hearing characteristics.

In FM bats with unspecialized tuning properties (*Myotis lucifugus*, Ramprashad et al. 1979; *Megaderma lyra*, Fiedler 1983; see Fig. 5.13), both the width and the thickness of the basilar membrane change gradually toward the apex. Such regular patterns are also seen in most nonecholocating species (cat [Fig. 5.13], rat, guinea pig, mouse), but the absolute gradients in BM width can be considerably larger than in bats (see also Table 5.1). These features are expected to correlate with differences in absolute hearing range and the species-specific course of frequency maps (see also Fig. 5.9).

In *Trachops cirrhosus* (Fig. 5.13), the frog-eating bat, which emits FM calls but largely depends on passive listening to frequencies below 5 kHz for catching its prey, a conspicuous BM thickening in the basal turn is followed by a significant decrease in BM thickness to about 60% distance from base. From there up to the apex, thickness is less (about 2 μm). The BM width gradually increases from base to apex with the greatest basoapical differences found in small mammals. Accordingly, *Trachops* is expected to have the largest frequency range among the species studied to date, and might have the best low-frequency hearing among bats (Bruns, Burda, and Ryan 1989).

The basoapical gradients in BM morphology of CF-FM bats deviate considerably from the patterns observed in other mammals. This feature appears related to specializations of the frequency map and tuning properties of single units. Because the observed patterns differ among species of CF-FM bats, they are described individually.

Rhinolophus: In *Rhinolophus* (Fig. 5.13; Bruns 1976a), the BM in the basal turn is considerably thickened up to a point located at about 25% distance from the base, where the thickness abruptly drops from 35 μm to 10 μm within only a few hundred micrometers. From this point to about 55% distance from the base, BM thickness remains almost constant. The pattern of change in BM width found up to about 55% distance from the base is also highly unusual. Coinciding with the transition in BM thickness, the BM width decreases and then remains almost constant up to about 50% distance from the base. A systematic increase in BM width and a regular decrease in BM thickness, typical for a nonspecialized cochlea, is found only in more apical regions.

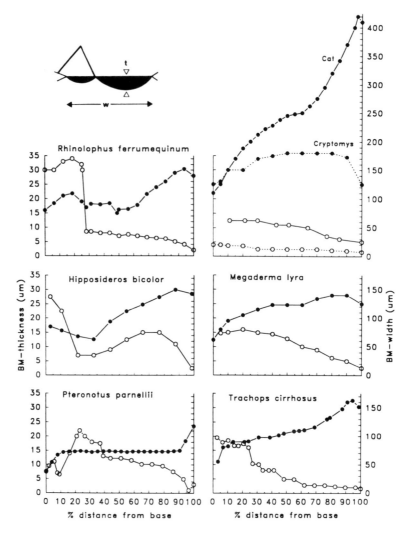

Figure 5.13. Basoapical gradients in BM width (*solid symbols*) and BM thickness (*open symbols*) represented on normalized BM length in different species. *Left column*: CF-FM bats (*Rhinolophus ferrumequinum*, after Bruns 1976a; *Hipposideros bicolor*, after Dannhof and Bruns 1991; *Pteronotus parnellii*, BM thickness after Kössl and Vater 1985b and BM width after Henson 1978). *Right column*: Nonecholocating mammals and FM bats (cat after Cabezudo 1978; *Cryptomys* after Müller et al 1992; *Megaderma lyra* after Fiedler 1983; *Trachops cirrhosus* after Bruns, Burda, and Ryan 1989).

Hipposideros: Some of the specialized features observed in *Rhinolophus* are also present in the cochlea of the related *Hipposideros* (Dannhof and Bruns 1991; see Fig. 5.13). The changes in BM width and thickness in the basal turn appear less abrupt than in *Rhinolophus*, but this feature probably results from a rather wide spacing of the measurement points. In contrast to other bats, *Hipposideros* possesses a second broad maximum in BM thickness located in the apical cochlea.

Pteronotus: In *Pteronotus* (Fig. 5.13; Kössl and Vater 1985b), low values of BM thickness are measured within the first 10% of cochlear length. These are followed by an apically directed increase in thickness between 10% and 23% of BM length. The area of increased BM thickness is apically terminated by an abrupt decrease of BM thickness at about 40% distance from the base. More apically, the decrease in BM thickness is very slight to about 80% distance from base. Contrasting with all other species, the BM width in *Pteronotus* appears to be constant throughout the cochlea, except for the very basal and apical portions (Henson 1978).

Although the exact gradients in BM morphology clearly differ among CF-FM bats, some common features are noted. First, in both *Rhinolophus* and *Pteronotus*, an abrupt decrease in BM thickness occurs just basal to the place of representation of the second-harmonic CF signal component. The functional role of these discontinuities is not fully understood, but they might play a role in enhancing tuning in a narrow frequency band (see Section 6.2.2). Second, areas of very little change in BM morphology, found just apical to the discontinuity, in both species include the expanded representation of the second harmonic CF component and coincide with a local maximum in innervation density (see also Section 5.4.5). In these regions, the stiffness gradient is probably very slight, thus leading to an expanded frequency mapping. Interestingly, the only other mammal in which expanded frequency mapping is found is the African mole rat (see Fig. 5.9). In the mole rat, long stretches of BM encompassing the acoustic fovea also have almost constant BM morphology (Fig. 5.13; Müller et al. 1992).

Thus, the presence of expanded mapping correlates with the presence of cochlear regions with almost constant BM stiffness. However, frequency expansion and enhanced tuning do not appear directly related. In *Pteronotus*, the representation places of CF_1 and CF_3 showing enhanced tuning do not coincide with BM specializations compatible to the CF_2 region. CF_1 is represented at about 90% distance from the base where a regular decrease in BM thickness occurs while CF_3 corresponds to a minimum in BM thickness at about 10% distance from base. The inverse pattern of BM thickness change within the most basal 20% of BM length in *Pteronotus* has been suggested to facilitate reverse traveling wave propagation, thus contributing to the exceptionally loud otoacoustic emissions in this species (see Section 6.2.2).

5.3 Tectorial Membrane

The receptor cell stereocilia are sheared by relative movement between the organ of Corti and the TM. The TM is either considered as a rigid beam providing stiffness and mass for shearing displacement or as a second resonator superimposed on the BM resonance (Zwislocki and Kletsky 1979; Zwislocki 1986). In most mammals, the TM increases in size from base to apex (e.g., Lim 1986). This pattern is also seen in *Hipposideros* (Dannhof and Bruns 1991) and *Trachops* (Bruns, Burda, and Ryan 1989) in which the cross-sectional area of the TM increases from base to apex with the largest changes observed in the upper half of the cochlea (Fig. 5.14). *Pteronotus* is most unusual because the cross-sectional area of the TM, and consequently its mass in the basal turn, is about 3.5-fold larger than in other species, and a distinct maximum of cross-sectional area is found at about 55% cochlear length. The basoapical changes in height of the spiral limbus reflect the pattern observed in TM dimensions. The peculiar TM morphology is thought to have substantial effects on hair cell stimulation in specific frequency bands (Henson and Henson 1991).

While it is generally accepted that OHC stereocilia are firmly embedded in the subsurface of the TM, the mode of linkage of IHC stereocilia is disputed: in the chinchilla, IHC stereocilia appear not embedded at all (Lim

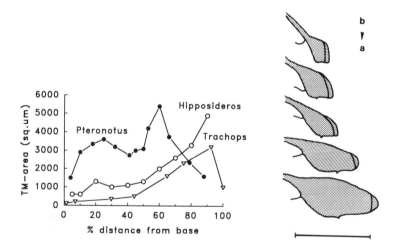

FIGURE 5.14. *Left:* Basoapical gradients in cross-sectional area of the TM in the CF-FM bats *Pteronotus parnellii* (after Henson and Henson 1991) and *Hipposideros bicolor* (after Dannhof and Bruns 1991), and the frog-eating bat, *Trachops cirrhosus* (after Bruns, Burda, and Ryan 1989). *Right:* Shape of the TM in cross-sections of the cochlea of *Hipposideros bicolor* obtained at different basoapical positions (from top to bottom at 0, 2, 4, 6, and 8 mm distance from base; after Dannhof and Bruns 1991). Note the prominent Hensen stripe at the subsurface of the TM. Calibration bar, 100 µm.

1986) whereas in the rat, imprints of IHC stereocilia in the subsurface of the TM are confined to the basal cochlea (Lenoir, Puel, and Pujol 1987). In the horseshoe bat *Rhinolophus rouxi*, imprints of IHC and OHC stereocilia have been observed throughout the cochlea (Vater and Lenoir 1992), and it was speculated that this feature is relevant for IHC stimulation in the high-frequency range of the mammalian audiogram. In all bat species studied so far, a prominent Hensen stripe is present on the subsurface of the TM above the region of the IHCs throughout most of the cochlea (Dannhof and Bruns 1991; Vater, Lenoir, and Pujol 1992). This structure may be relevant for IHC excitation at high frequencies, because in other mammals it is usually confined to basal cochlear regions (rat; Lenoir, Puel, and Pujol 1987).

5.4 Receptor Cells

5.4.1 Receptor Cell Arrangements

The structural organization of the receptor cells in the bat cochlea basically conforms to the patterns seen in most nonecholocating mammals: one single row of IHCs is located medial to three rows of OHCs. In the horseshoe bat and *Pteronotus*, the OHC arrangements are highly regular throughout the cochlea, and hair cell loss or additional hair cell rows typical for the rodent cochlea were not observed (Vater and Lenoir 1992; Vater, unpublished data). Such irregularities are, however, present in the apical cochlea of *Tadarida brasiliensis* and *Eptesicus fuscus* (Vater, unpublished data).

IHCs: Commonly, the elongated cuticular plates of IHCs form a tightly spaced single row in the longitudinal course of the organ of Corti (Lim 1986). A deviation from this bauplan was observed only in *Rhinolophus* (Bruns and Goldbach 1980; Vater and Lenoir 1992): in the lower basal turn, where the BM is considerably thickened, the spacing between neighboring IHCs is considerably enlarged because the size of cuticular plates and the number of stereocilia decrease. The transition from normal to specialized spacing coincides with the transition in BM dimensions (Vater and Lenoir 1992).

OHCs: Bat OHCs possess the typical W-shaped stereocilia bundles common to all mammals. As in most small mammals, three rows of stereocilia are present. The opening angle of the stereocilia bundles of bat OHCs is enlarged compared with other mammals (Vater and Lenoir 1992). This enlargement is paralleled by an increase in number of stereocilia per row and might produce a reinforced mechanical attachment of the TM relevant for high-frequency hearing.

Significantly, in *Rhinolophus* the transition in the morphology of IHC stereocilia bundles that parallels the change in BM morphology is not accompanied by a change in OHC organization.

5.4.2 Receptor Cell Ultrastructure

The ultrastructure of the receptor cell bodies of *Rhinolophus* is shown in Fig. 5.15 (Vater, Lenoir, and Pujol 1992). The salient cytological features closely conform to the typical mammalian pattern (Lim 1986). The IHC have a pearshaped body with an elongated neck portion carrying the cuticular plate. They are completely surrounded by supporting cells. An irregular arrangement of subsurface cisternae, specialized parts of the endoplasmic reticulum, is located along the lateral cell wall. The base of the IHC is contacted by numerous afferent endings of the dendrites of type I spiral ganglion cells. Each ending is opposed by a presynaptic bar or vesicle in the IHC. Efferent endings mainly synapse on the afferent dendrites.

The OHCs are cylindrical. Their cuticular plates are linked to the processes of supporting cells to form a continuous reticular lamina. The basal part of the OHC is attached to a specialized rigid cup formation of the Deiters cell. The lateral cell wall between the reticular lamina and the

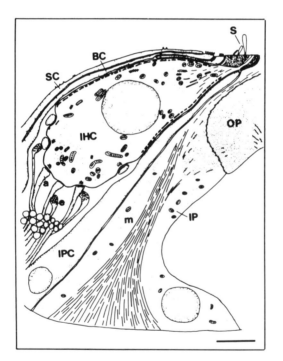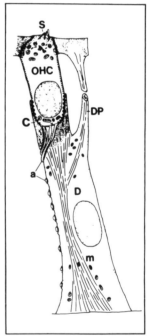

FIGURE 5.15. Ultrastructural organization of the organ of Corti of the horseshoe bat at the level of the IHC (left) and the level of the middle row of OHCs (right) (after Vater, Lenoir, and Pujol 1992). a, afferent endings; e; efferent endings; D; Deiters cell body; SC, inner sulcus cell; C, Deiters cup; DP, Deiters cell process; IHC, inner hair cell; IP, inner pillar cell; IPC, inner phalangeal cell; ISB; inner spiral bundle; m, microtubuli bundle; OP, head of outer pillar cell; S, stereocilia; ssc, subsurface cisternae. Calibration bar, 5 μm.

Deiters cup is surrounded by the fluid spaces of Nuel. A highly regular cytoskeleton is located along the lateral cell wall, composed of vesicle-shaped subsurface cisternae attached to the cell membrane via electron-dense pillars. Such a system is expected to play an integral role in mammalian OHC function: it forms a cytoskeletal spring maintaining the shape of the cylindrical cell, which is important for passive as well as active motion. Transmembrane motorproteins likely to underly the fast force generation mechanism in OHCs of other mammals (Kalinec et al. 1992) remain to be demonstrated in bats. The innervation of OHCs in the horseshoe bat is unusual because it consists of afferent endings (dendrites of type II ganglion cells) only (Bruns and Schmieszek 1980; Bishop and Henson 1988; Vater, Lenoir, and Pujol 1992; and see also Section 5.4.5).

5.4.3 Receptor Cell Dimensions

In mammals, apical OHCs and their stereocilia are much longer than basal ones (Spoendlin 1966). The absolute dimensions and gradients in OHC length and stereocilia size are related to the frequency representation of the organ of Corti: the gradations of stereocilia length along the cochlea follow a tonotopic order (Strelioff, Flock, and Minser 1985), and there is evidence that sound-induced motility of the OHCs is tuned and correlated with cell length (Brundin, Flock, and Canlon 1989); the smaller the cell, the higher its BF.

The dimensions of the receptor cells and their stereocilia in bats, in particular those of the OHCs, are considerably smaller than in other mammals (Dannhof and Bruns 1991; Vater and Lenoir 1992; Vater, Lenoir, and Pujol 1992). Figure 5.16 shows that the maximal length of OHCs in the apical cochlea of *Hipposideros* (Dannhof and Bruns 1991) is lower than the minimal length measured in the guinea pig cochlea (Pujol et al. 1992). In fact, it appears to represent a basally directed continuation of the gradient seen in the other species. The short size of OHCs in the basal turn of *Hipposideros* correlates with the expansion of its high-frequency limit of hearing to values of at least 160 kHz as compared to about 40 kHz in the guinea pig (Heffner, Heffner, and Masterton 1971). Because OHC and stereocilia size (Fig. 5.17) in the apex of the bat cochlea are much smaller than those observed in other mammals, it might be expected that the low-frequency limit of hearing in the bat is located at higher frequencies.

The OHC stereocilia in bats show a clear trend toward miniaturization (see Fig. 5.17). The comparison with the human cochlea (Wright 1984) again shows that the gradients observed in bats appear as a basally directed continuation of the gradients seen in other species. *Rhinolophus* (Vater and Lenoir 1992), however, possesses smaller OHC stereocilia in the apical cochlea than *Hipposideros* (Dannhof and Bruns 1991). As physiological data are few, we can only speculate that the lower frequency limit of hearing and/or the representation of low frequencies might differ between the two bats.

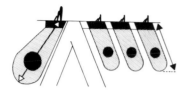

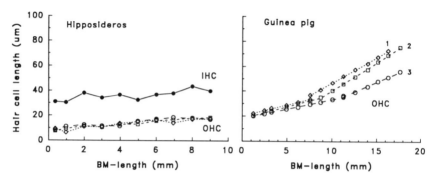

FIGURE 5.16. Basoapical gradients in the length of the cell bodies of IHCs and OHCs in *Hipposideros bicolor* (after Dannhof and Bruns 1991) and OHCs in the guinea pig (after Pujol et al. 1992). Different symbols and numbers denote measurements for innermost (first), middle (second), and outermost (third) row of OHCs.

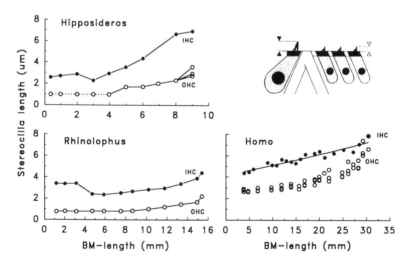

FIGURE 5.17. Basoapical gradients in sterocilia length for IHCs (*solid symbols*) and OHCs (*open symbols*) in the cochlea of CF-FM bats (*Hipposideros bicolor*, after Dannhof and Bruns 1991; *Rhinolophus rouxi*, after Vater and Lenoir 1992) and the human (after Wright 1984).

5.4.4 Receptor Cell Densities

The densities of OHCs in the cochlea of bats fall in the range reported for other mammals, but the density of IHCs appears slightly higher (Burda, Fiedler, and Bruns 1988). In particular, *Rhinolophus* has the highest density of IHC/mm of all bats studied (Bruns and Schmieszek 1980; Fig. 5.18°). Independent of the echolocation calls used, a trend for increasing OHC density towards the cochlear apex is observed while density of IHCs stays almost constant (Fig. 5.18).

5.4.5 Receptor Cell Innervation

Afferent innervation. The afferent innervation patterns of IHCs and OHCs have only been quantified in the horseshoe bat (Bruns and Schmieszek 1980) and are schematically illustrated in Figure 5.18. As in other mammals (Spoendlin 1973), the bulk of afferents contact the IHCs via highly convergent radially directed dendrites of type I spiral ganglion cells. Depending on cochlear location, between 80% and 90% of the afferent fibers terminate at the IHCs (Bruns and Schmieszek 1980). A small percentage of afferents contact the OHCs via outer spiral fibers that are probably derived from type II ganglion cells (Vater, Lenoir, and Pujol 1992; Zook and Leake 1989) coursing toward the cochlear base and making synaptic contacts with several sensory cells. The travel distance varies among turns (Bruns and Schmieszek 1980). As compared to other mammals in which only about 5% of the total afferents contact OHCs (Spoendlin 1973), the proportion of the afferent innervation of OHCs of the horseshoe bat is two- to fourfold higher.

Quantitative measurements of densities of spiral ganglion cells are available for several species (see Fig. 5.18), giving clues on variations of receptor cell innervation along the cochlea duct. However, only the study of Zook and Leake (1989) gives direct information on innervation densities of IHCs, because only type 1 ganglion cells were counted. In all species, distinct maxima and minima in innervation density are found. In *Pteronotus* (Henson 1973; Zook and Leake 1989), two maxima are located in the basal turn, separated by a sparsely innervated zone (see also Fig. 5.12). These innervation maxima correspond to the representation places of the second and third harmonic of the orientation call (Kössl and Vater 1985b; Zook and Leake 1989; also see Section 4). The sparsely innervated zone corresponds to the region of thickened BM. In *Rhinolophus* (Bruns and Schmieszek 1980), absolute densities of spiral ganglion neurons are much lower than in *Pteronotus*. The region of thickened BM is sparsely innervated, and the maximum of innervation density occurs in the upper basal turn, where according to Vater, Feng, and Betz (1985) the second-harmonic CF is represented. In *Taphozous* (Burda, Fiedler, and Bruns 1988), maxima and minima in innervation density are less pronounced than in CF-FM bats.

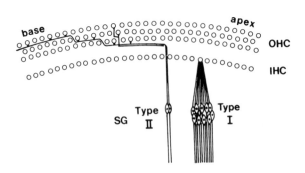

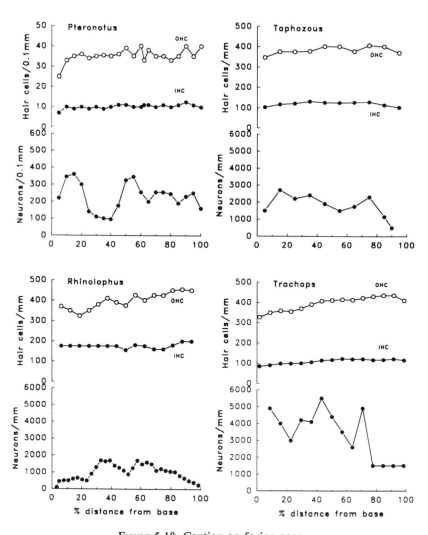

FIGURE 5.18. Caption on facing page.

The highest densities of spiral ganglion cells are found in the frog-eating bat *Trachops* (Bruns, Burda, and Ryan 1989).

The ratio of afferent fibers to IHCs in the region of innervation maxima amounts to 34–37:1 in *Pteronotus* (Zook and Leake 1989) and to 23.5:1 in *Rhinolophus* (Bruns and Schmieszek 1980). These values are in the range of maximal innervation densities reported for the cat at places in the cochlea representing frequencies above 2 kHz (Liberman, Dodds, and Pierce 1990). In the sparsely innervated zones of the cochlea of *Pteronotus* and *Rhinolophus*, only 8–12 afferent fibers contact IHCs. These values agree with the lowest values observed in the apex and extreme hook region of the cat cochlea (Liberman, Dodds, and Pierce 1990; Keithley and Schreiber 1986). In the FM bat *Myotis*, a ratio of 70:1 was calculated (Ramprashad et al. 1978), which is much higher than in CF-FM bats or other mammals. The exact numbers of afferent endings per OHC have not been quantified throughout the cochlear length in bats. Nevertheless, it is of interest to note that ratios of 4:1 in the basal turn of *Rhinolophus* (Vater and Lenoir 1992) and about 6:1 in the basal turn of *Pteronotus* (Bishop and Henson 1988) closely agree with values measured in basal cochlea regions of the cat (Liberman, Dodds, and Pierce 1990).

Efferent innervation. The efferent cochlear system in mammals is organized into two subsystems. The lateral olivocochlear system arises from small fusiform neurons located within or around the lateral superior olive (LSO) and synapses predominantly in the ipsilateral cochlea on afferent dendrites of type I ganglion cells contacting the IHCs. The medial olivocochlear system arises from large multipolar neurons located in the vicinity of the medial superior olive (MSO) and directly contacts OHC bodies of both cochleae with a distinct bias for the contralateral side (for review see: Warr 1992). Despite decades of intense research, the exact functions of the cochlear efferents remain unclear. It has been proposed that the lateral olivocochlear system exerts a tonic influence on primary auditory afferents (Liberman 1990). The medial olivocochlear efferents decrease cochlear sensitivity through a mechanism involving mechanical changes in the OHCs (Brown and Nuttal 1984) and might serve to protect the cochlea from sound-induced trauma (Rajan and Johnstone 1988), improve signal detectability in noise (Liberman 1988), or are involved in homeostasis (Johnstone, Patuzzi, and Yates 1986) by setting

←
___

FIGURE 5.18. *Top*: Schematic representation of basic innervation pattern of the organ of Corti in the horseshoe bat (*Rhinolophus ferrumequinum*, after Bruns and Schmieszek 1980). *Bottom*: Basoapical gradients of hair cell densities (upper graphs) and corresponding densities of spiral ganglion cells (lower graphs) in different bat species: *Pteronotus parnellii* (after Zook and Leake 1989); *Rhinolophus ferrumequinum* (after Bruns and Schmieszek 1980); *Taphozous kachhensis* (after Burda, Fiedler, and Bruns 1988); and *Trachops cirrhosus* (after Bruns, Burda, and Ryan 1989).

the operating point of OHCs compensating for a bias in BM position introduced by fluctuations of endolymphatic pressure.

The origin of olivocochlear fibers has been studied in CF-FM bats (*Pteronotus*, Bishop and Henson 1987, 1988; *Rhinolophus*, Aschoff and Ostwald 1987) and non-CF-FM bats (*Rhinopoma, Tadarida, Phylostomus*; Aschoff and Ostwald 1987). In all species, the lateral efferent system is present and arises in nuclei located close to the LSO (*Pteronotus:* interstitial nucleus; *Rhinolophus,* nucleus olivocochlearis; Fig. 5.19) or in cells located within the main body of LSO (*Tadarida, Rhinopoma, Phylostomus*). It is remarkable that this system only contains the ipsilateral component. Putative neurotransmitters in the horseshoe bat include acetylcholine (Bruns and Schmieszek 1980), γ-aminobutyric acid (GABA) (Vater, Kössl, and Horn 1992), and enkephalins (Tachibana, Senuma, and Kumamoto 1992). Ultrastructural data (Vater, Lenoir, and Pujol 1992) show that in the horseshoe bat, as in other mammals, efferent endings predominantly contact dendrites of type I ganglion cells.

The nuclei of origin of the medial olivocochlear system are present in all

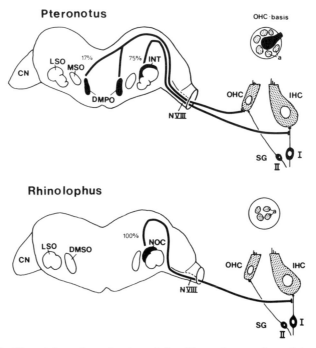

FIGURE 5.19. The origin and termination of the efferent innervation of the cochlea in *Pteronotus parnellii* (adapted from Bishop and Henson 1988) and *Rhinolophus* Spec. (adapted from Aschoff and Ostwald 1987; Vater, Lenoir, and Pujol 1992). CN, cochlear nucleus; DMPO, dorsomedial periolivary nucleus; DMSO, dorsomedial superior olive; INT, interstitial nucleus; LSO, lateral superior olive; MSO, medial superior olive; NOC, nucleus olivocochlearis; SG, spiral ganglion.

species (*Pteronotus:* dorsomedial periolivary nucleus, DMPO; Fig. 5.19) with the notable exception of *Rhinolophus* (Fig. 5.19). As in other mammals, the medial olivocochlear system of bats projects to the cochlea of both sides with the dominant component arising contralaterally. The lack of medial olivocochlear neurons in horseshoe bats agrees with reports that OHCs throughout the cochlea of horseshoe bats completely lack efferent synapses and thus considerably deviate from the typical mammalian scheme (Bruns and Schmieszek 1980; Bishop and Henson 1987, 1988; Vater, Lenoir, and Pujol 1992).

The medial olivocochlear system of *Pteronotus* has some specialized attributes. Each OHC only receives one efferent ending (Bishop and Henson 1988) centered at the OHC base and surrounded by afferent fibers (see Fig. 5.19), whereas in other mammals there are multiple efferent terminals on each OHC and their distribution varies along the cochlea (Liberman, Dodds, and Pierce 1990). The control of cochlear mechanics in *Pteronotus* appears very precise, because individual MOC fibers do not branch extensively and terminate on only one to as many as nine OHCs (Wilson, Henson, and Henson 1991), thus resembling the pattern in mice and deviating from the elaborate branching observed in the cat (Liberman, Dodds, and Pierce 1990). Acetylcholine esterase staining of efferent boutons in *Pteronotus* shows that they are particularly large within the auditory fovea (Xie et al. 1993).

The functional significance of the lack of the medial olivocochlear system in *Rhinolophus* and the presence of a uniform efferent system in *Pteronotus* is obscure as the two species are confronted with similar demands in analyzing echolocation signals. Explanations considering differences in cochlear mechanics, such as differences in overall damping (Bishop and Henson 1988), are intriguing for basal cochlear locations but of limited explanatory value because the lack of efferents is not confined to a particular cochlear place. Efferent terminals on OHCs are present in the cochleae of bats emitting FM calls (*Pteronotus suapurensis, Tadarida brasiliensis, Eptesicus fuscus*; Vater unpublished data). Interestingly, the only other mammal with no or considerably reduced efferent innervation of OHCs is the mole rat *Spalax ehrenbergi*, a species that is specialized to low-frequency hearing (Raphael et al. 1991). Again, no generally valid explanation for the presence or lack of this feature can be found.

6. Cochlear Mechanisms

6.1 Passive and Active Processes in the Cochlea of Mammals

The frequency-response characteristics of the mammalian cochlea are shaped to a large degree by cochlear macromechanics, that is, the more or less "passive" properties of the tectorial and basilar membranes and the supporting structures. In most nonspecialized mammals, a continuous basoapical decrease of BM thickness and an increase of BM width generate a gradient of decreasing membrane stiffness. Thus, sound energy of different frequencies can be dissipated along the BM at different locations enabling the frequency place transformation of the cochlea. However, to explain the sensitive threshold, the sharp tuning characteristics, and the occurrence of spontaneous otoacoustic emissions, cochlear models must include "active" components that introduce energy into the passive traveling wave, preferentially at cochlear locations located basal to the traveling wave maximum (de Boer 1983a,b; Neely and Kim 1983, 1986). The active components are likely to induce fast positive mechanical feedback, which results in negative damping or negative impedance of the BM, and a slower negative feedback system seems to be necessary to stabilize the traveling wave (Zweig 1991). The movement and sharp tuning of the BM also could be affected by different radial oscillation modes (Kolston et al. 1989; pars arcuata versus pars pectinata) or by radial differences in BM stiffness (Novoselova 1989). In addition, the TM could act as a second resonator and play a important role in a further filtering step to improve sharp tuning (Zwislocki and Kletsky 1979; Zwislocki 1986; Allen and Neely 1992).

The OHCs, which in vitro show motile responses to electrical and mechanical stimulation (e.g., Brownell et al. 1985; Zenner, Zimmermann, and Schmidt 1985; Ashmore 1987; Brundin, Flock, and Canlon 1989), seem to provide a cellular basis of the proposed active force generation in the cochlea. The extent to which movements of the OHC body are capable of following high-frequency sound stimuli and thus are able to act as mechanical amplifier on a cycle-by-cycle basis is still being investigated (Reuter et al. 1992; Santos-Sacchi 1992; He et al. 1993). Recent measurements of electrically induced traveling waves in the gerbil at about 30 kHz may indicate that fast movements of the OHC body are able to follow high frequencies (Xue, Mountain, and Hubbard 1993). In addition, slow contractions and elongations of OHCs that do not follow the frequency of the stimulus should significantly affect the geometry and interaction of the "passive" structures and may also contribute to a sharpening of the traveling wave.

6.2 Cochlear Structure and Function in Bats

To evaluate possible cochlear mechanisms from the viewpoint of a high-frequency echolocator, several features of structure and function in the cochlea of bats are discussed next.

6.2.1 Active Micromechanics

The strong spontaneous OAEs that occur in some individuals of *Pteronotus* at 60 kHz (up to 40 dB SPL) require a fast force generator which could be established by fast movements of the OHCs. In the foveal region of *Rhi-*

nolophus (Vater, Lenoir, and Pujol 1992), the OHCs are very short and are rigidly embedded in the Deiter cell cups and the reticular plate. This anatomical constellation, also valid for *Pteronotus,* should restrict the amplitude of fast movements of the OHC body. Indeed, in the FM bat *Carollia perspicillata,* the electrically induced movement in the isolated organ of Corti and of single OHCs of basal and apical cochlear turns is smaller than in the guinea pig and is below a noise level of 0.03 μm for frequencies above 1 kHz (Kössl et al. 1993). Therefore, the question is still open if the fast movements of the OHC body, which are likely to be very small at ultrasonic frequencies, can deliver enough energy to produce the large SOAEs. It has to be considered that the small OHC movements may feed into a nearly undamped resonator that depends on the passive macromechanics (see following).

One is not only looking for a mechanism to generate enhanced tuning and SOAEs at 60 kHz, but also for a more general process that ensures high sensitivity within a wide range of frequencies. The pronounced thickenings of the bat BM (pars pectinata, pars arcuata) should emphasize more or less passive mechanisms of sharp tuning that could be based on multiple oscillation modes in the radial direction (Kolston 1988; Kolston et al. 1989). Force generation in OHCs might influence the geometry of the organ of Corti and hence the possible oscillation modes.

It should be kept in mind that active force generation in mammalian OCHs also depends on transduction at the stereocilia. For low-level stimuli in the excitatory direction, the compliance of the mammalian hair bundle is increased (Russell, Kössl, and Richardson 1992), similar to the situation in the bullfrog sacculus (Howard and Hudspeth 1988). Such changes in stereocilia stiffness are a consequence of the asymmetric transduction mechanism at the tip links and may play a role in further increasing mechanical sensitivity. The large degree of tilt of the third-row stereocilia toward the second row in OHCs of *Rhinolophus* (Vater, Lenoir, and Pujol 1992) could enhance the excitatory versus inhibitory asymmetry of hair cells tuned to ultrasonic frequencies and also potentiate stiffness changes. The fact that in *Rhinolophus* the IHC stereocilia are embedded in the TM (Vater and Lenoir 1992) should enable a more direct communication between active processes in the body or stereocilia of the OHCs and the transduction site of the IHCs.

6.2.2 Macromechanics Exemplified in *Pteronotus*

In contrast to normal laboratory mammals, in *Pteronotus* and in other CF-FM bats the size and thickness of the BM and the TM vary considerably along the cochlear length.

i. At about 40% cochlear length the BM abruptly thickens toward the base (Fig. 5.13) which could provide a reflection zone for incoming waves. If the reflected waves reverberate between this discontinuity and the stapes, standing waves could occur and a more or less passive, but

226 Manfred Kössl and Marianne Vater

highly tuned resonator would be implemented (see also Kemp 1981; Duifhius and Vater 1985). The frequencies of the CF_2 range and of the OAEs are represented expandedly within a cochlear region just apical to the BM discontinuity. The OAEs are strongly suppressed by loud sound exposure with exposure frequencies between 62 and 70 kHz (Kössl and Vater 1985a, 1990b; see also Fig. 5.6). This range approximately matches the thickened BM region. The resonator would ensure the high sensitivity and sharp tuning just apically to the BM-thickness discontinuity.

ii. For frequencies at and just below CF_2, the cochlear resonance might, because of phase changes, produce an insensitivity. The neuronal on/off responses (Fig. 5.5) indicate that a delayed dampening or cancellation of the respective frequencies could be taking place. After switching off the tone burst stimulus, the resonator may oscillate on its own for a few cycles and produce the off response. Both OAEs and the ringing in the cochlear microphonic potentials can be evoked by slightly lower frequencies (Suga, Simmons, and Jen 1975; Kössl and Vater 1985a), that is, by stimuli within the on/off range. Therefore the on/off responses might arise from mechanical cancellations between the basilar membrane response to the stimulus and the concomitantly evoked resonance.

iii. The large size of the opening of the cochlear aqueduct and the "thick lining" of the scala tympani (see Section 5.1) could be adaptations to prevent damage from the reverberant oscillations. Apical to the 45% position, there is a local maximum of the TM area (Fig. 5.14; Henson and Henson 1991) whose functional significance is not yet understood. It could further dampen the 60-kHz response or act as a second resonator and enhance frequency tuning.

iv. At about 70 kHz there is a threshold maximum, the slopes of which do not coincide with enhanced tuning (see Fig. 5.7). Between 10% and 23% cochlear length a basoapical increase in BM thickness occurs (Fig. 5.13): This increase is just opposite to the normal mammalian gradient and may be responsible for the poor sensitivity. In addition, it could ensure that optimum reverse propagation of the OAEs does occur in the cochlea of *Pteronotus*, analogous to the "paradoxical waves" traveling toward the stapes in the mechanical model of von Békésy (1960), where the stiffness gradient was inverse.

v. It is still unclear whether the sharp neuronal tuning at about 90 and 30 kHz, which coincides with steep threshold slopes (Fig. 5.7), is generated by an harmonic effect of the 60-kHz resonator or by different mechanisms. After loud sound exposure at about 60 kHz, the 60-kHz threshold minimum of the N_1-off response from the auditory nerve is abolished but there are no significant effects on the thresholds at 30 and 90 kHz (Pollak, Henson, and Johnson 1979). This may suggest that independent cochlear mechanisms are acting at the three CF frequency ranges. The anatomical substrates for the specialized processing of CF_1 and CF_3 are largely unknown. There is a local decrease in BM thickness between about 6% and 9% of the cochlear length (Fig. 5.13) that coincides with the CF_3 area. It could be involved in producing another BM discontinuity responsible for 90-kHz reflections. However, OAEs have not yet been found in this frequency range.

7. Summary

The cochlea of bats is designed to yield high sensitivity at ultrasonic frequencies. For this purpose, the stiffness of the BM, as extrapolated from the thickness/width ratio, is increased. The size of the OHCs and their stereocilia is decreased, which may guarantee a high-speed micromechanical amplification of low-level signals by active OHC processes. In CF bats, the cochlear response is further shaped by macromechanical specializations of the BM, TM, and spiral ligament, and by the presence of an acoustic fovea, an expanded representation of the dominant CF frequency on the BM. A mechanical resonator that is evident in ringing of CM potentials and in otoacoustic emissions in the mustached bat is probably involved in creating extraordinarily sharp tuning to the CF frequency. In horseshoe bats, sharp tuning and high sensitivity in the CF range does not require the presence of the medial efferent system that is thought to affect OHC micromechanics.

References

Allen JB, Neely S T (1992) Micromechanical models of the cochlea. Physics Today, Vol. 45, July 1992, pp. 40–47.

Aschoff A, Ostwald J (1987) Different origins of cochlear efferents in some bat species, rats, and guinea pigs. J Comp Neurol 264:56–72.

Ashmore JF (1987) A fast motile response in guinea-pig outer hair cells: the cellular basis of the cochlear amplifier. J Physiol 288:323–347.

Bishop AL, Henson OW Jr (1987) The efferent cochlear projections of the superior olivary complex in the mustached bat. Hear Res 31:175–182.

Bishop AL, Henson OW Jr (1988) The efferent auditory system in Doppler-shift compensating bats. In: Nachtigall PE, Moore PWB (eds) Animal Sonar. New York: Plenum, pp. 307–311.

Brown AM, Gaskill SA (1990) Measurement of acoustic distortion reveals underlying similarities between human and rodent mechanical responses. J Acoust Soc Am 88:840–849.

Brown MC, Nuttall AL (1984) Efferent control of cochlear inner hair cell responses in the guinea pig. J Physiol 354:625–646.

Brownell WE, Bader CR, Bertrand D, Ribaupierre DEY (1985) Evoked mechanical responses of isolated cochlear outer hair cells. Science 227:194–196.

Brundin L, Flock A, Canlon B (1989) Sound induced motility of isolated cochlear outer hair cells is frequency selective. Nature 342:814–816.

Bruns V (1976a) Peripheral auditory tuning for fine frequency analysis by the CF-FM bat, *Rhinolophus ferrumequinum*. I. Mechanical specializations of the cochlea. J Comp Physiol 106:77–86.

Bruns V (1976b) Peripheral auditory tuning for fine frequency analysis by the CF-FM bat, *Rhinolophus ferrumequinum*. II. Frequency mapping in the cochlea.

J Comp Physiol 106:87-97.

Bruns V (1980) Basilar membrane and its anchoring system in the cochlea of the greater horseshoe bat. Anat Embryol 161:29-51.

Bruns V, Goldbach M (1980) Hair cells and tectorial membrane in the cochlea of the greater horseshoe bat. Anat Embryol 161:65-83.

Bruns V, Schmieszek E (1980) Cochlear innervation in the greater horseshoe bat; demonstration of an acoustic fovea. Hear Res 3:27-43.

Bruns V, Müller M, Hofer W, Heth G, Nevo E (1988) Inner ear structure and electrophysiological audiograms of the Subterranean mole rat, *Spalax ehrenbergi*. Hear Res 33:1-10.

Bruns V, Burda H, Ryan MJ (1989) Ear morphology of the frog-eating bat (*Trachops cirrhosus*, Family: Phyllostomidae): apparent specializations for low-frequency hearing. J Morphol 199:103-18.

Burda H, Fiedler J, Bruns V (1988) The receptor and neuron distribution in the cochlea of the bat, *Taphozous kachhensis*. Hear Res 32:131-136.

Cabezudo LM (1978) The ultrastructure of the basilar membrane in the cat. Acta Otolaryngol 86:160-175.

Dallos P, Corey M E (1991) The role of outer hair cell motility in cochlear tuning. Curr Opin Neurobiol 1:215-220.

Dannhof BJ, Bruns V (1991) The organ of Corti in the bat *Hipposideros bicolor*. Hear Res 53:253-268.

de Boer E (1983a) No sharpening? A challenge for cochlear mechanics. J Acoust Soc Am 73:567-573.

de Boer E (1983b) On active versus passive cochlear models – toward a generalized analysis. J Acoust Soc Am 73:574-576.

Dieler R, Shehata-Dieler WE, Brownell WE (1991) Concomitant salicylate-induced alterations of outer hair cell subsurface cisternae and electromotility. J Neurocytol 20:637-653.

Duifhuis H, Vater M (1985) On the mechanics of the horseshoe bat cochlea. In: Allen JB, Hall JL, Hubbard A, Neely ST, Tubis A (eds) Peripheral Auditory Mechanisms. Berlin: Springer, pp.89-96.

Ehret G (1974) Age-dependent hearing loss in normal hearing mice. Naturwissenschaften 11:506.

Ehret G, Frankenreiter M (1977) Quantitative analysis of cochlear structures in the house mouse in relation to mechanisms of acoustical information processing. J Comp Physiol 122:65-85.

Evans EF (1972) The frequency response and other properties of single fibres in the guinea pig cochlear nerve. J Physiol 226:263-287.

Evans EF (1975) Cochlear nerve and cochlear nucleus. In: Keidel WD, Neff WD (eds) Auditory Systems. (Handbook of Sensory Physiology, Vol.5/2.) Berlin: Springer, pp. 1-108.

Fernandez C (1952) Dimensions of the cochlea (guinea pig). J Acoust Soc Am 24:519-523.

Fiedler J (1983) Vergleichende Cochlea-Morphologie der Fledermausarten *Molossus ater, Taphozous nudiventris kachhensis* und *Megaderma lyra*. Ph.d. Thesis, University of Frankfurt, Germany.

Firbas W (1972) Über anatomische Anpassungen des Hörorgans an die Aufnahme höherer Frequenzen. Monatszeitschr Ohrenheilkd Laryngo-Rhinol 106:105-156.

Gaskill S A, Brown A M (1990) The behavior of the acoustic distortion product,

$f2_1$-f_2, from the human ear and its relation to auditory sensitivity. J Acoust Soc Am 88:821–839.

Grinnell A D (1963) The neurophysiology of audition in bats: intensity and frequency parameters. J Physiol 167:38–66.

Grinnell A D (1973) Rebound excitation (off-responses) following non-neural suppression in the cochleas of echolocating bats. J Comp Physiol 82:179–194.

Habersetzer J, Storch G (1992) Cochlea size in extant chiroptera and middle eozene microchiropterans from Messel. Naturwissenschaften 79:462–466.

Habersetzer J, Schuller G, Neuweiler G (1984) Foraging behavior and Doppler shift compensation in echolocating bats, *Hipposideros bicolor* and *Hipposiideros speoris*. J Comp Physiol A 155:559–567.

He ZZ, Evans BN, Clark B, Sziklai I, Dallos P (1993) Voltage-dependent frequency response properties of isolated outer hair cells. Abstracts of the 30th Inner Ear Biology Workshop, Budapest, Hungary, p. 49.

Heffner R, Heffner H, Masterton RB (1971) Behavioural measurement of absolute and frequency-difference threshold in guinea pig. J Acoust Soc Am 49:1888–1895.

Henson MM (1973) Unusual nerve-fiber distribution in the cochlea of the bat *Pteronotus p. parnellii* (Gray). J Acoust Soc Am 53:1739–1740.

Henson MM (1978) The basilar membrane of the bat *Pteronotus p. parnellii*. Am J Anat 153:143–159.

Henson MM, Henson OW Jr (1988) Tension fibroblasts and the connective tissue matrix of the spiral ligament. Hear Res 35:237–258.

Henson MM, Henson OW Jr (1991) Specializations for sharp tuning in the mustached bat: the tectorial membrane and spiral limbus. Hear Res 56:122–132.

Henson MM, Henson OW Jr, Goldman LJ (1977) The perilymphatic spaces in the cochlea of the bat, *Pteronotus p. parnellii* (Gray). Anat Rec 187:767.

Henson MM, Henson OW Jr, Jenkins DB (1984) The attachment of the spiral ligament to the cochlear wall: anchoring cells and the creation of tension. Hear Res 16:231–242.

Henson MM, Jenkins DB, Henson OW Jr (1982) The cells of Boettcher in the bat, *Pteronotus parnellii*. Hear Res 7:91–103.

Henson MM, Jenkins DB, Henson OW Jr (1983) Sustentacular cells of the organ of Corti—the tectal cells of the outer tunnel. Hear Res 10:153–166.

Henson OW Jr (1970) The ear and audition. In: Wimsatt WA (ed) Biology of Bats. New York: Academic, Press pp. 181–256.

Henson OW Jr, Schuller G, Vater M (1985) A comparative study of the physiological properties of the inner ear in Doppler shift compensating bats (*Rhinolophus rouxi, Pteronotus parnellii*). J Comp Physiol 157:587–597.

Henson OW Jr, Bishop A, Keating A, Kobler J, Henson MM, Wilson B, Hansen R (1987) Biosonar imaging of insects by *Pteronotus parnellii*, the mustached bat. Natl Geogr Res 3:82–101.

Henson OW Jr, Koplas PA, Kating AW, Huffman RF, Henson MM (1990) Cochlear resonance in the mustached bat: behavioral adaptations. Hear Res 50:259–274.

Howard J, Hudspeth A J (1988) Compliance of hair bundle associated with gating of mechanoelectrical transduction channels in the bullfrog's saccular hair cell. Neuron 1:189–199.

Huffman RF, Henson OW Jr (1991) Cochlear and CNS tonotopy: normal physiological shifts in the mustached bat. Hear Res 56:79–85.

Huffman RF, Henson OW Jr (1993a) Labile cochlear tuning in the mustached bat I. Concomitant shifts in biosonar emission frequency. J Comp Physiol A 171:725–734.

Huffman RF, Henson OW Jr (1993b) Labile cochlear tuning in the mustached bat II. Concomitant shifts in neural tuning. J Comp Physiol A 171:735–748.

Jenkins DB, Henson MM, Henson OW Jr (1983) Ultrastructure of the lining of the scala tympani of the bat, *Pteronotus parnellii*. Hear Res 11:23–32.

Johnstone BM, Patuzzi R, Yates GK (1986) Basilar membrane measurements and the travelling wave. Hear Res 22:147–153.

Kalinec F, Holley CM, Iwasa K H, Lim DJ, Kachar B (1992) A membrane-based force generation mechanism in auditory sensory cells. Proc Natl Acad Sci USA 89:8671–8675.

Keithley EN, Schreiber RC (1986) Frequency map of the spiral ganglion in the cat. J Acoust Soc Am 81:1036–1042.

Kelly JB, Masterton B (1977) Auditory sensitivity of the albino rat. J Comp Physiol Psychol 91:930–936.

Kemp DT (1978) Stimulated acoustic emissions from within the human auditory system. J Acoust Soc Am 64:1386–1391.

Kemp DT (1981) Physiologically active cochlear micromechanics—one source of tinnitus. In: Evered D, Lawrencson G (eds) Tinnitus. (CIBA Foundation Symposium 85.) London: Pitman, pp. 54–81.

Kolston PJ (1988) Sharp mechanical tuning in a cochlear model without negative damping. J Acoust Soc Am 83:1481–1487.

Kolston PJ, Viergever MA, de Boer E, Diependaal RJ (1989) Realistic mechanical tuning in a micromechanical model. J Acoust Soc Am 86:133–140.

Kössl M (1992a) High frequency distortion products from the ears of two bat species, *Megaderma lyra* and *Carollia perspicillata*. Hear Res 60:156–164.

Kössl M (1992b) High frequency two-tone distortions from the cochlea of the mustached bat *Pteronotus parnellii* reflect enhanced cochlear tuning. Naturwissenschaften 79:425–427.

Kössl M (1994) Otoacoustic emissions from the cochlea of the 'constant frequency' bats, *Pteronotus parnellii* and *Rhinolophus rouxi*. Hear Res 72:59–72.

Kössl M, Vater M (1985a) Evoked acoustic emissions and cochlear microphonics in the mustache bat, *Pteronotus parnellii*. Hear Res 19:157–170.

Kössl M, Vater M (1985b) The cochlear frequency map of the mustache bat, *Pteronotus parnelli*. J Comp Physiol A 157:687–697.

Kössl M, Vater M (1990a) Tonotopic organization of the cochlear nucleus of the mustache bat, *Pteronotus parnelli*. J Comp Physiol A 166:695–709.

Kössl M, Vater M (1990b) Resonance phenomena in the cochlea of the mustache bat and their contribution to neuronal response characteristics in the cochlear nucleus. J Comp Physiol A 166:711–720.

Kössl M, Reuter G, Hemmert W, Preyer S, Zimmermann U, Zenner H-P (1993) Motility of outer hair cells in the organ of Corti of the bat, *Carollia perspicillata*. In: Elsner N, Heisenberg M (eds) Gene-Brain-Behaviour. (Proceedings of the 21th Göttingen Neuroscience Conference.) Stuttgart: Georg Thieme Verlag, p. 264.

Kraus H (1983) Vergleichende und funktionelle Cochleamorphologie der Fledermausarten *Rhinopoma hardwickii*, *Hipposideros speoris* und *Hipposideros fulvus* mit Hilfe einer Computergestützten Rekonstruktionsmethode. Ph.D. Thesis, University of Frankfurt, Germany.

Lenoir M, Puel J-L, Pujol R (1987) Stereocilia and tectorial membrane development in the rat cochlea. A SEM study. Anat Embryol 175:477–487.

Liberman MC (1982) The cochlear frequency map for the cat: labeling auditory-nerve fibers of known characteristic frequency. J Acoust Soc Am 72:1441–1449.

Liberman MC (1988) Response properties of cochlear efferent neurons: monaural vs. binaural stimulation and the effects of noise. J Neurophysiol (Bethesda) 60:1779–1798.

Liberman MC (1990) Effects of chronic cochlear de-efferentation on auditory-nerve response. Hear Res 49:209–224.

Liberman MC, Dodds LW, Pierce S (1990) Afferent and efferent innervation of the cat cochlea: quantitative analysis with light and electron microscopy. J Comp Neurol 301:443–460.

Lim DJ (1986) Functional structure of the organ of Corti: a review. Hear Res 22:117–146.

Long GR, Schnitzler H-U (1975) Behavioural audiograms from the bat, *Rinolophus ferrumequinum*. J Comp Physiol 100:211–219.

Manley GA (1990) Peripheral hearing mechanisms in reptiles and birds. Berlin: Springer-Verlag.

Metzner W, Radtke-Schuller S (1987) The nuclei of the lateral lemniscus in the rufous horseshoe bat, *Rhinolophus rouxi*. J Comp Physiol 160:395–411.

Müller M (1991) Frequency representation in the rat cochlea. Hear Res 51:247–254.

Müller M, Burda H (1989) Restricted hearing range in a subterranean rodent, *Cryptomys hottentotus* (Bathyergidae). Naturwissenschaften 76:134–135.

Müller M, Laube B, Burda H, Bruns V (1992) Structure and function of the cochlea in the African mole rat (*Cryptomys hottentottus*): evidence for a low frequency acoustic fovea. J Comp Physiol A 171:469–476.

Nadol JB Jr (1988) Comparative anatomy of the cochlea and auditory nerve in mammals. Hear Res 34:253–266.

Neely ST, Kim DO (1983) An active cochlear model showing sharp tuning and high sensitivity. Hear Res 9:123–130.

Neely ST, Kim DO (1986) A model for active elements in cochlear biomechanics. J Acoust Soc Am 79:1472–1480.

Neff W, Hind J (1955) Auditory thresholds of the cat. J Acoust Soc Am 27:480–483.

Neuweiler G (1990) Auditory adaptations for prey capture in echolocating bats. Physiol Rev 70:615–641.

Neuweiler G, Bruns V, Schuller G (1980) Ears adapted for the detection of motion, or how echolocating bats have exploited the capacities of the mammalian auditory system. J Acoust Soc Am 68:741–753.

Novoselova SM (1989) A possibility of sharp tuning in a linear transversally inhomogeneous cochlear model. Hear Res 41:125–136.

Obrist MK, Fenton MB, Eger JL, Schlegel PA (1993) What ears do for bats: a comparative study of pinna sound pressure transformation in chiroptera. J Exp Biol 180:119–152.

Pollak G, Henson OW Jr, Johnson R (1979) Multiple specializations in the peripheral auditory system of the CF-FM bat, *Pteronotus parnellii*. J Comp Physiol 131:255–266.

Pollak GD, Henson OW Jr, Novick A (1972) Cochlear microphonic audiograms in the 'pure tone' bat, *Chilonycteris parnellii parnellii*. Science 176:66–68.

Probst R, Lonsbury-Martin BL, Martin GK (1991) A review of otoacoustic emissions. J Acoust Soc Am 89:2027–2067.

Pujol R, Lenoir M, Ladrech S, Tribillac F, Rebillard G (1992) Correlations between the length of outer hair cells and the frequency coding of the cochlea. In: Cazals Y., Demany L., Horner K (eds) Advances in Biosciences, Auditory Physiology and Perception. Carcans: Pergamon.

Rajan R, Johnstone BM (1988) Electrical stimulation of cochlear efferents at the round window reduces auditory desensitization in guinea pigs. I. Dependence on electrical stimulus parameters. Hear Res 36:53–74.

Ramprashad F, Money KE, Landolt JP, Laufer J (1978) A neuroanatomical study of the little brown bat (*Myotis lucifugus*). J Comp Neurol 178:347–363.

Ramprashad F, Landolt JP, Money KE, Clark D, Laufer J (1979) A morphometric study of the cochlea of the little brown bat (*Myotis lucifugus*). J Morphol 160:345–358.

Raphael Y, Lenoir M, Wroblewski R, Pujol R (1991) The sensory epithelium and its innervation in the mole rat cochlea. J Comp Neurol 314:367–382.

Reuter G, Gitter AH, Thurm U, Zenner H-P (1992) High frequency radial movements of the reticular lamina induced by outer hair cell motility. Hear Res 60:236–246.

Robertson D, Manley GA (1974) Manipulation of frequency analysis in the cochlear ganglion of the guinea pig. J Comp Physiol 91:363–375.

Roth B, Bruns V (1992) Postnatal development of the rat organ of Corti. I. General morphology, basilar membrane, tectorial membrane and border cells. Anat Embryol 185:559–569.

Russell IJ, Sellick PM (1978) Intracellular studies of hair cells in the mammalian cochlea. J Physiol 284:261–290.

Russell IJ, Kössl M, Richardson GP (1992) Nonlinear mechanical responses of mouse cochlear hair bundles. Proc R Soc Lond B Biol Sci 250:217–227.

Rübsamen R, Neuweiler G, Sripathi K (1988) Comparative collicular tonotopy in two bat species adapted to movement detection, *Hipposideros speoris* and *Megaderma lyra*. J Comp Physiol A 163:271–285.

Santos-Sacchi J (1992) On the frequency limit and phase of outer hair cell motility: effects of the membrane filter. J Neurosci 12:1906–1919.

Schnitzler H-U (1968) Die Ultraschall-Ortungslaute der Hufeisen-Fledermäuse (*Chiroptera – Rhinolophidae*) in verschiedenen Orientierungssituationen. Z Vgl Physiol 57:376–408.

Schnitzler H-U (1970) Echoortung bei der Fledermaus *Chilonycteris rubiginosa*. Z Vgl Physiol 68:25–39.

Schuller G (1980) Hearing characteristics and Doppler shift compensation in South Indian CF-FM bats. J Comp Physiol 139:349–356.

Schuller G, Beuter K, Schnitzler H-U (1974) Response to frequency shifted artifical echoes in the bat *Rhinolophus ferrumequinum*. J Comp Physiol A 89:275–286.

Spoendlin H (1966) The organization of the cochlear receptor. In: Ruedi L (ed) Advances in Otorhinolaryngology, Vol. 13. Basel: Karger.

Spoendlin H (1973) The innervation of the cochlear receptor. In: Moller AR (ed) Basic Mechanisms in Hearing. New York: Academic, pp. 185–235.

Steele CR (1976) Cochlear mechanics. In: Keidel WD, Neff WD (eds) Handbook of Sensory Physiology, Vol. 513. Berlin: Springer-Verlag, pp. 443–478.

Sterbing S, Rübsamen R, Schmidt U (1990) Auditory midbrain frequency-place code and audio-vocal interaction during postnatal development in the phyllostomid bat, *Carollia perspicillata*. In: Elsner N, RothG (eds) Proceedings of the 18th Göttingen Neurobiology Conference. Stuttgart: Georg Thieme Verlag, p. 140.

Strelioff D, Flock A, Minser KE (1985) Role of inner and outer hair cells in mechanical frequency selectivity of the cochlea. Hear Res 18:169–175.

Suga N (1964) Single unit activity in cochlear nucleus and inferior colliculus of echolocating bats. J Physiol 172:449–474.

Suga N, Jen P-HS (1977) Further studies on the peripheral auditory system of CF-FM bats specialized for fine frequency analysis of Doppler shifted echoes. J Exp Biol 69:207–232.

Suga N, Neuweiler G, Möller J (1976) Peripheral auditory tuning for fine frequency analysis by the CF-FM bat, *Rhinolophus ferrumequinum*. J Comp Physiol 106:111–125.

Suga N, Simmons JA, Jen P-HS (1975) Peripheral specialization for fine analysis of Doppler shifted echoes in the auditory system of the CF-FM bat *Pteronotus parnellii*. J Exp Biol 63:161–192.

Tachibana M, Senuma H, Kumamoto K (1992) Enkephaline-like immunoreactivity in the cochlear efferents in the bat, *Rhinolophus ferrumequinum*. Hear Res 59:14–16.

Vater M (1988) Cochlear physiology and anatomy in bats. In: Nachtigall PE, Moore PWB (eds) Animal Sonar. New York: Plenum, pp. 225–242.

Vater M, Lenoir M (1992) Ultrastructure of the horseshoe bat's organ of Corti. I. Scanning electron microscopy. J Comp Neurol 318:367–379.

Vater M, Feng A S, Betz M (1985) An HRP-study of the frequency-place map of the horseshoe bat cochlea: morphological correlates of the sharp tuning to a narrow frequency band. J Comp Physiol A 157:671–686.

Vater M, Kössl M, Horn AKE (1992) GAD- and GABA-immunoreactivity in the ascending auditory pathway of horseshoe and mustached bats. J Comp Neurol 325:183–206.

Vater M, Lenoir M, Pujol R (1992) Ultrastructure of the horseshoe bat's organ of Corti. II. Transmission electron microscopy. J Comp Neurol 318:380–391.

von Békésy G (1960) Experiments in Hearing. New York: McGraw-Hill.

Warr WB (1992) Organization of olivocochlear efferent systems in mammals. In: Webster DB, Popper AN, Fay RR (eds) Mammalian Auditory Pathway: Neuroanatomy. New York: Springer-Verlag, pp. 410–448.

Wilson JP (1985) The influence of temperature on frequency-tuning mechanisms. In: Allen JB, Hall JL, Hubbard A, Neely ST, Tubis A (eds) Peripheral Auditory Mechanisms. New York: Springer, pp. 229–236.

Wilson JL, Henson MM, Henson OW Jr (1991) Course and distribution of efferent fibers in the cochlea of the mouse. Hear Res 55:98–108.

Wit HP (1985) Diurnal cycle for spontaneous oto-acoustic emission frequency. Hear Res 18:197–199.

Wright A (1984) Dimensions of the stereocilia in man and the guinea pig. Hear Res 13:89–98.

Xie DH, Henson MM, Bishop AL, Henson OW Jr (1993) Efferent terminals in the cochlea of the mustached bat: quantitative data. Hear Res 66:81–90.

Xue S, Mountain DC, Hubbard AE (1993) Direct measurement of electrically-evoked basilar membrane motion. In: Biophysics of Hair Cell Sensory Systems. Groningen: Academisch Ziekenhuis, pp. 303–309.

Zenner H-P, Zimmermann U, Schmidt U (1985) Reversible contraction of isolated cochlear hair cells. Hear Res 18:127–133.

Zook JM, Leake PA (1989) Connections and frequency representation in the auditory brainstem of the mustache bat, *Pteronotus parnellii*. J Comp Neurol 290:243–261.

Zweig G (1991) Finding the impedance of the organ of Corti. J Acoust Soc Am 89:1229–1254.

Zwicker E, Schloth E (1984) Interrelation of different oto-acoustic emission. J Acoust Soc Am 75:1148–1145.

Zwislocki JJ (1986) Analysis of cochlear mechanics. Hear Res 22:155–169.

Zwislocki JJ, Kletsky EJ (1979) Tectorial membrane: a possible effect on frequency analysis in the cochlea. Science 204:639–641.

6

The Lower Brainstem
Auditory Pathways

ELLEN COVEY and JOHN H. CASSEDAY

1. Introduction

In the auditory system, more than any other sensory modality, extensive processing of incoming signals occurs in the brainstem. In all vertebrates, the auditory pathways below the inferior colliculus consist of a complex system of parallel pathways, each with its own centers for signal processing. The auditory structures of the lower brainstem act as filters to selectively enhance specific stimulus features and as computational centers to add, subtract, or compare signals in different channels. Some brainstem structures, such as the superior olive, have been studied extensively, and their function is at least partially understood. Others, such as the nuclei of the lateral lemniscus, have been largely ignored, and their functional roles are just beginning to be discovered.

In no vertebrate species are the brainstem auditory pathways larger or more highly differentiated than they are in echolocating bats. The large size and elegant organization of the lower brainstem auditory pathways in echolocating bats are not surprising when one considers the fact that these animals depend almost exclusively on their sense of hearing to navigate and capture prey in partial to complete darkness. It *is* surprising that the lower brainstem is the only part of the central auditory system that contains any clear anatomical specializations for echolocation (e.g., Baron 1974).

In this chapter, the structure and function of the auditory pathways of the lower brainstem in echolocating bats are compared and contrasted with those in other species of mammals; features peculiar to particular species of bats are discussed in terms of their possible role in echolocation behavior in each species.

2. Origins of Parallel Pathways in the Cochlear Nucleus

2.1 Structure of the Cochlear Nuclear Complex

In mammals, the auditory nerve branches as it enters the brainstem to innervate three major structures: the anteroventral cochlear nucleus

(AVCN), the posteroventral cochlear nucleus (PVCN), and the dorsal cochlear nucleus (DCN) (Ryugo 1992). This distribution of a common input to several distinct targets, each composed of a very different mixture of cell types (Cant 1992), forms the basis for the complex system of parallel pathways in the auditory brainstem.

The structure of each division of the cochlear nuclear complex has been described in several species of echolocating bats; these include the mustached bat, *Pteronotus parnellii* (Zook and Casseday 1982a; Kössl and Vater 1990), the greater horseshoe bat, *Rhinolophus ferrumequinum* (Poljak 1926; Schweizer 1981), and the rufous horseshoe bat, *Rhinolophus rouxi* (Feng and Vater 1985).

Figure 6.1 shows the cochlear nucleus of the mustached bat. In all species of echolocating bats that have been examined, the cochlear nucleus is extremely large relative to the size of the brainstem. This is true in species that have an expanded cochlear representation of certain spectral components of the echolocation call, such as the mustached and horseshoe bats, and also in those species that have a uniformly distributed frequency representation, such as the vespertilionids (Baron 1974).

In insectivorous echolocating bats, the AVCN is the largest division of the cochlear nuclear complex, comprising more than half of its entire volume (Baron 1974; Schweizer 1981; Zook and Casseday 1982a; Vater and Feng 1990). In addition to its large size, several cytoarchitectural features seem to be unique to and characteristic of the AVCN in all species of bats that have been studied. The first of these special features is the types of cells that are present in the AVCN and their distribution. In all mammals, the anterior subdivision of the AVCN contains mainly spherical cells, also known as bushy cells because they have a single thick dendrite that branches extensively at some distance from the cell body. In nonecholocating mammals, there are two populations of spherical bushy cells, distinguished by their size and location. Large spherical cells are found in the most rostral part of the AVCN, in an area where low frequencies are represented, whereas small spherical cells are distributed more caudally in areas responsive to high frequencies (see Cant 1992 for review). In echolocating bats, all the spherical cells in the AVCN are of the small type. In bats, small spherical cells are distributed throughout most of the rostrocaudal extent of the AVCN (Fig. 6.1A–C, G). These extend from the dorsal and medial part of the posterior subdivision, where they are intermixed with stellate cells, to the most rostral tip of the anterior subdivision, where they are the only cell type present (Zook and Casseday 1982a; Kössl and Vater 1990; Feng and Vater 1985). The absence of large spherical cells in bats, together with their rostral location in other mammals, suggests that large spherical cells may represent a special adaptation for processing low-frequency sounds. If so, this cell population in bats has either been lost or never developed during evolution. In spite of the difference in size, however, the spherical bushy cells of the bat have much in common with their counterparts in the cat. In both species, these cells receive end-bulbs of Held, large synaptic endings of

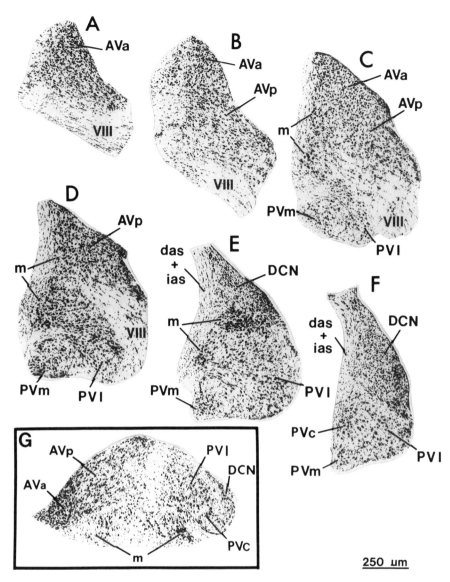

FIGURE 6.1A–G. Cytoarchitecture of the cochlear nucleus in the mustached bat as seen in Nissl-stained frontal sections arranged from rostral (A) to caudal (F) and in a horizontal section (G). The sections illustrate several structural features characteristic of echolocating bats: (1) the large size of the anteroventral cochlear nucleus (AVCN) and posteroventral cochlear nucleus (PVCN) relative to the dorsal cochlear nucleus (DCN); (2) the presence of small spherical bushy cells throughout the AVCN, even in the rostral part; (3) the presence of the marginal cell area medial to the AVCN and separating DCN from PVCN; and (4) poor lamination in the DCN. AVa, anterior division of AVCN; AVp, posterior division of AVCN; AVm, medial division of AVCN; PVm, medial division of PVCN; PV, lateral division of PVCN; PVc, caudal division of PVCN; das + ias, dorsal acoustic stria and intermediate acoustic stria (From Zook and Casseday © 1982 J. Comp. Neurol. 207, reprinted by permission of Wiley-Liss, a division of John Wiley and Sons Inc.)

auditory nerve fibers that surround the cell bodies. This conservation of input morphology indicates that whatever the functional significance of the end-bulbs, it is not restricted to preserving phase locking to low-frequency sounds.

The caudal part of the AVCN in echolocating bats resembles that of other mammals in that a number of different cell types are present (see Fig. 6.1B–E). These include multipolar and globular cells in the ventral region and granule cells along the lateral edge and caudally at the border with the DCN.

A feature of the AVCN that appears to be peculiar to echolocating bats is the presence of a distinctive population of very large multipolar cells along the medial edge of the AVCN and posteroventrally at the border with the PVCN (Fig. 6.1C–E). The cells that make up this population are the largest in the cochlear nucleus. This cell group has been termed the "marginal" subdivision of the AVCN (Zook and Casseday 1982a; Kössl and Vater 1990) and has further been subdivided into a medial part at the medial border of the AVCN and PVCN and a lateral part that lies along the common border between AVCN, PVCN, and DCN (Kössl and Vater 1990).

The cytoarchitecture of the PVCN in bats is similar to that of other mammals (Schweizer 1981; Zook and Casseday 1982a; Vater and Feng 1990; Cant 1992) (see Fig. 6.1D, F–G). The PVCN has been separated into two subdivisions in some bat species and three in others. Throughout most of the PVCN there is a mixture of stellate, small multipolar, elongate, and small round cells. As in other mammals (Osen 1969a,b), the caudal part of the PVCN contains octopus cells, large neurons with two or more long thick dendrites that extend orthogonal to the fibers of the descending branch of the auditory nerve. In most bats the octopus cell area is prominent, and in some horseshoe bats the octopus cells are arranged in a particularly distinctive U-shaped formation (Poljak 1926; Schweizer 1981; Pollak and Casseday 1989).

Compared to the AVCN and PVCN, the DCN in most bat species is relatively small and poorly differentiated (Poljak 1926; Henson 1970; Baron 1974; Schweizer 1981; Zook and Casseday 1982a; Feng and Vater 1985) (see Fig. 6.1E–G). In many mammals, the chief characteristic of the DCN is the laminar arrangement of its cells (Cant 1992). Prominent features of the lamination are that the fusiform cells are surrounded by many granule cells, and further that the fusiform cells are aligned in a single layer with their long axes perpendicular to the surface of the nucleus so that one dendrite extends into the deep layer and the other into the molecular layer. In bats, the granule cells surrounding the fusiform cells are few in number, and the fusiform cells are not uniformly aligned perpendicular to the surface of the nucleus. In all species of bats that have been described, the lamination of the DCN is poorly developed, and the fusiform cells are not arranged in an orderly fashion. The near absence of lamination in DCN is a feature shared by primates, including humans (Moore 1987).

In the horseshoe bat, two divisions of DCN can be distinguished on the basis of the pattern of lamination. The part of the DCN that represents frequencies below the constant-frequency (CF) range is laminated, whereas the part of the DCN that represents frequencies at and above the CF is not. Because this division of DCN into laminated and unlaminated divisions is not seen in the mustached bat, it appears to be a feature peculiar to the horseshoe bat and not a specialization common to all bats that use a CF component in their echolocation calls.

2.2 Inputs to the Cochlear Nucleus

In bats, the pattern of connectivity of the cochlear nucleus is basically the same as in other mammals, with a few exceptions. The main input is via the afferent fibers of the auditory nerve. These fibers bifurcate to form an ascending branch that terminates in the AVCN and a descending branch that terminates in the PVCN and DCN. Anterograde labeling of auditory nerve fibers entering the cochlear nucleus shows that in both the rufous horseshoe bat (Feng and Vater 1985) and the mustached bat (Zook and Leake 1989) these fibers terminate in each division of the cochlear nucleus as thick bands or "slabs" which correspond to isofrequency contours. Structural and functional evidence of a slablike connectional organization of afferents has also been seen in nonecholocating mammals (Cajal 1909; Rose, Galambos, and Hughes 1959; Moskowitz and Liu 1972; Noda and Pirsig 1974; Bourke, Mielcarz, and Norris 1981; Ryan, Woolf, and Sharp 1982). Thus, the basic pattern of innervation of the cochlear nucleus in bats follows the standard mammalian pattern. Descending inputs to the cochlear nucleus originate mainly in the periolivary cell groups and project to all three divisions. Within the cochlear nucleus itself, there are intrinsic connections that link one division with another.

The connectional basis for tonotopy in each division of the cochlear nucleus and the pattern of innervation from the cochlea has been elegantly demonstrated in the mustached bat (Zook and Leake 1989). In these experiments, very small horseradish peroxidase (HRP) injections were placed in physiologically characterized sites in the cochlear nucleus. Retrograde transport from the injection sites labeled cells in the spiral ganglion, and thus identified the portion of the cochlea projecting to a given site in the cochlear nucleus. In addition, anterograde transport identified the target regions of each site within the superior olivary complex (Fig. 6.2). The distribution of labeled ganglion cells in the cochlea was always proportional to injection size, even within the expanded 60- to 63-kHz region. This finding supports and adds to physiological evidence for two ideas concerning frequency representation: first, the proportion of tissue devoted to processing any given frequency range is approximately the same at the level of the cochlear nucleus as it is at the cochlea, and second, the expanded

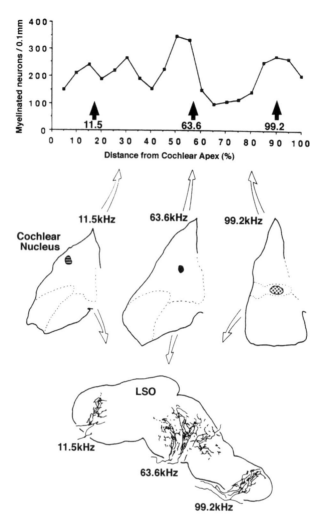

FIGURE 6.2. Patterns of connections of small regions in the AVCN as seen by anterograde and retrograde transport of horseradish peroxidase (HRP). In the cochlea, shown at the *top*, the location and extent of labeled ganglion cells varies with the frequency range and size of the injection sites, shown in *middle*. In the lateral superior olive (LSO), shown at the *bottom*, the region of terminal label is much larger for the injection at 63.6 kHz than for the injections at higher and lower frequencies, indicating a greater degree of divergence in this frequency range. (Adapted from Zook and Leake © 1988 J. Comp. Neurol. 290, reprinted by permission of Wiley-Liss, a division of John Wiley and Sons, Inc.)

representation of certain frequency ranges in the cochlea of CF-FM bats is maintained in a one-to-one projection to the cochlear nucleus.

2.3 Intrinsic Connections of the Cochlear Nucleus

Small HRP injections confined to subdivisions of the cochlear nucleus have been used to determine the patterns of branching and termination of auditory nerve fibers and patterns of intrinsic connections between cochlear nucleus subdivisions in the rufous horseshoe bat (Feng and Vater 1985) and in the mustached bat (Zook and Leake 1988; Kössl and Vater 1990). These results demonstrate frequency-specific reciprocal connections between the DCN and AVCN and between the DCN and PVCN. Like the afferent arborizations, the intrinsic connections between divisions of the cochlear nucleus are tonotopically organized and take the form of slabs. In the mustached bat, the intrinsic projection from the AVCN to the DCN molecular layer is especially large (Zook and Casseday 1985).

2.4 Physiology of the Cochlear Nucleus

The response properties of neurons in the cochlear nucleus have been studied in two FM species, the little brown bat (*Myotis lucifugus*) (Suga 1964) and the big brown bat (Haplea, Covey, and Casseday 1994), and in two CF-FM species, the mustached bat (Suga, Simmons, and Jen 1975) and the horseshoe bat (Neuweiler and Vater 1977; Feng and Vater 1985).

2.4.1 Tonotopy and Frequency Tuning

The tonotopic organization in the cochlear nucleus of bats is basically the same as that found in other mammals such as the cat (Bourke, Mielcarz, and Norris 1981). Thus, in bats, the tonotopic axis from high to low frequencies extends from dorsal to ventral in the DCN (Feng and Vater 1985), from dorsomedial to ventrolateral in the PVCN, and from caudal to rostral in the AVCN (Feng and Vater 1985; Haplea, Covey, and Casseday 1994).

The cochleas of CF-FM bats are specialized so that a very large proportion of the basilar membrane is tuned to the CF_2 frequency range, about 61 kHz (see Kössl and Vater, Chapter 5, this volume). As mentioned in Section 2.2, the cochlear frequency expansion is preserved in the tonotopic organization of each division of the cochlear nucleus. The result is a prominent central expansion of the same small frequency range that is expanded on the basilar membrane. Thus, more than half of all neurons in the cochlear nucleus are tuned to frequencies about 61 kHz in the mustached bat (Suga, Simmons, and Jen 1975), 81–88 kHz in the greater horseshoe bat (Neuweiler and Vater 1977), and 77–79 kHz in the rufous horseshoe bat (Feng and Vater 1985).

Figure 6.3 compares the distribution of best frequencies (BFs) and sharpness of tuning in the cochlear nucleus of a CF-FM bat with that of a bat that uses a purely frequency-modulated (FM) echolocation call. The cochleas of FM bats, so far as is known, have no specializations for fine frequency tuning nor do they show an expansion of any frequency range. Correspondingly, FM bats have a uniform representation of frequencies within the cochlear nucleus (Suga 1964; Suga, Simmons, and Jen 1975; Haplea, Covey, and Casseday 1994). These observations suggest that the frequency representations seen in the cochlear nucleus reflect the cochlear frequency representation in approximately a one-to-one relationship. In horseshoe bats, however, the expansion of the CF_2 frequency range is more pronounced in the AVCN and PVCN than in the DCN (Feng and Vater 1985). This finding suggests that although the cochlear frequency expansion is largely responsible for establishing frequency representation in the cochlear nucleus of CF-FM bats, this one-to-one correspondence may not be preserved at all levels. There is evidence that specific frequency ranges are progressively expanded in the central auditory system, especially in FM species which have no cochlear frequency expansion. Figure 6.4 shows that the BFs of neurons in the cochlear nucleus of the big brown bat are approximately evenly distributed across the entire audible range just as they are in the little brown bat (Fig. 6.3B); this distribution is probably a direct reflection of the cochlear frequency map. However, at the nuclei of the

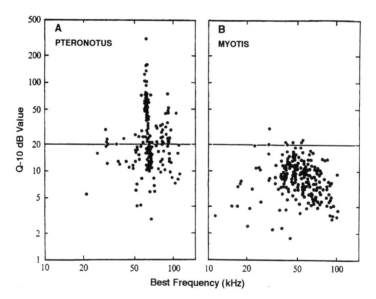

FIGURE 6.3A,B. Comparison of the best frequency (BF) distribution and sharpness of frequency tuning in the cochlear nucleus of (A) a constant frequency, frequency-modulated (CF-FM) bat (the mustached bat) and (B) an FM bat (the little brown bat). (From Suga, Simmons, and Jen 1975.)

FIGURE 6.4A–C. Comparison of the BF distribution and sharpness of frequency tuning in an FM species, the big brown bat, at the cochlear nucleus (C), nuclei of the lateral lemniscus (B), and inferior colliculus (A). At the cochlear nucleus and nuclei of the lateral lemniscus, the distribution of BFs is approximately uniform throughout the audible range, and Q_{10dB}s are nearly all below 20; at the inferior colliculus, there is an expanded representation of the frequency range between 20 and 30 kHz, and the Q_{10dB}s within this range are high, with a large proportion greater than 20. (From Haplea, Covey, and Casseday 1994.)

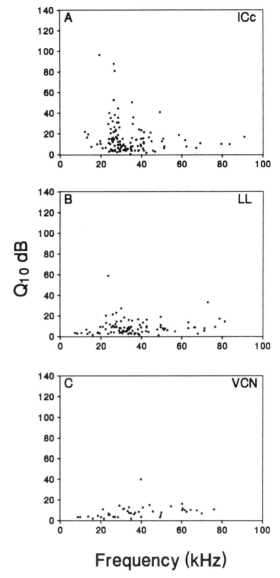

Frequency (kHz)

lateral lemniscus, the proportion of neurons with BFs in the frequency range between 20 and 30 kHz is slightly increased. In the inferior colliculus, there is a large expansion of this frequency range, so that about one-third of all BFs are in the range between 20 and 30 kHz. Thus, it seems selective convergence or divergence of projections results in portions of the frequency map being selectively contracted or expanded.

In CF-FM bats, the expanded representation of the CF_2 frequency range is accompanied by specializations in frequency tuning. In the cochlear nucleus of horseshoe bats and mustached bats, most neurons with BFs outside the CF_2 frequency range have ordinary V-shaped tuning curves similar to those in nonecholocating mammals; most units with BFs in the CF_2 range have extremely narrow tuning curves (Suga, Simmons, and Jen 1975; Neuweiler and Vater 1977; Feng and Vater 1985). For the narrowly tuned neurons in all these species, the quality factor at 10 dB above threshold (Q_{10dB}) is above 20, with some as high as 400.

Neurons with frequency selectivity comparable to that of the narrowly tuned units in CF-FM bats have not been found in the cochlear nucleus of FM bats. In the cochlear nucleus of the little brown bat and the big brown bat, neurons with BFs throughout the entire audible frequency range have V-shaped tuning curves comparable in breadth to those in cats or other nonecholocating mammals (Suga 1964; Haplea, Covey, and Casseday 1994).

Although it is likely that the fine frequency tuning in CF-FM bats is largely the result of cochlear mechanical specializations (see Kössl and Vater, Chapter 5, this volume), frequency tuning may be further modified in the central auditory system through neural inhibition. Inhibitory sidebands are associated with the tuning curves of some cochlear nucleus neurons in both CF-FM bats and FM bats, although they do not appear to be very common at this level (Suga 1964; Neuweiler and Vater 1977). Because inhibitory sidebands are found in both narrowly tuned and broadly tuned neurons, there is no reason to believe that the presence or absence of inhibitory sidebands is correlated with BF or narrow cochlear tuning. Suga (1964) proposed that the abrupt transition between excitation and inhibition in neurons with inhibition at only one side (Fig. 6.5) could play a role in creating directional selectivity for FM sweeps, a function that could be common to both FM and CF-FM bats because both use FM echolocation signals.

It is important to note that FM bats do have neurons at the level of the inferior colliculus (IC) that are narrowly tuned across a wide amplitude range, with Q_{10dB}, Q_{30dB} and Q_{40dB} values of 20 and higher. In the big brown bat, the BFs of all the narrowly tuned neurons fall within a range (20–30 kHz) that corresponds to the portion of the FM call that is lengthened to a "quasi-CF" during the search phase of echolocation (Casseday and Covey 1992). This finding suggests that extremely narrow frequency tuning is necessary at some level for any echolocating bat that must analyze echoes of a CF or quasi-CF call; furthermore, narrow tuning can be produced entirely by neural mechanisms. Thus, narrow tuning may have evolved independently through very different mechanisms in the two types of bats.

In the big brown bat, fine frequency tuning first appears at the IC and is created through neural inhibitory mechanisms (Casseday and Covey 1992; Covey et al. 1993; Haplea, Covey, and Casseday 1994). In the mustached bat, the already fine frequency tuning present at lower levels is sharpened at

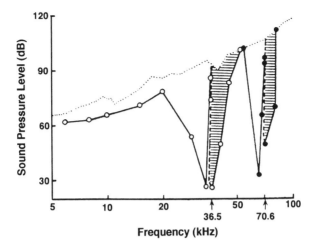

FIGURE 6.5. Turning curves of two spontaneously active neurons in the cochlear nucleus of the little brown bat have excitatory areas (*unshaded*) in which discharge increases above spontaneous rate, and inhibitory areas (*shaded*) in which discharge decreases below spontaneous rate. For both these neurons, the excitatory area occupies a range of frequencies below the inhibitory area. (From Suga 1964.)

the IC (Yang, Pollak, and Resler 1992). In either species it is unknown whether the inhibitory input responsible for modifying frequency tuning arises from interneurons in the IC or from projection neurons in lower brainstem pathways. The finding that narrow frequency tuning first appears at the level of the IC in FM bats suggests that even though fine frequency tuning is present throughout the brainstem auditory system of CF-FM bats, it may not be exploited until the level of the IC. The fact that the IC is an early stage of output to systems for motor control (Covey, Hall, and Kobler 1987; Schuller, Covey, and Casseday 1991) suggests that fine frequency tuning might be important for motor tasks such as Doppler-shift compensation (see Chapter 3, this volume).

2.4.2 Discharge Patterns and Timing Characteristics

The discharge patterns of cochlear nucleus units in both CF-FM and FM bats are basically similar to those in nonecholocating mammals such as the cat (Rose, Galambos, and Hughes 1959; Pfeiffer 1966). Transient responses of one or a few spikes may be correlated with either the onset or offset of sound. Sustained responses include primary-like, primary-like with notch, tonic (nonadapting), pauser, chopper, and buildup. In most species, including the little brown bat (Suga 1964), the big brown bat (Haplea, Covey, and Casseday 1994), the rufous horseshoe bat (Feng and Vater 1985), and the mustached bat (Suga, Simmons, and Jen 1975), sustained responses are reported to be the predominant type. In the greater horseshoe bat (Neuweiler and Vater 1977), the most common discharge pattern was reported to be transient, probably because in this study recording was limited to the PVCN and DCN. In addition, in this study a significant number of neurons were reported to have complex response patterns that changed depending on frequency or intensity of the stimulus. This phenomenon has also been seen in the DCN of nonecholocating mammals (e.g., Goldberg and Brownell 1973; Godfrey, Kiang, and Norris 1975; Adams 1976; Rhode and Kettner 1987).

Several lines of evidence suggest that some initial analysis of temporal information occurs in the cochlear nucleus. Calbindin-like immunoreactivity is thought to provide a marker for neural pathways that preserve timing information with great precision (Carr 1986; Takahashi et al. 1987). In the cochlear nucleus of the mustached bat, calbindin reactivity is prevalent; it is found in the end-bulbs of Held, auditory nerve root neurons, multipolar and globular cells in AVCN, multipolar and octopus cells in PVCN, and small- and medium-sized cells in DCN (Zettel, Carr, and O'Neill 1991). There is functional evidence that at least some of these cell types do play a role in transmitting precise information about the temporal structure of sounds.

In CF-FM bats, many cells that respond with a transient onset discharge also have a transient discharge correlated in time with stimulus offset. Usually the onset response is present throughout the excitatory frequency range, but the offset response appears only within a restricted frequency range. *On/off* responses are most commonly seen in cells with BFs around the CF_2 frequency (Suga, Simmons, and Jen 1975; Neuweiler and Vater 1977). In the mustached bat, the off responses at 60–61 kHz do not appear to be rebounds from neural inhibition because they are not preceded by suppression of spontaneous activity. Instead, the off discharges are correlated with cochlear microphonic off or "after" activity, suggesting that on/off responses in CF-FM bats are a result of cochlear mechanical specializations (Suga, Simmons, and Jen 1975; Suga and Jen 1977; Bruns 1976ab; Suga, Neuweiler, and Möller 1976).

It is not known whether on/off responses serve some purpose in neural processing of echolocation sounds. However, they persist at some higher levels of the auditory system such as the medial superior olive (MSO) (Covey, Vater, and Casseday 1991; Grothe et al. 1992) and IC (Pollak and Bodenhamer 1981; Lesser et al. 1990), but are rare or absent at other levels such as the lateral superior olive (LSO) (Covey, Vater, and Casseday 1991).

An unusual response pattern common to both FM and CF-FM bats is the so-called afterdischarge, a prolonged period of neural activity lasting up to 600 msec after the cessation of a 5- or 10-msec stimulus. Afterdischarges are especially common following sounds at high amplitudes (Suga 1964; Neuweiler and Vater 1977; Haplea, Covey, and Casseday 1994). Presentation of a second, identical tone during the afterdischarge causes brief inhibition of firing (Suga 1964). This finding suggests that the sequence of

synaptic events leading to afterdischarge is inhibition followed by prolonged excitation, although cochlear mechanisms have not been ruled out as
a possible cause of afterdischarge.

Some cochlear nucleus neurons in horseshoe bats are suppressed rather
than excited for a period of several tens of milliseconds following their
response to a sound (Neuweiler and Vater 1977). It seems likely that long
periods of excitation or suppression following the presentation of a stimulus
could provide a neural trace that marks the occurrence of a stimulus for a
prolonged time period, thus facilitating or suppressing the response to
subsequent sounds that occur during the period of the afterdischarge or
suppression. Afterdischarges thus might play a role in echoranging or in the
analysis of long sequences of sounds such as the rapid and continuous train
of pulses and echoes that commonly occurs just before prey capture in
many bat species.

Additional evidence that cells in the bat cochlear nucleus may be an initial
stage in analyzing timing information comes from experiments in the
greater horseshoe bat and big brown bat in which a population of neurons
in the PVCN fire a single spike correlated precisely in time with stimulus
onset. Virtually no spikes occur at any other time. These neurons maintain
the same latency and probability of firing over a wide range of stimulus
intensities (Neuweiler and Vater 1977; Haplea, Covey, and Casseday 1994).
Neurons with similar timing characteristics have also been found in the
marginal division of AVCN in the mustached bat (Kössl and Vater 1990).

In the mustached bat, most transient responses seen in the cochlear
nucleus are recorded from neurons in the marginal division of AVCN. Most
neurons in the marginal division have BFs within the frequency range of the
first FM harmonic (FM_1), from 24 to 32 kHz (Kössl and Vater 1990). The
finding of precisely timed transient responses in a population of neurons
tuned to the FM_1 may be important for understanding how target range is
determined by the mustached bat. The thalamic and cortical neurons that
are thought to encode information about target range respond to the FM_2
or FM_3 only if it is preceded by the FM_1, and the maximal response occurs
when the two sounds are separated by a specific time interval (O'Neill and
Suga 1979, 1982; Olsen and Suga 1991a, b). It seems likely that FM_1-
selective neurons in the marginal division of AVCN provide a sharp and
precise timing marker for the occurrence of the FM_1, which could optimize
time delay measurements at higher levels of the auditory system. The
marginal division of AVCN in the mustached bat contains a high density of
noradrenergic fibers that originate mainly in the locus coeruleus (Kössl,
Vater, and Schweizer 1988). It has been shown that noradrenaline enhances
the phasic nature and precise timing of discharges of cells in this region
(Kössl and Vater 1989). This finding suggests that attention may play a role
in regulating the responses of cochlear nucleus neurons, and that behavioral
attention is especially important in the system that originates in the
marginal division of the AVCN.

Besides providing timing information from which to derive target range, cochlear nucleus neurons are capable of providing highly accurate information about the fine timing characteristics of rapidly changing complex stimuli. Transient responders in both CF-FM and FM bats have recovery times as short as 0.3 msec, and this is reflected in their ability to follow amplitude modulations up to about 3000 Hz (Suga, Simmons, and Jen 1975; Neuweiler and Vater 1977).

Neurons in the cochlear nucleus of horseshoe bats respond with synchronized discharges to both sinusoidal amplitude modulations and sinusoidal frequency modulations. For amplitude-modulated sounds, the minimum modulation depth needed to produce synchronization appears to be independent of BF; in contrast, for frequency-modulated sounds, minimum modulation depth is related to BF range (Vater 1982). Although all neurons follow sinusoidal frequency modulations, those with BFs in the CF_2 range are much more sensitive to small frequency modulations than are any others. Neurons in the CF_2 range can synchronize their discharges to a modulation depth of ± 20 Hz applied to a carrier frequency of more than 80 kHz. This represents a modulation of only 0.00025%. This exquisite sensitivity of the CF_2 filter neurons to sinusoidal frequency modulations is a neural correlate of the behavioral observations that CF-FM bats can discriminate periodic Doppler shifts of the CF_2 frequency caused by the beating wings of an insect.

All these findings suggest that neurons in the cochlear nucleus of echolocating bats transmit information about the temporal pattern of sound in the form of discharges which are highly synchronized to sound onset, amplitude modulations, or frequency modulations. At higher brainstem levels, this synchronized pattern of neural activity undergoes various transformations such as bandpass or lowpass filtering at the MSO (Grothe 1990), further sharpening of time locking in the nuclei of the lateral lemniscus (Covey and Casseday 1991), or conversion to pattern selectivity at the inferior colliculus (Covey et al. 1993; Casseday, Ehrlich, and Covey 1994).

2.5 Central Connections of the Cochlear Nucleus

In all mammals, the three divisions of the cochlear nucleus are the origin of multiple parallel pathways (e.g., Warr 1982). In the bat, the pattern of ascending projections from the cochlear nucleus is similar to the basic mammalian plan, with some exceptions that may be specializations for localization of high-frequency sounds, adaptations for target range determination, or adaptations for identification of targets using echolocation. Figure 6.6 shows the basic pattern of outputs from each division of the cochlear nucleus in echolocating bats. Three major groups of targets receive projections from the cochlear nucleus: the superior olivary complex (SOC), the monaural nuclei of the lateral lemniscus, and the central nucleus of the IC.

In all bat species that have been studied, the AVCN supplies the predominant input to the principal nuclei of the SOC (Feng and Vater 1985;

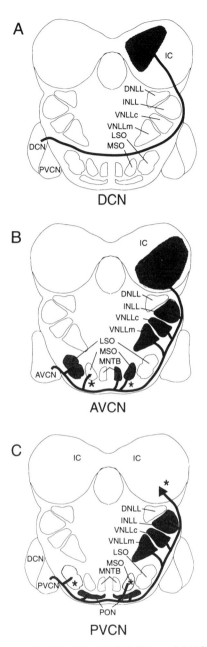

FIGURE 6.6A–C. Outputs of DCN (A), AVCN (B), and PVCN (C) shown in sche-
matized frontal sections from the brain of an echolocating bat. *Shaded areas* are
targets of the different divisions of the cochlear nucleus; *heavy stippling* indicates
major projections; *light stippling* indicates sparse projections; and *asterisks* indicate
pathways that have been described only in echolocating bats, or that differ significantly between bats and other mammals. DNLL, dorsal nucleus of lateral lemniscus;
INLL, intermediate nucleus of lateral lemniscus; VNLLc, columnar division of
ventral nucleus of lateral lemniscus; VNLLm, multipolar cell division; MSO, medial
superior olive; MNTB, medial nucleus of trapezoid body; PON, periolivary nuclei.

Zook and Casseday 1985, 1987; Covey and Casseday 1986; Vater and Feng 1990). However, in several species of bats, the PVCN supplies a significant projection to the LSO and MSO (Zook and Casseday 1985; Vater and Feng 1990). Because a projection from the PVCN to the SOC has not been described in other mammals, it may be restricted to echolocating bats. Surrounding the principal nuclei of the SOC are a number of different periolivary cell groups. The periolivary nuclei are highly variable among all mammalian species, especially among different species of bats. All divisions of the cochlear nucleus contribute projections to the periolivary cell groups, but the largest proportion of input comes from the PVCN, followed by the DCN (Zook and Casseday 1985).

The second major set of targets of the cochlear nucleus are the intermediate and ventral nuclei of the lateral lemniscus (see Fig. 6.6B,C). In all mammals, most of the input to the intermediate and ventral nuclei of the lateral lemniscus arises from the contralateral cochlear nucleus, from both AVCN and PVCN (Glendenning et al. 1981; Zook and Casseday 1982b, 1985; Covey and Casseday 1986).

The third major target of direct output from the cochlear nucleus is the contralateral central nucleus of the inferior colliculus (ICc) (Fig. 6.6). The direct projections to the ICc originate from two main sources, the AVCN and DCN. However, in the species of echolocating bats that have been studied, projections to the ICc also originate in the PVCN (Zook and Casseday 1982b; Vater and Feng 1990). Because projections from the PVCN to the IC have not been described in other mammals, this pathway may be a feature peculiar to and characteristic of echolocating bats. Like the pathways to the SOC and nuclei of the lateral lemniscus, the direct projections from the cochlear nucleus to the ICc are tonotopically organized (Zook and Casseday 1982b; Casseday and Covey 1992). In all mammals that have been studied, including several species of bats, the projections from the cochlear nucleus to the ICc terminate in a banded pattern (e.g., Zook and Casseday 1985, 1987; Casseday and Covey 1992; Oliver and Huerta 1992).

In the mustached bat and the big brown bat, small spherical cells of rostral AVCN project directly to the anterolateral, low-frequency area of the ICc; cells in the caudal, high-frequency part of AVCN project directly to the ventromedial, high-frequency part of the ICc, and small multipolar, globular, elongate, ovoid, and spherical cells at intermediate locations and frequencies project to corresponding intermediate positions in the ICc (Zook and Casseday 1982b, 1987). In noncholocating mammals, the projection to the ICc originates mainly in the multipolar, or stellate, cell population of the AVCN (see Oliver and Huerta 1992). The fact that in the bat not only stellate cells but also spherical and globular cells project to the ICc may indicate that a different and larger subset of inputs are integrated in the IC of the echolocating bat than in noncholocating mammals. The projection from the PVCN to the ICc does not originate in the octopus cell area but from other cell types (Zook and Casseday 1982b).

In the bat species that have been examined, the projection from the DCN to the ICc originates in fusiform and giant cells, just as it does in other mammals. This projection is tonotopically organized so that the dorsal DCN projects to the ventromedial, high-frequency area of the ICc while the ventral DCN projects to the anterolateral, low-frequency area of the ICc. The target of the DCN within the ICc extends more medially and dorsally than the AVCN target (Zook and Casseday 1987), suggesting that there is partial segregation of the two systems of projections.

3. Superior Olivary Complex

3.1 Structure of the Superior Olivary Complex

In all species of echolocating bats that have been studied, the SOC is large and well developed. It appears to contain all the same structures that are present in nonecholocating mammals, including the LSO, MSO, medial nucleus of the trapezoid body (MNTB), and various periolivary cell groups, although there is some controversy as to whether the large and specialized MSO of bats can be considered equivalent to MSO in nonecholocating animals. The SOC of the mustached bat is illustrated in Figure 6.7.

3.1.1 Principal Nuclei

In all bats that have been examined, the LSO and MNTB are unusually large relative to the size of the brainstem (Baron 1974). In some bat species, the LSO has more convolutions than does the LSO of most nonecholocating mammals. Nevertheless, in terms of their cytoarchitecture and patterns of connections, these nuclei in bats are virtually identical to the LSO and MNTB of other mammals (Schweizer 1981; Zook and Casseday 1982a; Schwartz 1992). The LSO is composed mainly of elongate cells oriented with their long axis perpendicular to the outer contour of the nucleus. Surrounding the LSO is a dense fiber plexus (Fig. 6.7B). The MNTB in bats and other mammals is a conspicuous group of densely packed cells, most of which are principal cells. These are round or oval, with an eccentric nucleus, and are contacted on a one to one basis by large calyces of Held that originate in the AVCN (Zook and Casseday 1982a; Kuwabara and Zook 1991; Kuwabara, DiCaprio, and Zook 1991).

In the mustached bat and other bats that have been examined, there is a large cell group in the position normally occupied by the MSO. This structure is cytoarchitecturally similar to MSO in other mammals in that a dense fiber plexus surrounds a column of cells (Fig. 6.7A,B) that are mainly elongate with their long axis oriented medial to lateral (Zook and Casseday 1982a). However, there is one structural aspect in which the bat MSO differs from that of most nonecholocating mammals. The MSO of a primate, for example, is many cells thick in the rostrocaudal and dorsoven-

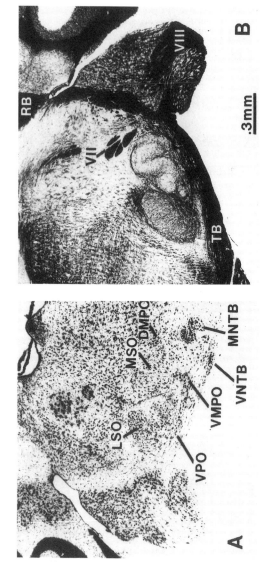

FIGURE 6.7A,B. Subdivisions of the superior olivary complex (SOC) in the mustached bat, *Pteronotus parnellii*. (A) Nissl-stained frontal section through the left side of the brainstem; (B) adjacent frontal section through the right side of the brainstem stained to show fibers. DMPO, dorsomedial periolivary nucleus; VPO, ventral periolivary nucleus; VMPO, ventromedial periolivary nucleus; VNTB, ventral nucleus of trapezoid body; TB, trapezoid body; RB, restiform body; VIII, eighth cranial nerve. (From Zook and Casseday © 1982a. J. Comp Neurol. 207, reprinted by permission of Wiley-Liss a division of John Wiley and Sons, Inc.)

tral dimensions, but only a few cells thick in the mediolateral dimension; in most echolocating bats, the MSO is many cells thick in all these dimensions. In the mustached bat, the MSO is convoluted to form a large dorsal limb and a small ventral limb. In horseshoe bats there are two structures, each of which is more like MSO in cytoarchitecture and connections than like any of the periolivary cell groups (Schweizer 1981). These have been termed the dorsal MSO and the ventral MSO (Casseday, Covey, and Vater 1988).

3.1.2 Periolivary Nuclei

The most detailed description of the cytoarchitecture and organization of the periolivary cell groups is in the mustached bat (Zook and Casseday 1982a), but some information is also available for a number of additional bat species (Aschoff and Ostwald 1987). Most bats have a prominent group of large multipolar neurons dorsomedial to the MSO. Although this cell group has been called the dorsomedial periolivary nucleus in the mustached bat (Zook and Casseday 1982a), it may be the same as the superior paraolivary nucleus (SPN) of rodents, which occupies a similar location and is made up of large multipolar neurons (e.g., Ollo and Schwartz 1979). Ventral to the principal nuclei of the SOC are the ventral nucleus of the trapezoid body and the ventral and ventromedial periolivary nuclei (Fig. 6.7A) (Zook and Casseday 1982a). In at least some bat species a large group of neurons is located lateral and anterolateral to the LSO. In the mustached bat and the horseshoe bat, these cells have been called the lateral nucleus of the trapezoid body (LNTB) (Zook and Casseday 1982a; Casseday, Covey, and Vater 1988; Kuwabara and Zook 1992), although it is not clear whether this is the same structure that is called LNTB in the cat. Rostral to the LSO is a group of very large multipolar neurons that are prominent in the mustached bat and in horseshoe bats, but are not well developed in vespertilionid bats. This cell group has been called the anterolateral periolivary nucleus (ALPO: Zook and Casseday 1982a; Covey, Hall, and Kobler 1987; Kobler, Isbey, and Casseday 1987) and more recently, the nucleus of the central acoustic tract (NCAT), because it is the source of the central acoustic tract, an extralemniscal pathway to the superior colliculus and thalamus (Casseday et al. 1989). It is highly likely that the NCAT in echolocating bats is a more robust homolog of a group of cells in the ventromedial lateral lemniscus of the cat, which has been shown to have sparse projections to the thalamus (Henkel 1983).

3.2 Variability of SOC Organization

Because of differences in the structure of MSO and the periolivary nuclei, the overall organization of the SOC varies considerably from one family of bats to another. Some of these differences are illustrated in Figure 6.8, in which the general structural features of the SOC are compared for the

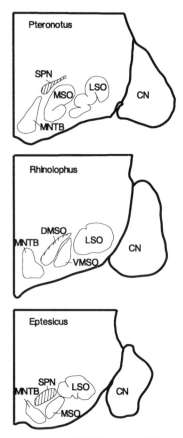

FIGURE 6.8. Structural features of the SOC in three different species of echolocating bats. The LSO and MNTB are similar in all three species, but the medial cell groups are variable. The *shaded area* (SPN) indicates a population of large multipolar neurons just dorsal to MSO. SPN is especially prominent in vespertilionid bats such as the big brown bat, intermediate in size in the mustached bat, and nearly absent in horseshoe bats, where there are only a few multipolar neurons intermingled with the cells along the dorsal border of the DMSO. The MSO in the mustached bat is large, prominent, and convoluted to form a dorsal and a ventral limb. In the big brown bat and other vespertilionids, the MSO is an ovoid structure of intermediate size. In horseshoe bats, the MSO consists of two distinct subdivisions, the dorsal MSO (DMSO) and the ventral MSO (VMSO).

mustached bat, a horseshoe bat, and a vespertilionid bat. It is not known whether these differences in organization of the SOC are systematically correlated with differences in echolocation behavior or hearing capabilities. The differences are systematically correlated with taxonomy, and they may provide as reliable a criterion for classification as do any of the external morphological features commonly used, such as tooth structure or forearm length.

Because of the controversial status of the MSO in bats, the MSO connectional pattern is described here in detail. In the best-studied example, the mustached bat, there are connectional data from both retrograde and anterograde transport of neural tracers as well as from single-fiber injections.

3.3 Inputs to the SOC

The pattern of input from the cochlear nucleus to the SOC in bats is in most respects similar to that in nonecholocating mammals. The LSO receives direct excitatory input from the ipsilateral AVCN (Casseday, Covey, and Vater 1988; Zook and DiCaprio 1988; Covey, Vater, and Casseday 1991; Kuwabara and Zook 1991), mainly from spherical bushy cells (Zook and Casseday 1985; Casseday, Covey, and Vater 1988; Kuwabara and Zook 1991). The input to the LSO from the contralateral AVCN is relayed via neurons in the MNTB that provide glycinergic inhibitory input. The input to MNTB is mainly from globular neurons in AVCN. The axons of the globular cells terminate in large calyces of Held, which synapse in a one-to-one manner on principal cells of the MNTB (Zook and DiCaprio 1988; Casseday, Covey, and Vater 1988; Zook and Leake 1989; Kuwabara and Zook 1991; Kuwabara, DiCaprio, and Zook 1991).

The one major respect in which the brainstem auditory system of echolocating bats appears to diverge from the general mammalian plan is in the pattern of inputs to the MSO. In nonecholocating animals, the MSO receives direct input from the AVCN of both sides, and the relative timing of these two inputs provides an important cue for sound localization (see Irvine 1992 for review). Although there is no unique input to the MSO in any bat that has been studied, bats do differ from nonecholocating mammals in the relative proportion of ipsilateral to contralateral inputs from the AVCN. In the horseshoe bat and mustached bat, for example, the contralateral input is robust while the ipsilateral input is sparse. Furthermore, the ipsilateral and contralateral projections to MSO seem to originate in different populations of cells in the AVCN (Casseday, Covey, and Vater 1988; Covey, Vater, and Casseday 1991; Vater, Casseday, and Covey 1995).

In all mammals, including bats, the projections to the principal nuclei of the SOC are tonotopically organized. Within this organization, the CF harmonic ranges, especially the CF_2, are greatly expanded in the horseshoe and mustached bats (Casseday, Covey, and Vater 1988; Covey, Vater, and Casseday 1991). Although these are the same frequency ranges that are expanded at the cochlea and cochlear nucleus, they appear to be even further expanded at the level of the SOC. Zook and Leake (1989) showed that in the mustached bat the extent of anterograde transport from the cochlear nucleus to the principal nuclei of the SOC is not proportional to the size of the injection site, but rather is differentially expanded in each target structure (see Fig. 6.2). Thus, the representations of certain frequency ranges, especially the CF_2 range from 60 to 63 kHz, are further expanded at the SOC. The relative amount by which each range is expanded differs in the different nuclei. This pattern of expanded CF harmonic frequency representation is reflected in the distribution of BFs in the SOC of the mustached bat (Covey, Vater, and Casseday 1991). In the LSO, approximately one-third of the total volume is devoted to representation of the frequency range from 60 to 63 kHz, and much of the remaining volume to the CF_1, CF_3, and CF_4 frequency ranges. In the MSO, an even larger proportion of the total volume, nearly one-half, is devoted to the CF_2; additionally, there is a large expansion of the CF_3 frequency range, which occupies most of the ventral limb of the MSO.

As would be expected from observations of most other mammals (Friauf and Ostwald 1988; Smith et al. 1991), the projections from the cochlear nucleus to the LSO are ipsilateral only, while those to the MNTB are exclusively contralateral (Kuwabara, DiCaprio, and Zook 1991). The projections to the MSO are unusual in two respects. First, instead of an equal distribution to the two sides, most of the projections are to the contralateral MSO (Casseday, Zook, and Kuwabara 1993; Vater, Casseday, and Covey 1995). Second, in most mammals, the projections from AVCN terminate in the half of MSO nearest their side of origin (Stotler 1953; Strominger and Strominger 1971 Warr 1982). In the mustached bat, there are only sparse projections to the ipsilateral MSO; these are confined to the lateral edge as would be expected. On the contralateral side, however, heavy projections are distributed throughout the entire mediolateral extent of MSO (Casseday, Zook, and Kuwabara 1993; Vater, Casseday, and Covey 1995).

Figure 6.9A shows the results of intraaxonal injection of a fluorescent dye into two fibers of the trapezoid body. The fiber that arises from a spherical bushy cell projects only to the contralateral MSO. The fiber that arises from a stellate cell projects bilaterally. These projection patterns are examples of a rule: most of the projections to MSO arise from spherical bushy cells in the contralateral AVCN, but a small proportion of the projections arise from stellate cells in the AVCN, mainly from the ipsilateral side (Fig. 6.10).

In nonecholocating mammals with good low-frequency hearing, the input to MSO originates mainly from the large spherical cells of the AVCN bilaterally (Cant and Casseday 1986; Casseday and Covey 1987; Schwartz 1992). However, large spherical cells are absent in the AVCN of echolocating bats, and the projection to the MSO originates mainly from the small spherical cells that occupy the rostral AVCN (Covey, Vater, and Casseday 1991; Casseday, Zook, and Kuwabara 1993). Despite these differences in projection pattern, the details of innervation of individual cells in bat MSO seem to be like those of other mammals (Clark 1969a,b; Schwartz 1980; Kiss and Majorossy 1983; Schwartz 1984; Zook and Leake 1989). The fibers from the AVCN form varicosities on the cell body and along one of the two main dendrites, usually the one nearest the origin of the fiber (Zook and Leake 1989).

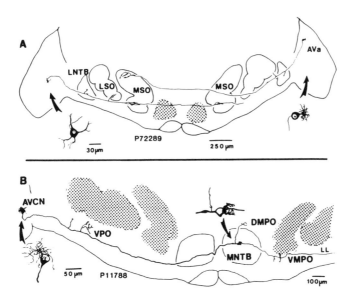

FIGURE 6.9A,B. Projections from intracellularly labeled axons of AVCN neurons in the mustached bat. (A) The axon of a spherical bushy cell (on *right*) projects to the ipsilateral LNTB and the contralateral MSO (*arrows*). The axon of a labeled stellate cell (on *left*) projects to the ipsilateral LNTB, ipsilateral LSO, ipsilateral MSO, and contralateral MSO. (B) The axon of a globular bushy cell projects to periolivary cell groups bilaterally and to the MNTB contralaterally (*large arrows*). In MNTB, the axon terminates in a large calyx of Held. (From Casseday, Zook, and Kuwabara 1993.)

The inputs to the bat MNTB are virtually identical to those in other mammals (Warr 1972; Friauf and Ostwald 1988). In the mustached bat (Kuwabara, DiCaprio, and Zook 1991), the calyces of Held in the MNTB arise from globular bushy cells in the AVCN (Fig. 6.9B). The cells that receive the calyces provide inhibitory projections to the LSO (Moore and Caspary 1983; Spangler, Warr, and Henkel 1985; Zook and DiCaprio 1988; Adams and Mugnaini 1990; Bledsoe et al. 1990; Kuwabara and Zook 1991). Recently it has been shown that MNTB cells also project to MSO (cat, Adams and Mugnaini 1990; bat, Kuwabara and Zook 1991) and that this projection follows the tonotopic arrangement just as it does in LSO (Kuwabara and Zook 1991). Thus, the MNTB is a major source of input to MSO from the contralateral AVCN. The input to MSO, like that to LSO, is inhibitory (Grothe 1990; Grothe et al. 1992).

Another recently recognized potential source of inhibitory input to both LSO and MSO is the LNTB. In bats and other mammals, LNTB cells are contacted by fibers from the ipsilateral cochlear nucleus (Warr 1992; Tolbert, Morest, and Yurgelun-Todd 1982; Roullier and Ryugo 1984; Spirou, Brownell, and Zidanic 1990; Smith et al. 1991; Kuwabara, DiCa

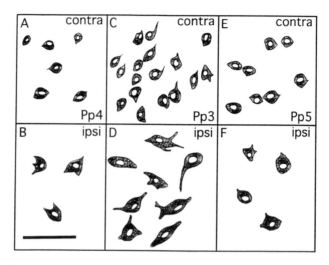

FIGURE 6.10 A–F. Labeled cells in AVCN of the mustached bat following three MSO injections that resulted in bilateral transport. In all three cases, most labeled cells on the contralateral side (*contra*) are small and oval or round in shape, probably spherical bushy cells (A,C,E). Most of the cells labeled on the ipsilateral side *(ipsi)* are larger and multipolar in shape (B,D,F). Calibration bar = 100 μm. (From Vater, Casseday, and Covey © 1995. J. Comp. Neurol. 351, reprinted by permission of Wiley-Liss a division of John Wiley and Sons, Inc.)

prio, and Zook 1991). Labeling of single axons of LNTB cells shows that they project ipsilaterally to both LSO and MSO. Because cells in the LNTB stain for glycine (Peyret, Geffard, and Aran 1986; Peyret et al. 1987; Wenthold et al. 1987; Helfert et al. 1989; Bledsloe et al. 1990), it seems likely that the LNTB cells transform excitatory input from the cochlear nucleus to ipsilateral inhibitory input to the LSO and MSO. At present there is no clue as to the function of an ipsilateral inhibitory input to LSO in bats or any other animal.

3.4 Physiology of the SOC

The functional properties of the principal nuclei of the SOC have been studied extensively in nonecholocating mammals. The resulting view is that information about differences in sound level at the two ears is first encoded in the LSO, whereas information about differences in binaural timing is first encoded in the MSO. The anatomical and physiological bases for binaural hearing in nonecholocating mammals have been reviewed elsewhere (Casseday and Covey 1987; Kuwada and Yin 1987; Irvine 1992). The responses of neurons in the SOC have been studied in an FM bat, *Molossus ater* (Harnischfeger, Neuweiler, and Schlegel 1985) and in two CF-FM bats, the rufous horseshoe bat (Casseday, Covey, and Vater 1988) and the mustached bat (Covey, Vater, and Casseday 1991).

3.4.1 Tonotopy and Frequency Tuning

In echolocating bats, just as in other mammals, the LSO and MSO are tonotopically organized, with a progressive increase in BF going from lateral to medial in the LSO and dorsal to ventral in the MSO. However, it seems clear, based on the observed connectional patterns and results of electrophysiological studies, that both the LSO and MSO in bats are adapted for the specific echolocation behavior of each species. CF-FM bats have clear specializations in the pattern of frequency representation in LSO and MSO (Covey, Vater, and Casseday 1991). Figure 6.11B shows that, in the mustached bat, the CF_2 range from about 60 to 63 kHz has a very large representation in both LSO and MSO. In the LSO, the CF_2 representation occupies the middle one-third of the nucleus, and in MSO, more than half of the nucleus. The representation of the third harmonic, CF_3, although only slightly expanded in the LSO, is greatly expanded in the MSO, where it includes nearly the entire ventral limb of the nucleus. In both LSO and MSO, only a very small proportion of volume is devoted to the range of frequencies in the FM portion of the echolocation signal. It appears that not only are the CF harmonic ranges expanded, but the FM ranges are contracted. The great expansion of the CF representation in CF-FM bats suggests that whatever the respective roles of the LSO and MSO in echolocation may be, both MSO and LSO are concerned mainly with processing the CF components of the call rather than the FM components.

In FM bats, the pattern of frequency representation in the SOC is similar to that in the cochlear nucleus in that no frequency range is expanded. Figure 6.11A shows that in the FM bat *Molossus*, BFs of neurons in MSO are rather uniformly distributed across frequency throughout the audible range (Harnischfeger, Neuweiler, and Schlegel 1985). This uniform frequency representation is not only very different from that in CF-FM bats, it is also very different from the expanded representation of low frequencies seen in the MSO of nonecholocating mammals with good low-frequency hearing (e.g. Guinan, Norris, and Guinan 1972). It is possible that the MSO in FM bats is in some sense less specialized than the MSO of animals that perform low-frequency sound localization, in which a major portion of the MSO contains an expanded low-frequency representation.

Thus, although the general tonotopic sequence in the LSO and MSO follows the standard mammalian plan, the relative representation of specific frequency ranges within the tonotopic sequence is highly variable and species dependent. The result is that the LSO and MSO of each species have an expanded representation of biologically important sounds.

3.4.2 Binaural Response Characteristics

Some clues as to how biologically important sounds may be processed can be seen in the responses of cells in the LSO and MSO to tests with binaural stimuli. In the bat species that have been studied, the responses of neurons in LSO are like those in other mammals. That is, nearly all LSO cells are

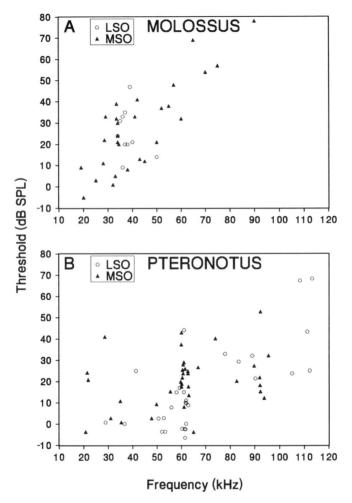

FIGURE 6.11. Distribution of BFs in the LSO and MSO of an FM bat, *Molossus ater* (A) compared with the distribution of BFs in the MSO of a CF-FM species, the mustached bat (*Pteronotus*) (B). (Data for the mustached bat are from Covey, Vater, and Casseday 1991; data for *Molossus* are from Harnischfeger, Neuweiler, and Schlegel 1985.)

excited by sounds at the ipsilateral ear, and this excitatory response is inhibited by sound at the contralateral ear. Thus, the LSO in echolocating bats presumably processes information about relative sound level at the two ears. In CF-FM species, this processing utilizes mainly the CF components of the call.

In contrast, the responses of neurons in MSO appear to vary considerably across bat species and, at least in CF-FM bats, to differ substantially from the responses of MSO cells in nonecholocating mammals. In the dog or cat

MSO, approximately two-thirds of neurons are excited by a sound at either ear, a small percentage are excited by sound at one ear and inhibited by a simultaneous sound at the other ear, and only a very few neurons are monaural (Goldberg and Brown 1968; Yin and Chan 1990). In the rat, MSO neurons are approximately evenly divided between those excited by sound at either ear and those excited by a contralateral sound but inhibited by a simultaneous ipsilateral sound (Inbody and Feng 1981). Functionally, the MSO of the FM bat *Molossus* more closely resembles that of the rat than that of the dog or cat. In *Molossus*, approximately one-fourth of MSO neurons are excited by sound at either ear, and about one-third are excited by a contralateral sound and inhibited by an ipsilateral sound. *Molossus* differs from the rat, however, in that monaural neurons are found in the MSO. If neurons excited by either ear are primarily responsible for encoding interaural phase differences of low-frequency sounds, perhaps their lower incidence in MSO of the rat and their even lower incidence in an FM bat reflects the decreased importance of interaural time or phase differences in sound localization for animals with poor low-frequency hearing, regardless of echolocation capability.

In CF-FM bats, the response properties of MSO neurons are even farther removed from those of dogs, cats, and rats than are those of FM bats. In horseshoe bats, the neurons in the dorsal MSO are approximately evenly divided into three groups: (1) those excited by sound at either ear, (2) those excited by a contralateral sound and inhibited by a simultaneous ipsilateral sound, and (3) monaural cells, responsive to sound at the contralateral ear only. In the ventral MSO, virtually all neurons are monaural, excited by the contralateral ear (Casseday, Covey, and Vater 1988). In the mustached bat, more than three-fourths of MSO neurons are monaural, responding only to a contralateral sound; the remainder are binaural, with the largest binaural class those excited by a contralateral sound and inhibited by a simultaneous ipsilateral sound. Only a very small percentage of neurons in the MSO of the mustached bat are excited by sound at either ear (Covey, Vater, and Casseday 1991; Grothe et al. 1992).

3.4.3 Discharge Patterns and Timing Characteristics

Temporal response properties reinforce the view that in bats LSO neurons are like those in other mammals whereas MSO neurons are specialized. In both FM and CF-FM bats, LSO neurons typically fire in a sustained, "fast chopper" pattern (Harnischfeger, Neuweiler, and Schlegel 1985; Casseday, Covey, and Vater 1988; Covey, Vater, and Casseday 1991). However, the discharge patterns of MSO neurons in both FM and CF-FM bats are somewhat different from those seen in mammals with low-frequency hearing. In the cat, nearly 90% of MSO units respond in a sustained pattern (Yin and Chan 1990). As in the case of binaural responses, the discharge patterns of MSO cells in FM bats more closely resemble those in nonecholocating species than do those of CF-FM bats. In the FM bat *Molossus*, about three-fourths of MSO units respond in a sustained pattern, but the remaining one-fourth respond transiently at the onset of a sound (Harnisch-feger, Neuweiler, and Schlegel 1985). In both species of CF-FM bat that have been studied, the percentage of transient responses in MSO is even higher, 40% in the horseshoe bat (Casseday, Covey, and Vater 1988) and about 70% in the mustached bat (Covey, Vater, and Casseday 1991). For many MSO neurons with transient responses in these species, the discharge switches from onset to offset with small changes in frequency. This change from on to off is seen most commonly in neurons with a BF near the CF_2 or CF_3.

The transient on and off responses in the MSO of the mustached bat have been shown to arise through the interaction of a sustained excitatory input with a sustained inhibitory input. Whether the response is to the onset or offset of the sound is determined by the relative timing of the two inputs (Grothe et al. 1992). The direct excitatory input is almost certainly from spherical bushy cells in the contralateral AVCN. The inhibitory input probably arises from the ipsilateral MNTB, which in turn receives its input from globular bushy cells in the contralateral AVCN. A similar type of interaction could account for the responses of MSO units that are excited by a contralateral sound and inhibited by an ipsilateral sound. Ipsilateral inhibitory input, relayed by LNTB cells, could reach MSO cells before or concurrent with the direct excitatory input from the contralateral cochlear nucleus.

3.5 Possible Function of the MSO in Echolocation

The MSO of the bat differs from that of other mammals in two important respects. Its inputs are altered to emphasize the contralateral ear, and its responses are altered to emphasize transients in the CF frequency range. Of what significance are these alterations for the bat? Although there is no single obvious answer, the data suggest several possibilities. The first involves localization in the vertical plane. Because of the aerial acrobatics that bats perform to pursue and capture flying prey, they must be as adept at localizing sound in the vertical plane as in the horizontal plane. The external ears of bats and other animals impose elevation-dependent changes in the intensity of specific spectral components of sounds, and it has been suggested that these systematic changes in the spectrum provide cues that can be analyzed in the central auditory system to compute the vertical location of a sound source (Simmons and Lawrence 1982; Fuzessery and Pollak 1985). Spectral changes might provide a particularly robust cue for elevation in echolocating bats, as the outgoing pulse could provide a reference for comparison.

In the IC of the mustached bat, the thresholds of all binaural neurons are lowest for sound at the midline in the horizontal dimension. This sensitivity pattern is independent of the neuron's frequency selectivity. In the vertical dimension, however, neurons tuned to the different harmonics of the echolocation call are differentially sensitive to a sound, depending on signal elevation. For example, neurons tuned to the CF_2 are most sensitive to about $0°$ elevation, whereas neurons tuned to the CF_2 are most sensitive to elevations of about $-40°$ (Fuzessery and Pollak 1985). In the MSO and LSO of the mustached bat, units tuned to the CF_2 frequency range, about 60 kHz, have the lowest thresholds and the widest range of thresholds; units tuned to the CF_3 frequency range, about 90 kHz, have higher thresholds (Covey, Vater, and Casseday 1991). As there is an increase in intensity for the frequencies about 90 kHz at lower elevations, the units in the LSO and MSO with BFs around 90 kHz would be more likely to be activated at lower elevations of the echo source. Thus, one function of MSO might be to transmit spectral cues that could be integrated at the IC with interaural intensity difference cues to derive the vertical and horizontal coordinates of an object in space.

A second possible function of the MSO in bats concerns localization of objects in the third dimension, that is, estimation of the distance to an object. This potential function for the MSO would by no means preclude the simultaneous transmission of spectral cues for elevation. When the bat flies toward a stationary or slowly moving object, the returning echoes are Doppler shifted to a frequency slightly higher than the emitted pulse. Through a reflex compensatory mechanism, the bat lowers the frequency of its own emitted call to keep the frequency of the echoes within the narrow excitatory range of filter neurons with BFs about 60 kHz. The response pattern of many MSO neurons in the mustached bat changes from on to off as sound frequency decreases. This on/off pattern is seen mainly in neurons tuned to the intense CF_2 harmonic of the echolocation call, about 60 kHz. Thus, on/off neurons in MSO could respond at the offset of the lower frequency emitted pulse and at the onset of the higher frequency Doppler-shifted echo; presumably, if the two overlapped, the response would be facilitated. Field studies have shown that the bat systematically decreases the duration of the CF component of its call as it approaches its prey (Novick and Vaisnys 1964). Thus, the facilitated response could provide a mechanism to signal the bat to adjust the duration of its echolocation signal in relationship to the distance from its target, thus acting as a sort of vocal yardstick (Casseday, Zook, and Kuwabara 1993).

A third possibility is that MSO neurons encode information about the fine temporal structure of the envelope of sounds. Nearly all MSO neurons in the mustached bat are capable of following amplitude modulations (AM) with on or off responses. Approximately one-third of MSO neurons exhibit bandpass selectivity for AM rate, while the remainder have low-pass filter characteristics with upper limits between 100 and 500 Hz, with most between 200 and 300 Hz (Grothe 1990, 1994). The upper AM filter limits for MSO neurons are considerably lower than those found at the level of the cochlear nucleus, where individual neurons show synchronized firing up to 1000 Hz (Vater 1982). Blocking glycinergic inhibitory input to MSO shows that the upper AM filter limit is created through an interplay of excitatory and inhibitory inputs, offset in time relative to one another. This mechanism for producing AM filter characteristics is almost certainly related to, or identical with, the mechanism by which frequency-specific on and off responses are produced in the bat MSO (Grothe et al. 1992; Grothe, 1994).

Behavioral studies have shown that CF-FM bats are capable of discriminating among different species of insects on the basis of the pattern of frequency or amplitude modulations imposed on the CF portion of the echo by an insect's wingbeats (Schnitzler et al. 1983; Von der Emde and Schnitzler 1986, 1990; Kober and Schnitzler 1990). An as-yet-untested hypothesis is that MSO neurons are selective for the wingbeat rates of insects that are attractive to the bat and that their activity could signal the presence of a prey worth pursuing.

A fourth possibility, given that bat MSO neurons exhibit phase locking to amplitude modulations of a high-frequency carrier tone, is that monaural phase information from the two MSOs could ultimately come together in some form at a higher level, the dorsal nucleus of the lateral lemniscus (DNLL) or the IC, for example, to provide localization cues based on interaural timing differences of low-frequency amplitude modulations of a high-frequency carrier.

Finally, echoes from three-dimensional objects are known to contain characteristic interference patterns or "glints" that can be seen as amplitude modulations in the envelope of the sound (Simmons 1989). Because each of the bat's ears is at a slightly different angle with respect to the object that produces the echo, the only conditions under which the echo would be the same at both ears would be for a perfectly symmetrical object located at the midline. Thus, another untested hypothesis about the function of MSO is that it could convey information about binaural disparities in the glint pattern of echoes, that could then be integrated at a higher level to derive information about an object's three-dimensional structure much as binocular disparity is used in the visual system to perceive depth.

3.6 Do Bats Really Have a Medial Superior Olive?

Animals with very small heads and high-frequency hearing are thought to have little or no usable range of binaural phase or envelope time differences (Masterton et al. 1975). This limitation is consistent with the idea, proposed several decades ago, that bats do not have an MSO, or at best have one that is extremely small and rudimentary (Harrison and Irving 1966; Irving and Harrison 1967). A different point of view, suggested by more recent studies of the superior olive in bats, is that the unusual requirements for localizing high-frequency sound reflected from flying prey in three-dimensional space have imposed an evolutionary pressure that has resulted in a highly developed and differentiated MSO homolog that is functionally different from the MSO in animals with good low-frequency hearing. This functional difference appears to result from a difference in the strength of projections from the ipsilateral cochlear nucleus. Otherwise, the cytoarchitecture and connections basically resemble those of MSO in mammals with low-frequency hearing.

If the MSO is defined as being a structure that receives direct excitatory input in equal measure from the two ears and compares interaural phase differences, then the bat "MSO" is clearly not an MSO. If, on the other hand, the MSO is defined as being a population of elongate cells between the LSO and the MNTB that receives direct or indirect input in equal or unequal measure from the two ears, and that provides tonotopically organized projections to the ipsilateral DNLL and IC, then the bat "MSO" clearly is an MSO. Unfortunately, no developmental data are available to answer the obvious question of whether the bat MSO is derived from the same embryonic precursor as the MSO in the cat or other large mammals. However, even if this question were resolved, the problem of the monaurality of MSO in the bat would remain. The fact that there seems to be a continuum of increasing monaurality from cat to rat to FM bat to CF-FM bat suggests that small changes in the relative strength of one input to a common precursor structure can result in very different functional properties. These properties customize the structure to fit the particular needs of each species. Just as the bat's head, pinnae, and hands have undergone evolutionary changes, the MSO has also undergone changes to adapt it to the special requirements of aerial navigation and hunting guided by echolocation.

3.7 Outputs of the SOC

The SOC is the source of three very different classes of outputs. First, and most thoroughly studied, are the pathways that ascend within the *lemniscal system*. The contribution of the SOC to the lemniscal system includes all the ascending projections that originate in the LSO and MSO and that terminate in the DNLL and the IC. The major lemniscal outputs of the SOC are summarized in Figure 6.12. Second is the *efferent system*, which includes all the projections that originate in the periolivary cell groups or principal nuclei of the SOC and terminate in the cochlea or cochlear nucleus. The third output of the SOC is a small but distinct *extralemniscal pathway* that originates in the NCAT and bypasses the IC to terminate in the superior colliculus and auditory thalamus.

3.7.1 SOC Lemniscal System: LSO, MSO, and DNLL

The lemniscal system as a whole includes all the pathways that ascend within the fiber tract of the lateral lemniscus. These pathways originate in the cochlear nucleus, the SOC, and the nuclei of the lateral lemniscus. All the lemniscal pathways terminate, directly or indirectly, in the IC. The main sources of input to the DNLL are the LSO and MSO; the DNLL in turn projects to the IC. Thus, the LSO, MSO, and DNLL together make up an integral system of direct and indirect pathways to the auditory midbrain, largely concerned with binaural processing. The MNTB does not project directly to the IC. However, in bats and other mammals, the MNTB provides dense input to the intermediate nucleus of the lateral lemniscus (INLL) and the ventral nucleus of the lateral lemniscus (VNLL) (Fig. 6.12C). This input is probably inhibitory, as is the input from the MNTB to the LSO and MSO.

As in all other mammals that have been studied, the projections from the LSO to the IC in echolocating bats are bilateral (Fig. 6.12A) (Schweizer 1981; Zook and Casseday 1982b, 1987; Casseday, Covey, and Vater 1988; Ross, Pollak, and Zook 1988; Ross and Pollak 1989; Vater, Casseday, and Covey 1995). The target of the projection from the LSO occupies roughly the ventral two-thirds of the IC, and the terminal fields take the form of bands or slabs oriented roughly parallel to the orientation of disk-shaped cells, and also roughly parallel to isofrequency contours (Zook and Casseday 1985, 1987; Casseday, Covey, and Vater 1988).

The ascending projections from the MSO in echolocating bats (see Fig. 6.12B) resemble those of other mammals in that they terminate only in the ipsilateral DNLL and IC (Adams 1979; Brunso-Bechtold, Henkel, and Linville 1990; Schweizer 1981; Zook and Casseday 1982b, 1987; Casseday, Covey, and Vater 1988; Ross, Pollak, and Zook 1988; Ross and Pollak 1989; Vater, Casseday, and Covey 1995). Like the LSO projections, the MSO projections extend throughout the ventral two-thirds of the IC and are tonotopically organized. The MSO projections, again like the LSO projections, appear to terminate in a pattern of bands or slabs. The monaural nature of MSO responses in the mustached bat appears to be preserved up through the level of the midbrain. Electrophysiological experiments combined with retrograde transport of HRP show that in this species the MSO is the main source of input to the monaural region of the enlarged 60-kHz contour of the inferior colliculus (Ross and Pollak 1989). Thus, although the overall target areas of MSO and LSO are similar, there still appears to be functional segregation according to monaural or binaural response properties. Nevertheless, there is considerable overlap between the targets of LSO and those of MSO, so it is highly likely that many individual cells in DNLL and IC receive input from both LSO and MSO (Vater, Casseday, and Covey 1995).

The DNLL: structure and function. Although published information about the DNLL in echolocating bats is not extensive, it is sufficient to indicate that its structure, function, and connections are essentially the same as they are in nonecholocating mammals. In both FM bats and CF-FM bats, the

FIGURE 6.12 A–C. Outputs of the SOC.
The *shaded areas* indicate the targets
of LSO (A), MSO (B), and MNTB (C).
These pathways are essentially the
same in echolocating bats as they are
in other mammals.

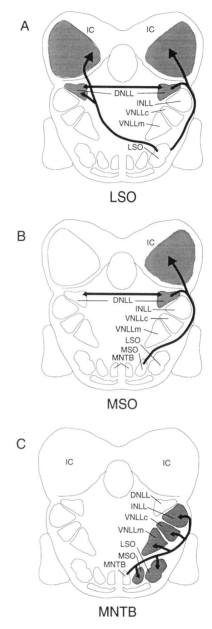

DNLL is a wedge-shaped structure composed mainly of medium to large
fusiform cells arranged in clusters between the dense fascicles of ascending
fibers of the lateral lemniscus (see Fig. 6.15, later in this chapter). The long
axes of the cells are typically oriented perpendicular to the ascending fibers.
The DNLL receives bilateral projections from the LSO and ipsilateral
projections from the MSO. Each DNLL projects to the opposite DNLL via

the commissure of Probst, and to the IC bilaterally, where its target region is approximately coextensive with that of the AVCN, LSO, and MSO (Zook and Casseday 1987; Casseday, Covey, and Vater 1988). In addition, in both bats and nonecholocating mammals, there is a projection from the rostral region of the DNLL and the adjacent auditory paralemniscal tegmentum to the deep layers of the superior colliculus (Casseday, Jones, and Diamond 1979; Kudo 1981; Henkel 1983; Tanaka et al. 1985; Covey, Hall, and Kobler 1987). As in other mammals, most DNLL neurons in the bat use γ-aminobutyric acid (GABA) as a neurotransmitter, indicating that the DNLL provides mainly inhibitory input to its targets (Adams and Mugnaini 1984; Covey 1993a).

As would be expected in a structure that receives bilateral input from the SOC, the responses of most DNLL neurons are binaural. This is true in the big brown bat (Covey 1993a), the mustached bat (Markovitz and Pollak 1994), and the cat (Aitkin, Anderson, and Brugge 1970; Brugge, Anderson, and Aitkin 1970), the three species in which binaural responses in DNLL have been studied. Most commonly, DNLL neurons are excited by sound at the contralateral ear and inhibited by sound at the ipsilateral ear. In the big brown bat, some DNLL neurons are excited by sound at either ear while others are facilitated by ipsilateral sound at some levels and inhibited at other levels (Covey 1993a). Because so few data have been obtained in nonecholocating mammals, it is impossible to say whether these response properties are common to the DNLL of all mammals or whether they are peculiar to FM bats. In either case, binaural neurons in the DNLL clearly play a role in shaping, through GABAergic inhibition, the binaural responses of neurons at the IC (Li and Kelly 1992); this must also be the case in echolocating bats.

In the cat, the DNLL contains a complete tonotopic representation that is clearly separate from the tonotopic representation in the more ventral parts of the lateral lemniscal nuclei (Aitkin, Anderson, and Brugge 1970; Brugge, Anderson, and Aitkin 1970). In horseshoe bats and big brown bats, the tonotopic representation in the DNLL is not as highly organized as in the cat (Metzner and Radtke-Schuller 1987; Covey and Casseday 1991). Nevertheless, even in these species, the tonotopy in the DNLL is distinct from that in the intermediate and ventral nuclei of the lateral lemniscus.

Although the role of DNLL in binaural hearing is almost certainly the same in echolocating bats as it is in nonecholocating mammals, there are several features of DNLL responses that may be specializations for echolocation. In the CF-FM species, the rufous horseshoe bat, some DNLL neurons have multiple BFs that are harmonically related. Because the BFs of these multiple-tuned neurons are not within the CF harmonic ranges, it has been suggested that they may play a role in range determination, which depends on the FM frequency ranges, or in social communication between individuals (Metzner and Radtke-Schuller 1987).

In the FM species, the big brown bat, some DNLL neurons are delay tuned in that they respond in a highly facilitated manner to the second of two identical sounds when it occurs at a specific time after the first. The best delays of these neurons are all within the range that would be relevant for echolocation behavior (Covey 1993a). Similar facilitated responses to the second of two identical sounds have also been described in the lateral lemniscus of another FM species, the little brown bat (Suga and Schlegel 1973). Although recording sites were not localized in the latter study, it seems likely, given the data in the big brown bat, that the facilitated neurons were in the DNLL. Delay-tuned neurons in the DNLL of FM bats may represent the first stage of neural selectivity for echoes that originate from objects at specific distances from the bat.

3.7.2 Efferent System

Although the ascending lemniscal pathways are the largest and most prominent outputs from cell groups in the SOC, there is also a system of decending pathways that are thought to provide feedback to modulate neural activity in the cochlea and cochlear nucleus. Although the organization of the efferent system differs greatly even among nonecholocating mammals, a few generalizations can be made. In most mammals, two distinct populations of neurons project to the cochlea. These are large olivocochlear neurons that terminate on outer hair cells and small olivo-cochlear neurons that terminate on inner hair cells (Warr and Guinan 1979; Guinan, Warr, and Norris 1983; Aschoff and Ostwald 1987; Warr 1992). The projections of olivocochlear neurons have been described in a number of different species of echolocating bats, including the greater horseshoe bat (Bruns and Schmieszek 1980), the mustached bat (Bishop and Henson 1987, 1988), the rufous horseshoe bat (Aschoff and Ostwald 1987; Bishop and Henson 1988), *Hipposideros lankadiva* (Bishop and Henson 1988), and *Rhinopoma*, *Tadarida*, and *Phyllostomus discolor* (Aschoff and Ostwald 1987). The pattern of distribution of large and small olivocochlear neurons in different species of bats is summarized in Figure 6.13. The reason for this seemingly inordinate amount of comparative data is the finding that certain species of bats lack efferent projections to the outer hair cells (see Kössl and Vater, Chapter 5, this volume). These species include the greater horseshoe bat, the rufous horseshoe bat, and *Hipposideros lankadiva*. All these species are horseshoe bats that use CF-FM calls. In the other unrelated CF-FM species, the mustached bat, each outer hair cell receives a single large efferent ending. Corresponding to the lack of outer hair cell innervation, horseshoe bats have only one population of olivocochlear neurons. These are small cells located in a region between the LSO and MSO, and they project only to the ipsilateral cochlea (Aschoff and Ostwald 1987).

In the mustached bat, two populations of neurons project to the cochlea. First, a population of small olivocochlear neurons located ipsilaterally in a region between the LSO and MSO probably innervate the inner hair cells.

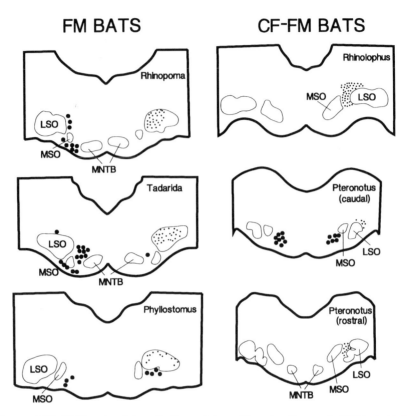

FIGURE 6.13. Different origins of olivocochlear efferent pathways in FM bats (*left*) and CF-FM bats (*right*). In FM bats, as in rodents, small olivocochlear neurons are located within the ipsilateral LSO. In CF-FM bats, they are ipsilateral, but outside the LSO, mainly in the region between the LSO and MSO. Large olivocochlear neurons are mostly located medial and ventral to the MSO. In some species they are bilateral, but in others they are contralateral only. In horseshoe bats, large olivocochlear neurons are absent. (Data for *Rhinopoma*, *Tadarida*, *Phyllostomus*, and the horseshoe bat are redrawn from Aschoff and Ostwald 1987; data for the mustached bat are redrawn from Bishop and Henson 1988.)

Second, a population of large olivocochlear neurons located bilaterally in a region caudal to the MSO probably innervate the outer hair cells (Bishop and Henson 1987, 1988). In the FM bats *Rhinopoma* and *Tadarida*, and in the whispering FM bat *Phyllostomus*, the organization of the olivocochlear efferents is very different. In all three species, small olivocochlear neurons are found inside the ipsilateral LSO, where they make up a significant proportion of the total cells. Large olivocochlear neurons are present bilaterally in the ventral and medial periolivary regions. Except for the lack of small olivocochlear neurons in the contralateral LSO, the organization in these bat species resembles that in rodents such as the guinea pig (Aschoff and Ostwald 1987). Thus, there is a clear difference between CF-FM bats and FM bats. The efferent systems of the FM bats are similar to those of rodents while those of the CF-FM bats are highly specialized and deviate considerably from what seems to be the general mammalian plan of organization.

Decending projections from the SOC terminate not only in the cochlea, but also in the cochlear nucleus. In the rufous horseshoe bat, efferents to all three divisions of the cochlear nucleus originate in the ventral periolivary cell groups bilaterally and in the LNTB ipsilaterally (Vater and Feng 1990). This pattern of efferent projections is essentially the same as that seen in the FM species, the big brown bat (Covey, unpublished data), suggesting that whereas the origins of efferents to the cochlea in CF-FM bats are very different from those in FM bats, the origins of efferents to the cochlear nucleus are similar.

3.7.3 The Central Acoustic Tract: An Extralemniscal Pathway to the Thalamus

In the lemniscal pathway, a variety of signals from the auditory brainstem reach the thalamus indirectly via one or more synapses in the inferior colliculus, which is often thought of as an obligatory relay (Goldberg and Moore 1967; Aitkin and Phillips 1984). However, the central acoustic tract, an extralemniscal pathway from the auditory medulla to the thalamus, was first described in nonecholocating mammals, including man (Papez 1929a,b). Perhaps because of the small size of this pathway relative to the lemniscal system, it has largely been ignored. However, in some echolocating bats, the central acoustic tract is large and robust. In the mustached bat it originates in a group of large multipolar neurons just rostral to the principal nuclei of the SOC, the nucleus of the central acoustic tract (NCAT) (Kobler, Isbey, and Casseday 1987; Casseday et al. 1989). The NCAT receives bilateral input from the cochlear nucleus and gives rise to a fiber bundle that courses medial to the lateral lemniscus and ventral to the inferior colliculus. It terminates in the deep layers of the superior colliculus and in the suprageniculate nucleus of the thalamus, a group of large cells just medial to the medial geniculate body. The suprageniculate nucleus, in turn, projects to the auditory cortex and frontal cortex. These connections are summarized in Figure 6.14. The fact that the central acoustic tract is especially highly developed in some species of echolocating bats suggests that it may play a role in specialized motor behaviors such as Doppler-shift compensation (Casseday et al. 1989).

3.7.4 The Paraleminscal Tegmentum and Auditory–Vocal Interactions

In echolocating bats, there is evidence for interaction between the auditory pathways of the lower brainstem and the motor systems for vocalization. In experiments on the gray bat, Suga and Schlegel (1972) and Suga and

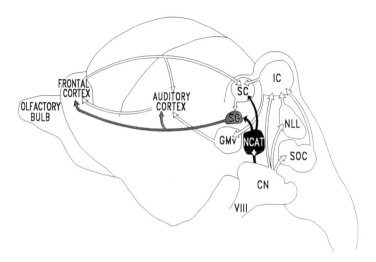

FIGURE 6.14. The central acoustic tract in the mustached bat and its relationship with the lemniscal auditory pathways and pathways to the cortex. The components of the central acoustic tract are shown by the *black arrows* and the lemniscal system by the *white arrows*. The *grey arrows* indicate further projections beyond the thalamic target of the central acoustic tract. NCAT, nucleus of central acoustic tract; IC, inferior colliculus; GMv, central division of medial geniculate body. (From Casseday et al. © 1989. J. Comp. Neurol. 287 reprinted by permission of Wiley-Liss, a division of John Wiley and Sons, Inc.)

Shimozawa (1974) measured evoked potentials in response to self-vocalized FM sounds and tape recordings of the same sounds. They found that in the region of the lateral lemniscus, the responses to self-vocalized sounds were considerably smaller than the responses to artificial sounds and suggested that the nuclei of the lateral lemniscus are the site of neural attenuation of vocalized sounds. Experiments on the rufous horseshoe bat (Metzner 1989; 1993) showed that the responses of many neurons in the paralemniscal tegmentum are affected by vocalization; about half of these neurons were inhibited when the bat vocalized. Thus, it is possible that the origin of the neural attenuation described in the early studies on the gray bat originated in the paralemniscal tegmentum.

Other auditory-vocal neurons in the paralemniscal tegmentum of the horseshoe bat show a variety of complex interactions between sound-evoked responses and vocal activity. Metzner (1989, 1993) has suggested that at least some of these neurons may play an active role in Doppler-shift compensation. Electrical stimulation of the paralemniscal tegmentum in several bat species elicits species-specific echolocation sounds (Suga et al. 1973; Schuller and Radtke-Schuller 1990). Thus it appears that the paralemniscal tegmentum acts as an interface between the sensory pathways of the auditory system and the motor pathways involved in vocalization.

4. Monaural Nuclei of the Lateral Lemniscus

In all mammals, groups of cell bodies located among the ascending fibers of the lateral lemniscus receive input from the cochlear nucleus and project to the IC. In echolocating bats these nuclei are hypertrophied and highly differentiated. Like the nuclei of the SOC, the nuclei of the lateral lemniscus represent a stage of specialized neural processing that is accomplished before the outputs of the different brainstem pathways converge at the IC. With the exception of the DNLL, which is clearly part of the binaural system, all the cell groups of the lateral lemniscus receive the bulk of their input from the contralateral cochlear nucleus and project densely and almost exclusively to the ipsilateral IC (for review, see Schwartz 1992). Thus, the monaural nuclei of the lateral lemniscus together make up a system of pathways that is organized in parallel to those from the SOC to the IC.

4.1 Structure

In nonecholocating animals such as cats and gerbils, the monaural nuclei of the lateral lemniscus are at least as large in size as the SOC, and their projections to the IC are at least equal in magnitude to those from the SOC (Adams 1979). In echolocating bats, the monaural nuclei of the lateral lemnsicus are extraordinarily large relative to the rest of the brainstem, and are exquisitely differentiated into morphologically distinct regions (Poljak 1926; Baron 1974; Zook and Casseday 1982a; Covey and Casseday 1986).

In all species of echolocating bats that have been examined, there are at least four separate cell groups embedded among the fibers of the lateral lemniscus: the dorsal nucleus (DNLL), the intermediate nucleus (INLL), and two parts of the ventral nucleus (VNLL), the columnar division (VNLLc) and the multipolar cell division (VNLLm). All of these cell groups are easily distinguished from one another on the basis of the arrangement and morphology of their neurons. Figure 6.15 shows the overall organization of the nuclei of the lateral lemniscus in the mustached bat. The INLL is a thick triangular wedge-shaped structure, which in many bats is so large that it forms a conspicuous protrusion on the side of the brainstem. It is bounded medially and laterally by thick bundles of ascending fibers. Fine fascicles of fibers course throughout the INLL, some parallel to the ascending fibers and others orthogonal to them. The principal cell type in the INLL is elongate, with dendrites that are oriented orthogonal to the ascending fibers of the lateral lemniscus. The VNLLc of echolocating bats is especially prominent because of the unusually high packing density and distribution of cells in columns separated by thin fiber bundles. Virtually every neuron in VNLLc is of the same type, small and round, with one thick dendrite that branches extensively at some distance from the cell

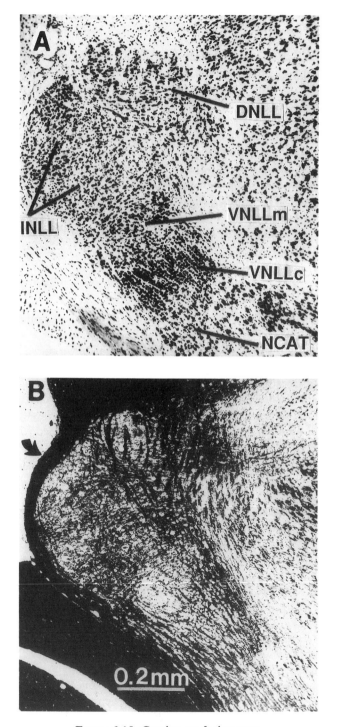

FIGURE 6.15. Caption on facing page.

body. They are very similar in appearance to spherical bushy cells in the AVCN. They appear to receive terminals mainly or exclusively on the cell body, some as large, calyx-like endings (Zook and Casseday 1985; Covey and Casseday 1986). As its name implies, the principal cell type in VNLLm is multipolar in shape. The dendrites of the multipolar cells do not appear to have any preferred orientation.

All echolocating bats have the same cell populations in INLL and VNLL, but the cells in the VNLL are distributed somewhat differently, according to species. Examples of these different patterns of organization are shown in Figure 16.6. In the mustached bat, the VNLLc is most ventral and the VNLLm lies between the INLL and the VNLLc. In the big brown bat and other vespertilionid bats, the locations of the two ventral nuclei are reversed, with the VNLLc located in the middle and the VNLLm in the ventralmost position. In horseshoe and hipposiderid bats, the VNLLc and VNLLm are side by side, with the VNLLc medial and the VNLLm lateral. Thus, just as in the case of SOC organization, there is a strong relationship between taxonomy and VNLL morphology.

In nonecholocating mammals there is no clear segregation of cell types within the INLL and VNLL, as there is in the bat, although the same basic cell types seem to be present (Adams 1979). Echolocating dolphins resemble bats in that the MNTB and monaural nuclei of the lateral lemniscus are hypertrophied and the VNLL is differentiated into columnar and multipolar cell regions (Zook et al. 1988). The similarity between bats and dolphins suggests that the neural processing performed by the monaural nuclei of the lateral lemniscus must be of crucial importance for echolocation. The variability in the location of the two cell populations of VNLL among different taxonomic groups of bats suggests that segregation of bushy cells from multipolar cells is somehow important for echolocation, but this segregation is accomplished in different ways in different species.

4.2 Inputs to the Monaural Nuclei of the Lateral Lemniscus

Anatomical studies in echolocating bats and nonecholocating mammals show that the INLL and VNLL are major targets of projections from the ventral cochlear nucleus and that this pathway is almost entirely contralateral (Warr 1966, 1969, 1982; Glendenning et al. 1981; Zook and

←——————————————————

FIGURE 6.15A,B. The nuclei of the lateral lemniscus in the mustached bat. (A) Nissl-stained frontal section through the left brainstem; (B) fiber-stained frontal section through the left brainstem. (From Zook and Casseday © 1982. J. Comp. Neurol, 207 reprinted by permission of Wiley-Liss, a division of John Wiley and Sons, Inc.)

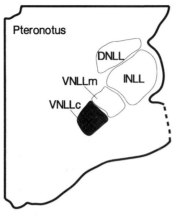

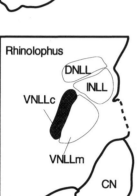

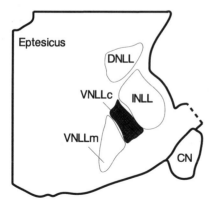

FIGURE 6.16. Schematic drawings of the organization of the nuclei of the lateral lemniscus in three different species of echolocating bats. The DNLL and INLL are similar in all three species, but the organization of the VNLL is variable. In the mustached bat (*top*), the VNLLm is dorsal to the VNLLc; in the horseshoe bat (*middle*), the two divisions are side by side with VNLLm lateral and VNLLc medial, and in the big brown bat (*lower*), the VNLLc is dorsal to the VNLLm.

Casseday 1985). In all mammals, the projections to the INLL and VNLL originate in both the AVCN and PVCN, probably from several different cell types with different response properties (Zook and Casseday 1985; Covey and Casseday 1986; Friauf and Ostwald 1988; Covey 1993b). Additionally, the INLL and VNLL in all mammals receive indirect projections from the contralateral cochlear nucleus via the MNTB and periolivary nuclei (Glendenning et al. 1981; Zook and Casseday 1985, 1987).

In the mustached bat, injections of retrograde tracers placed in the INLL label many cells in the AVCN, MNTB, and periolivary cell groups; only a few cells are labeled in the PVCN. Retrograde tracers placed in the VNLL label many cells in the AVCN and PVCN; only a few cells are labeled in MNTB and periolivary cell groups (Zook and Casseday 1985). A similar pattern of projections is seen in the cat (Glendenning et al. 1981; see Covey 1993b for review of connections in nonecholocating mammals). These results suggest that in all mammals including bats the AVCN projects densely throughout the entire region of the INLL and VNLL, but the MNTB and PVCN project with a dorsal-to-ventral density gradient. Thus, the most dense MNTB projection is to the INLL and the most dense PVCN projection is to the VNLLm.

Injections of anterograde tracers in the cochlear nucleus of both cats and bats result in multiple patches or bands of anterograde label in the region ventral to DNLL (Covey 1993b). Similarly, injection of HRP in axons of cells in the cochlear nucleus of the rat shows that each axon provides several collaterals that terminate in the region below the DNLL (Friauf and Ostwald 1988). Taken together, this evidence strongly suggests that the region below the DNLL contains multiple tonotopic representations; these multiple representations originate, at least partly, from collaterals of single axons.

4.3 Physiology of the Monaural Nuclei of the Lateral Lemniscus

Despite a convincing body of anatomical evidence to indicate that the monaural nuclei of the lateral lemniscus provide a major system of parallel pathways to the midbrain, there have to date been only four published studies of single unit response properties in the INLL and VNLL of any animal. Two of these are in the cat (Aitkin, Anderson, and Brugge 1970; Guinan, Norris, and Guinan 1972), and two are in the echolocating species, the rufous horseshoe bat (Metzner and Radtke-Schuller 1987) and the big brown bat (Covey and Casseday 1991). An abstract is also available on the mustached bat (O'Neill, Holt, and Gordon 1992). Despite a few inconsistencies among the results reported in these studies, the data present a reasonably coherent picture of the response properties of neurons in the mammalian INLL and VNLL, and provide functional evidence for the importance of these structures in echolocation.

4.3.1 Monaurality

As would be expected from the connections, the responses of most neurons in the INLL and VNLL are monaural, driven by sound at the contralateral ear. It was on the basis of monaural versus binaural responses in the cat lateral lemniscus that the distinction was first made between the DNLL, where responses are binaural, and the auditory region ventral to DNLL, where responses are nearly all monaural (Aitkin, Anderson, and Brugge 1970; Brugge, Anderson, and Aitkin 1970). In the big brown bat, all units in the VNLLc and VNLLm and nearly all units in the INLL are monaural, excited by sound at the contralateral ear. The few binaural units found in the INLL are at marginal locations, mainly near the border with the DNLL (Covey and Casseday 1991; Covey 1993a).

4.3.2 Tonotopy and frequency tuning.

In the CF-FM horseshoe bat as well as the FM species, the big brown bat, the INLL, VNLLc, and VNLLm each has a separate and distinct tonotopic representation. The INLL is organized so that low frequencies are represented laterally and high frequencies medially. In the VNLLc of both the big brown bat and the horseshoe bat, there is a well-defined tonotopy in which low frequencies are dorsal and high frequencies are ventral. However, there is one major difference between the two species. Whereas the VNLLc of the big brown bat contains a complete tonotopic sequence, in the horseshoe bat the CF_2 frequency range is absent (Metzner and Radtke-Schuller 1987). This finding is particularly striking in view of the fact that the CF_2 frequency range is greatly expanded in most other auditory nuclei in the horseshoe bat. It has also been reported that in the mustached bat, another CF-FM species, neurons in VNLLc do not respond to the CF harmonics of the echolocation call (O'Neill, Holt, and Gordon 1992). The absence of CF representation suggests that the VNLLc is specialized for processing the FM components of the echolocation call. This idea is supported by evidence that neurons in the VNLLc are the source of an extemely precise timing signal that provides the information necessary to determine the delay between the emission of the FM component of the echolocation signal and the return of the FM component of the echo. This delay information is essential for target range determination (Covey and Casseday 1991).

The VNLLm in the big brown bat has only a rough tonotopic organization. A central core of neurons tuned to high frequencies appears to be surrounded by concentric layers of neurons tuned to progressively lower frequencies. In the rufous horseshoe bat, the situation is much more-straightforward: all neurons in the VNLLm have BFs within the CF_2 range around 79 kHz. This highly specialized frequency representation indicates that the role of the VNLLm in CF-FM bats must be in processing the CF component of the echolocation call. The lack of a precise tonotopic map in either species suggests that a highly organized tonotopy is not important for the function of VNLLm.

Neurons in the INLL and VNLLm are not sharply tuned for sound frequency, and many neurons in the VNLLc in the big brown bat have extremely broad tuning curves. The frequency tuning of the majority of neurons in the INLL and VNLL basically reflects that of neurons in the cochlear nucleus (Neuweiler and Vater 1977; Metzner and Radtke-Schuller 1987; Covey and Casseday 1991; Haplea, Covey, and Casseday 1994). There is no evidence in any species to indicate that sharpness of frequency tuning increases in the nuclei of the lateral lemniscus; in the big brown bat, frequency tuning may actually become broader in the VNLLc. For example, in the big brown bat, frequency tuning has been compared at several brainstem levels (Fig. 6.17). The broadest tuning curves are found in the

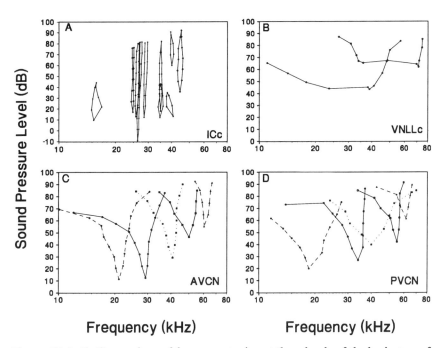

FIGURE 17.A–D Comparison of frequency tuning at three levels of the brainstem of the big brown bat. (A) The level-tolerant narrow-band tuning curves and closed tuning curves are found in the IC but not in the nuclei of the lateral lemniscus or cochlear nucleus. (B) Neurons in VNLLc are responsive to a very broad range of frequencies at only a few decibels above threshold. (C) Neurons in the AVCN have V-shaped tuning curves that resemble those found in AVCN of other mammals. (D) Likewise, neurons in PVCN have V-shaped tuning curves. (Modified from Haplea, Covey, and Casseday 1994).

VNLLc (Fig. 6.17B). These curves can be compared with the V-shaped tuning curves of neurons in the AVCN and PVCN (Fig. 6.17C,D) and with the "filter" and "closed" type tuning curves seen in the inferior colliculus (Fig. 6.17A). The broad frequency tuning of VNLLc neurons probably reflects integration across frequency to increase precision in the temporal domain (Covey and Casseday 1991; Covey 1993b).

Broad frequency tuning in the VNLLc may be a feature common to mammalian auditory systems rather than a specific adaptation for echolocation. In the cat, most units in the VNLL also have V-shaped tuning curves, but some "wide" tuning curves have been reported, possibly similar to those seen in VNLLc of the big brown bat (Guinan, Norris, and Guinan 1972). Thus, broad frequency tuning may be a correlate of the processing mechanisms common to the VNLL of all mammals. However, in echolocating bats, this processing mechanism has acquired an increased degree of importance.

4.3.3 Discharge Patterns and Timing Characteristics

Although neurons in the INLL, VNLLc, and VNLLm of the big brown bat exhibit a variety of response properties, they share certain features that make them ideally suited to transmit information about the timing of auditory events (Covey and Casseday 1991; Covey 1993b). Because neurons throughout these nuclei have little or no spontaneous activity, the occurrence of spikes that are synchronized to temporal features of the sound provides an unambiguous timing signal. Our main examples are from studies in the big brown bat, the species for which the most physiological information is available.

The discharge patterns of neurons in the INLL and VNLL include both sustained and transient responses. This is true in horseshoe bats and big brown bats, as well as in the cat (Aitkin, Anderson, and Brugge 1970; Metzner and Radtke-Schuller 1987; Covey and Casseday 1991). Based on regularity of firing, the sustained responses include choppers, which have regular interspike intervals, and nonchoppers, which have irregular interspike intervals. On the basis of the amount by which discharge rate decreases over the course of a response, sustained discharges can be classified as adapting or nonadapting. In the big brown bat, at least, there is some segregation of response types.

Transient responses in the INLL, VNLLc, and VNLLm typically consist of one or a few spikes. The most striking class of transient responding cells, found mainly in the VNLLc, are those with "phasic constant latency" responses (Covey and Casseday 1991). Virtually all neurons in the VNLLc of the big brown bat, and probably also in the mustached bat (O'Neill, Holt, and Gordon 1992), are phasic constant latency responders. These neurons discharge only one spike per stimulus, with a standard deviation in latency less than 0.1 msec. In addition, the latency of the first spike is virtually independent of stimulus level and frequency. Figure 6.18 shows a comparison of the timing properties of a neuron in VNLLc with those of neurons in AVCN and PVCN. A phasic constant latency neuron provides an extremely precise marker of the time of onset of a sound, either for the onset of a tone at any frequency within its range of sensitivity, or for the time when a frequency- or amplitude-modulated stimulus enters its range of sensitivity (Suga 1970; Pollak et al. 1977; Bodenhamer, Pollak, and Marsh 1979; Bodenhamer and Pollak 1981; Covey and Casseday 1991). Thus, phasic constant latency responses may be important in providing timing markers for specific portions of the bat's echolocation call and the returning echoes.

Neurons in the VNLLc are not the only ones that exhibit constant latency. These properties are also seen in the responses of some neurons in the INLL, VNLLm, and DNLL of the big brown bat, although they are not common in these areas (Covey 1993a). Not all transiently responding neurons are constant latency; in fact, the majority of transient responses in INLL and VNLLm have a variable first-spike latency that shifts with changes in sound amplitude and frequency much as in the auditory nerve and cochlear nucleus (see Fig. 6.18C–F).

In the INLL and VNLLm of the big brown bat, about two-thirds of neural responses are sustained. Most of the sustained responses in both INLL and VNLLm are the nonadapting type, which means that a neuron continues to fire at a fairly constant rate as long as the frequency and amplitude of a sound remain within its range of sensitivity. Sustained, nonadapting responses may be important in transmitting information about the duration of specific components of the echolocation call and the returning echo.

As in the cochlear nucleus, nearly all neurons in the monaural nuclei of the lateral lemniscus, whether transient or sustained responders, have very short integration times. These neurons respond robustly to stimuli a millisecond or less in duration, so that they should be capable of responding to transient auditory events such as an FM sound that sweeps rapidly through their range of frequency sensitivity, or to rapid frequency or amplitude modulations. In fact, neurons in the VNLLc of the mustached bat do synchronize their discharge to follow high rates of sinusoidal amplitude modulations, but they are unresponsive to modulation rates of less than about 600 Hz (O'Neill, Holt, and Gordon 1992). This selectivity of VNLLc neurons for high rates of amplitude modulation is in contrast to neurons in the cochlear nucleus, which have no lower limit for synchronizing to (SAM) (Vater 1982), and neurons in the MSO, which have an upper limit of about 200–300 Hz (Grothe 1990, 1994). This comparison further suggests that VNLLc neurons are specialized to respond to stimuli having very rapid changes in amplitude or frequency.

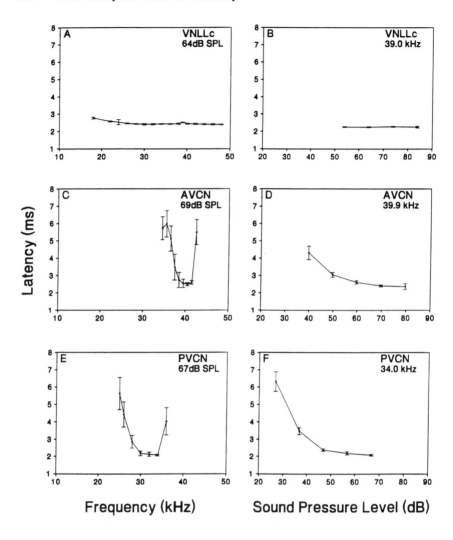

Frequency (kHz)   Sound Pressure Level (dB)

FIGURE 6.18A–F Comparison of the effects of changing sound frequency and amplitude on the latency variability of neurons in the VNLLc, AVCN, and PVCN of the big brown bat. In all graphs, mean first-spike latency is plotted as a function of frequency or amplitude; *vertical bars* represent standard deviations. (A) For neurons in the VNLLc, there is little change in mean latency across a wide range of sound frequencies; latency variability is uniformly low. (B) In the VNLLc, latency values and variability also remain constant across a wide range of sound amplitudes. (C) Responses of a neuron in AVCN to a sound 30 dB above threshold as frequency was varied. At BF, latency variability was as low as for the neuron in VNLLc. However, as frequency deviated from BF, mean first-spike latency of the AVCN neuron increased by several milliseconds and latency variability also increased. (D) Responses of the same neuron in AVCN to a sound at BF as amplitude was varied. Mean latency and latency variability progressively decreased as sound level was

4.4 Selection and Enhancement of Temporal Features of Sound: Two Streams of Processing

Both connectional and physiological evidence indicate that the INLL, VNLLc and VNLLm receive convergent parallel inputs from cell populations in the cochlear nucleus and other brainstem structures such as the MNTB and certain periolivary nuclei. These inputs are then transformed in a way that not only preserves the temporal pattern of sounds, but selects and enhances certain features of this pattern. Like auditory nerve fibers and certain neurons in the ventral cochlear nucleus, neurons throughout the INLL and VNLL respond robustly to transient stimuli of short duration. Therefore, they can follow very rapid frequency or amplitude modulations. It has been shown that low-BF globular cells in the cochlear nucleus are more precise phase lockers than are auditory nerve fibers (Smith et al. 1991); it therefore seems likely that the enhancement of temporal features is a multistage process that begins in the cochlear nucleus and culminates in the nuclei of the lateral lemniscus. In echolocating bats, the important temporal features may be rapid frequency or amplitude modulations or a rapid sequence of echolocation pulses and their echoes.

As summarized in Figure 6.19, the neurons in INLL and VNLL can be divided into two broad populations. The first population consists of the phasic constant latency neurons of the VNLLc, which signal very precisely the onset time of a sound. The second population includes the sustained nonadapting neurons of the INLL and VNLLm, which respond continuously throughout the time a sound is present and thus can transmit information about its intensity and duration. At least in the bat, these two streams of processing can clearly be distinguished from one another on the basis of their neuronal morphology, the type of terminals supplied by axons originating in the ventral cochlear nucleus and breadth of frequency tuning. In addition, they differ in their patterns of termination in the IC.

The population of onset encoders includes virtually all the neurons in the VNLLc. These cells have shorter response latencies than any others in the INLL, VNLL, or the nuclei of the SOC, averaging just a little over 3 msec. They appear to be glycinergic and provide dense and widespread projections to the IC, possibly organized so that a single sheet of cells in the VNLLc sends divergent projections to a broad frequency range within the IC

FIGURE 6.18A–F. (*continued*) increased. (E) Responses of a neuron in PVCN to a sound 40 dB above threshold as frequency was varied. At BF, latency variability was low, but as frequency deviated from BF, mean first-spike latency and latency variability increased. (F) Responses of the same neuron in PVCN to a sound at BF as amplitude was varied. Mean latency and latency variability progressively decreased as sound level was increased. (From Haplea, Covey and Casseday 1994)

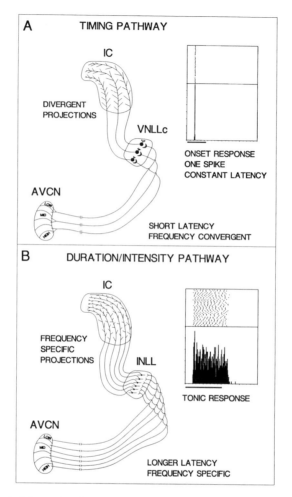

FIGURE 6.19A,B Schematic diagram of hypothesis that there are two different streams of processing within the monaural auditory pathways of the brainstem, each transmitting a different class of information about temporal features of sounds. (A) Constant latency neurons in VNLLc respond with one spike locked to the onset of a sound. (B) Neurons with sustained responses in INLL respond continously throughout the duration of a sound. The insets show representative post-stimulus time histograms (PSTHs) (*lower parts*) and dot raster displays (*upper parts*) of spikes. The bar below each horizontal axis represents the stimulus duration.

(Covey and Casseday 1986). Thus, the output of the VNLLc is short latency, frequency tolerant and level tolerant, and is distributed widely in the inferior colliculus. The VNLLc provides the IC with very precise information about the onset of sound and about frequency and intensity transitions of sound. The VNLLc may be the source of transient inhibition that precedes excitatory responses in the IC. (Covey et al. 1993)

The population of intensity/duration encoders include the nonadapting sustained units in the INLL and VNLLm. These units probably vary somewhat in their morphology, but all have multiple large dendrites that are contacted by punctate terminals of axons originating in the ventral cochlear nucleus. Many of the sustained units have a relatively broad dynamic range over which firing rate changes in response to sound pressure level. Their responses would therefore provide a signal either of stimulus duration for a tone, or for an FM signal, the dwell time within their range of frequency-sensitivity. They respond with latencies that are 1 to several milliseconds longer than those of VNLLc neurons, averaging about 4–5 msec, and spanning a considerably wider range of latencies. Neurons in INLL and VNLLm appear to project in a tonotopic manner to the IC. The outputs of both timing pathways probably play an important role in shaping the responses of IC neurons to temporal features such as the duration of sounds (Casseday, Ehrlich, and Covey 1994).

4.5 Outputs of the Monaural Nuclei of the Lateral Lemniscus

The INLL, VNLLc, and VNLLm give rise to parallel pathways that terminate in the ipsilateral IC. Both anterograde and retrograde tracing experiments show that all these pathways terminate densely in the ventral two-thirds of the IC in a region that is largely if not entirely coextensive with the targets of the AVCN, LSO, MSO, and DNLL. In at least some species of echolocating bats, this region is also the target of projections from the PVCN (Schweizer 1981; Zook and Casseday 1982b, 1987; Covey and Casseday 1986).

5. Summary: Convergence and Integration of Parallel Pathways at the Inferior Colliculus

The auditory system described up to this point is really a set of very separate subsystems. Each subsystem has its own tonotopy, and each provides some separate and unique transformation of the auditory stimulus. However, the outputs of the subsystems converge at the inferior colliculus in a manner that reassembles the separate components back into one tonotopy. Because it is clear that a major reorganization of the input signals occurs at the IC, we conclude by reviewing the variety of these signals.

Inputs to the IC arrive directly from the cochlear nucleus and indirectly via one or more stages of processing. Inputs are both monaural and binaural. Monaural inputs arise from the three divisions of the cochlear nucleus, and from the INLL and VNLL. In addition, the MSO provides mainly monaural input in bats. Binaural inputs arise from the LSO and DNLL. Either monaural or binaural inputs may be excitatory or inhibitory.

For example, the DNLL provides almost exclusively inhibitory inputs, whereas the LSO provides excitatory and inhibitory inputs. The cochlear nuclei provide mainly excitatory inputs but may provide some inhibitory input. The INLL and VNLL contribute substantial inhibitory inputs.

When one considers the temporal characteristics of the responses in the different pathways, it becomes clear that much of the processing in the IC concerns the integration of inputs with different temporal patterns. First, different pathways have different latencies. The various inputs arrive at the IC at different times, depending on the latency of the pathway through which they are transmitted. This range of input latencies has the potential to create a prolonged and complex sequence of inhibitory and excitatory synaptic events at the IC, a sequence that is highly labile and can change depending on which systems are active and the order in which they are activated. Second, in a manner analogous to the XYZ system in visual pathways, the different auditory pathways in the lower brainstem have different temporal patterns of response. The temporal patterns range from a single spike locked to the onset or offset of a sound, to chopper or pauser patterns, to a continuous train of spikes that persists for the entire duration of a sound or even longer in the case of an afterdischarge. These differences in discharge pattern mean that a variety of excitatory and inhibitory inputs with activity distributed differentially over time could interact at the IC to produce complex sequences of postsynaptic excitation and inhibition (Casseday, Ehrlich, and Covey 1994). These sequences could last for tens of milliseconds or longer, thereby modulating the response of a neuron to subsequently occurring sounds. The resulting time-varying changes in the state of the neuron might be manifested in the simplest case as windows of facilitation or suppression, or in a more complex case as neural filters for biologically important temporal sequences of sounds.

Acknowledgments The authors were supported by U.S National Institutes of Health (NIH) grants DC-00607 (E.C.) and DC-00287 (J.H.C.) during the preparation of this chapter. Special thanks to Boma Rosemond for help in preparing the illustrations.

References

Adams JC (1976) Single unit studies on the dorsal and intermediate acoustic striae. J Comp Neurol 170:97–106.

Adams JC (1979) Ascending projections to the inferior colliculus. J Comp Neurol 183:519–538.

Adams JC, Mugnaini E (1984) Dorsal nucleus of the lateral lemniscus: a nucleus of GABAergic projection neurons. Brain Res Bull 13:585–590.

Adams JC, Mugnaini E (1990) Immunocytochemical evidence for inhibitory and disinhibitory circuits in the superior olive. Hear Res 49:281–298.

Aitkin LM, Phillips SC (1984) Is the inferior colliculus an obligatory relay in the cat auditory system? Neurosci Lett 44:259–264.

Aitkin LM, Anderson DJ, Brugge JF (1970) Tonotopic organization and discharge characteristics of single neurons in nuclei of the lateral lemniscus of the cat. J Neurophysiol 33:421–440.

Aschoff A, Ostwald J (1987) Different origins of cochlear efferents in some bat species, rats, and guinea pigs. J Comp Neurol 264:56–72.

Baron G (1974) Differential phylogenetic development of the acoustic nuclei among chiroptera. Brain Behav Evol 9:7–40.

Bishop AL, Henson OW (1987) The efferent cochlear projections of the superior olivary complex in the mustached bat. Hear Res 31:175–182.

Bishop AL, Henson OW (1988) The efferent auditory system in Doppler-shift compensating bats. In: Nachtigall PE, Moore PWB (eds) Animal Sonar: Processes and Performance. New York: Plenum, pp. 307–310.

Bledsoe SC, Snead CR, Helfert RH, Prasad V, Wenthold RJ, Altschuler RA(1990) Immunocytochemical and lesion studies support the hypothesis that the projection from the medial nucleus of the trapezoid body to the lateral superior olive is glycinergic. Brain Res 517:189–194.

Bodenhamer RD, Pollak GD (1981) Time and frequency domain processing in the inferior colliculus of echolocating bats. Hear Res 5:317–335.

Bodenhamer RD, Pollak GD, Marsh DS (1979) Coding of fine frequency information by echoranging neurons in the inferior colliculus of the Mexican free-tailed bat. Brain Res 171:530–535.

Bourke TR, Mielcarz JP, Norris BE (1981) Tonotopic organization of the anteroventral cochlear nucleus of the cat. Hear Res 4:215–241.

Brugge JF, Anderson DJ, Aitkin LM (1970) Responses of neurons in the dorsal nucleus of the lateral lemniscus of cat to binaural tonal stimulation. J Neurophysiol (Bethesda) 33:441–458.

Bruns V (1976a) Peripheral auditory tuning for fine frequency analysis by the CF-FM bat, *Rhinolophus ferrumequinum*. I. Mechanical specializations of the cochlea. J Comp Physiol 106:77–86.

Bruns V (1976b) Peripheral auditory tuning for fine frequency analysis by the CF-FM bat, *Rhinolophus ferrumequinum*. J Comp Physiol 107:87–97.

Bruns V, Schmieszek ET (1980) Cochlear innervation in the greater horseshoe bat: demonstration of an acoustic fovea. Hear Res 3:27–43.

Brunso-Bechtold JK, Henkel CK, Linville C (1990) Synaptic organization in the adult ferret medial superior olive. J Comp Neurol 294:389–398.

Cajal Ramon SY (1909) Histologie du systeme nerveux de l'homme et des vertebres. Tome I. Madrid: Instituto Ramon y Cajal (1952), pp. 778–848.

Cant NB (1992) The cochlear nucleus: neuronal types and their synaptic organization. In: Popper AN, Fay RR (eds) The Mammalian Auditory Pathway: Neuroanatomy. New York: Springer-Verlag, pp. 66–116.

Cant NB, Casseday JH (1986) Projections from the anteroventral cochlear nucleus to the lateral and medial superior olivary nuclei. J Comp Neurol 247:457–476.

Carr CE (1986) Time coding in electric fish and barn owls. Brain Behav Evol 28:122–134.

Casseday JH, Covey E (1987) Central auditory pathways in directional hearing. In: Yost W, Gourevitch G (eds) Directional Hearing. New York: Springer-Verlag, pp. 109–145.

Casseday JH, Covey E (1992) Frequency tuning properties of neurons in the inferior colliculus of an FM bat. J Comp Neurol 319:34–50.

Casseday JH, Covey E, Vater M (1988) Connections of the superior olivary complex in the rufous horseshoe bat, *Rhinolophus rouxi*. J Comp Neurol 278:313–329.

Casseday JH, Ehrlich D, Covey E (1994) Neural tuning for sound duration: role of inhibitory mechanisms in the inferior colliculus. Science 264:847–850.

Casseday JH, Jones DR, Diamond IT (1979) Projections from cortex to tectum in the tree shrew, *Tupaia glis*. Comp Neurol 185:253–292.

Casseday JH, Zook JM, Kuwabara N (1993) Projections of cochlear nucleus to superior olivary complex in an echolocating bat: relation to function. In: Merchan MA, Juiz JM, Godfrey DA (eds) The Mammalian Cochlear Nuclei: Organization and Function. New York: Plenum, pp. 303–319.

Casseday JH, Kobler JB, Isbey SF, Covey E (1989) The central acoustic tract in an echolocating bat: an extralemniscal auditory pathway to the thalamus. J Comp Neurol 287:247–259.

Clark GM (1969a) The ultrastructure of nerve endings in the medial superior olive of the cat. Brain Res 14:298–305.

Clark GM (1969b) Vesicle shape versus type of synapse in the nerve endings of the cat medial superior olive. Brain Res 15:548–551.

Covey E (1993a) Response properties of single units in the dorsal nucleus of the lateral lemniscus and paralemniscal zone of an echolocating bat. J Neurophysiol (Bethesda) 69:842–859.

Covey E (1993b) The monaural nuclei of the lateral lemniscus: parallel pathways from cochlear nucleus to midbrain. In: Merchan MA, Juiz JM, Godfrey DA (eds) The Mammalian Cochlear Nuclei: Organization and Function. New York: Plenum, pp. 321–334.

Covey E, Casseday JH (1986) Connectional basis for frequency representation in the nuclei of the lateral lemniscus of the bat, *Eptesicus fuscus*. J Neurosci 6:2926–2940.

Covey E, Casseday JH (1991) The monaural nuclei of the lateral lemniscus in an echolocating bat: parallel pathways for analyzing temporal features of sound. J Neurosci 11:3456–3470.

Covey E, Hall WC, Kobler JB (1987) Subcortical connections of the superior colliculus in the mustache bat, *Pteronotus parnellii*. J Comp Neurol 263:179–197.

Covey E, Vater M, Casseday JH (1991) Binaural properties of single units in the superior olivary complex of the mustached bat. J Neurophysiol (Bethesda) 66:1080–1094.

Covey E, Johnson BR, Ehrlich D, Casseday JH (1993) Neural representation of the temporal features of sound undergoes transformation in the auditory midbrain: evidence from extracellular recording, application of pharmacological agents and *in vivo* whole cell patch clamp recording. Neurosci Abstr 19:535.

Feng AS, Vater M (1985) Functional organization of the cochlear nucleus of rufous horseshoe bats (*Rhinolophus rouxi*): frequencies and internal connections are arranged in slabs. J Comp Neurol 235:529–553.

Friauf E, Ostwald J (1988) Divergent projections of physiologically characterized rat ventral cochlear nucleus neurons as shown by intra-axonal injection of horseradish peroxidase. Exp Brain Res 73:263–284.

Fuzessery ZM, Pollak GD (1985) Determinants of sound location selectivity in bat inferior colliculus: A combined dichotic and free-field stimulation study. J Neurophysiol 54:757–781.

Glendenning KK, Brunso-Bechtold JK, Thompson GC, Masterton R B (1981) Ascending auditory afferents to the nuclei of the lateral lemniscus. J Comp Neurol 197:673–704.

Godfrey DA, Kiang NYS, Norris BA (1975) Single unit activity in the dorsal cochlear nucleus of the cat. J Comp Neurol 162:269–284.

Goldberg JM, Brown PB (1968) Functional organization of the dog superior olivary complex: an anatomical and electrophysiological study. J Neurophysiol (Bethesda) 31:639–656.

Goldberg JM, Brownell WE (1973) Discharge characteristics of neurons in antero-ventral and dorsal cochlear nuclei of cat. Brain Res 64:35–54.

Goldberg JM, Moore RY (1967) Ascending projections of the lateral lemniscus in the cat and monkey. J Comp Neurol 129:143–156.

Grothe B (1990) Versuch einer Definition des medialen Kernes des oberen Olivenkomplexes bei der Neuweltfledermaus *Pteronotus parnellii*. Ph.D. dissertation, Ludwig-Maximilians Universität, Munich, Germany.

Grothe B (1994) Interaction of excitation and inhibition in processing of pure tone and amplitude-modulated stimuli in the medial superior olive of the mustached bat. J Neurophysiol 71:706–721.

Grothe B, Vater M, Casseday JH, Covey E (1992) Monaural interaction of excitation and inhibition in the medial superior olive of the mustached bat: an adaptation for biosonar. Proc Natl Acad Sci USA 89:5108–5112.

Guinan JJ, Norris BE, Guinan SS (1972) Single auditory units in the superior olivary complex. II: Locations of unit categories and tonotopic organization. Int J Neurosci 4:147–166.

Guinan JJ, Warr WB, Norris BE (1983) Differential olivocochlear projections from lateral vs. medial zones of the superior olivary complex. J Comp Neurol 221:358–370.

Haplea S, Covey E, Casseday JH (1994) Frequency tuning and response latencies at three levels in the brainstem of the echolocating bat, *Eptesicus fuscus*. J Comp Physiol A 174:671–683.

Harnischfeger G, Neuweiler G, Schlegel P (1985) Interaural time and intensity coding in superior olivary complex and inferior colliculus of the echolocating bat, *Molossus ater*. J Neurophysiol 53:89–109.

Harrison JM, Irving R (1966) Visual and nonvisual auditory systems in mammals. Science 154:738–743.

Helfert RH, Bonneau JM, Wenthold RJ, Altschuler RA (1989) GABA and glycine immunoreactivity in the guinea pig superior olivary complex. Brain Res 501:269–286.

Henkel CK (1983) Evidence of sub-collicular projections to medial geniculate nucleus in the cat: an autoradiographic and horseradish peroxidase study. Brain Res 259:21–30.

Henson OW (1970) The central nervous system. In: Wimsatt WA (ed) Biology of Bats, Vol. 2. New York: Academic, pp. 57–152.

Inbody SB, Feng AS (1981) Binaural response characteristics of single neurons in the medial superior olivary nucleus of the albino rat. Brain Res 210:361–366.

Irvine DRF (1992) Physiology of the auditory brainstem. In: Popper AN, Fay RR (eds) The Mammalian Auditory Pathway: Physiology. New York: Springer-Verlag, pp. 153–231.

Irving R, Harrison JM (1967) Superior olivary complex and audition: A comparative study. J Comp Neurol 130:77–86.

Kiss A, Majorossy K (1983) Neuron morphology and synaptic architecture in the medial superior olivary nucleus. Exp Brain Res 52:15–327.

Kober R, Schnitzler H-U (1990) Information in sonar echoes of fluttering insects available for echolocating bats. J Acoust Soc Am 87:874–881.

Kobler JB, Isbey SF, Casseday JH (1987) Auditory pathways to the frontal cortex of the mustache bat, *Pteronotus parnellii*. Science 236:824–826.

Kössl M, Vater M (1989) Noradrenaline enhances temporal auditory contrast and neuronal timing precision in the cochlear nucleus of the mustached bat. J Neurosci 9:4169–4178.

Kössl M, Vater M (1990) Tonotopic organization of the cochlear nucleus of the mustache bat, *Pteronotus parnellii*. J Comp Physiol A 166:695–709.

Kössl M, Vater M, Schweizer H (1988) Distribution of catecholamine fibers in the cochlear nuclei of horseshoe bats and mustache bats. J Comp Neurol 269:523–535.

Kudo M (1981) Projections of the lateral lemniscus in the cat: an autoradiographic study. Brain Res 221:57–69.

Kuwabara N, Zook JM (1991) Classification of the principal cells of the medial nucleus of the trapezoid body. J Comp Neurol 314:707–720.

Kuwabara N, Zook JM (1992) Projections to the medial superior olive from the medial and lateral nuclei of the trapezoid body in rodents and bats. J Comp Neurol 324:522–538.

Kuwabara N, DiCaprio RA, Zook JM (1991) Afferents to the medial nucleus of the trapezoid body and their collateral projections. J Comp Neurol 314:684–706.

Kuwada S, Yin TCT (1987) Physiological studies of directional hearing. In: Yost WA, Gourevitch G (eds) Directional Hearing. New York: Springer-Verlag, pp. 146–176.

Lesser HD, O'Neill WE, Frisina RD, Emerson RC (1990) On-off units in the mustached bat inferior colliculus are selective for transients resembling "acoustic glint" from fluttering insect targets. Exp Brain Res 82:137–148.

Li L, Kelly JB (1992) Inhibitory influence of the dorsal nucleus of the lateral lemniscus on binaural responses in the rat's inferior colliculus. J Neurosci 12:4530–4539.

Markovitz NS, Pollak G.D. (1994) Binaural processing in the dorsal nucleus of the lateral lemniscus Hear. Res. 73:121–140.

Masterton RB, Thompson GC, Bechtold JK, RoBards MJ (1975) Neuroanatomical basis of binaural phase-difference analysis for sound localization: a comparative study. J Comp Physiol Psychol 89:379–386.

Metzner W (1989) A possible neuronal basis for Doppler-shift compensation in echo-locating horseshoe bats. Nature 341:529–532.

Metzner, W (1993) An audio-vocal interface in echolocating horseshoe bats. J. Neurosci. 13:1899–1915.

Metzner W, Radtke-Schuller S (1987) The nuclei of the lateral lemniscus in the rufous horseshoe bat, *Rhinolophus rouxi*. J Comp Physiol 160:395–411.

Moore JK (1987) The human auditory brain stem: a comparative view. Hear Res 29:1–32.

Moore MM, Caspary DM (1983) Strychnine blocks binaural inhibition in lateral superior olivary neurons. J Neurosci 3:237–242.

Moskowitz N, Liu J-C (1972) Central projections of the spiral ganglion of the squirrel monkey. J Comp Neurol 144:335–344.

Neuweiler G, Vater M (1977) Response patterns to pure tones of cochlear nucleus units in the CF-FM bat, *Rhinolophus ferrumequinum*. J Comp Physiol A 115:119–133.

Noda Y, Pirsig W (1974) Anatomical projection of the cochlea to the cochlear nuclei of the guinea pig. Arch Otolaryngol 208:107–120.

Novick A, Vaisnys JR (1964) Echolocation of flying insects by the bat *Chilonycteris parnellii*. Biol Bull 127:478–488.

Oliver DL, Huerta MF (1992) Inferior and superior colliculi. In: Popper AN, Fay RR (eds) The Mammalian Auditory Pathway: Neuroanatomy. New York: Springer-Verlag, pp. 168–221.

Ollo C, Schwartz I (1979) The superior olivary complex in C5BL/6 mice. Am J Anat 155:349–374.

Olsen JF, Suga N (1991a) Combination sensitive neurons in the medial geniculate body of the mustached bat: encoding of relative velocity information. J Neurophysiol (Bethesda) 65:1254–1274.

Olsen JF, Suga N (1991b) Combination sensitive neurons in the medial geniculate body of the mustached bat: encoding of target range information. J Neurophysiol 65:1275–1296.

O'Neill WE, Suga N (1979) Target range-sensitive neurons in the auditory cortex of the mustache bat. Science 203:69–73.

O'Neill WE, Suga N (1982) Encoding of target range and its representation in the auditory cortex of the mustached bat. J Neurosci 2:17–31.

O'Neill WE, Holt JR, Gordon M (1992) Responses of neurons in the intermediate and ventral nuclei of the lateral lemniscus of the mustached bat to sinusoidal and pseudorandom amplitude modulations. Assoc Res Otolaryngol Abstr 15:140.

Osen KK (1969a) Cytoarchitecture of the cochlear nuclei in the cat. J Comp Neurol 136:453–484.

Osen KK (1969b) The intrinsic organization of the cochlear nuclei in the cat. Acta Otolaryngol 67:352–359.

Papez JW (1929a) Central acoustic tract in cat and man. Anat Rec 42:60.

Papez JW (1929b) Comparative Neurology. New York: Crowell, pp. 270–293.

Peyret D, Geffard M, Aran J-M (1986) GABA immunoreactivity in the primary nuclei of the auditory central nervous system. Hear Res 23:115–121.

Peyret D, Campistron G, Geffard M, Aran J-M (1987) Glycine immunoreactivity in the brainstem auditory and vestibular nuclei of the guinea pig. Acta Otolaryngol 104:71–76.

Pfeiffer RR (1966) Classification of response patterns of spike discharges for units in the cochlear nucleus: tone-burst stimulation. Exp Brain Res 1:220–235.

Poljak S (1926) Untersuchungen am Oktavussystem der Säugetiere und an den mit diesem koordinierten motorischen Apparaten des Hirnstammes. J Psychol Neurol 32:170–231.

Pollak GD, Bodenhamer R (1981) Specialized characteristics of single units in the inferior colliculus of mustache bats: frequency representation, tuning and discharge patterns. J Neurophysiol 46:605–620.

Pollak GD, Casseday JH (1989) The Neural Basis of Echolocation in Bats. Berlin: Springer-Verlag.

Pollak GD, Marsh DS, Bodenhamer R, Souther A (1977) Echo-detecting characteristics of neurons in inferior colliculus of unanesthetized bats. Science 196:675–678.

Rhode WS, Kettner RE (1987) Physiological study of neurons in the dorsal and posteroventral cochlear nucleus of the unanesthetized cat. J Neurophysiol 57:414–442.

Rose JE, Galambos R, Hughes JR (1959) Microelectrode studies of the cochlear nuclei of the cat. Bull Johns Hopkins Hosp 104:211–251.

Ross LS, Pollak GD (1989) Differential ascending projections to aural regions in the 60-kHz contour of the mustache bat's inferior colliculus. J Neurosci 9:2819–2834.

Ross LS, Pollak GD, Zook JM (1988) Origin of ascending projections to an isofrequency region of the mustache bat's inferior colliculus. J Comp Neurol 270:488–505.

Rouiller EM, Ryugo DK (1984) Intracellular marking of physiologically character-ized cells in the ventral cochlear nucleus of the cat. J Comp Neurol 225:167–186.

Ryan AF, Woolf NK, Sharp FR (1982) Tonotopic organization in the central auditory pathway of the mongolian gerbil: a 2-deoxyglucose study. J Comp Neurophysiol 207:369–380.

Ryugo DK (1992) The auditory nerve: peripheral innervation, cell body morphol-ogy, and central projections. In: Popper AN, Fay RR (eds) The Mammalian Auditory Pathway: Neuroatomy. New York: Springer-Verlag, pp. 23–65.

Schnitzler H-U, Menne D, Kober R, Heblich K (1983) The acoustical image of fluttering insects in echolocating bats. In: Huber F, Markl H (eds) Neuroethology and Behavioral Ethology: Roots and Growing Points. Berlin: Springer-Verlag, pp. 235–250.

Schuller G, Radtke-Schuller S (1990) Neural control of vocalization in bats: mapping of brainstem areas with electrical microstimulation eliciting species-specific echolocation calls in the rufous horseshoe bat. Exp. Brain Res. 79: 192–206.

Schuller G, Covey E, Casseday JH (1991) Auditory pontine grey: connections and response properties in the horseshoe bat. Eur J Neurosci 3:648–662.

Schwartz IR (1980) The differential distribution of synaptic terminal classes on marginal and central cells in the cat medial superior olivary nucleus. Am J Anat 159:25–31.

Schwartz IR (1984) Axonal organization in the cat medial superior olivary nucleus. In: Neff WD (ed) Contributions to Sensory Physiology, Vol. 8. New York: Academic, pp. 99–129.

Schwartz IR (1992) The superior olivary complex and lateral lemniscal nuclei. In: Popper AN, Fay RR (eds) The Mammalian Auditory Pathway: Neuroanatomy. New York: Springer-Verlag, pp. 117–167.

Schweizer H (1981) The connections of the inferior colliculus and the organization of the brainstem auditory system in the greater horseshoe bat (*Rhinolophus ferrumequinum*). J Comp Neurol 201:25–49.

Simmons JA (1989) A view of the world through the bat's ear: the formation of acoustic images in echolocation. Cognition 33:155–199.

Simmons JA, Lawrence BD (1982) Echolocation in bats: The external ear and perception of the vertical positions of targets. Science 218:481–483.

Smith PH, Joris PX, Carney LH, Yin TCT (1991) Projections of physiologically characterized globular bushy cell axons from the cochlear nucleus of the cat. J Comp Neurol 304:387–407.

Spangler KM, Warr RB, Henkel CK (1985) The projections of principal cells of the medial nucleus of the trapezoid body in the cat. J Comp Neurol 238:249–262.

Spirou GA, Brownell WE, Zidanic M (1990) Recordings from cat trapezoid body and HRP labeling of globular bushy cell axons. J Neurophysiol (Bethesda) 63:1169–1190.

Stotler WA (1953) An experimental study of the cells and connections of the superior olivary complex of the cat. J Comp Neurol 98:401–432.

Strominger NL, Strominger AI (1971) Ascending brain stem projections of the anteroventral cochlear nucleus in the rhesus monkey. J Comp Neurol 143:217–242.

Suga N (1964) Single unit activity in cochlear nucleus and inferior colliculus of echo-locating bats. J Physiol (Lond) 172:449–474.

Suga N (1970) Echo-ranging neurons in the inferior colliculus of bats. Science 170:449–452.

Suga N, Jen PHS (1977) Further studies on the peripheral auditory system of "CF-FM" bats specialized for fine frequency analysis of Doppler-shifted echoes. J Exp Biol 69:207–232.

Suga N, Schlegel P (1972) Neural attenuation of responses to emitted sounds in echolocating bats. Science 177:82–84.

Suga N, Schlegel P (1973) Coding and processing in the auditory systems of FM-signal-producing bats. J Acoust Soc Am 54:174–190.

Suga N, Neuweiler G, Möller J (1976) Peripheral auditory tuning for fine frequency analysis by the CF-FM bat *Rhinolophus ferrumequinum*. IV. Properties of peripheral auditory neurons. J Comp Physiol 106:111–125.

Suga N, Shimozawa T (1974) Site of neural attenuation of responses to self-vocalized sounds in echolocating bats Science 183:1211–1213.

Suga N, Simmons JA, Jen PHS (1975) Peripheral specializations for fine frequency analysis of Doppler-shifted echoes in the CF-FM bat, *Pteronotus parnellii*. J Exp Biol 63:161–192.

Takahashi TT, Carr CE, Brecha N, Konishi M (1987) Calcium binding protein-like immunoreactivity labels the terminal field of nucleus laminaris of the barn owl. J Neurosci 7:1843–1856.

Tanaka K, Otani K, Tokunaga A, Sugita S (1985) The organization of neurons in the nucleus of the lateral lemniscus projecting to the superior and inferior colliculi in the rat. Brain Res 341:252–260.

Tolbert LP, Morest DK, Yurgelun-Todd DA (1982) The neuronal architecture of the anteroventral cochlear nucleus in the cat in the region of the cochlear nerve root: horseradish peroxidase labeling of identified cell types. Neuroscience 7:3031–3052.

Vater M (1982) Single unit responses in cochlear nucleus of horseshoe bats to sinusoidal frequency and amplitude modulated signals. J Comp Physiol A 149:369–388.

Vater M, Feng AS (1990) Functional organization of ascending and descending connections of the cochlear nucleus of horseshoe bats. J Comp Neurol 292:373–395.

Vater M, Casseday JH, Covey E (1995) Convergence and divergence of ascending binaural and monaural pathways from the superior olives of the mustached bat. J Comp Neurol 351:632–646.

Von der Emde G, Schnitzler H-U (1986) Fluttering target detection in hipposiderid bats. J Comp Physiol 159:765–772.

Von der Emde G, Schnitzler H-U (1990) Classification of insects by echolocating greater horseshoe bats. J Comp Physiol A 167:423–430.

Warr WB (1966) Fiber degeneration following lesions in the anterior ventral cochlear nucleus of the cat. Exp Neurol 14:453–474.

Warr WB (1969) Fiber degeneration following lesions in the posteroventral cochlear nucleus of the cat. Exp Neurol 23:140–155.

Warr WB (1982) Parallel ascending pathways from the cochlear nucleus: neuroanatomical evidence of functional specialization. In: Neff WD (ed) Contributions to Sensory Physiology, Vol. 7. New York: Academic, pp. 1–38.

Warr WB (1992) Organization of olivocochlear efferent systems in mammals. In: Popper A N, Fay R R (eds) The Mammalian Auditory Pathway: Neuroanatomy. New York: Springer-Verlag, pp. 410–448.

Warr WB, Guinan JJ (1979) Efferent innervation of the organ of Corti: two separate systems. Brain Res 173:152–155.

Wenthold RJ, Huie D, Altschuler RA, Reeks KA (1987) Glycine immunoreactivity localized in the cochlear nucleus and superior olivary complex. Neuroscience 22:897–912.

Yang L, Pollak GD Resler C (1992) GABAergic circuits sharpen tuning curves and modify response properties in the mustache bat inferior colliculus. J Neurophysiol (Bethesda) 68:1760–1774.

Yin TCT, Chan JCK (1990) Interaural time sensitivity in medial superior olive of cat. J Neurophysiol (Bethesda) 64:465–488.

Zettel ML, Carr CE, O'Neill WE (1991) Calbindin-like immunoreactivity in the central auditory system of the mustached bat, *Pteronotus parnelli*. J Comp Neurol 313:1–16.

Zook JM, Casseday JH (1982a) Cytoarchitecture of auditory system in lower brainstem of the mustache bat, *Pteronotus parnellii*. J Comp Neurol 207:1–13.

Zook JM, Casseday JH (1982b) Origin of ascending projections to inferior colliculus in the mustache bat, *Pteronotus parnellii*. J Comp Neurol 207:14–28.

Zook JM, Casseday JH (1985) Projections from the cochlear nuclei in the mustache bat, *Pteronotus parnellii*. J Comp Neurol 237:307–324.

Zook JM, Casseday JH (1987) Convergence of ascending pathways at the inferior colliculus in the mustache bat, *Pteronotus parnellii*. J Comp Neurol 261:347–361.

Zook JM, DiCaprio RA (1988) Intracellular labeling of afferents to the lateral superior olive in the bat, *Eptesicus fuscus*. Hear Res 34:141–148.

Zook JM, Leake PA (1989) Connections and frequency representation in the auditory brainstem of the mustache bat, *Pteronotus parnellii*. J Comp Neurol 290:243–261.

Zook JM, Jacobs MS, Glezer I, Morgane PJ (1988) Some comparative aspects of auditory brainstem cytoarchitecture in echolocating mammals: speculations on the morphological basis of time-domain signal processing. In: Nachtigall PE, Moore PWB (eds) Animal Sonar: Processes and Performance. New York: Plenum, pp. 311–316.

Abbreviations

ALPO	anterolateral periolivary nucleus
AN	auditory nerve
AV, a, d, m, p	anteroventral cochlear nucleus, anterior division, dorsal division, medial division, posterior division
AVCN	anteroventral cochlear nucleus

BP	brachium pontis
CB, CER	cerebellum
CG	central grey
CP	cerebral peduncle
DCN, d, v	dorsal cochlear nucleus, dorsal, ventral
DMPO	dorsomedial periolivary nucleus
DMSO	medial superior olive, dorsal division
DNLL	dorsal nucleus of lateral lemniscus
DPO	dorsal periolivary nucleus
GM, v	medial geniculate body, ventral division
ICc	inferior colliculus, central nucleus
INLL	intermediate nucleus of the lateral lemniscus
LNTB	lateral nucleus of the trapezoid body
LSO	lateral superior olive
MNTB	medial nucleus of the trapezoid body
MSO	medial superior olive
NCAT	nucleus of the central acoustic tract
PV, a, c, l, m	posteroventral cochlear nucleus, anterior division, caudal division, lateral division, medial division
PVCN, d, v	posteroventral cochlear nucleus, dorsal division, ventral division
Pyr, Py	pyramidal tract
RB	restiform body
SC	superior colliculus
TB	trapezoid body
VII	seventh cranial nerve
VIII	eighth cranial nerve
VMPO	ventromedial periolivary nucleus
VMSO	medial superior olive, ventral division
VNLL, c, m	ventral nucleus of the lateral lemniscus, columnar division, multipolar cell division
VNTB	ventral nucleus of the trapezoid body
VPO	ventral periolivary nucleus

7

The Inferior Colliculus

George D. Pollak and Thomas J. Park

1. Introduction

The mammalian inferior colliculus sits as a protuberance on the dorsal surface of the midbrain (Fig. 7.1) and is composed of several subdivisions (see Oliver and Huerta 1992 for a discussion of various subdivisions). The largest division is the central nucleus of the inferior colliculus (ICc), which is the target of the ascending auditory projections of the lateral lemniscus (Adams 1979; Schweizer 1981; Zook and Casseday 1982; Aitkin 1986; Irvine 1986; Ross, Pollak, and Zook 1988; Ross and Pollak 1989; Frisina, O'Neill, and Zettel 1989; Vater and Feng 1990; Casseday and Covey 1992; Oliver and Huerta 1992). In most echolocating bats, the ICc is especially large while the other divisions of the inferior colliculus are considerably smaller than in other animals (e.g., Pollak and Casseday 1989). The ICc is truly a nexus in the mammalian auditory system where the ascending fibers from 10 or more of the lower auditory nuclei make an obligatory synaptic connection (Fig. 7.2). The ICc neurons, in turn, send a large projection to the medial geniculate (Olsen 1986; Frisina, O'Neill, and Zettel 1989; also see review by Winer 1992, and Wenstrup, Chapter 8) and lesser projections to other divisions of the inferior colliculus as well as to the superior colliculus (Covey, Hall, and Kobler 1987; Oliver and Huerta 1992). In addition, the ICc sends a large descending projection to numerous lower nuclei, including the lateral superior olive, cochlear nucleus, and pontine nuclei (Schuller, Covey, and Casseday 1991; also see reviews by Aitkin 1986; Oliver and Huerta 1992).

In this chapter the principal features that characterize the ICc are discussed. Particular attention is given to those features that are clearly related to the processing of information required for particular aspects of echolocation. However, the general theme is that the structure, immunocytochemistry, and physiology of the bat's inferior colliculus are strikingly similar to the inferior colliculus of other, less specialized mammals. Throughout the text, we point out how the particular features of the ICc of echolocating bats compare with those reported for other mammals.

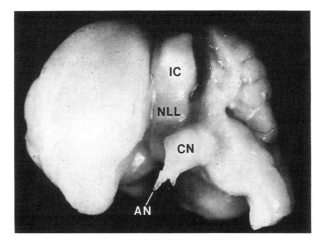

FIGURE 7.1. Photograph of brain of mustache bat. The cerebellum has been partially removed to more clearly reveal auditory structures. AN, auditory nerve; CN, cochlear nucleus; IC, inferior colliculus; NLL, nuclei of the lateral lemniscus.

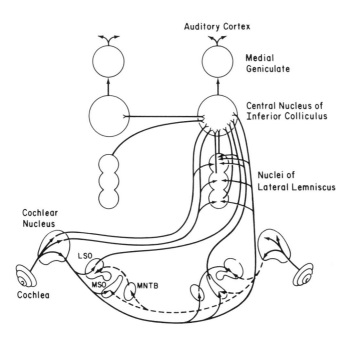

FIGURE 7.2. Wiring diagram shows principal connections of the mammalian auditory pathway. LSO, lateral superior olive; MNTB, medial nucleus of the trapezoid body; MSO, medial superior olive.

1.1 Features of the Projections from Lower Centers

The massive projections from most lower auditory nuclei terminate in the ICc in an orderly manner (Ross and Pollak 1989; also see Oliver and Huerta 1992). The magnitude of the projections, in combination with their orderly termination, underlie two other key characteristics of the inferior colliculus: (1) the remarkable amount of processing that transforms the properties of the converging inputs into new response properties expressed by ICc neurons (Faingold, Gelhbach, and Caspary 1989; Faingold, Boersma-Anderson, and Caspary 1991; Vater, Kössl, and Horn 1992; Yang, Pollak, and Resler 1992; Park and Pollak 1993a,b; Pollak and Park 1993); and (2) the orderly arrangement of those response properties in the ICc (Roth et al. 1978; Wenstrup, Ross, and Pollak 1985, 1986a; Schreiner and Langner 1988; Pollak and Casseday 1989; Park and Pollak 1993b).

The projections from the lower nuclei that terminate in the ICc are both excitatory and inhibitory. Excitatory projections originate from several lower nuclei. Among these are the cochlear nucleus (Semple and Aitkin 1980) and lateral superior olive (LSO) (Saint Marie et al. 1989; Glendenning et al. 1992). It is also likely that excitatory projections originate from the medial superior olive and the intermediate nucleus of the lateral lemniscus.

The inhibitory projections are as large as, if not larger than, the excitatory projections and are both glycinergic and γ-aminobutyric acid (GABA)ergic (Figs. 7.3 and 7.4) (Vater et al. 1992; Winer, Larue, and Pollak 1995). There are three prominent examples of nuclei that provide inhibitory inputs to their targets in the ICc. The first is the columnar division of the ventral nucleus of the lateral lemniscus (VNLLv), whose neurons are predominantly glycinergic (Fig. 7.3) (Wenthold and Hunter 1990; Pollak and Winer 1989; Larue et al. 1991; Winer, Larue, and Pollak 1995) and project heavily to the ICc (Schweizer 1981; Zook and Casseday 1982, 1987; Ross, Pollak, and Zook 1988; O'Neill, Frisina, and Gooler 1989; Vater and Feng 1990; Casseday and Covey 1992).

The second example is the LSO. The LSO projects bilaterally to the ICc (Beyerl 1978; Roth et al. 1978; Schweizer 1981; Zook and Casseday 1982; Shneiderman and Henkel 1987; Ross, Pollak, and Zook 1988) and contains a large number of glycinergic neurons (Saint Marie et al. 1989; Park et al. 1991; Glendenning et al. 1992; Winer, Larue, and Pollak 1995). Recent studies indicate the projection from the LSO to the ipsilateral ICc is largely, and possibly entirely, glycinergic (see Fig. 7.3), whereas the projection to the contralateral ICc is most likely excitatory (Saint Marie et al. 1989; Park et al. 1991; Glendenning et al. 1992; Park and Pollak 1993a).

The third example is the bilateral projections from the dorsal nucleus of the lateral lemniscus (DNLL). The neurons of the DNLL are predominantly

FIGURE 7.3. Diagram of auditory brainstem of the mustache
bat shows ascending γ-aminobutyric acid (GABA)ergic
(*solid lines*) and glycinergic
(*dashed lines*) inhibitory projections to the inferior colliculus. DNLL, dorsal nucleus of
the lateral lemniscus; DpD,
dorsoposterior division of the
inferior colliculus (the region
containing neurons sharply
tuned to 60 kHz); INLL, intermediate nucleus of the lateral
lemniscus; LSO, lateral superior olive; MSO, medial superior olive; VNLLv; ventral or
columnar division of the ventral nucleus of the lateral lemniscus; VNLLd, dorsal division
of the ventral nucleus of the lateral lemniscus; VNTB; ventral
nucleus of the trapezoid body.

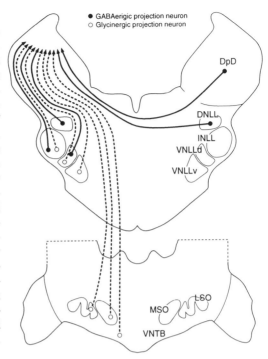

GABAergic (Adams and Mugniani 1984; Thompson, Cortez, and Lam
1985; Pollak et al. 1992; Vater et al. 1992; Winer, Larue, and Pollak 1995)
and provide a substantial inhibitory input to the ICc (Fig. 7.3) (Roth et al.
1978; Brunso-Bechtold, Thompson, and Masterton 1981; Schweizer 1981;
Zook and Casseday 1982, 1987; Covey and Casseday 1986; Ross, Pollak,
and Zook 1988; Shneiderman, Oliver, and Henkel 1988; O'Neill, Frisina,
and Gooler 1989; Ross and Pollak 1989; Shneiderman and Oliver 1989). In
addition, many of the neurons in the ICc are GABAergic (Adams and
Wenthold 1979; Mugniani and Oertel 1985; Roberts and Ribak 1987;
Oliver, Nuding, and Beckius 1988; Pollak and Winer 1989; Vater et al.
1992; Winer, Larue, and Pollak 1995), and thus may act locally to
complement the inhibition provided by the incoming fibers from lower
centers (Oliver et al. 1991; also see Oliver and Huerta 1992). The inhibitory
projections and intrinsic inhibitory neurons act to modify the response
features of the excitatory projections to produce the coding properties of
collicular cells (Faingold, Gelhbach, and Caspary 1989; Faingold, Boersma-
Anderson, and Caspary 1991; Yang, Pollak, and Resler 1992; Vater, Kössl,
and Horn 1992; Park and Pollak 1993a,b; Pollak and Park 1993). As we
explore in the following section, the ordering of projection systems creates,
in turn, an arrangement of response properties within the tonotopic
framework of the colliculus.

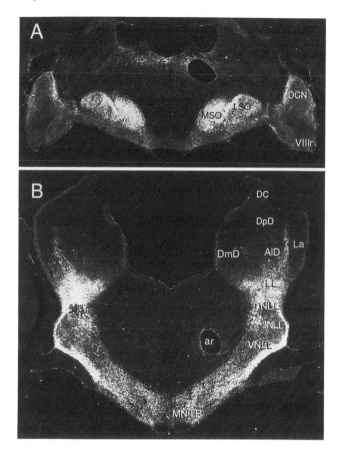

FIGURE 7.4A,B. Sections of the medullary (A) and pontine and midbrain auditory regions (B) of the mustache bat. Sections were reacted with antibodies against conjugated glycine and were photographed under darkfield illumination. Notice the massive projections of glycine-positive fibers in B that terminate in the inferior colliculus. Mark made by investigator during sectioning is indicated by *ar*. ALD, anterolateral division of the inferior colliculus (region representing frequencies below 60 kHz); DmD, medial division of the inferior colliculus (region representing frequencies above 60 kHz); DpD, dorsoposterior division of the inferior colliculus (region representing neurons sharply tuned to 60 kHz); DC, dorsal cortex of the inferior colliculus; DCN, dorsal cochlear nucleus; DNLL, dorsal nucleus of the lateral lemniscus; INLL, intermediate nucleus of the lateral lemniscus; LL, lateral lemniscus; LSO, lateral superior olive; MSO, medial superior olive; VNLL, ventral nucleus of the lateral lemniscus; VIII, auditory nerve. (Adapted from Winer, Larue, and Pollak, 1995.)

1.2 Organization of Response Properties

The results from a wide variety of studies during the past decade have shown that the projections to the ICc, as well as the response properties of its neurons, are remarkably well organized. This organization is expressed on different levels. The most fundamental and well-known is tonotopic organization, the remapping of each point of the cochlear surface on sheets of collicular cells (Fig. 7.5) (e.g., Aitkin 1976; Irvine 1986; Pollak and Casseday 1989). Thus, the cells that compose each sheet of cells in the colliculus are tuned to the same frequency as the cells in the lower nuclei from which they receive their innervation. Because all the cells in a sheet have the same best frequency, the sheet is said to be isofrequency. The sheets in turn are stacked one on top of another dorsoventrally, so that dorsal sheets represent low frequencies (the apical cochlea) and ventral sheets represent progressively higher frequencies and thus the more basal cochlear regions. This general form of tonotopy is seen in all common laboratory mammals, as well as in most of the bat species that have been studied.

The strict tonotopic organization of the lower auditory centers that is preserved in their projections to the ICc is conceptually significant. The importance of this feature is that it provides justification for thinking about the auditory system as having a modular organization. The concept of a modular organization derives from studies of sensory cortices, in which a module is an elementary unit of neuronal organization that encompasses the total processing from one segment of the sensory surface (Hubel and Wiesel 1977; Mountcastle 1978; Sur, Merzenich, and Kass 1980). Thus, the total processing from a given segment of the cochlear surface receives representation in the corresponding isofrequency contour of the ICc. An isofrequency contour is, then, the unit of neuronal organization, or the unit module, and like a cortical module each contour is, in principle, an iterative version of any other contour. In the following sections, we consider the next level of organization, the orderly arrangement of projections and arrangement of response properties within isofrequency contours.

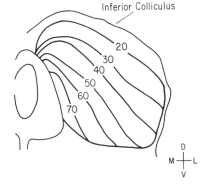

FIGURE 7.5. Drawing of transverse section through the inferior colliculus of the big brown bat, *Eptesicus fuscus*, showing its tonotopic arrangement. The orientation of isofrequency contours is shown by the *solid lines*, and the frequency in kilohertz represented in each contour is indicated. (Reproduced from Pollak and Casseday 1989.)

1.3 The 60-kHz Isofrequency Contour of the Mustache Bat's Inferior Colliculus Is a Model Contour

Both horseshoe bats (*Rhinolophus ferrumequinum* and *R. rouxi*) and mustache bats (*Pteronotus parnellii*) display a special variation of the general tonotopic pattern (Pollak and Schuller 1981; Zook et al. 1985; Rübsamen 1992). As described by Fenton in Chapter 2 and Schnitzler in Chapter 3, the dominant component of the mustache bat's orientation call is the 60-kHz constant-frequency component, and that of the horseshoe bat is the 83-kHz constant-frequency component. The dominant constant-frequency component of their call is the critical element that these bats use for Doppler-shift compensation and for the detection and recognition of insects, features discussed in the next section. As a consequence of their reliance on the dominant constant-frequency component, that frequency is greatly overrepresented throughout their auditory system, including the ICc. The overrepresentation, in turn, has distorted the general tonotopic arrangement in the ICc.

This distortion is well illustrated by the tonotopy of the mustache bat's ICc (Fig. 7.6). As described by Kössl and Vater in Chapter 5, the cochlear region representing 60 kHz is greatly expanded in the mustache bat and has adaptations that create exceptionally sharp tuning curves in the auditory

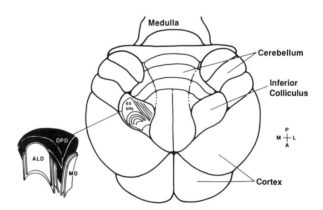

FIGURE 7.6. Schematic of dorsal view of the mustache bat's brain to show location of the 60-kHz region in the inferior colliculus. The hypertrophied inferior colliculi protrude between the cerebellum and cerebral cortex. In the left colliculus are shown the isofrequency contours, as determined in anatomical and physiological studies. A three-dimensional representation of the laminar arrangement is shown on the far left. Low-frequency contours, representing an orderly progression of frequencies from about 59 kHz to about 10 kHz, fill the anterolateral division (ALD). High-frequency contours, representing frequencies from about 64 to more than 120 kHz, occupy the medial division (MD). The 60-kHz region is the dorsoposterior division (DPD), and is the only representation of the filter units in the bat's colliculus. (Reproduced from Pollak and Casseday 1989.)

nerve fibers which innervate that region (Suga, Simmons, and Jen 1975; Suga and Jen 1977; Kössl and Vater 1985, 1990; Zook and Leake 1989). The auditory nerve fibers have $Q_{10 \text{ dB}}$ values that range from about 50 to more than 300 and are, on average, an order of magnitude sharper than the fibers tuned just a few hundred hertz higher or lower in frequency. This overrepresentation of sharply tuned 60-kHz neurons is manifest in each nucleus of the auditory pathway, from cochlear nucleus to cortex (e.g., Suga and Jen 1976; Schlegel 1977; Pollak and Bodenhamer 1981; Kössl and Vater 1990; Covey, Vater, and Casseday 1991), and is expressed in the ICc as an isofrequency contour composed of sharply tuned 60-kHz cells that occupy roughly a third of its volume (Figs. 7.6 and 7.7) (Zook et al. 1985; Wenstrup, Ross, and Pollak 1985, 1986a; Ross, Pollak, and Zook 1988). Indeed, the central nucleus of the mustache bat's inferior colliculus can be divided into three divisions (see Fig. 7.6): an anterolateral division, in which the sheets of isofrequency cells are stacked more or less vertically and create an orderly representation of frequencies from about 59 kHz to about 20 kHz; a medial division, where the isofrequency sheets are also stacked vertically and create an orderly representation of frequencies from about 64 kHz to about 120 kHz; and, wedged between these, the dorsoposterior division, in which all the units are sharply tuned to a small frequency band around 60 kHz. In short, the inferior colliculus of this animal has a greatly expanded isofrequency contour that is easily distinguished both by the uniformity of the best frequencies of its neuronal population and by their exceptionally sharp tuning curves. We use here the 60-kHz isofrequency contour of the mustache bat's colliculus as a model to illustrate the ways in which the ascending projections and response properties are organized within a collicular contour.

1.4 Projections to the 60-kHz Contour of the Mustache Bat's Inferior Colliculus

In the previous section we emphasized that the inferior colliculus possesses a tonotopic organization and receives projections from a large number of lower auditory nuclei. Each isofrequency region of the colliculus, then, presumably receives a complement of projections from corresponding isofrequency regions of the lower brainstem nuclei. This is also true of the 60-kHz contour in the mustache bat's inferior colliculus. Figure 7.8 shows the results of a study in which the afferent inputs to the 60-kHz contour were determined by initially mapping its boundaries physiologically, and then making injections of horseradish peroxidase (HRP) in the contour. The 60-kHz contour receives projections from the same set of lower auditory nuclei that project to the entire central nucleus of the inferior colliculus (Ross, Pollak, and Zook 1988). These include: (1) projections from the three divisions of the contralateral cochlear nucleus; (2) projec-

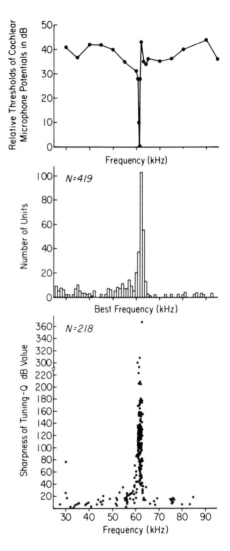

FIGURE 7.7. Distribution of best frequencies and Q_{10dB} values of neurons recorded from the inferior colliculus of the mustache bat. *Top panel*, a cochlear microphonic audiogram recorded from the inner ear of a mustache bat, shows that the cochlea is sharply tuned to 60 kHz. *Middle panel* shows that about a third of the neurons in the inferior colliculus have best frequencies around 60 kHz and correspond closely to the frequency to which the cochlea is tuned. *Lower graph* shows that the tuning curves of the 60-kHz neurons have large Q_{10dB} values, and thus their tuning curves are much sharper than are those of neurons tuned to higher or lower frequencies.

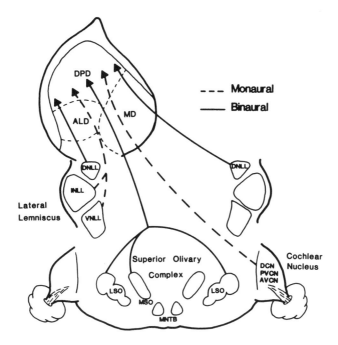

FIGURE 7.8. Ascending projections to the 60-kHz region (DPD) of the mustache bat's inferior colliculus. Projections from binaural auditory nuclei that receive innervation from the two ears are shown as *solid lines*; projections from monaural nuclei that receive innervation from only one ear are shown as *dashed lines*. Note that the 60-kHz region receives projections from all lower auditory nuclei, and thus receives the same types of inputs as the entire inferior colliculus. Although not shown, the neurons that project to the 60-kHz region in each lower nucleus are restricted to regions that presumably represent 60 kHz. (Drawing based on Ross, Pollak, and Zook 1988.)

tions from the ipsilateral medial superior olive; (3) bilateral projections from the LSO; (4) ipsilateral projections from the ventral and intermediate nuclei of the lateral lemniscus; (5) bilateral projections from the DNLL; and (6) projections from the contralateral inferior colliculus (not shown). The projections arise from discrete segments of the various projecting nuclei, each of which represents 60 kHz.

1.5 Aural Types Are Topographically Arranged in the 60-kHz Contour of the Inferior Colliculus

The inferior colliculi of all mammals have three types of neurons that are classified on the basis of aural preference (e.g., Semple and Aitkin 1979; Roth et al. 1978; Semple and Kitzes 1985; Fuzessery and Pollak 1985;

Wenstrup, Ross, and Pollak 1986a; Wenstrup, Fuzessery, and Pollak 1988a). These are monaural neurons (EO) and two types of binaural neurons. One binaural type receives excitatory inputs from both ears; these are called excitatory-excitatory (EE) neurons. The second type, called excitatory-inhibitory (EI) neurons, receives excitation from one ear and inhibition from the other ear. The noteworthy feature is that monaural and binaural neurons are arranged topographically within the 60-kHz contour of the inferior colliculus (Fig. 7.9) (Wenstrup, Ross, and Pollak 1985, 1986a; Ross and Pollak 1989). Monaural units are located along the dorsal and lateral portions of the contour. EE cells occur in two regions, one in the ventrolateral region and the other in the dorsomedial region. EI neurons also are restricted to two regions. The main population is in the ventromedial region, and a second population occurs along the very dorsolateral margin of the contour, which is most likely in the external nucleus of the inferior colliculus.

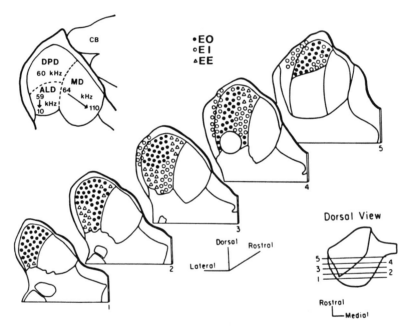

FIGURE 7.9. Schematic shows segregation of neurons with different aural preferences in the 60-kHz region (DPD) of the mustache bat's inferior colliculus. Transverse sections of the inferior colliculus are arranged in a caudal-to-rostral direction. Each section is separated by 180 μm, and the rostrocaudal position of each section is illustrated on the dorsal view of the inferior colliculus in the lower right. The medial division (MD), where high frequencies are represented, is the large area to the right of the DPD in each section. The anterolateral division (ALD), representing frequencies below 60 kHz, is shown below the DPD in sections 4 and 5. EO, monaural neurons; EI, excitatory-inhibitory neurons; EE, excitatory-excitatory neurons. (Adapted from Wenstrup, Ross, and Pollak 1986a.)

1.6 Each Aural Subregion of the 60-kHz Contour Receives Afferent Projections from a Unique Set of Lower Auditory Nuclei

The topographic arrangement of neurons with particular monaural and binaural properties in the 60-kHz contour suggests that the aural properties are a consequence of the projections that terminate in each subregion. By making small iontophoretic deposits of HRP within each of the physiologically defined aural regions, it was shown that each monaural and binaural region within the 60-kHz contour receives its chief inputs from a different subset of nuclei in the lower brainstem (Ross and Pollak 1989). The principal connections of the EI and both EE aural subregions are shown in Fig. 7.10. For most regions, the response properties reflect the subset of

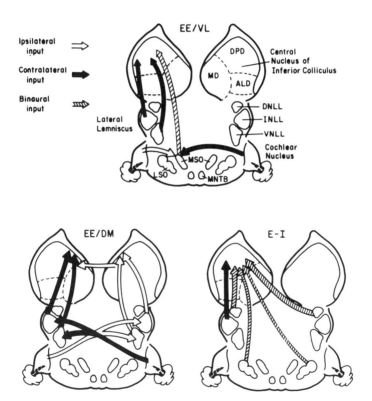

FIGURE 7.10. Schematic shows some of the major projections to the ventrolateral EE region *(top panel)*, dorsomedial EE region *(lower left panel)*, and ventromedial EI region *(lower right panel)* of the 60-kHz contour of the mustache bat's inferior colliculus. Notice that the EI region receives bilateral projections from the LSO and DNLL, and that those nuclei do not project to either of the EE regions. (Adapted from Ross and Pollak 1989.)

inputs. Of particular significance are the projections to the ventromedial EI region. The afferents to this area arise largely from binaural nuclei, especially the DNLL (bilaterally) and from the LSO (bilaterally). A major input also originates in the intermediate nucleus of the lateral lemniscus, which is a monaural nucleus. The robust inputs from the DNLL and LSO distinguish the EI region from all other aural regions of the 60-kHz contour, which receive few or no projections from either of these nuclei.

1.7 Rate-Intensity Functions and Temporal Discharge Patterns Are Other Characteristic Features of Auditory Neurons

Two additional properties that characterize all auditory neurons are the way in which spike counts change with increasing sound intensity, and the temporal discharge patterns that neurons display to tone bursts presented at their best frequencies. In response to increasing intensity, spike counts of auditory neurons change in one of two principal ways. In some neurons the spike counts increase as intensity is raised, and then either reach a plateau or continue to increase even with the highest intensities presented. Such neurons are said to have monotonic spike-count functions, that are either saturated or nonsaturated. In the majority of ICc neurons, however, spike counts at first increase with intensity and then decrease, sometimes to zero spikes, as intensity is increased further. These neurons are said to have nonmonotonic rate-intensity functions. In many neurons, the discharge rate falls to zero at higher intensities. These neurons not only have a lower threshold but an upper threshold as well. In the 60-kHz contour of the mustache bat, about 60% of the neurons have nonmonotonic rate-intensity functions, whereas 40% have monotonic functions (Pollak and Park 1993).

Another characteristic feature of auditory neurons is their temporal discharge pattern evoked by tone bursts at the neuron's best frequency. Discharge patterns can be complex, and in lower nuclei the number of different types is quite large (e.g., Rhode and Greenberg 1992). For purposes of simplicity, we classify collicular neurons into one of two types; those that fire with sustained discharges for the duration of the tone burst, and those that fire only to the onset, or to both the onset and offset of the signal. The discharge patterns of neurons in most lower nuclei are sustained (Goldberg and Brown 1968; Brugge, Anderson, and Aitkin 1970; Guinan, Guinan, and Norris 1972; Harnischfeger, Neuweiler, and Schlegel 1985; Covey, Vater, and Casseday 1991; Rhode and Greenberg 1992) and onset patterns are less common. In the ICc of bats, the population of onset neurons exceeds the population of sustained neurons (Pollak et al. 1978; Vater, Schlegel, and Zöller 1979; Jen and Suthers 1982; O'Neill 1985; Harnischfeger, Neuweiler, and Schlegel 1985; Park and Pollak 1993b). In the 60-kHz contour, for example, about 66% of the neurons have onset patterns and only about 33% have sustained patterns (Park and Pollak 1993b).

Neurons with monotonic and nonmonotonic rate-intensity functions, and neurons with phasic and sustained discharge patterns, occur throughout the dorsoventral extent of the 60-kHz contour, and thus at all levels of the EI, EE and monaural subregions (Park and Pollak 1993b).

1.8 Inhibitory Thresholds of EI Neurons are Topographically Organized in the Ventromedial Region of the 60-kHz Collicular Contour

The population of EI neurons is of particular interest because they differ in their sensitivities to interaural intensity disparities (IIDs). These neurons compare the sound intensity at one ear with the intensity at the other ear by subtracting the activity generated in one ear from that in the other, and thus play an important role in coding sound location. The way in which EI neurons code for sound location is discussed in detail in a later section. Suprathreshold sounds delivered to the excitatory (contralateral) ear evoke a certain discharge rate that is unaffected by low-intensity sounds delivered simultaneously to the inhibitory (ipsilateral) ear. However, when the ipsilateral intensity reaches a certain level, and thus generates a particular IID, the discharge rate declines sharply, and even small increases in the ipsilateral intensity will in most cases completely inhibit the cell (Fig. 7.11). Thus, each EI neuron has a steep IID function and reaches a criterion inhibition at a specified IID that remains relatively constant over a wide range of absolute intensities. We refer to the IID that produces a 50% reduction in discharge rate as the neuron's "inhibitory threshold" (Wenstrup, Ross, and Pollak 1986a; Wenstrup, Fuzessery, and Pollak 1988a,b). An inhibitory threshold is assigned a negative value if the criterion inhibition occurred when the ipsilateral signal was more intense than the contralateral signal, and was assigned a positive value if the intensity at the ipsilateral ear was lower than the contralateral ear when the criterion inhibition was achieved. The inhibitory threshold of the EI cell in Fig. 7.11 is 0 dB. Inhibitory thresholds among EI neurons vary from about +15 dB to about −25 dB, encompassing much of the range of IIDs that the bat would normally experience (Wenstrup, Fuzessery, and Pollak 1988a).

Particularly noteworthy is that the inhibitory thresholds of EI neurons are arranged in an orderly fashion in the EI subregion of the 60-kHz contour (Fig. 7.12). EI neurons with negative inhibitory thresholds (i.e., neurons requiring a louder ipsilateral stimulus than contralateral stimulus to produce 50% inhibition) are located in the dorsal EI region. Subsequent EI neurons display a progressive shift to more positive inhibitory thresholds. The most ventral EI neurons have the most positive inhibitory thresholds: they are suppressed by ipsilateral sounds equal or less intense than the contralateral sounds.

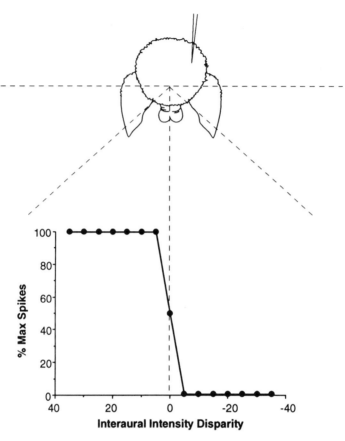

FIGURE 7.11. Schematic of interaural intensity function of an EI neuron. The bat's head is shown in the *upper portion* with a microelectrode that would record the activity of the EI neuron in the colliculus on the side shown. The IID function is obtained by presenting sound through earphones situated in the bat's ears (not shown). A 60-kHz sound is presented to the excitatory ear, the ear contralateral to the colliculus from which the recording is taken. The sound intensity is fixed at some intensity above threshold and evokes a discharge that is represented as 100% maximum spikes on the ordinate. A low-intensity sound is then presented simultaneously at the ipsilateral (inhibitory) ear, and has no effect on the discharge rate evoked by the sound presented to the excitatory ear. As the sound intensity at the ipsilateral ear is increased, indicated by the decreasing IIDs, the discharge rate remains unaffected until the IID reaches a certain value, in this case about 0 dB, at which point the spike count declines markedly. Further increases of the intensity at the ipsilateral ear (more negative IIDs) result in a complete inhibition of discharges. The bat's head is drawn to indicate how the IIDs translate into spatial locations and how the cell might respond to free-field sounds. The IID function suggests that discharges should be inhibited by 50% when the sound is directly in front of the bat, at 0° azimuth. The cell should not be inhibited by sounds in the hemifield contralateral to the recording electrode, but the cell should be completely inhibited by sounds from the hemifield ipsilateral to the electrode.

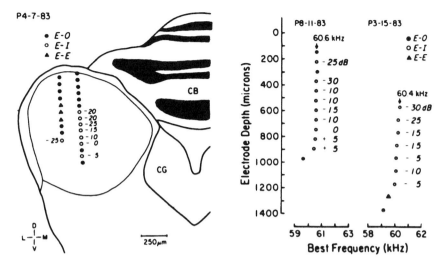

FIGURE 7.12. *Left:* Systematic shifts in the inhibitory thresholds (IIDs at which the discharge rate declines by 50%) of EI unit clusters shown in a transverse section of the mustache bat's inferior colliculus. Numbers indicate IIDs of the inhibitory thresholds recorded at those locations. All units were sharply tuned to frequencies around 60 kHz. *Right:* Systematic increase of inhibitory thresholds of EI neurons with depth in the dorsoposterior division of the mustache bat's inferior colliculus recorded in two dorsoventral penetrations from different bats. (Adapted from Wenstrup, Ross, and Pollak 1985).

1.9 Latencies Are Systematically Arranged Within the Aural Subregions of the 60-kHz Contour

Latency is another response feature that is arranged in an orderly fashion within isofrequency contours of the ICc (Schreiner and Langner 1988; Park and Pollak 1993b). Figure 7.13 shows plots of latency as a function of depth in the 60-kHz contour and reveals two salient features. The first is that neurons in the dorsal colliculus have, on the average, longer latencies than neurons located ventrally; the average latency in the more dorsal part of the colliculus is about 15.4 msec, which decreased to an average of about 9.8 msec in the more ventral region. The second feature is that the range of latencies changes markedly with depth. Dorsally there is a broad distribution of latencies while deeper regions have a much narrower range of latencies. Long-latency cells are found only dorsally whereas short-latency cells are found at all depths. The latencies of dorsal neurons, within 200–400 μm of the collicular surface, range from 30 to 6 msec, whereas the latencies at depths from 1000 to 1300 μm range from 5 to 10 msec. Thus, dorsal regions are characterized by a wide distribution of latencies and a relatively high average latency, while ventral neurons have a narrower range of relatively short average latencies.

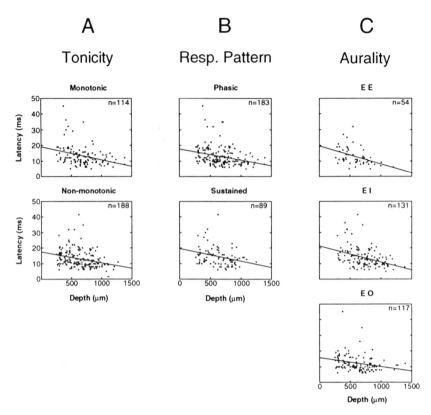

FIGURE 7.13A–C. Plots of latency as a function of depth for cells with different response types. (A) Latency distributions for cells with monotonic and nonmonotonic rate-intensity functions. (B) Latency distributions for cells with phasic and sustained response patterns. (C) Distributions for cells with different binaural and monaural properties. (Adapted from Park and Pollak 1993b.)

The foregoing features are also seen in each of the aural subregions. As shown in Fig. 7.13C, a wide distribution of latencies and a relatively high average latency are found dorsally in each of the aural subregions, where either monaural, EI, or EE units predominate. In more ventral portions of each aural subregion, the neuronal populations have a narrower range of latencies and shorter average latencies.

Finally, the neuronal populations that have monotonic and nonmonotonic rate-intensity functions, and those that have phasic and sustained discharge patterns, also express the same characteristic pattern of latency change with depth. This is illustrated in Figure 7.13A,B by the characteristic decrease in average latency and latency range with depth for neurons with either monotonic or nonmonotonic rate-intensity functions, and by neurons with either a sustained or phasic discharge pattern.

1.10 Overall View of the Organization of the Auditory Brainstem

The general picture of the auditory brainstem is that it is constructed from a hierarchical series of organizational principles. Once an organization is established at a certain level of the system, that organization is maintained throughout the system. At each level, a new organization is created and becomes embedded within the framework of the organization that was created at the lower level. Thus, the primary level of ordering, tonotopy, is initially established in the cochlea. The tonotopy is maintained throughout the brainstem auditory pathway, and results in the isofrequency contours of each nucleus. The projections from the isofrequency regions in each of the lower centers are further ordered and target regional areas of the same isofrequency contours in the inferior colliculus. The regional targeting of projection systems creates regions within each collicular contour dominated by monaural or one of the two types of binaural neurons (Roth et al. 1978; Wenstrup, Ross, and Pollak 1986a; Ross and Pollak 1989). We refer to these regions as aural subregions. Imposed on the neurons in each aural subregion are additional organizational features (Wenstrup, Ross, and Pollak 1986a; Schreiner and Langner 1988; Ross and Pollak 1989; Pollak and Casseday 1989; Park and Pollak 1993b).

Because sustained and phasic neurons occur throughout the contour, they can be considered to be arranged as parallel arrays that run dorsoventrally within each aural subregion of the 60-kHz contour. Furthermore, neurons that have a sustained discharge pattern and a monotonic rate-intensity function are also found throughout the contour, and may be arranged parallel to sustained neurons that have nonmonotonic rate-intensity functions. These populations, in turn, are arranged in parallel with neurons that have phasic, monotonic and phasic, nonmonotonic response types. Thus, aural type, temporal discharge pattern, and rate-intensity function are qualitative features that appear to be constant within a dorsoventral "column" or array of cells. Other features are then arranged orthogonal to the dorsoventral axis and change quantitatively along this spatial dimension; specifically, a characteristic range of latencies and a characteristic average latency are present at each dorsoventral level (Park and Pollak 1993b), and in the case of EI neurons, a characteristic interaural intensity disparity that causes the discharge rate to decline by 50% as well (inhibitory threshold) (Wenstrup, Ross, and Pollak 1985, 1986a). The values of both the average latency and range of latencies and the inhibitory thresholds (the interaural intensity disparities of the 50% cutoffs) change systematically with depth. Thus, an array of cells is conceptually analogous to a cortical column, and the complement of arrays may be analogous to a cortical hypercolumn (Hubel and Wiesel 1977).

In the following sections the way in which biologically relevant acoustic cues are encoded by collicular neurons are discussed. Where possible, we show how the organization of the response features within an isofrequency contour contributes to the representation of a particular cue, or to the establishment of new features at the next level. Finally, we will also show how, for certain features, the projection patterns from lower regions act to shape new response features, and how such transformations contribute to one or another of the different organizational features of the inferior colliculus.

2. Response Properties of Collicular Neurons That Underlie Doppler-Shift Compensation and the Coding of Target Features

One of the benefits of using echolocating bats for neurophysiological investigations of the auditory system is that the way in which they manipulate their orientation calls is known in considerable detail, as are the acoustic cues from which these animals derive information about objects in the external world (see Grinnell, Chapter 1; Fenton, Chapter 2; and Schnitzler, Chapter 3, this volume). This knowledge allows investigators to generate signals that mimic the types of signals that bats emit and receive during echolocation, as well as the sorts of acoustic cues which are contained in the echoes reflected from flying insects. In addition, it permits the investigator to make informed inferences about how the nervous system encodes and represents a particular cue from the discharge properties of the cells that are monitored. In this section, we briefly discuss one of the most remarkable behaviors exhibited by horseshoe and mustache bats, Doppler-shift compensation. We then consider the acoustic cues these bats use for the detection and recognition of insects, and why Doppler-shift compensation is important for the perception of those cues. Finally, we discuss some of the neural features that are important both for Doppler-shift compensation and for the coding of the acoustic cues these bats use for the identification of their targets.

2.1 Some Aspects of Doppler-Shift Compensation

Doppler-shift compensation is the expression of the bat's extreme sensitivity for motion. One way that relative motion is created is by the difference in flight speed between the bat and background objects in its environment. Because a flying bat approaches an object at a certain speed, the echo from that object will be Doppler shifted upward, and will therefore have a higher frequency than the emitted signal. Long constant-frequency, frequency-modulating (CF/FM) bats are exquisitely sensitive to upward Doppler shifts in the CF component of the echo. They compensate for the Doppler shifts by lowering the frequency of the next emitted CF component by an amount that is nearly equivalent to the Doppler shift in the preceding echo (Fig. 7.14) (Schnitzler 1970; Schuller, Beuter, and Schnitzler 1974; Simmons 1974; Henson et al. 1982; Trappe and Schnitzler 1982; Gaioni, Riquima-roux, and Suga 1990).

The flight of the bat, however, is not the only motion in the night sky, and thus is not the only source of Doppler shifts. Flying insects beat their wings to stay aloft, and the motion of the wings creates periodic Doppler shifts, which impose periodic frequency modulations on the CF component of the echo (Fig. 7.15) (Schnitzler and Ostwald 1983; Schuller 1984; Kober 1988; Ostwald, Schnitzler, and Schuller 1988). In addition, the wingmotion, or flutter, presents a reflective surface whose area alternates with wing position, thereby also creating periodic amplitude fluctuations (i.e., ampli-tude modulations) in the CF component of the echo. The modulations imposed upon the echo CF component by the fluttering wings of an insect are of critical importance to long CF/FM bats, because it is from these modulations that the bats discriminate flying insects from background objects and recognize the particular insect that has wandered into their acoustic space (Goldman and Henson 1977; Kober 1988; Ostwald, Sch-nitzler, and Schuller 1988; von der Emde 1988).

One of the key adaptations that underlies the perception of both Doppler shifts and modulation patterns is the large population of sharply tuned neurons. In the following sections we first consider some properties of these neurons and the mechanisms that create the sharp tuning curves. We then describe how the sharp tuning permits the processing of Doppler-shifted

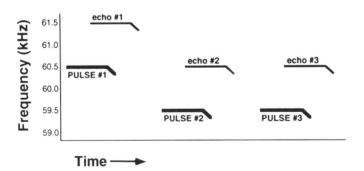

FIGURE 7.14. Schematic of Doppler-shift compensation in a flying mustache bat. The first pulse is emitted at about 60.5 kHz, and the Doppler-shifted echo returns with a higher frequency, at about 61.5 kHz. The mustache bat detects the difference in frequency between the pulse and echo, and lowers the frequency of its subsequent emitted pulses by an amount almost equal to the Doppler shift. Thus, the frequencies of the subsequent echoes are held constant and return at a frequency very close to that of the first emitted signal. Also notice that because of the length of the emitted CF component, the echoes return to the ear while the bat is still emitting the pulse, resulting in substantial periods of pulse–echo overlap.

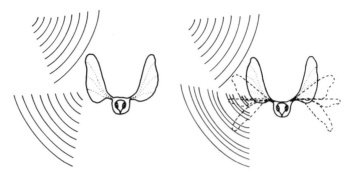

FIGURE 7.15. Drawings illustrate the echoes of tone bursts reflected from an insect in which the wings are stationary *(left)* and from an insect whose wings are in motion *(right)*. Echoes reflected from the insect whose wings are stationary will also be have the same frequency as the tone burst, while the echoes from the insect whose wings are beating will be frequency modulated, as indicated by the alternate spacing of the reflected waves. The frequency modulations are created by the Doppler effect caused by the velocity of wing movements. (Reproduced from Pollak and Casseday 1989.)

echoes and its significance for coding the variety of modulation patterns these bats receive from flying insects.

2.2 Long CF-FM Bats Have Large Populations of Neurons Sharply Tuned to the CF Components of Their Biosonar Signals

Perhaps the most dramatic neuronal response feature of the long CF/FM bats is the overrepresented population of neurons that have very sharp tuning curves and best frequencies that are close to the dominant CF component of the bat's orientation calls (see Fig. 7.7) (Suga, Simmons, and Jen 1975; Suga and Jen 1976, 1977; Schuller and Pollak 1979; Pollak and Bodenhamer 1981; Ostwald 1984; Vater, Feng, and Betz 1985; Kössl and Vater 1990). Because of the high degree of frequency selectivity, these neurons are also referred to as "filter neurons" (Neuweiler and Vater 1977).

The narrow tuning of filter neurons is a consequence of at least two processes. The first is the cochlear specializations that were mentioned earlier and which are considered in detail by Kössl and Vater in Chapter 5 (also see Suga, Simmons, and Jen 1975; Suga and Jen 1977; Kössl and Vater 1990). The second process is neural inhibition (Suga and Tsuzuki 1985; Yang, Pollak, and Resler 1992; Vater et al. 1992; Casseday and Covey 1992). In higher auditory nuclei, inhibitory circuits often modify the shape of tuning curves in at least two ways. The first way is to further sharpen the tuning curve with lateral inhibition; either the high- or low-frequency skirts of the tuning curve, or both, are trimmed by inhibitory inputs from neurons whose best frequencies are on the flanks of the excitatory tuning curve. In filter units, the inhibition most often affects the width of the tuning curve at intensities 20 dB or more above threshold, and has little or no effect on the width of the curve at 10 dB above threshold (Suga and Tsuzuki 1985; Vater et al. 1992; Yang, Pollak, and Resler 1992). Such sharpening is illustrated by the neuron in the top panel of Fig. 7.16. This figure shows the expansion of the tuning curve of a filter neuron in the mustache bat's inferior colliculus that occured when GABAergic inhibition was blocked by the drug bicuculline. The origin of the GABAergic inhibition is unknown, but it seems likely that the GABAergic inhibition that shapes tuning curves is caused by local GABAergic collicular neurons,

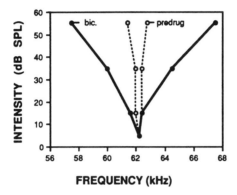

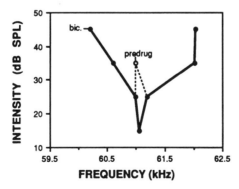

FIGURE 7.16. Two 60-kHz units from the mustache bat's inferior colliculus whose tuning curves broadened from the iontophoretic application of bicuculline. The unit in the *top panel* originally had a very sharp tuning curve (predrug), which opened substantially when GABAergic inhibition was blocked with bicuculline (bic). The cell in the *lower panel* originally had an upper-threshold tuning curve, which was transformed into a much wider, open tuning curve with bicuculline. (Adapted from Yang, Pollak, and Resler 1992.)

although the involvement of GABAergic projection neurons cannot be ruled out.

The second way that inhibition can modify a tuning curve is to suppress discharges at high intensities, thereby producing a closed tuning curve that has both a minimum threshold and an upper threshold (Vater et al. 1992; Yang, Pollak, and Resler 1992). The bottom panel in Figure 7.16 shows the closed tuning curve of a filter unit in the mustache bat's inferior colliculus that was changed to a more conventional "open" tuning curve when GABAergic inhibition was blocked with bicuculline. Although neurons with closed tuning curves are common in the ICc and at higher levels of the auditory system (Grinnell 1963; Suga 1964; Jen and Schlegel 1982; Jen and Suthers 1982; O'Neill 1985; Olsen and Suga 1991a,b; Suga and Tsuzuki 1985; Casseday and Covey 1992; Vater et al. 1992; Yang, Pollak, and Resler 1992), the significance of these tuning curves for echolocation is not entirely clear. One suggestion is that these neurons might be inhibited by the relatively loud emitted call, but would be left in a responsive condition to respond to the fainter echo that arrives shortly thereafter (Grinnell 1963; Casseday and Covey 1992). A mechanism of this sort could be important for processing the FM components of biosonar signals, but it apparently does not play the same role in the processing of the dominant CF components. The reason is that the CF components of the pulse and echo are processed by different neuronal populations, as explained next.

2.3 The Sharp Tuning of Filter Neurons Is Relevant for Doppler-Shift Compensation

The long CF/FM bats derive two major benefits by having sharply tuned filter units. The first benefit is that it allows the echo CF component to be perceived during periods of pulse–echo overlap, when the emitted and echo CF components are stimulating the ear at the same time. The second benefit is that the sharp tuning imparts exceptional sensitivity for encoding frequency modulation patterns that are important for detecting and characterizing insects (Suga and Jen 1977; Schuller 1979; Pollak and Schuller 1981; Schnitzler and Ostwald 1983; Bodenhamer and Pollak 1983; Suga, Niwa, and Taniguchi 1983; Ostwald 1988; von der Emde 1988; Lesser et al. 1990). We first consider the issue of pulse–echo overlap, and then turn to the detection and characterization of targets.

2.4 Sharp Tuning Allows Spatial Segregation of Activity Evoked by the Pulse and Echo During Periods of Pulse-Echo Overlap

Because the emitted CF components are as long as 30 msec in mustache bats (Novick and Vaisnys 1964) and up to 90 msec in horseshoe bats (Griffin and

Simmons 1974), the CF component of the echo always returns to the bat's ear while the animal is still emitting its pulse (see Fig. 7.14) (Novick 1971). This creates relatively long periods of pulse–echo overlap at the ear, a condition that seemingly would interfere with the bat's perception of the echo CF component. The sharp tuning of filter neurons is significant in this regard, because it segregates the neuronal populations that respond to the emitted CF component from the filter neurons that respond to the CF component of the Doppler-shifted echo (Schuller and Pollak 1979). Consider, for example, that the flight speed of mustache bats can create Doppler shifts up to about 1500 Hz in the echoes from stationary objects. When a mustache bat compensates for a 1500-Hz Doppler shift, its emitted CF component is about 58.5 kHz and the echo CF component is about 60 kHz. Under these conditions, the filter neurons tuned to 60 kHz would respond to the Doppler-shifted echo, but the extreme narrowness of their tuning curves would prevent them from discharging to the lower frequency of the emitted CF component. Especially important in this regard are the very sharp slopes on the low-frequency sides of the tuning curves, which are, in part, a consequence of lateral inhibition (Suga and Jen 1977; Suga and Tsuzuki 1985; Kössl and Vater 1990; Yang, Pollak, and Resler 1992). Conversely, because the high-frequency skirts of the tuning curves of units tuned to 58 and 59 kHz are steep (Suga and Jen 1977; O'Neill 1985; Kössl and Vater 1990), the majority of units in the 58-kHz and 59-kHz collicular contours would discharge to the emitted CF component, but would be largely unaffected by the higher frequency of the Doppler-shifted echo. Thus, the activity evoked by the emitted CF would be spatially segregated from the activity evoked by the Doppler-shifted echo. Such an arrangement effectively prevents the emitted pulse from masking the discharge activity in filter units evoked by the echo CF during periods of pulse–echo overlap, a feature of clear importance for the effective operation of a Doppler-based sonar system.

2.5 Advantage of Doppler-Shift Compensation

The adaptive advantage of Doppler-shift compensation is that it enhances the ability of long CF/FM bats to readily detect and recognize a fluttering insect in the presence of background echoes (Neuweiler 1983, 1984; Pollak and Casseday 1989). All bats that utilize this type of biosonar system hunt under the forest canopy (Neuweiler 1983, 1984; Bateman and Vaughan 1974). This habitat provides a rich source of food for which there is probably little other competition, but it is also an environment that requires the bat to be able to both detect and perceive prey in the midst of the echoes from background objects. Long CF/FM bats cope with the clutter by compensating for the Doppler shifts in the echoes from the background objects, and thereby clamp those echoes at a constant frequency that corresponds closely to the best frequencies of the overrepresented filter neurons. When a small insect crosses the path of a hunting horseshoe or mustache bat, the presence of modulations in the echo are clear and unambiguous cues that a flying insect has invaded the bat's acoustic space, and the modulation pattern provides the information for characterizing the insect (Goldman and Henson 1977; Ostwald, Schnitzler, and Schuller 1988; von der Emde 1988). Thus, by manipulating the frequencies of their emitted CF/FM components, long CF/FM bats confine echoes to a very narrow frequency band, and this behavior ensures the bat that modulated echoes will be processed by an exceptionally large number of sharply tuned filter neurons. For reasons explained here, the sharp tuning curves are also important because they impart an exceptional sensitivity and effectiveness for encoding the patterns of the periodic frequency modulations.

2.6 Filter Neurons Are Exceptionally Sensitive to Modulated Signals

The periodic amplitude and frequency modulations created by the wingbeat pattern of a fluttering insect are complex (Schnitzler and Ostwald 1983: Schuller 1984; Kober 1988). The way in which filter units encode these patterns is usually studied with electronically generated signals that mimic the natural echoes (Suga and Jen 1977; Schuller 1979; Pollak and Schuller 1981; Vater 1982; Ostwald, Schnitzler, and Schuller 1988; Bodenhamer and Pollak 1983; Suga, Niwa, and Taniguchi 1983; Lesser et al. 1990). The response features of filter units evoked by simulated echoes have proven to be a good approximation to the way they respond to natural echoes from a fluttering insect (Schuller 1984; Ostwald 1988). The echoes are mimicked by modulating the frequency or amplitude of a carrier tone with low-frequency sinusoids of 50–500 Hz (Fig. 7.17). The low-frequency sinusoid is referred to as the modulating waveform to distinguish it from the carrier or center frequency. This arrangement creates either sinusoidally amplitude modulated (SAM) or sinusoidally frequency modulated (SFM) signals. With SFM signals, the depth of modulation, the amount by which the frequency varies around the carrier, mimics the degree of Doppler shift created by the wing motion of a flying insect. The other modulating parameter, the modulation rate, simulates the insect's wingbeat frequency.

When SFM or SAM signals are presented to the ear of any mammal, be it a bat (e.g., Schuller 1979), rat (M;daller 1972), or cat (Schreiner and Langner 1988), neurons in the mammalian auditory system respond with discharges that are tightly locked to the modulating waveform. This is also the case for both filter and nonfilter collicular neurons in long CF/FM bats. The feature that distinguishes filter neurons from all other neurons is that they are far more sensitive to very small periodic modulations than are more broadly tuned units (Schuller 1979; Bodenhamer and Pollak 1983; Vater 1982). The filter neurons fire only when the frequencies sweep into the

Constant Frequency Tone Burst

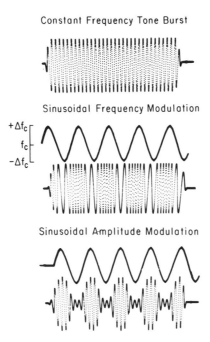

Sinusoidal Frequency Modulation

Sinusoidal Amplitude Modulation

FIGURE 7.17. The fine structure of a tone burst *(upper record)*, a sinusoidally frequency-modulated (SFM) burst *(middle record)*, and a sinusoidally amplitude-modulated (SAM) burst *(lower record)*. The tone burst in the upper record is a shaped sine wave in which the frequency, or fine structure, of the signal remains constant. The fine structure of the SFM burst changes sinusoidally in frequency. F_c indicates the center or carrier frequency; the degree to which the frequency changes around the carrier is indicated by ΔF. The modulation waveform is shown above the record of the signal's fine structure. In the SAM burst, the fine structure of the signal is a constant frequency, but the amplitude varies with the sinusoidal modulating waveform. (Reproduced from Pollak and Casseday 1989.)

narrow confines of the tuning curve and remain silent when the signal is outside of the tuning curve (Suga and Jen 1977; Schuller 1979; Pollak and Schuller 1981; Bodenhamer and Pollak 1983; Ostwald 1988). The truly spectacular synchronization of discharges of filter neurons to the modulation waveform, even when the depth of modulation is as small as ± 10 Hz, is illustrated in Figure 7.18.

2.7 Selectivity of Filter Neurons for Parameters of Modulated Signals

Although filter neurons having firing patterns tightly locked to the modulating waveform are commonly encountered in the colliculus, most are selective for some parameter of the modulated signal to which they will

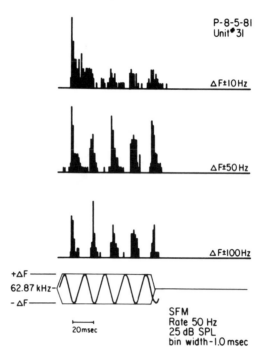

FIGURE 7.18. Peristimulus time histograms of a sharply tuned 60-kHz neuron in the mustache bat's inferior colliculus that phase locked to sinusoidally frequency-modulated (SFM) signals. The signal envelope and modulating waveform are shown below. The depth of modulation was varied in the three records around a carrier frequency of 62.87 kHz. This unit was unusually sensitive, and phase locked when the frequency swings (Δf) were as small as ±10 Hz around the 62.87-kHz center frequency. (Reproduced from Bodenhamer and Pollak 1983.)

synchronously discharge (Schuller 1979; Pollak 1980; Pollak and Schuller 1981; Bodenhamer and Pollak 1983; Ostwald 1988). This feature is well illustrated by considering how discharge synchrony is influenced by signal intensity, modulation depth and rate, and carrier frequency.

Many filter neurons respond with discharges that are about equally well synchronized to the modulating waveform at all intensities above threshold. Other filter neurons, however, discharge in registry with the modulation waveform only over a preferred range of intensities. Typically, the sharpest locking and most vigorous responses are evoked only at low or moderately low intensities, and at higher intensities the discharge registration either declines significantly or disappears completely. Four neurons exhibiting this feature are shown in (Fig. 7.19). It should also be pointed out that the preferred range of intensities of the four neurons in Figure 7.19 differs slightly from neuron to neuron. The population, then, forms a continuous gradient in which some neurons encode the modulation pattern at all intensities above threshold, whereas others are more selective and only

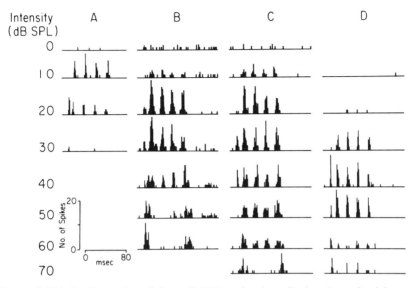

FIGURE 7.19A–D. Examples of four 60-kHz units that displayed a selectivity, or tuning, of their locked discharges for intensity. Each of the units phase locked to the SFM signals only over a narrow intensity range that was specific to each neuron. Modulation depths: ±100 Hz (A), ±50 Hz (B), ±200 Hz (C), and ±100 Hz (D). Modulation rate was 50 Hz for all units. Signal duration was 80 msec. (Reproduced from Bodenhamer and Pollak 1983.)

encode the modulation waveform if the echo falls within a narrow intensity slot.

Filter neurons also exhibit preferences for the range of modulation depths to which they will phase lock. As was the case for intensity, most filter neurons exhibit a continuous gradation in their abilities to encode modulation depth. This feature is illustrated by the three filter neurons in Figure 7.20. The neuron on the far left of Figure 7.20 locked best for modulations depths between ±100 Hz to ±400 Hz. In contrast, the neuron in the middle panel locked best to modulation depths that ranged from ±50 Hz to ±200 Hz, while the neuron in the far right panel locked best to ±50 Hz. Notice that the modulation *rate* presented to all three neurons was 50 Hz.

As is the case for intensity and modulation depth, filter neurons also display a selectivity for modulation rate, which simulates the various wingbeat frequencies of different insects. A few neurons accurately encode modulation rates from 25 Hz to 200–300 Hz, whereas most other units show a preference for a range of rates.

The effects of various SFM parameters are generally evaluated when the carrier frequency of the modulated signal is set at the neuron's best frequency. Under natural conditions, however, the center frequency of the echo will not correspond to the best frequencies of all the neurons excited by the signal. For example, consider a situation where a Doppler compen-

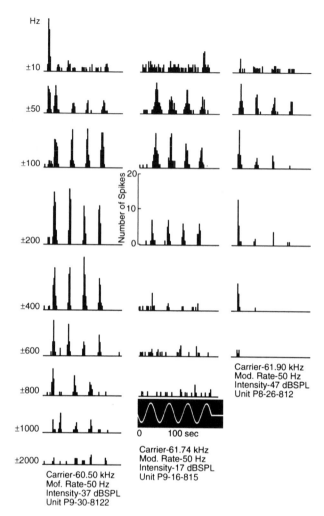

FIGURE 7.20. Three 60-kHz units from the mustache bat's inferior colliculus that had preferences for different SFM modulation depths. The modulation depth used to generate each histogram is shown at the far left. The modulation rate was 50 Hz and the signal duration was 80 msec for each unit. (Reproduced from Bodenhamer and Pollak 1983.)

sating mustache bat receives an echo in which the frequency modulations vary by ±500 Hz around a 60-kHz carrier. Under these conditions, units having a best frequency of 60 kHz will be excited, but so will other units having best frequencies slightly above or slightly below 60 kHz. Given these conditions, it becomes of some interest to assess how filter units encode modulation patterns for a variety of different carrier frequencies.

Carrier frequency affects synchronized discharges to SFM signals in two general ways. The first way is that many units lock in a symmetrical manner for carrier frequencies around the unit's best frequency (Fig. 7.21, left panel). Whether the carrier frequency is above or below the neuron's best frequency is unimportant to such units. Rather, the magnitude of the locked discharge rate depends largely on the extent to which the SFM signal encroaches upon the tuning curve. The second way is displayed by other units that respond in an asymmetric manner for different carrier frequencies. In some of these units the magnitude of the synchronized discharges is clearly greater when the carrier frequency is below the neuron's best frequency, or, in other units, when the carrier frequency is above the best frequency. An asymmetric unit in which the neuron favored carrier frequencies above the best frequency is shown in the right panel of Figure 7.21.

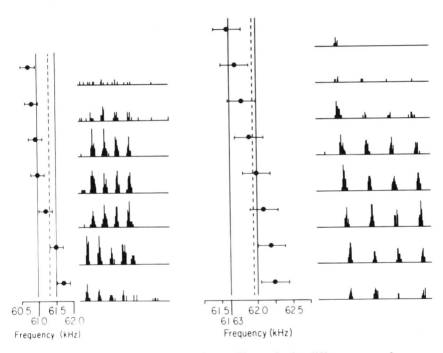

FIGURE 7.21. Phase-locked discharges in two filter units for different center frequencies. *Left:* Neuron that phase locked regardless of carrier frequency, so long as a portion of the SFM signal encroached upon the tuning curve. *Right:* Neuron displaying a marked preference, or asymmetry, for SFM carrier frequencies on the high-frequency side of its tuning curve. *Solid lines* to left of histograms show the limits of the unit's tuning curve at the same intensity as the SFM signal. *Dashed line* indicates the unit's best frequency. *Horizontal lines* indicate the position of the SFM signal. SFM signal in left panel was ±100 Hz at 20 dB SPL (sound pressure level), and SFM signal in right panel was ±200 Hz at 30 dB SPL. (Reproduced from Bodenhamer and Pollak 1983.)

2.8 Representation of an Insect's Signature in the Filter Region of the Inferior Colliculus

Insights into the spatial extent of activity evoked by an echo from a fluttering insect can be obtained from considerations of the sharp tuning curves and the way in which filter units respond to SFM signals. For example, on detecting a target, a mustache bat can determine if the target is inanimate or animate, for example, a falling leaf or a flying insect, on the basis of the temporal patterns of the responding neurons: an echo from a falling leaf will be encoded in a manner similar to a tone burst whereas a fluttering insect will elicit synchronized discharges locked to the modulation pattern of the echo. The active region of the colliculus will have boundaries sharply confined to the 60-kHz contour because of the narrow tuning curves of the filter units. The response characteristics evoked by SFM signals suggest that the discharge rate and synchrony should be maximal in units whose best frequencies coincide with the echo center frequency and which respond symmetrically to various SFM carrier frequencies. However, the activity should also be maximal in the asymmetric units whose best frequencies are slightly above or below the center or carrier frequency of the echo. In these neurons, the synchronized discharges are evoked only when the carrier frequency is below, or in other units, above, the neuron's best frequency.

Changes in position, orientation, and speed of either the bat or its target will cause subsequent echoes to differ more or less in carrier frequency, modulation pattern, and intensity from the previous echo(es). In principle, each echo will be encoded in a similar manner. However, the preferences of many filter units for selective ranges of intensity as well as modulation rate and depth suggest that the changes in echo parameters will cause some neural elements to drop out and new elements to be recruited, while others will simply change response vigor or firing registration to reflect the changes in echo characteristics. In short, the properties of filter neurons endow the system with the ability to encode the features of the echo CF component. Thus, the sum total of neural activity is a dynamic pattern that differs from echo to echo.

3. Sound Localization and the Processing of Localization Cues by the EI Circuit

The ability to localize a sound source in both azimuth and elevation is of obvious importance to a nocturnal predator that hunts flying insects in the night sky. The cues animals use to associate a sound with its position in space are interaural disparities in time or intensity (Erulkar 1972; Mills

1972; Gourevitch 1980). Animals that hear "high" frequencies, such as echolocating bats, rely on interaural intensity disparities (IIDs) to localize sounds composed of those frequencies (Erulkar 1972; Mills 1972).

The intensity disparities are generated by acoustic shadowing and the directional properties of the ear (Grinnell and Grinnell 1965; Blauert 1969/1970; Erulkar 1972; Flannery and Butler 1981; Musicant and Butler 1984; Fuzessery and Pollak 1984, 1985; Makous and O'Neill 1986; Jen and Sun 1984; Jen and Chen 1988; Pollak and Casseday 1989). Once produced, the intensity disparities are conveyed into the central nervous system where the information from the two ears is initially "compared" in the LSO (Stotler 1953; Boudreau and Tsuchitani 1968; Warr 1982; Cant and Casseday 1986; Harnischefeger, Neuweiler, and Schlegel 1985; Covey, Vater, and Casseday 1991; Finlayson and Caspary 1989). The comparison is a subtractive process, whereby signals from the ipsilateral ear excite and signals from the contralateral ear inhibit LSO cells (Boudreau and Tsuchitani 1968; Brownell, Manis, and Ritz 1979; Caird and Klinke 1983; Harnischefer, Neuweiler, and Schlegel 1985; Sanes and Rubel 1988; Covey, Vater, and Casseday 1991). These so-called EI cells are sensitive to intensity disparities and express the comparison of IIDs in their firing rates.

The comparisons of IIDs must be performed on a frequency-by-frequency basis, a requirement permitted by tonotopic organization. The frequency-by-frequency comparison is necessary because the directional properties of the ear amplify sounds emanating from certain positions in space in a frequency-dependent manner (Grinnell and Grinnell 1965; Blauert 1969/1970; Flannery and Butler 1981; Fuzessery and Pollak 1984, 1985; Makous and O'Neill 1986; Jen and Sun 1984; Jen and Chen 1988; Musicant and Butler 1984). Thus, a particular frequency emanating from a given region of space off the midline will generate an IID of a particular value, but a different frequency originating from the same location will create a different IID. This feature is illustrated in the top panel of Figure 7.22 for the IIDs generated at the mustache bat's ear by three harmonics of its orientation calls, at 30 kHz, 60 kHz, and 90 kHz.

Consistent with the requirement that binaural cues be processed on a frequency-by-frequency basis is the finding that LSO cells are binaurally innervated from comparable regions of the cochlear surface, and thereby compare the intensity of a given frequency from one ear with the intensity of the same frequency from the other ear (Boudreau and Tsuchitani 1968; Caird and Klinke 1983; Semple and Kitzes 1985; Sanes and Rubel 1988). Once the computation is made by the LSO, the encoded information is conveyed bilaterally to the inferior colliculus and DNLL (see Fig. 7.2). The DNLL, in turn, projects bilaterally to the inferior colliculus (Casseday, Covey, and Vater 1988; Glendenning et al. 1981; Irvine 1986; Roth et al. 1978; Ross, Pollak, and Zook 1988). Both the DNLL (Brugge, Anderson, and Aitkin 1970; Covey 1993; Markovitz and Pollak 1993) and inferior

Distribution of IIDS

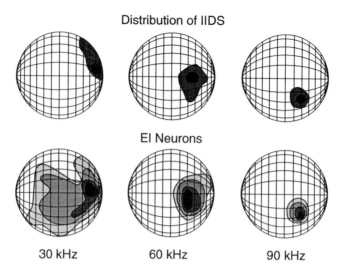

El Neurons

30 kHz   60 kHz   90 kHz

FIGURE 7.22. Interaural intensity disparities (IIDs) generated by 30-, 60-, and 90-kHz sounds are shown in *top row*. Each sphere represents the bat's frontal sound field. The bat's head would lie in the center of each sphere. Each horizontal line represents a 20° increment in elevation, and each vertical line represents a 13° increment in azimuth. The *blackened areas* in each panel indicate the spatial locations of the largest IIDs (20 dB or greater), and the *grey areas* indicate those locations where the IIDs are at least 10 dB. The panels in the *lower row* show the spatial selectivity of three EI units, one tuned to 30 kHz, one to 60 kHz, and one to 90 kHz. The *blackened areas* indicate the spatial locations where the lowest thresholds were obtained. Isothreshold contours are drawn for threshold increments of 5 dB. Areas not contained within isothreshold contours indicate spatial locations from which sounds failed to evoke discharges. (Adapted from Fuzessery 1986.)

colliculus (Roth et al. 1978; Schlegel 1977; Semple and Aitkin 1979; Wenstrup, Fuzessery and Pollak 1988a,b; Irvine and Gago 1990) have large populations of EI cells.

Thus, the task of the auditory system is to encode each of the IIDs generated by the various frequencies of a complex sound. The location of a sound source, in both azimuth and elevation, must then be represented in the inferior colliculus by the pattern of activity among the population of EI neurons within each of the tonotopically organized isofrequency contours.

In the following sections, the 60-kHz contour of the mustache bat's inferior colliculus is used as a model to illustrate the manner in which IIDs are represented within an individual contour. Consideration of how the IIDs generated by other frequencies are represented in their contours leads to a hypothesis of how sound location, in both azimuth and elevation, is represented in the inferior colliculus. Finally, the role played by the inhibition from lower auditory nuclei for creating the representation of IIDs in the colliculus is discussed.

3.1 Directional Properties of the Ear and IIDs
Generated by 60 kHz Along the Azimuth

Before describing the binaural and spatial properties of 60 kHz EI units, we consider how 60-kHz sounds are affected by the ear at different spatial locations. For simplicity, only spatial locations along the azimuth, at 0° elevation, are considered. The directional properties of the ear for 60 kHz are shown in the left panel of Figure 7.23. The noteworthy feature of the directional pattern is that the ear is most sensitive to sounds at about 26–40° from the midline, and that thresholds increase progressively as the source moves toward the midline and then into the sound field on the opposite side of the bat's head. The IIDs generated by 60 kHz were estimated by taking the difference in threshold between the two ears at mirror image locations in the ipsilateral and contralateral sound fields (right panel of Fig. 7.23). The largest IIDs originate at about 40° azimuth, the location at which the sound is most intense in one ear and least intense in the other. Notice that the IIDs change roughly linearly, by about 0.75 dB/degree, for locations ±40° around the midline. The importance of this linearity is that specifying the IID automatically specifies azimuthal location, but this is only true within ±40° around the midline and only when elevation is held constant. Beyond about 40°, the magnitude of the IIDs decline. In the following sections, we discuss the coding of IIDs within 40° of the midline. In later sections, we discuss the significance of the nonlinear IID changes that occur beyond 40° and the influence of elevation.

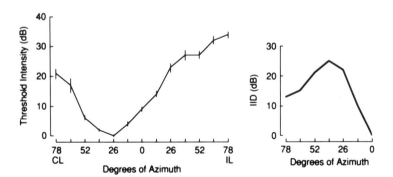

FIGURE 7.23. Measures of directional properties of the mustache bat's ear at 60 kHz *(left panel)* and the interaural intensity disparities (IIDs) generated by 60-kHz sounds *(right panel)*. Directional properties describe the changes in threshold of either the cochlear microphonic potential or monaural single units for a 60-kHz sound at different azimuthal positions in the acoustic hemifield. (Adapted from Wenstrup, Fuzessery, and Pollak 1988b.)

3.2 The Medial Border of Each 60-kHz EI Neuron is Determined by the Disparities Generated by the Ears and the Neuron's Inhibitory Threshold

When sounds are presented from various regions of space, EI neurons display a spatial selectivity, in that they discharge only when the sounds emanate from some regions of space and not from others. The regions of space from which discharges can be evoked constitute the neuron's receptive field (right panel of Fig. 7.24). The receptive fields of EI collicular units are characterized by a medial border in space. Sounds emanating from one side of the border, which correspond to more intense sound at the contralateral (excitatory) ear, evoke a more or less strong discharge rate depending on the signal intensity. Sounds emanating from the other side of the border, corresponding to locations closer to the ipsilateral (inhibitory) ear, evoke few or no discharges. The border is the location in space at which the discharge rate changes from one that is fairly vigorous to a much lower rate, or a complete inhibition.

To determine how binaural properties shape a neuron's receptive field, the binaural properties of 60-kHz EI cells were first determined with loudspeakers inserted into the ear canals, and subsequently the spatial properties of the same neurons were evaluated with free-field stimulation delivered from loudspeakers located around the bat's hemifield (Fuzessery and Pollak 1984, 1985; Wenstrup, Fuzessery, and Pollak 1988b; Fuzessery, Wenstrup, and Pollak 1990). By correcting for the directional properties of the ear at 60 kHz and the IIDs generated at each location, the quantitative aspects of the binaural responses could be associated with, and thus predict, the neuron's spatially selective properties (Wenstrup, Fuzessery, and Pollak 1988b). These studies showed that the receptive field of a 60-kHz EI neuron could be constructed from a family of IID functions. Each IID function was obtained when the intensity at the contralateral (excitatory) ear was held constant and the intensity at the ipsilateral (inhibitory) ear was progressively increased, as shown in the left panel of Figure 7.24. The neuron's receptive field was then constructed from a family of such IID functions, each obtained with a different sound intensity at the contralateral ear. The receptive field that was constructed from the IID functions was an accurate representation of the azimuthal receptive field of the same unit determined with free-field stimulation (Wenstrup, Fuzessery, and Pollak 1988b). These studies also showed: (1) that the medial borders and the interaural intensity disparities that are generated at the locations of those borders differ among EI cells; and (2) that the medial border and the corresponding IID correlate closely with, and thus can be predicted by, the neuron's inhibitory threshold (the IID at which the response declines by 50%) (Fuzessery and Pollak 1985; Wenstrup, Fuzessery, and Pollak 1988b).

That the medial border can be predicted by the cell's inhibitory threshold

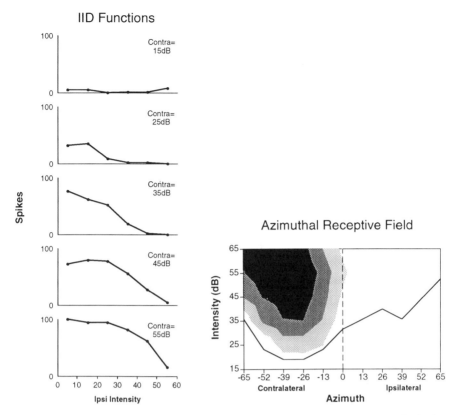

FIGURE 7.24. Azimuthal receptive field of a 60-kHz EI neuron in the mustache bat's inferior colliculus is shown on the *right*. *Blackened regions* show discharge rates ranging from 75% to 100% of the maximum rate. *Dark grey area* indicates discharge rates from 50% to 74% of maximum; *light grey areas*, regions of discharge rates from 25% to 49% of maximum. The *solid line* below the receptive field indicates the regions of space from which discharges could be evoked in a monaural neuron. The receptive field was constructed from the family of IID functions shown on the *left*. Each IID function was obtained by holding the intensity at the contralateral ear constant while changing the intensity at the ipsilateral ear. The algorithm described in Wenstrup, Fuzessery, and Pollak (1988b) was used to construct the receptive field.

is illustrated in Figure 7.25. This figure shows the receptive fields of three 60-kHz EI units, each having a different inhibitory threshold, and the receptive field of a monaural unit. As a starting point, consider the receptive field of the monaural neuron (Fig. 7.25A). Monaural neurons tuned to 60 kHz are always most sensitive at the same spatial position, about 26°–40° from the midline, which is a consequence of the directional properties of the ear for 60 kHz. Thus, a 60-kHz tone presented at about 40° in the contralateral sound field will have the lowest threshold, and a

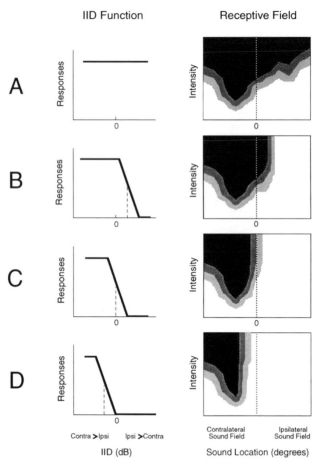

FIGURE 7.25A–D. Schematic shows influence of inhibitory thresholds of the IID functions on the azimuthal receptive fields of 60-kHz EI neurons. *Black area* in each receptive field shows locations and intensities that evoke discharge rates between 75% and 100% of maximum. *Dark grey* contours, firing rates between 50% and 74% of maximum; *light grey contour*, firing rates 25% to 49% of maximum. (A) The receptive field of a monaural neuron extends throughout the acoustic hemifield. The thresholds are lowest in the contralateral sound field and highest in the ipsilateral sound field because of the directional properties of the ear, together with the shadowing produced by the head and ears. (B) An EI neuron with a negative inhibitory threshold (sound at the ipsilateral ear has to be more intense than at the contralateral ear to produce a 50% reduction of discharge rate), The receptive field no longer extends throughout the ipsilateral sound field, because sounds presented from those regions are sufficiently intense at the ipsilateral ear to completely inhibit any discharges. (C) and (D) Two EI units with progressively more positive inhibitory thresholds have receptive fields that are correspondingly more limited to the contralateral sound field. (Based on model in Wenstrup, Fuzessery, and Pollak 1988b.)

low-intensity sound at that location will evoke a low discharge rate, as indicated by the lightly shaded area. Progressively louder sounds at that location will, in turn, elicit progressively higher discharge rates as indicated by the darker shading. If the sound source is moved to 40° in the ipsilateral sound field, a much more intense sound is needed to reach threshold because of the shadowing effect of the head and ears. To elicit the same range of discharge rates from the ipsilateral sound field as were evoked from the contralateral sound field, the sound in the ipsilateral field need only be made more intense. Thus, a monaural neuron can discharge over its entire dynamic range to sounds presented from any azimuthal location. However, the sound intensity required to evoke a given discharge rate varies with location because of the acoustic shadow cast by the head and ears and the directional properties of the ear. It is for these reasons that physical properties dictate the shape of the receptive field.

The receptive fields of EI neurons have the same general shape as monaural neurons, but with one difference; a portion of the receptive field in the ipsilateral sound field is cut off. The particular spatial location at which the cell's receptive field is cut off, that is, its medial border, is largely predicted by the cell's inhibitory threshold (Fig. 7.25B–D). The explanation is simply that the inhibitory projection that has been added from the other ear has a certain threshold, and when that threshold is exceeded, the inhibition begins to dominate. Thus, as the sound is moved around the acoustic hemifield, from the contralateral to the ipsilateral side, the sound initially drives only the excitatory ear, and thus evokes a discharge rate similar to the rate that would be evoked only by monaural stimulation. However, when the sound moves into the ipsilateral field, the increasing intensity at the ipsilateral ear will, at a certain location, exceed the threshold of the inhibitory neurons driven by sound at the ipsilateral ear. The activity evoked from the ipsilateral ear will then suppress the discharges evoked by sound at the contralateral ear. Thus, 60-kHz units with high inhibitory thresholds, such as the unit in Figure 7.25B, require a more intense stimulation of the inhibitory ear for complete inhibition, and therefore the medial borders of these units are in the ipsilateral sound field. In 60-kHz units with positive inhibitory thresholds, as in Figure 7.25 B,C, their medial borders occur along the midline or are even in the contralateral sound field. For these units, sounds presented ipsilateral to their borders are incapable of eliciting discharges, even with high-intensity stimulation. The important spatial feature of each EI unit receptive field is its medial border, and the particular azimuth of that border is determined principally by its inhibitory threshold.

3.3 The Systematic Arrangement of Inhibitory Thresholds Creates a Representation of IIDs

The finding that an EI neuron's inhibitory threshold determines where along the azimuth the cell's medial border is located, coupled with the orderly arrangement of inhibitory thresholds in the ventromedial EI region as presented previously, has implications for the representation of the azimuthal position of a sound in the mustache bat's inferior colliculus (Fuzessery and Pollak 1985; Wenstrup, Fuzessery, and Pollak 1988b; Pollak, Wenstrup, and Fuzessery 1986; Pollak and Casseday 1989). Specifically, the value of an IID is represented in the ventromedial 60-kHz contour as a border separating a region of discharging from a region of inhibited cells, as shown in Figure 7.26. Consider, for instance, the pattern of activity in the 60-kHz contour of one inferior colliculus generated by a 60-kHz sound that is 15 dB louder in the ipsilateral ear than in the contralateral ear. The IID in this case is −15 dB. Because neurons with positive inhibitory thresholds are situated ventrally, the high relative intensity in the ipsilateral (inhibitory) ear will inhibit all the EI neurons in the ventral portion of the contour. The same sound, however, will not be sufficiently intense at the ipsilateral ear to inhibit the EI neurons in the more dorsal portion of the contour, where neurons require an even more intense ipsilateral sound for inhibition. The topology of inhibitory thresholds and the steep IID functions of EI neurons, then, can create a border between excited and inhibited cells within the ventromedial region of the contour. The locus of the border, in turn, should shift with changing IID, and therefore should shift correspondingly with changing sound location (Fig. 7.26, lower panel).

3.4 Spatial Properties of Neurons Tuned to Other Frequencies

The binaural processing found in the 60-kHz contour appears to be representative of binaural processing within other isofrequency contours (Fuzessery and Pollak 1984, 1985). This feature is most readily appreciated by considering the spatial properties of EI units tuned to 90 kHz, the third harmonic of the mustache bat's orientation calls (see Fig. 7.22, lower panel

———→

FIGURE 7.26. *Top panel:* Schematic of systematic changes in the medial borders of the receptive fields of EI neurons in the 60-kHz contour of the inferior colliculus. Shifts in the medial borders result from the progressive shifts of the inhibitory thresholds of EI neurons along the dorsoventral axis of the 60-kHz contour. *Bottom panel*: Stylized illustration of relationship between the value of the interaural intensity disparity produced by a sound source at a given location (moths at the top of the panel) and the pattern of activity in the ventromedial EI region of the left colliculus, where IID sensitivities are topographically organized. The activity in this region, indicated by the *blackened area*, spreads ventrally as a sound source moves from the ipsilateral (−15 dB IID) to the contralateral sound field (+15 dB IID).

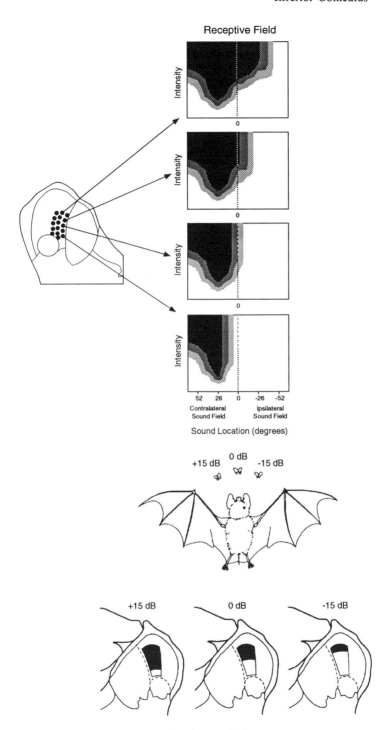

FIGURE 7.26. Caption on facing page.

on right). The significant feature is that the maximal interaural intensity disparities generated by 90-kHz tones occur in a region of space that is different from the region that generates the maximum disparity with 60-kHz tones. The largest disparities obtained for 90 kHz are at about 40° along the azimuth and −40° in elevation (Fig. 7.22, top panel on right). The 90-kHz EI neurons, like those tuned to 60 kHz, are most sensitive to sounds presented from the same spatial location at which the maximal interaural intensity disparities are generated. Additionally, the inhibitory thresholds of these neurons determine the azimuthal border defining the region in space from which 90-kHz sounds can evoke discharges from the region where sounds are incapable of evoking discharges (Fuzessery and Pollak 1985). The population of 90-kHz EI neurons has a variety of inhibitory thresholds that appear to be topographically arranged within that contour (Wenstrup, Ross, and Pollak 1986b). Therefore, a particular interaural intensity disparity will be encoded by a population of 90-kHz EI cells, having a border separating the inhibited from the excited neurons, in a fashion similar to that shown for the 60-kHz contour. The same argument can be applied to the 30-kHz cells, but in this case the maximal interaural intensity disparity is generated from the very far lateral regions of space (see Fig. 7.22, top panel on left).

3.5 The Representation of Auditory Space in the Mustache Bat's Inferior Colliculus

We can now begin to see how the cues that define both the azimuth and elevation of a sound source are derived. The directional properties of the ears generate different interaural intensity disparities among frequencies, an interaural spectral difference, when a broadband sound emanates from a particular location. The interaural spectral difference will evoke a specific pattern of activity across each of the isofrequency contours in the bat's midbrain that are driven by the spectral differences received at the ears. Figure 7.27 shows a stylized illustration of the interaural intensity disparities generated by 30, 60, and 90 kHz within the mustache bat's hemifield, and below is shown the loci of borders in the EI region of each contour that would be generated by a biosonar signal containing the three harmonics emanating from different regions of space. Consider first a sound emanating from 40° along the azimuth and 0° elevation (Fig. 7.27, white circle in top panel). This position creates a maximal interaural intensity disparity at 60 kHz, a lesser interaural intensity disparity at 30 kHz, and yet a different interaural intensity disparity at 90 kHz. The borders created within each of the isofrequency contours by these interaural intensity disparities are shown in the bottom panel of Figure 7.27. Next, consider the interaural intensity disparities created by the same sound, but from a slightly different position in space, at about 60° azimuth and −20°

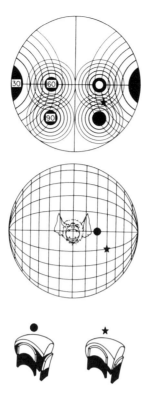

FIGURE 7.27. Loci of borders in the 30-, 60-, and 90-kHz EI regions generated by biosonar signals emanating from two regions of space. The *top panel* is a schematic representation of IIDs produced by 30-, 60-, and 90-kHz sounds that occur in the mustache bat's acoustic field. *Blackened areas* indicate the regions in space where the maximum IID is generated for each harmonic. The *circular lines* surrounding each blackened region indicate the spatial locations that produce iso-IIDs for that frequency. Each successive circle indicates a progressively smaller IID. The *middle panel* depicts the bat's head and the spatial positions of two sounds (*star* and *circle*), where each sound is composed of the three harmonics that have equal intensities. The *lower panel* shows the borders separating the regions of excited from regions of inhibited neurons in the 30-, 60-, and 90-kHz contours that would result from the IIDs generated by the sounds at the two locations. (Reproduced from Pollak and Casseday 1989.)

elevation (Fig. 7.27, star in top panel). In this case, there is a decline in the 60-kHz interaural intensity disparity, and an increase in the 90-kHz interaural intensity disparity, but the 30-kHz interaural intensity disparity will be the same as it was when the sound emanated from the previous position. The fact that the interaural intensity disparity at 30 kHz did not change although the location of the sound source changed is a crucial point. It illustrates that the interaural intensity disparity generated by a particular frequency is not uniquely associated with one position in space, but rather can be generated from a variety of positions. It is for this reason that sound localization with only one frequency is ambiguous (Blauert 1969/1970; Butler 1974; Musicant and Butler 1984; Fuzessery 1986). However, spatial location, *in both azimuth and elevation*, is rendered unambiguous by the simultaneous comparison of three interaural intensity disparities, because their values in combination are associated with a unique spatial location in the bat's acoustic hemifield, except for one special region of space.

The exception is the vertical midline, at 0° azimuth. The representation by EI regions becomes ineffective along the vertical midline because the interaural intensity disparities will be 0 dB at all frequencies and all elevations. The borders among the EI populations, therefore, will not change with elevation because the interaural intensity disparities remain constant. Several types of binaural cells may be important for the encoding of elevation along the midline (Fuzessery and Pollak 1985; Fuzessery 1986; Fuzessery, Wenstrup, and Pollak 1990). One type is the EI/f cell (Fig. 7.28). The distinguishing feature of EI/f cells is that as sound intensity increases at the ipsilateral (inhibitory) ear, the increase initially causes the neuron's firing rate to *increase* by at least 25% above the rate evoked by the sound at the contralateral (excitatory) ear alone (Fig. 7.28, left panel). Additional intensity increases at the ipsilateral ear then result in a marked decline in

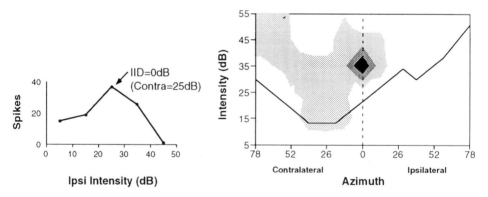

FIGURE 7.28. IID function *(left)* and azimuthal receptive field *(right)* of an 60-kHz EI/f unit recorded from the mustache bat's inferior colliculus. The IID function on the left was obtained when the contralateral intensity was fixed at 25 dB SPL and the intensity at the ipsilateral ear was varied from 5 to 45 dB. With low intensities at the ipsilateral ear (5 and 15 dB), the neuron's discharge rate was equal to that evoked when the 25-dB signal was presented only to the contralateral ear (not shown). When the ipsilateral intensity increased to 25 dB, generating an IID of 0 dB, the discharge rate increased markedly. However, additional intensity increments at the ipsilateral ear then resulted in a progressively greater inhibition. Note that the highest discharge rates could be evoked only from limited regions of space close to the midline. These regions generate the IIDs at which the facilitation is expressed in the IID functions. The receptive field was constructed from a family of IID functions (not shown), each generated with a different intensity at the contralateral ear.

response rate, as in regular EI neurons. The receptive fields of EI/f cells are variations of the fields of EI cells (Fuzessery and Pollak 1985; Fuzessery, Wenstrup, and Pollak 1990). Like EI cells, they respond poorly or not at all to sounds located in the ipsilateral acoustic field, and their borders are determined by the cell's inhibitory threshold.

Unlike EI cells, the discharge rates of EI/f cells are enhanced for sound locations that generate the IIDs at which they are facilitated. The IIDs producing facilitation are usually at, or close to, 0 dB and thus occur around the midline. EI/f cells, then, respond most vigorously to sounds located directly ahead as a consequence of the binaural facilitation. However, the elevation at which they discharge maximally depends on their best frequency. EI/f units tuned to different frequencies exhibit selectivities for different elevations; EI/f units tuned to 60 kHz are maximally sensitive at about $0°$ to $-10°$ elevation, whereas units tuned to 90 kHz are most sensitive at about $-40°$ (Fuzessery and Pollak 1985; Fuzessery, Wenstrup, and Pollak 1990). The elevation selectivity is a consequence of the directional properties of the ear for 30, 60, and 90 kHz. One can, therefore, visualize that as the elevation of a sound source along the vertical meridian shifts, the response magnitude also shifts among the EI/f units in different isofrequency contours. In combination then, these two binaural types could provide a neuronal representation of sound located anywhere within the bat's acoustic hemifield.

A striking feature of the foregoing scenario is its similarity, in principle, to the ideas about sound localization proposed previously by Pumphery (1948) and Grinnell and Grinnell (1965). What the recent studies provide are the details of how interaural disparities are encoded and how they are topologically represented in the acoustic midbrain.

3.6 A Wiring Diagram of the Excitatory and Inhibitory Projections to the Inferior Colliculus Can Be Constructed

The previous sections showed how the spatial properties of 60-kHz EI neurons are shaped by both the directional properties of the ear and the inhibition provided by the projections activated by stimulation of the ipsilateral ear. In the following sections, the nuclei that provide the inhibition evoked by the ipsilateral ear are identified, and the role that the inhibitory neurons in each of those nuclei could play in shaping the binaural properties and receptive fields of EI neurons in the inferior colliculus are discussed. This issue is addressed in two stages. In the first stage, the origins of the GABAergic and glycinergic inhibitory neurons that innervate the EI region of the colliculus are determined. In the second stage, the changes in binaural properties and receptive fields of EI neurons are described after inhibitory inputs are blocked by pharmacological antagonists applied iontophoretically.

The identification of inhibitory projections takes advantage of the fact that the majority of auditory nuclei have large numbers of projection neurons that are inhibitory (Adams and Mugniani 1984; Mugniani and Oertel 1985; Saint Marie et al. 1989; Park et al. 1991; Vater et al. 1992; Winer, Larue, and Pollak 1995). The prevalence of long-distance inhibitory projections is illustrated in Figure 7.4, which shows a section through the mustache bat's auditory brainstem that has been reacted with antibodies against the inhibitory neurotransmitter glycine. It is apparent from this photomicrograph that there is a massive glycinergic projection system ascending in the lateral lemniscus which innervates the inferior colliculus. As we pointed out previously, there are also strong GABAergic projections to the inferior colliculus (see Fig. 7.4). Determining which nuclei send glycinergic or GABAergic cells to the EI region of the colliculus was accomplished by injecting HRP in the 60-kHz isofrequency contour and then colocalizing the HRP reaction product from retrograde transport in sections that were stained alternately with antibodies against GABA and glycine (Larue et al. 1991; Park et al. 1991). Data of this sort show not only which cells project to the 60-kHz region, but also whether the cells are GABAergic, glycinergic, or neither, and thus are probably excitatory. If these findings are considered together with those from earlier studies, which revealed the particular subset of nuclei that project to the EI region, not only can the wiring diagram of the EI region be constructed but in addition an excitatory or inhibitory role can also be assigned to each projection. This neurotransmitter wiring diagram is shown in Figure 7.29.

Because the excitatory/inhibitory projection system involves both contralateral and ipsilateral structures, as well as crossed and uncrossed projections, we hereafter refer to structures ipsilateral to the left colliculus as being on the left, and structures contralateral to that colliculus as being on the right. Hence, EI cells in the left colliculus are excited by sound at the right (contralateral) ear and inhibited by sound at the left (ipsilateral) ear. The relevant features of Figure 7.29 concern the inhibitory projections that are activated by the (left) ipsilateral ear, because it is the inhibition evoked by stimulation of that ear which shapes the receptive fields of EI neurons in the left colliculus. Four nuclei are especially important: (1) the left (ipsilateral) LSO; (2) the right (contralateral) LSO; (3) the left (ipsilateral) DNLL; and (4) the right (contralateral) DNLL.

Turning first to the right (contralateral) LSO, this nucleus receives excitation from the right ear and inhibition from the left ear, which generates the IE properties of the cells in this nucleus. Indeed, LSO neurons in all mammals are overwhelming inhibitory/excitatory (Boudreau and Tsuchitani 1968; Caird and Klinke 1983; Sanes and Rubel 1988: Covey, Vater, and Casseday 1991). The right (contralateral) LSO, however, sends a putative excitatory projection to the contralateral (left) colliculus (Saint Marie et al. 1989; Park et al. 1991; Glendenning et al. 1992), and through this excitatory projection it can impose EI properties upon its targets in the

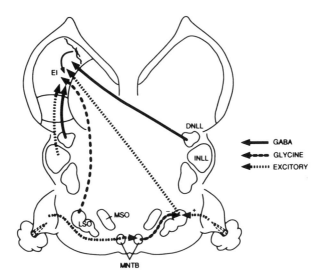

FIGURE 7.29. Wiring diagram shows sources of GABAergic, glycinergic, and excitatory projections to the 60-kHz EI region of the mustache bat's inferior colliculus. DNLL, dorsal nucleus of the lateral lemniscus; INLL, intermediate nucleus of the lateral lemniscus; LSO, lateral superior olive; MSO, medial superior olive; MNTB, medial nucleus of the trapezoid body.

left colliculus. The left (ipsilateral) LSO, in contrast, receives excitation from the left ear and inhibition from the right ear. However, the left LSO sends a glycinergic projection to the colliculus on the same side (Saint Marie et al. 1989; Park et al. 1991; Glendenning et al. 1992). It, therefore, can provide inhibition to the left colliculus when the left ear is stimulated. The DNLL, like the LSO, has mostly excitatory/inhibitory neurons (Brugge, Anderson, and Aitkin 1970; Covey 1993; Markovitz and Pollak 1993). Thus, stimulation of the left (ipsilateral) ear will provide an excitatory drive to neurons of the right DNLL. Because these neurons are GABAergic, they can, in turn, provide a potent inhibition to the left colliculus when the left (ipsilateral) ear is stimulated. This is in contrast to the left (ipsilateral) DNLL, which would be inhibited by sound presented to the left ear and excited by sound presented to the right ear. Thus, the left DNLL could provide inhibition to the left colliculus when sound was presented to the right ear. The effects of each of these complex connections are explained in greater detail in the following sections.

3.7 Effects of GABAergic and Glycinergic Innervation on Spatial Receptive Fields Can Be Evaluated

Armed with this information, we assessed the influences of inhibitory inputs on the binaural response properties of 60-kHz collicular EI neurons with microiontophoretic application of the GABA$_A$ receptor antagonist bicuculline and the glycine receptor antagonist strychnine. The rationale is that if EI properties are created in the colliculus by the convergence of excitatory and inhibitory inputs, then removing the influence of the inhibitory inputs should reduce or eliminate the acoustically evoked inhibition. On the other hand, if EI response properties are created in a lower nucleus and imposed upon the collicular cell via an excitatory projection, then the blockade of inhibitory inputs at the colliculus should have no effect on the expression of those response properties. We constructed the receptive field from the binaural properties obtained before blocking inhibition, and then compared it to the receptive field constructed from the binaural properties obtained when GABAergic inhibition was blocked by the application of bicuculline, or when glycinergic inhibition was blocked by the application of strychnine. The feature of interest is how inhibition evoked by the left (ipsilateral) ear limits the degree to which the receptive field extends into the left (ipsilateral) sound field. In this way, the effect of GABAergic inhibition on the spatial properties of an EI collicular neuron can be compared to the effects of glycinergic inhibition, and informed inferences can then be made as to which of the lower nuclei produced each of the effects.

3.8 Binaural Properties of EI Neurons Are Formed in at Least Five Ways

These studies show that a variety of EI properties are formed in the inferior colliculus by the mixing and matching of the various excitatory and inhibitory circuits that converge upon these binaural cells. The EI properties, and hence the features of receptive fields, are formed in at least five ways. The first way is through the complete generation of EI properties in the right (contralateral) LSO that are then imposed, without modification, on the left (ipsilateral) inferior colliculus via the crossed, excitatory projection from the LSO to the inferior colliculus. Because neither GABAergic nor glycinergic innervation of the inferior colliculus contributes to the binaural properties of the collicular cell, blocking either of these inhibitory transmitters should have no effect on the neuron's receptive field. An example of a neuron exhibiting these features, and the circuitry that could account for the absence of any changes in the receptive field of the cell, are shown in Figure 7.30.

The second way is more subtle, and involves a shift of the neuron's inhibitory threshold caused by GABAergic inhibition. Such shifts were seen in cells in which blocking GABAergic inhibition did not reduce the degree of inhibition evoked by the left (ipsilateral) ear, but rather the blockage caused the neuron's inhibitory threshold to shift to a more positive value (Park and Pollak 1993a). Figure 7.31 shows an example of a unit whose

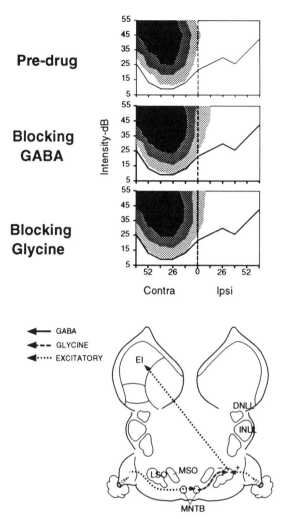

FIGURE 7.30. Example of a 60-kHz EI neuron whose receptive field was unaffected by either GABAergic or glycinergic inputs. *Top panel* shows the receptive field before drugs were applied; the *lower panels* show receptive fields obtained after blocking GABAergic and glycinergic inputs. The receptive field properties were most likely created in the lateral superior olive on the opposite side, and imposed upon the collicular neuron via the excitatory projection to the colliculus, as shown in the wiring diagram at the bottom of the figure. Thus, blocking either the GABAergic or glycinergic inputs had no effect on the cell's spatial receptive field.

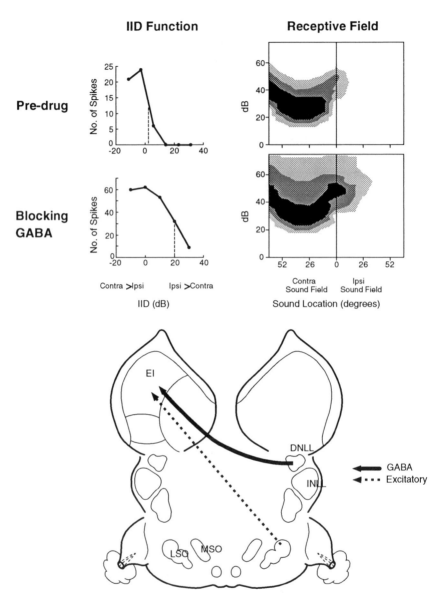

FIGURE 7.31. Example of a neuron in which blocking GABAergic inhibition had little effect on the magnitude of the ipsilaterally evoked inhibition, but caused a shift of 14 dB in its inhibitory threshold. The predrug IID function and the IID function obtained when GABAergic inhibition was blocked are shown in the *left panels*. Inhibitory thresholds are indicated by *dashed lines*. The changes in the azimuthal receptive fields are shown in the *right panels*. Note the expansion of the receptive field into the ipsilateral sound field by blockage of GABAergic inhibition. The *lower panel* shows a possible circuit that could generate these features. The simplest interpretation of such shifts is that in the normal condition there are at least two

inhibitory threshold changed by 14 dB because of the blockage of GA-BAergic inhibition and the effects of those changes on the unit's receptive field.

The simplest explanation for these effects is that the right (contralateral) LSO imparts its EI properties, and thus its inhibitory threshold, on the collicular cell, as was the case for the neuron in Figure 7.30. However, the inhibitory threshold of the LSO cell is shifted in the colliculus by the GABAergic inhibition evoked by stimulation of the left (ipsilateral) ear. The most likely source of that inhibition is the right (contralateral) DNLL, because its cells are excited by sound at the left (ipsilateral) ear and inhibited by sound at the right ear (Fig. 7.31, lower panel). This explanation is also supported by a recent study by Li and Kelly (1992a) in the rat. They blocked the DNLL pharmacologically while recording from EI cells in the colliculus and noted that the inhibitory thresholds (50% point IIDs) of many EI cells shifted because of the inactivation of the DNLL.

The third way is that the binaural properties are created entirely in the inferior colliculus by a monaural, excitatory projection from the right (contralateral) ear that is sculpted by GABAergic inhibition evoked from stimulation of the left (ipsilateral) ear (Faingold, Gelhbach, and Caspary 1989; Faingold, Boersma-Anderson, and Caspary 1991; Li and Kelly 1992a,b; Park and Pollak 1993a). These features are illustrated by the EI neuron in Figure 7.32. This neuron had a fairly sharply confined spatial receptive field before the application of any drug, but after blocking GABAergic inputs with bicuculline all the inhibitory inputs from the

---

FIGURE 7.31. (continued). projections that form the IID function in these collicular cells, one of which is the GABAergic projection from the contralateral DNLL. The other projection is from the contralateral LSO where the EI property is initially created. The LSO projection establishes both the strength of the inhibition and an inhibitory threshold is achieved when the intensity at the ipsilateral ear is substantially greater than the intensity at the contralateral ear. This projection is excitatory and therefore is not affected by iontophoresis of bicuculline. The second projection is GABAergic and innervates the same collicular cell as the LSO projection. The GABAergic circuit from the contralateral DNLL is driven only by stimulation of the ear ipsilateral to the IC, and has a lower absolute threshold than the inhibitory input to the LSO. Thus, the effect of the GABAergic circuit is to change the IID function at the colliculus, where it causes the discharge rate to decline with lower intensities at the ipsilateral ear than did the IID function created in the LSO. In short, the resultant IID function, caused by the summation of these circuits, has an inhibitory threshold that is shifted to a more positive IID (requires a less intense stimulus at the ipsilateral ear to inhibit the cell) but has the same maximum inhibition that was initially established in the LSO cell. The effect of bicuculline is to remove the influence of the GABAergic projection from the DNLL, thereby allowing the maximum inhibition and inhibitory threshold of the projection from the LSO to be expressed by the collicular cell.

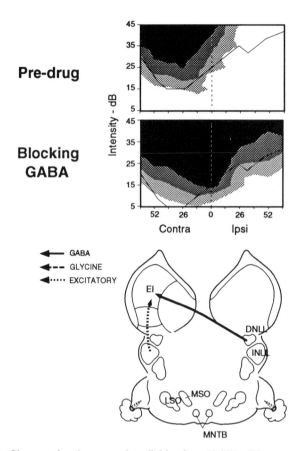

FIGURE 7.32. Changes in the receptive field of a 60-kHz EI neuron caused by blockage of GABAergic inputs. Notice that after blocking GABAergic inputs the receptive field expanded into the ipsilateral sound field and was similar to the receptive fields seen in monaural units. The wiring diagram (*below*) shows the circuit that could account for the expansion of the field. (See text for further explanation.)

ipsilateral ear were eliminated and the cell was rendered monaural. These charges are seen in the neuron's binaural properties (not shown) as well as its receptive field. Thus, the binaural properties of this cell were created entirely in the inferior colliculus by the convergence of excitatory projections from the right (contralateral) ear and GABAergic projections that most likely emanated from the right DNLL (Fig. 7.32). Because all inhibition evoked by stimulation of the ipsilateral ear was eliminated by bicuculline, no role for glycinergic projections needs to be proposed.

The EI neuron in Figure 7.33 illustrates the fourth way in which inhibitory circuits that converge on a collicular cell can shape its receptive field. In this case, the binaural properties and the receptive field properties were formed in the colliculus by both GABAergic and glycinergic inputs.

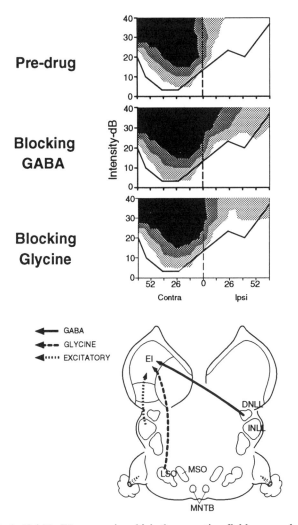

FIGURE 7.33. A 60-kHz EI neuron in which the receptive field was sculpted by both GABAergic and glycinergic inputs. (See text for further explanation.)

This neuron, like the previous one, had a receptive field sharply confined to the contralateral sound field. Blocking GABAergic inputs with bicuculline reduced the efficacy of the inhibition evoked by the ipsilateral ear, although it did not completely eliminate it. Because the inhibition evoked by the ipsilateral ear was reduced, the receptive field expanded into the ipsilateral acoustic hemifield. The unit, however, did not become completely monaural, because the highest discharge rates (shown in black) were still somewhat restricted to one side of the sound field. Effects similar to those produced by blocking GABA were observed when glycinergic inputs were blocked with strychnine. The receptive fields for both drug conditions show that discharges could be evoked from sound locations that were inhibitory in the predrug condition (Fig. 7.33). For neurons like this, GABAergic and glycinergic inputs appear to work in tandem to sculpt a receptive field from a monaural excitatory input from the right (contralateral) ear. The most likely source of the GABAergic inhibition is again the contralateral DNLL, and the source of the glycinergic inhibition is almost certainly the ipsilateral (left) LSO, because the cells in both these nuclei are excited by stimulation of the left (ipsilateral) ear and thus could provide the appropriate inhibition to their targets in the left colliculus. The circuit likely to produce the receptive field of this neuron is shown in the lower panel of Figure 7.33.

The fifth way that inhibition shapes receptive fields is one of the most interesting, and concerns the binaural facilitation of EI/f cells. Blocking GABA eliminated the binaural facilitation in many EI/f units independent of changes in maximum inhibition (Park and Pollak 1993a). Thus, binaural facilitation was reduced or lost in some cells in which the inhibition evoked by the left (ipsilateral) ear was also reduced or abolished after blocking GABA, whereas in other cells the ipsilaterally evoked inhibition was affected minimally or not at all. An example of an EI/f cell in which blocking GABAergic inhibition hardly affected the inhibition evoked by the left (ipsilateral) ear but abolished the facilitation is shown in Figure 7.34. For purposes of clarity, a circuit in which GABAergic inhibition could shape the facilitated response of an EI/f cell whose ipsilaterally evoked inhibition and inhibitory threshold are not affected by bicuculline is shown in Figure 7.35. The circuit has two components: a projection from the right LSO, which is responsible for generating the inhibition evoked by the left, ipsilateral ear and thus the cell's receptive field border, and a GABAergic cell in the *left* (ipsilateral) DNLL that generates the facilitation. One of the interesting features of this explanation is the inhibition is the proposed agent for producing the "facilitation." The explanation of how this circuit could produce binaural facilitation in an EI/f cell of the inferior colliculus is presented in Figure 7.35.

3.9 Summary

The studies reviewed here are beginning to reveal how neurons in the ICc integrate the signals from a number of parallel pathways to produce an output that is a synthesis of the information conveyed from those sources. The specific form of the neuron's output, in this case its binaural properties, is a reflection of the particular subset of inputs that converge upon that neuron. By mixing and matching different combinations of inputs, a variety of effects can be achieved. For example, the convergence of the inputs from the contralateral DNLL and contralateral LSO can adjust the expression of

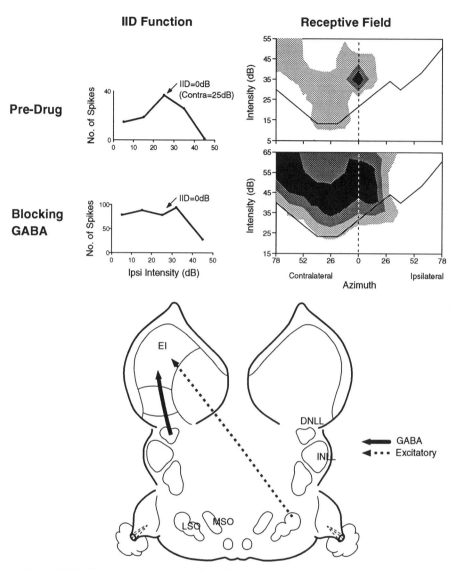

FIGURE 7.34. Change in the receptive field of an EI/f cell from blockage of GABAergic inhibition. The IID functions in the *left panels* show that blocking GABA eliminated the ipsilaterally evoked facilitation, but had only minimal effect on the magnitude of the ipsilaterally evoked inhibition. Thus, the effect of blocking GABAergic inhibition was to transform both the facilitated IID function and the focused activity of the spatial receptive field into those of a conventional EI cell. The simplest circuit that could generate these features would involve two projections: one from the contralateral LSO that generates the EI property, and another from the ipsilateral DNLL that generates the facilitation. A detailed description of how this could work is given in Figure 7.35.

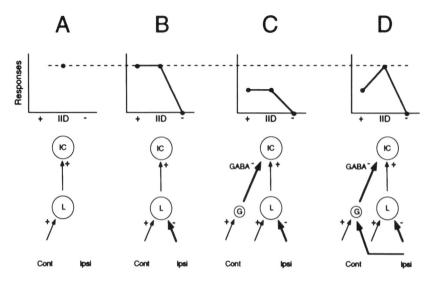

FIGURE 7.35. Construction of a circuit that could create binaural facilitation. (A) An excitatory input from the contra ear that makes a synaptic connection with cell L in a lower nucleus that drives the colliculus cell (IC) via an excitatory projection. The discharge rate of the IC cell is shown in the top graph. (B) An inhibitory input from the ipsi ear is added to L, making it EI. The IC cell thus also becomes EI (IID function in top panel). (C) A GABAergic input to the IC cell that originates from cell G is added next. G receives excitation from the contra ear. Sound at contra ear simultaneously evokes an excitation, via L, and an inhibition in the IC cell via G. Adding sound to the ipsi ear generates an IID function with a lower overall response rate than L, but has the same maximum inhibition. (D) Lastly, an inhibitory input evoked by the ipsi ear is added to G, making it EI. G has a *lower* 50% point than L. Sound at the contra ear alone at 10 dB above threshold, or with a subthreshold ipsi intensity, evokes the same reduced discharge rate as in panel C. When the intensity at the ipsi ear is increased, so that the inhibitory input is above threshold at G but is still below threshold for L, L still excites the IC cell. Since G is inhibited, and thus no longer imparts an inhibition at the IC, the discharge rate of the IC cell increases and expresses facilitation. Higher ipsi intensities now also suppress discharges in L. As the discharge rate of L falls to zero, so does the discharge rate of the IC cell. (Adapted from Park and Pollak 1993a.)

the inhibitory threshold of the LSO cell in its collicular target. An entirely different effect is achieved by the convergence of projections from the ipsilateral DNLL and contralateral LSO. Such a convergence does not shift the inhibitory threshold of the collicular target cell, but rather creates a new binaural property, a "facilitation" as expressed by the EI/f neurons. Other types of convergence create the binaural property in ICc neurons by combining inputs from a monaural, excitatory projection and an inhibitory projection from the contralateral DNLL. Such combinations are almost

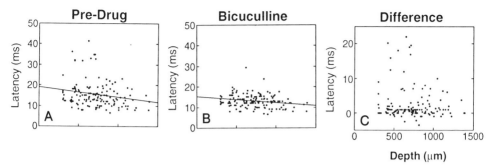

FIGURE 7.36A–C. Effects of blocking GABAergic inhibition with bicuculline on the pattern of latency with depth for 142 cells recorded from the 60-kHz region of the mustache bat inferior colliculus. (A) Plot of latency as a function of depth before bicuculline application. (B) Latencies of the same cells during application of bicuculline. (C) Difference in latency for each unit before and during blockage of GABAergic inhibition as a function of depth. (Adapted from Park and Pollak 1993b.)

certainly further affected by glycinergic input from the ipsilateral LSO, whose influence on binaural processing is not well understood at the present time.

4. Mechanisms That Create the Latency Distribution in the 60-kHz Contour and Some Functional Implications of That Distribution

In the first section of this chapter we showed that discharge latencies have a characteristic distribution in the 60-kHz contour of the inferior colliculus. Neurons in the dorsal colliculus have, on average, longer latencies than neurons located ventrally, and the range of latencies changes markedly with depth (Figs. 7.13 and 7.36) (Park and Pollak 1993b). Dorsally there is a broad distribution of latencies while deeper regions have a much narrower range of latencies. Long-latency cells are found only dorsally whereas short-latency cells are found at all depths. An orderly arrangement of latencies is apparently a general feature of collicular isofrequency contours, because it occurs not only in the mustache bat's inferior colliculus but in isofrequency contours of the cat's inferior colliculus as well (Schreiner and Langer 1988). The significance of such a distribution for information processing must be considerable, although at the present time it is not possible to exactly specify how different latencies affect binaural or other forms of processing. On the other hand, the ordering of latencies in collicular isofrequency contours could provide the topographical substrate for the generation of the combinatorial properties of neurons in the mustache bat's medial geniculate body. Before discussing this issue, we show that GABAergic inhibition is an important mechanism for producing the wide range of latencies expressed by collicular neurons, and that this inhibition apparently creates the particular form of the latency arrangement within the 60-kHz contour.

4.1 Effects of GABAeric Inhibition on Latency

GABAergic inhibition modifies the latencies of many, although not all, collicular neurons (Pollak and Park 1993; Park and Pollak 1993b). In neurons whose latencies are affected by inhibition, the effect is to increase latency thereby lengthening the time period between the presentation of a stimulus and the appearance of discharges. The change in latency is revealed when GABAergic inhibition is blocked by the application of bicuculline. In more than half of the neurons in the 60-kHz contour, blocking GABAergic inhibition with bicuculine caused latencies to shorten by 1–5 msec, and in about 20% of the population, latencies shortened by 5–30 msec. Examples of each type are shown in Figure 7.37.

Of significance is that the majority of neurons that had large (>8 msec) changes in latency with bicuculline were located in the dorsal region of the 60-kHz contour (Park and Pollak 1993b). Very few neurons in the more ventral regions of the colliculus had large bicuculline-induced latency changes. These features are illustrated by the four neurons in Figure 7.37 and by the three graphs in Figure 7.36. The top panel of Figure 7.36 shows the distribution of latencies as a function of depth before bicuculline was applied. The graph in the middle panel of Figure 7.36 shows the latency distribution during the application of bicuculline. The most apparent feature is the absence of neurons in the dorsal colliculus, between 200–600 μm from the surface, that had latencies longer than 20 msec during the application of bicuculline. There was also a reduction in the range of latencies at more ventral levels. At levels between 600 and 1000 μm, there was a smaller number of neurons with long latencies than in the predrug graph, although two neurons still had latencies well above 20 msec. In the most ventral colliculus, at depths of 1000–1400 μm, there was a reduction in the range of latencies, but the overall latency change from bicuculline was smaller than in the more dorsal regions.

The difference between the latencies before and during the application of bicuculline for neurons at the various levels of the colliculus is plotted graphically in the bottom panel of Figure 7.36. Here the largest changes in latency are clearly apparent for the neurons in the dorsal colliculus (200–600 μm from the surface). The smaller and moderate latency changes can also be seen for neurons at midlevels of the colliculus, while the neuronal population located ventrally, from about 1000–1400 μm, had, on average, the smallest latency change. It is also noteworthy that the latencies of large

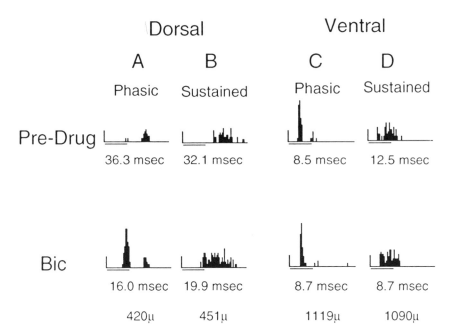

FIGURE 7.37A–D. Peristimulus time histograms of two neurons with sustained response patterns and two units with phasic discharge patterns from different dorsoventral locations in the 60-kHz region of the mustache bat's inferior colliculus. For each unit, the *top graph* shows responses before bicuculline (Bic) application; the *bottom graph* shows responses during the application of bicuculline. Numbers below the graphs are the median latency to the first spike in milliseconds and the depth of the unit (distance from the dorsal surface) in micrometers. (A) Phasic cell from dorsal colliculus. (B) Sustained cell from dorsal colliculus. (C) Phasic cell from ventral colliculus. (D) Sustained cell from ventral colliculus. Notice that the latencies of both the phasic and sustained neurons recorded from the dorsal colliculus (A and B) changed substantially with bicuculline, whereas the latencies of the phasic and sustained cells in the ventral colliculus changed very little. (Adapted from Park and Pollak 1993b.)

numbers of neurons at all levels of the colliculus were not changed by bicuculline.

The neurons in lower nuclei that send excitatory projections to the midbrain appear to have a much narrower range of latencies than the population of collicular neurons (Markovitz and Pollak 1993; Covey and Casseday 1991; Covey, Vater, and Casseday 1991 Haplea, Covey and Casseday 1994; Yang and Pollak 1994). The reason for the amplified range of collicular latencies is that some of the cells in the dorsal region of the contour, which receive fairly short latency excitatory innervation, are transformed into longer latency cells by GABAergic inhibition. Thus, the latencies of the cells receiving the appropriate inhibitory inputs are changed from the short-latency cells, seen after GABA is blocked, to the longer latencies which the cells normally express because of inhibitory innervation by their GABAergic neighbors.

4.2 Orderly Arrangement of Latencies Has Implications for the Creation of Combination-Sensitive Neurons in the Medial Geniculate

As discussed previously, the orderly arrangement of response properties seems to be a general feature of collicular isofrequency contours. The orderly arrangement of response features suggests that a neuron's location within a contour could specify its response properties. This arrangement may be important for the mustache bat because of the remarkable response properties that are generated in the medial and dorsal divisions of the medial geniculate body by the convergence of projections from at least two frequency contours (Olsen and Suga 1991a,b; Buttman 1992). The convergence results in the emergence of combination-sensitive neurons in the medial geniculate, a feature that is not present in the inferior colliculus (O'Neill 1985). If the organization of response properties in the other frequency contours is comparable to those in the 60-kHz contour, as we have argued they are, then all the response features required to generate combinatorial properties are already established and arranged by locus in the various frequency contours of the inferior colliculus. These considerations led us to suggest that the appropriate combinatorial properties could be created by the convergence of projections from a given locus in one isofrequency contour and the projections from a different locus in another isofrequency contour. We present here two hypothetical circuits that could generate the combinatorial properties in geniculate neurons.

The circuits can potentially account for the two general types of combination-sensitive neurons that have been found in the medial geniculate (Olsen and Suga 1991a,b). The characteristic features of these neurons are described in detail by Wenstrup in Chapter 8, and by O'Neill in Chapter 9. One type is selective for the frequencies in the constant frequency (CF) components of the bat's orientation calls, the so-called CF/CF neurons. The CF/CF neurons are characterized by their weak discharges to individual tone bursts and the vigorous firing when two tone bursts having different frequencies are presented together. These neurons are also distinguished by their relative insensitivity to the temporal interval separating the presentation of the two frequencies. CF/CF neurons are tuned to two of the harmonics of the CF components of the bat's orientation calls; thus, CF_1/CF_2 neurons are driven only when the fundamental and second harmonic of the CF components are combined, and CF_1/CF_3 neurons by the fundamental and third harmonic. The exact frequency requirements for optimally driving each type of CF/CF neuron are slightly different among the population, in that each neuron is tuned to a small deviation from an exact harmonic relationship. Thus, the population of CF_1/CF_2 neurons is optimally driven by the range of CF_1 frequencies the bat would emit during Doppler-shift compensation, combined with the CF_2 frequencies that correspond to the stabilized echoes it receives from Doppler-shift compensation. Similar precise frequency pairings are expressed by the population of CF_1/CF_3 neurons.

The other type, called FM/FM or delay-tuned neurons, are also most effectively driven by two harmonically related frequencies of the frequency-modulated (FM) components. Thus, there are FM_1/FM_2 neurons, as well as FM_1/FM_3 and FM_1/FM_4 neurons. For these neurons, the frequency specificity is not as strict as it is for the CF/CF neurons, but rather the neurons require that there be a specific delay between the two signals. For this reason, they are also referred to as delay-tuned neurons. The delay-tuned properties are a consequence of the latency differential with which the inputs evoked by each of the two frequencies arrive at the geniculate neuron (Olsen and Suga 1991b). We first propose a circuit for the generation of CF/CF neurons, and then suggest a slightly different circuitry that could lead to the formation of delay-tuned neurons.

The circuit for CF/CF neurons must account for the facilitated response from two specific frequencies and for the finding that the facilitation can be evoked over a wide range of temporal intervals separating the two tone bursts. To construct the appropriate circuit, we visualize first that a large number of 60-kHz cells project to a common neuron in the medial geniculate (Fig. 7.38, top panel). Furthermore, the projections arise from a dorsoventral column of cells in the contour. Thus the geniculate cell receives projections from a population of 60-kHz collicular cells in which each has a different latency. We next add projections from a population of topographically ordered cells in the 30-kHz contour, the fundamental frequency of the CF component, which then would also provide a wide distribution of latencies. The net result of this arrangement is that the presentation of a 30-kHz and a 60-kHz tone would evoke activity from the two frequency contours, and that the excitation evoked in the the geniculate neuron by the convergence of inputs from the two contours would be independent of the interval separating the two tones. If, for instance, the 30-kHz tone was presented 10 msec before the 60-kHz tone, the 30-kHz tone would evoke a response in the geniculate cell over a wide range of intervals, because each projection neuron would be activated sequentially. The 60-kHz tone, arriving 10 msec later, would also evoke activity distributed over time. The important feature is that because the population of 30-kHz cells and 60-kHz cells have distributed latencies, the discharges from at least some of the 30-kHz neurons will arrive simultaneously with the discharges from some 60-kHz cells and thereby evoke a facilitated response. Because the latency distribution of both the 30- and 60-kHz projection cells is large, there will be a convergence of excitation from some cells of both frequencies regardless of the time interval separating the two signals.

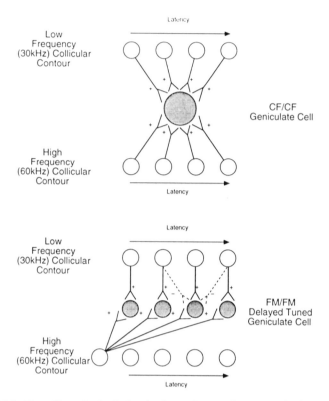

FIGURE 7.38. *Top*: Hypothetical circuit shows inputs from two isofrequency contours that could generate properties appropriate for CF/CF neurons in the mustache bat forebrain. *Bottom*: Hypothetical circuit shows inputs to an FM/FM neuron that could generate delay-tuned properties. (See text for discussion.) (Adapted from Park and Pollak 1993b.)

The projections from the two frequency contours that could generate the FM/FM delay-tuned neurons must be somewhat different. The circuit has to satisfy the additional criterion that the facilitated response occurs only for a small range of intervals between the two signals. Such properties could be achieved if projections from a region of the 30-kHz contour, where latencies are long, converged with the projections from the ventral 60-kHz contour, where latencies are relatively short (Fig. 7.38, lower panel). Under these conditions, the two tones presented simultaneously would arrive at their targets in the medial geniculate at different times and elicit a small response. However, if the 60-kHz signal was delayed relative to the 30-kHz signal by an interval equal to the latency differential evoked by the two frequencies in the colliculus, then the two inputs should converge simultaneously on the geniculate cell, resulting in a facilitated discharge.

An arrangement of the sort just described can be extended and thereby account for two additional features of delay-tuned neurons. The first feature is the distribution of delay-tuned values. This could be if a 60-kHz cell, having a particular latency, sent projections to several geniculate cells. These geniculate cells would also receive innervation from a population of topographically ordered 30-kHz cells, where the locus of each 30-kHz cell defined its latency. Each 30-kHz cell, however, would project to only one geniculate cell. Thus every geniculate cell in this array would be innervated with a common latency from the 60-kHz cell, but any given geniculate cell would be innervated with a different latency by the particular 30-kHz cell that projects to it. An arrangement of this sort would create a population of cells, each endowed with a different delay-tuned value. Another dividend derived from this arrangement is that it provides a simplified mechanism for generating a map of delay tuning. The ordered projection system described would at once generate a diversity of delay-tuned values and an orderly change in the delay-tuned values among the population of target cells in the medial geniculate. Assuming an orderly arrangment of latencies in other frequency contours, then this way of creating delay-tuned neurons can be generalized to any combination of two frequencies and does not have to include 60-kHz cells.

The second feature that requires explanation concerns the difference between neurons tuned to short best delays and those tuned to long best delays (Olsen and Suga 1991b). Medial geniculate neurons tuned to short delays are coincidence detectors and are thought to be generated only by the convergence of two excitatory inputs from the sort of circuit described here. Medial geniculate neurons tuned to long delays, however, are more complex and are created by interactions of excitation and inhibition. In these neurons, delays shorter or longer than the optimal delay inhibit the cell. Thus, the delay value to which the unit is tuned is sandwiched between periods of inhibition. The circuitry we proposed for generating delay-tuned neurons can easily incorporate inhibitory as well as excitatory projections (Fig. 7.38, lower panel). The mammalian inferior colliculus contains a large population of GABAergic principal neurons that presumably provide inhibitory inputs to the medial geniculate. One can postulate, for example, three topographically arranged projections from the 30-kHz contour; the first is an inhibitory projection with a short latency, the second is an excitatory projection with a slightly longer latency, and the third is another inhibitory projection with a latency somewhat longer than the second. The latency of the second neuron could generate an excitatory coincidence with the 60-kHz projection at a particular delay that would be sandwiched between the inhibitory periods produced by the shortest and longest latencies of the 30-kHz projections, thereby sharpening the neuron's delay tuning. These features satisfy many of the properties seen in long-delay geniculate neurons by Olsen and Suga (1991b).

The fact that an equivalent latency distribution occurs in monaural, as well as in the EE and EI subregions of the 60-kHz contour (see Fig. 7.13C), may also be important for creating combinatorial properties. An issue that has not previously been raised is that a mismatch in the aural properties of the convergent projections could confound the latency specificity of the delay-tuned projection system. The reason is that if the projection from one of the frequency contours were monaural and the projection from the other frequency were binaural, then interactions from the two ears could change the latency of the binaural projection. If this occurred, then the delay tuning of a given geniculate neuron would also vary with the interaural intensity disparity and hence with location of the sound source in space. The finding that the population of neurons in each of the 60-kHz aural subregions express a similar latency ordering seems significant in this regard. If a comparable arrangement exists in each of the other frequency contours, and we argue that it probably does, then a combinatorial projection system segregated according to aural type could be readily achieved. In such an arrangement, monaural projection neurons from the two frequency contours would converge on a common geniculate cell, and similarly for the other binaural types. If this proves to be correct, it would simplify what otherwise would appear to be a seriously confounded system.

In conclusion, the results indicate that the full range of delays, as well as other response properties required to form combinatorial neurons, are present in the inferior colliculus, and that the delays are arranged with a characteristic pattern in at least one frequency contour. We have also proposed a simplified model by which the convergence of projections from two frequency contours of the colliculus could theoretically generate the properties of the various types of combinatorial neurons in the medial geniculate. Whether the combinatorial properties are, in fact, created by such convergences remains for future studies to determine.

Abbreviations

ALD	anterolateral division of the inferior colliculus (region representing frequencies below 60 kHz)
AN	auditory nerve
CF	constant frequency
DC	dorsal cortex of the inferior colliculus
DCN	dorsal cochlear nucleus
DmD or MD	medial division of the inferior colliculus (region representing frequencies above 60 kHz)
DpD	dorsoposterior division of the inferior colliculus (the region containing neurons sharply tuned to 60 kHz)
EE	binaural neurons that are driven by acoustic stimulation presented to either ear
EI	binaural neurons that are driven by stimulation at one ear and inhibited by stimulation at the other ear

EO	monaural neurons that are are driven by stimulation at one ear and are unaffected by stimulation of the other ear
FM	frequency modulated
IC	inferior colliculus
ICc	central nucleus of the inferior colliculus
IID	interaural intensity disparity
INLL	intermediate nucleus of the lateral lemniscus
LL	lateral lemniscus
LSO	lateral superior olive
MNTB	medial nucleus of the trapezoid body
MSO	medial superior olive
NLL	nuclei of the lateral lemniscus
SAM	sinusoidal amplitude modulation
SFM	sinusoidal frequency modulation
SPL	sound pressure level
$VNLL_v$	ventral or columnar division of the ventral nucleus of the lateral lemniscus
VNLLd	dorsal division of the ventral nucleus of the lateral lemniscus
VNTB	ventral nucleus of the trapezoid body

References

Adams JC (1979) Ascending projections to the inferior colliculus. J Comp Neurol 183:519–538.

Adams JC, Mugniani E (1984) Dorsal nucleus of the lateral lemniscus: a nucleus of GABAergic projection neurons. Brain Res Bull 13:585–590.

Adams JC, Wenthold RJ (1979) Distribution of putative amino acid transmitters, choline acetyltransferase and glutamate decarboxylase in the inferior colliculus. Neuroscience 4:1947–1951.

Aitkin LM (1976) Tonotopic organization at higher levels of the auditory pathway. Intr Rev Physiol Neurophysiol 10:249–279.

Aitkin LM (1986) The Auditory Midbrain: Structure and Function in the Central Auditory Pathway. Clifton, NJ: Humana.

Bateman GC, Vaughan (1974) Nightly activities of mormoopid bats. J Mammal 55:45–65.

Beyerl BD (1978) Afferent projections to the central nucleus of the inferior colliculus in the rat. Brain Res 145:209–223.

Blauert J (1969/1970) Sound localization in the median plane. Acoustica 22:205–213.

Bodenhamer RD, Pollak GD (1983) Response characteristics of single units in the inferior colliculus of mustache bats to sinusoidally frequency modulated signals. J Comp Physiol A 153:67–79.

Boudreau JC, Tsuchitani C (1968) Binaural interaction in the cat superior olive S-segment. J Neurophysiol (Bethesda) 31:442–454.

Brownell WE, Manis PB, Ritz LA (1979) Ipsilateral inhibitory responses in the cat lateral superior olive. Brain Res 177:189–193.

Brugge JF, Anderson DJ, Aitkin LM (1970) Responses of neurons in the dorsalnucleus of the lateral lemniscus to binaural tonal stimuli. J Neurophysiol (Bethesda) 33:441–458.

Brunso-Bechtold JK, Thompson GC, Masterton RB (1981) HRP study of the organization of auditory afferents ascending to the central nucleus of the inferior colliculus in the cat. J Comp Neurol 97:705–722.

Butler RA (1974) Does tonotopy subserve the perceived elevation of a sound? Fed Proc 33:1920–1923.

Buttman JA (1992) Inhibitory and excitatory mechanisms of coincidence detection in delay-tuned neurons of the mustache bat. In: Proceedings of the 3rd International Congress of Neuroethology, p. 34.

Caird D, Klinke R (1983) Processing of binaural stimuli by cat superior olivary complex neurons. Exp Brain Res 52:385–399.

Cant NB, Casseday JH (1986) Projections from the anteroventral cochelar nucleus to the lateral and medial superior olivary nuclei. J Comp Neurol 247:457–476.

Casseday JH, Covey E (1992) Frequency tuning properties of neurons in the inferior colliculus of an FM bat. J Comp Neurol 319:34–50.

Casseday JH, Covey E, Vater M (1988) Connections of the superior olivary complex in the rufous horseshoe bat, *Rhinolophus rouxi*. J Comp Neurol 278:313–329.

Covey E (1993) Response properties of single units in the dorsal nucleus of the lateral lemniscus and paralemniscual zone of an echolocating bat. J Neurophysiol (Bethesda) 69:842–859.

Covey E, Casseday JH (1986) Connectional basis for frequency representation in the nuclei of the lateral lemniscus of the bat, *Eptesicus fuscus*. J Neurosci 6:2926–2940.

Covey E, Casseday JH (1991) The monaural nuclei of the lateral lemniscus in an echolocating bat: parallel pathways for analyzing temporal features of sound. J Neurosci 11:3456–3470.

Covey E, Hall WC, Kobler JB (1987) Subcortical connections of the superior colliclus in the mustache bat, *Pteronotus parnellii*. J Comp Neurol 263:179–197.

Covey E, Vater M, Casseday JH (1991) Binaural properties of single units in the superior olivary complex of the mustached bat. J Neurophysiol (Bethesda) 66:1080–1093.

Erulkar S (1972) Comparative aspects of sound localization. Physiol Rev 52:237–360.

Faingold CL, Boersma-Anderson CA, Caspary DM (1991) Involvment of GABA in acoustically evoked inhibition in inferior colliculus. Hear Res 52:201–216.

Faingold CL, Gehlbach G, Caspary DM (1989) On the role of GABA as an inhibitory neurotransmitter in inferior colliculus neurons: iontophoretic studies. Brain Res 500:302–312.

Finlayson PG, Caspary DM (1989) Synaptic potentials of chinchilla lateral superior olivary neurons. Hear Res 38:221–228.

Flannery R, Butler RA (1981) Spectral cues provided by the pinna for monaural localization in the horizontal plane. Percept Psychophys 29:438–444.

Frisina RD, O'Neill WE, Zettel ML (1989) Functional organization of mustached bat inferior colliculus: II. Connections of the FM$_2$ region. J Comp Neurol 284:85–107.

Fuzessery ZM (1986) Speculations on the role of frequency in sound localization. Brain Behav Evol 28:95–108.

Fuzessery ZM, Pollak GD (1984) Neural mechanisms of sound localization in an echolocating bat. Science 225:725–728.

Fuzessery ZM, Pollak GD (1985) Determinants of sound location selectivity in the bat inferior colliculus: a combined dichotic and free-field stimulation study. J Neurophysiol 54:757–781.

Fuzessery ZM, Wenstrup JJ, Pollak GD (1990) Determinants of horizontal sound location selectivity of binaurally excited neurons in an isofrequency region of the mustache bat inferior colliculus. J Neurophysiol (Bethesda) 63:1128–1147.

Gaioni SJ, Riquimaroux H, Suga N (1990) Biosonar behavior of mustached bats swung on a pendulum prior to cortical ablation. J Neurophysiol (Bethesda) 64:1801–1817.

Glendenning KK, Brunso-Bechtold JK, Thompson GC, Masterton RB (1981) Ascending auditory afferents to the nuclei of the lateral lemniscus. J Comp Neurol 197:673–703.

Glendenning KK, Baker BN, Hutson KA, Masterton RB (1992) Acoustic chiasm V: inhibition and excitation in the ipsilateral and contralateral projections of LSO. J Comp Neurol 319:100–122.

Goldberg JM, Brown PB (1968) Functional organization of the dog superior olivary complex: an anatomical and electrophysiological study. J Neurophysiol (Bethesda) 31:639–656.

Goldman LJ, Henson OW Jr (1977) Prey recognition and selection by the constant frequency bat, *Pteronotus parnellii*. Behav Ecol Sociobiol 2:411–419.

Gourevitch G (1980) Directional hearing in terrestrial mammals. In: Popper AN, Fay RR (eds) Comparative Studies of Hearing in Vertebrates. New York: Springer-Verlag, pp. 357–374.

Griffin DR, Simmons JA (1974) Echolocation of insects by horseshoe bats. Nature 250:731–732.

Grinnell AD (1963) The neurophysiology of audition in bats: intensity and frequency parameters. J Physiol 167:38–66.

Grinnell AD, Grinnell VS (1965) Neural correlates of vertical localization by echolocating bats. J Physiol 181:830–851.

Guinan JJ Jr, Guinan SS, Norris BE (1972) Single auditory units in the superior olivary complex. I. Responses to sounds and classifications based on physiological properties. Int J Neurosci 4:101–120.

Haplea S, Covey E, Casseday JH (1994) Frequency tuning and response latencies at three levels in the brainstem of the echolocating bat, *Eptesicus fuscus* J Comp Physiol 174:671–684.

Harnischfeger G, Neuweiler G, Schlegel P (1985) Interaural time and intensity coding in superior olivary complex and inferior colliculus of the echolocating bat, *Molossusater*. J Neurophysiol (Bethesda) 53:89–109.

Henson OW Jr, Pollak GD, Kobler JB, Henson MM, Goldman LJ (1982) Cochlear microphonics elicited by biosonar signals in flying bats. Hear Res 7:127–147.

Hubel DH, Wiesel TN (1977) Functional architecture of macaque monkey visual cortex (Ferrier lecture). Proc R Soc Lond 198:1–59.

Irvine DRF (1986) The auditory brainstem. In: Autrum H, Ottoson D (eds) Progress in Sensory Physiology, vol. 7. Berlin-Heidelberg: Springer-Verlag.

Irvine DRF, Gago G (1990) Binaural interaction in high-frequency neurons in inferior colliculus of the cat: effects of variations in sound pressure level on sensitivity to interaural intensity differences. J Neurophysiol (Bethesda) 63:570–591.

Jen PH-S, Chen D (1988) Directionality of sound pressure transformation at the pinna of echolocating bats. Hear Res 34:101–118.

Jen PH-S, Schlegel PA (1982) Auditory physiological properties of the neurons in the inferior colliculus of the big brown bat, *Eptesicus fuscus.* J Comp Physiol A 147:351–363.

Jen PH-S, Sun X (1984) Pinna orientation determines the maximal directional sensitivity of bat auditory neurons. Brain Res 301:157–161.

Jen PH-S, Suthers RA (1982) Responses of inferior collicular neurons to acoustic stimuli in certain FM and CF-FM, paleotropical bats. J Comp Physiol A 146:423–434.

Kober R (1988) Echoes of fluttering insects. In: Nachtigall PE, Moore PWB (eds) Animal Sonar: Processes and Performance. New York: Plenum, pp. 477–482.

Kössl M, Vater M (1985) The frequency place map of the bat, *Pteronotus parnellii.* J Comp Physiol 157:687–697.

Kössl M, Vater M (1990) Resonance phenomena in the cochlea of the mustache bat and their contribution to neuronal response characteristics in the cochlear nucleus. J Comp Physiol A 166:711–720.

Larue DT, Park TJ, Pollak GD, Winer JA (1991) Glycine and GABA immuno-staining defines functional subregions of the lateral lemniscal nuclei in the mustache bat. Proc Soc Neurosci 17:300.

Lesser HD, O'Neill WE, Frisina RD, Emerson RC (1990) ON-OFF units in the mustached bat inferior colliculus are selective for transients resembling "acoustic glint" from fluttering insect targets. Exp Brain Res 82:137–148.

Li L, Kelly JB (1992a) Inhibitory influences of the dorsal nucleus of the lateral lemniscus on binaural responses in the rat's inferior colliculus. J Neurosci 12:4530–4539.

Li L, Kelly JB (1992b) Binaural responses in rat inferior colliculus following kainic acid lesions of the superior olive: interaural intensity difference functions. Hear Res 61:73–85.

Makous JC, O'Neill WE (1986) Directional sensitivity of the auditory midbrain in the mustached bat to free-field tones. Hear Res 24:73–88.

Markovitz NS, Pollak GD (1993) The dorsal nucleus of the lateral lemniscus in the mustache bat: Monaural properties. Hearing Res 71:51–63. Abstracts, 16th Annual Meeting of the Association for Research in Otolarynology, p. 110.

Mills A W (1972) Auditory localization. In: Tobias JV (ed) Foundations of Modern Auditory Theory, Vol II. New York: Academic, pp. 303–348.

M;daller AR (1972) Coding of amplitude and frequency modulated sounds in the cochlear nucleus of the rat. Acta Physiol Scand 86:223–238.

Mountcastle VB (1978) An organizing principle for cerebral function: the unit module and the distributed system. In: The Mindful Brain. Cambridge: MIT Press.

Mugniani E, Oertel WH (1985) An atlas of the distribution of GABAergic neurons and terminals in rat CNS as revealed by GAD immunocytochemistry. In: Bjorlund A, Hokfelt T (eds) Handbook of Chemical Neuroanatomy, Vol. 4: GABA and Neuropeptides in the CNS, Part I. Amsterdam: Elsevier, pp. 436–608.

Musicant AD, Butler RA (1984) The psychophysical basis of monaural localization. Hear Res 14:185–190.

Neuweiler G (1983) Echolocation and adaptivity to ecological constraints. In: Huber F, Markl H (eds) Neuroethology and Behavioral Physiology: Roots and Growing Pains. Berlin: Springer-Verlag, pp. 280–302.

Neuweiler G (1984) Foraging, echolocation and audition in bats. Naturwissenschaften 71:446–455.

Neuweiler G, Vater M (1977) Response patterns to pure tones of cochlear nucleus units in the CF-FM bat, *Rhinolophus ferrumequinum*. J Comp Physiol A 115:119–133.

Novick A (1971) Echolocation in bats: some aspects of pulse design. Am Sci 59:198–209.

Novick A, Vaisnys JR (1964) Echolocation of flying insects by the bat, *Chilonycteris parnellii*. Biol Bull 127:478–488.

Oliver DL, Huerta MF (1992) Inferior and superior colliculi. In: Webster DB, Popper AN, Fay RR (eds) The Mammalian Auditory Pathway: Neuroanatomy, Vol 1. New York: Springer-Verlag, pp. 168–221.

Oliver DL, Nuding SC, Beckius G (1988) Multiple cell types have GABA immunoreactivity in the inferior colliculus of the cat. Proc Soc Neurosci 14:490.

Oliver DL, Kuwada S, Yin TCT, Haberly LB, Henkel CK (1991) Dendritic and axonal morphology of HRP-injected neurons in the inferior colliculus of the cat. J Comp Neurol 303:75–100.

Olsen JF (1986) Processing of biosonar information by the medial geniculate body of the mustached bat, *Pteronotus parnellii*. Ph.D. dissertation, Washington University, St. Louis, MO.

Olsen JF, Suga N (1991a) Combination sensitive neurons in the medial geniculate body of the mustached bat: encoding of relative velocity information. J Neurophysiol (Bethesda) 65:1254–1274.

Olsen JF, Suga N (1991b) Combination sensitive neurons in the medial geniculate body of the mustached bat: encoding of target range information. J Neurophysiol (Bethesda) 65:1275–1296.

O'Neill WE (1985) Responses to pure tones and linear FM components of the CF/FM biosonar signals by single units in the inferior colliculus of the mustached bat. J Comp Physiol A 157:797–815.

O'Neill WE, Frisina RD, Gooler DM (1989) Functional organization of mustached bat inferior colliculus: I. Representation of FM frequency bands important for target ranging revealed by C-2-deoxyglucose autoradiography and single unit mapping. J Comp Neurol 284:60–84.

Ostwald J (1984) Tontopical organization and pure tone response characteristics of single units in the auditory cortex of the greater horseshoe bat. J Comp Physiol A 155:821–834.

Ostwald J (1988) Encoding of natural insect echoes and sinusoidally modulated stimuli by neurons in the auditory cortex of the greater horseshoe bat, *Rhinolophus ferrumequinum*. In: Nachtigall PE, Moore PWB (eds) Animal Sonar: Processes and Performance. New York: Plenum, pp. 483–487.

Ostwald J, Schnitzler H-U, Schuller G (1988) Target discrimination and target classification in echolocating bats. In: Nachtigall PE, Moore PWB (eds) Animal Sonar: Processes and Performance. New York: Plenum, pp. 413–434.

Park TJ, Pollak GD (1993a) GABA shapes sensitivity to interaural intensity disparities in the mustache bat's inferior colliculus: implications for encoding sound location. J Neurosci 13:2050–2067.

364   George D. Pollak and Thomas J. Park

Park TJ, Pollak GD (1993b) GABA shapes a topographic organization of response latency in the mustache bat's inferior colliculus. J Neurosci 13:5172-5187.

Park TJ, Larue DT, Winer JA, Pollak GD (1991) Glycine and GABA in the superior olivary complex of the mustache bat: projections to the central nucleus of the inferior colliculus. Soc Neurosci Abstr 17:300.

Pollak GD (1980) Organizational and encoding features of single neurons in the inferior colliculus of bats. In: Busnel GR (ed) Animal Sonar Systems. New York: Plenum, pp. 549-587.

Pollak GD, Bodenhamer RD (1981) Specialized characteristics of single units in inferior colliculus of mustache bat: frequency representation, tuning, and discharge patterns. J Neurophysiol (Bethesda) 46:605-619.

Pollak GD, Casseday JH (1989) The Neural Basis of Echolocation in Bats. Berlin: Springer-Verlag.

Pollak GD, Park TJ (1993) The effects of GABAergic inhibition on monaural response properties of neurons in the mustache bat's inferior colliculus. Hear Res 65:99-117.

Pollak GD, Schuller G (1981) Tonotopic organization and encoding features of single units in inferior colliculus of horseshoe bats: functional implications for prey identification. J Neurophysiol (Bethesda) 45:208-226.

Pollak GD, Winer JA (1989) Glycinergic and GABAergic auditory brain stem neurons and axons in the mustache bat. Soc Neurosci Abstr 15:1115.

Pollak GD, Wenstrup JJ, Fuzessery ZM (1986) Auditory processing in the mustache bat's inferior colliculus. Trends Neurosci 9:556-561.

Pollak GD, Marsh DS, Bodenhamer R, Souther A (1978) A single-unit analysis of inferior colliculus in unanesthetized bats: response patterns and spike-count functions generated by constant frequency and frequency modulated sounds. J Neurophysiol (Bethesda) 41:677-691.

Pollak GD, Park TJ, Larue DT, Winer JA (1992) The role inhibitory circuits play in shaping receptive fields of neurons in the mustache bat's inferior colliculus. In: Singh RN (ed) Principles of Design and Function in Nervous Systems. New Delhi: Wiley, pp. 271-290.

Pumphery RJ (1948) The sense organs of birds. Ibis 90:171-190.

Rhode WS, Greenberg S (1992) Physiology of the cochlear nuclei. In: Webster DB, Popper AN, Fay RR (eds) The Mammalian Auditory Pathway: Neurophysiology, Vol. 2. New York: Springer-Verlag, pp. 94-152.

Roberts RC, Ribak CE (1987) GABAergic neurons and axon terminals in the brainstem auditory nuclei of the gerbil. J Comp Neurol 258:267-280.

Ross LS, Pollak GD (1989) Differential projections to aural regions in the 60-kHz isofrequency contour of the mustache bat's inferior colliculus. J Neurosci 9:2819-2834.

Ross LS, Pollak GD, Zook JM (1988) Origin of ascending projections to an isofrequency region of the mustache bat's inferior colliculus. J Comp Neurol 270:488-505.

Roth GL, Aitkin LM, Andersen RA, Merzenich MM (1978) Some features of the spatial organization of the central nucleus of the inferior colliculus of the cat. J Comp Neurol 182:661-680.

Rübsamen R (1992) Postnatal development of central auditory frequency maps. J Comp Physiol A 170:129-143.

Saint Marie RL, Ostapoff ME, Morest DK, Wenthold RJ (1989) Glycine-immunoreactive projection of the cat lateral superior olive: possible role in midbrain dominance. J Comp Neurol 279:382-396.

Sanes DH, Rubel EW (1988) The ontogeny of inhibition and excitation in the gerbil lateral superior olive J Neurosci 8:682-700.

Schlegel P (1977) Directional coding by binaural brainstem units of the CF-FM bat, *Rhinolophus ferrumequinum*. J Comp Physiol A 118:327-352.

Schnitzler H-U (1970) Comparison of echolocation behavior in *Rhinolophus ferrumequinum* and *Chilonycteris rubiginosa*. Bijdr Dierkd 40:77-80.

Schnitzler H-U, Ostwald J (1983) Adaptations for the detection of fluttering insects by echolocation in horseshoe bats. In: Ewert J-P, Capranica RR, Ingle DJ (eds) Advances in Vertebrate Neuroethology. New York: Plenum, pp. 801-828.

Schreiner CE, Langner G (1988) Periodicity coding in the inferior colliculus of the cat. II. Topographic organization. J Neurophysiol (Bethesda) 60:1823-1840.

Schuller G (1979) Coding of small sinusoidal frequency and amplitude modulations in the inferior colliculus of the CF-FM bat, *Rhinolophus ferrumequinum*. Exp Brain Res 34:117-132.

Schuller G (1984) Natural ultrasonic echoes from wing beating insects are coded by collicular neurons in the long CF-FM bat, *Rhinolophus ferrumequinum*. J Comp Physiol A 155:121-128.

Schuller G, Pollak GD (1979) Disproportionate frequency representation in the inferior colliculus of Doppler-compensating greater horseshoe bats: evidence for an acoustic fovea. J Comp Physiol A 132:47-54.

Schuller G, Beuter K, Schnitzler H-U (1974) Responses to frequency shifted artificial echoes in the bat, *Rhinolophus ferrumequinum*. J Comp Physiol A 89:275-286.

Schuller G, Covey E, Casseday JH (1991) Auditory pontine grey: connections and response properties in the horseshoe bat. Eur J Neurosci 3:648-662.

Schweizer H (1981) The connections of the inferior colliculus and organization of the brainstem auditory system in the greater horseshoe bat, *Rhinolophus ferrumequinum*. J Comp Neurol 201: 25-49.

Semple MN, Aitkin LM (1979) Representation of sound frequency and laterality by units in the central nucleus of the cat's inferior colliculus. J Neurophysiol (Bethesda) 42:1626-1639.

Semple MN, Aitkin LM (1980) Physiology of pathway from dorsal cochlear nucleus to inferior colliculus revealed by electrical and auditory stimulation. Exp Brain Res 41:19-28.

Semple MN, Kitzes LM (1985) Single-unit responses in the inferior colliculus: different consequences of contralateral and ipsilateral auditory stimulation. J Neurophysiol (Bethesda) 53:1467-1482.

Shneiderman A, Henkel CK (1987) Banding of lateral superior olivary nucleus afferents in the inferior colliculus: a possible substrate for sensory integration. J Comp Neurol 266:519-534.

Shneiderman A, Oliver DL (1989) EM autoradiographic study of the projections from the dorsal nucleus of the lateral lemniscus: a possible source of inhibitory inputs to the inferior colliculus. J Comp Neurol 286:28-47.

Shneiderman A, Oliver DL, Henkel CK (1988) The connections of the dorsal nucleus of the lateral lemniscus. An inhibitory parallel pathway in the ascending auditory system? J Comp Neurol 276:188-208.

Simmons JA (1974) Response of the Doppler echolocation system in the bat, *Rhinolophus ferrumequinum*. J Acoust Soc Am 56:672-682.

Stotler WA (1953) An experimental study of the cells and connections of the superior olivary complex of the cat. J Comp Neurol 98:401–432.

Suga N (1964) Single unit activity in the cochlear nucleus and inferior colliculus of echolocating bats. J Physiol 172:449–474.

Suga N, Jen PH-S (1976) Disproportionate tontopic representation for processing species specific CF-FM sonar signals in the mustache bat auditory cortex. Science 194:542–544.

Suga N, Jen PH-S (1977) Further studies on the peripheral auditory system of "CF-FM" bats specialized for the fine frequency analysis of Doppler-shifted echoes. J Exp Biol 69:207–232.

Suga N, Tsuzuki K (1985) Inhibition and level-tolerant frequency tuning in the auditory cortex of the mustached bat. J Neurophysiol (Bethesda) 53:1109–1145.

Suga N, Niwa H, Taniguchi I (1983) Representation of biosonar information in the auditory cortex of the mustached bat, with emphasis on representation of target velocity information. In: Ewert J-P, Capranica RR, Ingle DJ (eds) Advances in Vertebrate Neuroethology. New York: Plenum, pp. 829–870.

Suga N, Simmons JA, Jen PH-S (1975) Peripheral specializations for fine frequency analysis of Doppler-shifted echoes in the CF-FM bat, *Pteronotus parnellii*. J Exp Biol 63:161–192.

Sur M, Merzenich MM, Kass JH (1980) Magnification, receptive field area and "hypercolumn" size in areas 3b and 1 of somatosensory cortex in owl monkeys. J Neurophysiol (Bethesda) 44:295–311.

Thompson GC, Cortez AM, Lam DM-K (1985) Localization of GABA immuno-reactivity in the auditory brainstem of guinea pigs. Brain Res 339:119–122.

Trappe M, Schnitzler H-U (1982) Doppler-shift compensation in insect-catching horseshoe bats. Naturwissenschaften 69:193–194.

Vater M (1982) Single unit responses in the cochlear nucleus of horseshoe bats to sinusoidal frequency and amplitude moduated signals. J Comp Physiol A 149:369–388.

Vater M, Feng AS (1990) The functional organization of ascending and descending connections of the cochlear nucleus of horseshoe bats. J Comp Neurol 292:373–395.

Vater M, Feng AS, Betz M (1985) An HRP study of the frequency-place map of the horseshoe bat cochlea: morphological correlates of the sharp tuning to a narrow frequency band. J Comp Physiol A 157:671–686.

Vater M, Kössl M, Horn AKE (1992) GAD- and GABA-immunoreactivity in the ascending auditory pathway of horseshoe and mustache bats. J Comp Neurol 325:183–206.

Vater M, Schlegel P, Zöller H (1979) Comparative auditory neurophysiology of the inferior colliculus of two molossid bats, *Molossus ater* and *Molossus molossus*. I. Gross evoked potentials and single unit responses to pure tones. J Comp Physiol A:131–137.

Vater M, Habbicht H, Kössl M, Grothe B (1992) The functional role of GABA and glycine in monaural and binaural processing in the inferior colliculus of horseshoe bats. J Comp Physiol A 171:541–553.

von der Emde G (1988) Greater horseshoe bats learn to discriminate simulated echoes of insects fluttering with different wingbeat patterns. In: Nachtigall PE, Moore PWB (eds) Animal Sonar: Processes and Performance. New York: Plenum, pp. 495–500.

Warr WB (1982) Parallel ascending pathways from the cochlear nucleus: neuro-anatomical evidence of functional specialization. In: Neff WD (ed) Contributions to Sensory Physiology. New York: Academic, pp. 1–38.

Wenstrup JJ, Ross LS, Pollak GD (1985) A functional organization of binaural responses in the inferior colliculus. Hear Res 17:191–195.

Wenstrup JJ, Ross LS, Pollak GD (1986a) Binaural response organization within a frequency-band representation of the inferior colliculus: implications for sound localization. J Neurosci 6:962–973.

Wenstrup JJ, Ross LS, Pollak GD (1986b) Organization of IID sensitivity in isofrequency representations of the mustache bat's inferior colliculus. In: IUPS Satellite Symposium on Hearing, University of California, San Francisco, CA (Abstr. 415).

Wenstrup JJ, Fuzessery ZM, Pollak GD (1988a) Binaural neurons in the mustache bat's inferior colliculus: I. Responses of 60 kHz E-I units to dichotic sound stimulation. J Neurophysiol (Bethesda) 60:1369–1383.

Wenstrup JJ, Fuzessery ZM, Pollak GD (1988b) Binaural neurons in the mustache bat's inferior colliculus: II. Determinants of spatial responses among 60 kHz E-I units. J Neurophysiol (Bethesda) 60:1384–1404.

Wenthold RJ, Hunter C (1990) Immunocytochemistry of glycine and glycine receptors in the central auditory system. In: Ottersen OP, Storm-Mathisen J (eds) Glycine Neurotransmission. Chichester: Wiley, pp. 391–417.

Winer JA (1992) The functional architecture of the medial geniculate body and the primary auditory cortex. In: Webster DB, Popper AN, Fay RR (eds) The Mammalian Auditory Pathway: Neuroanatomy, Vol. 1. New York: Springer-Verlag, pp. 222–409.

Winer JA, Larue DT, Pollak GD (1995) GABA and glycine in the central auditory system of the mustache bat: structural substrates for inhibitory neuronal organization. J Comp Neurol (in press).

Yang L, Pollak GD, Resler C (1992) GABAergic circuits sharpen tuning curves and modify response properties in the mustache bat inferior colliculus. J Neurophysiol (Bethesda) 68:1760–1774.

Yang L, Pollak GD (1994) GABA and glycine have different effects on monaural response properties in the dorsal nucleus of the lateral lemniscus of the mustache bat. J Neurophysiol (Bethesda) 71:2014–2024.

Zook JM, Casseday JH (1982) Origin of ascending projections to inferior colliculus in the mustache bat, Pteronotus parnellii. J Comp Neurol 207:14–28.

Zook JM, Casseday JH (1987) Convergence of ascending pathways at the inferior colliculus of the mustache bat, Pteronotus parnellii. J Comp Neurol 261:347–361.

Zook JM, Leake PA (1989) Connections and frequency representation in the auditory brainstem of the mustache bat, Pteronotus parnellii. J Comp Neurol 290:243–261.

Zook JM, Winer JA, Pollak GD, Bodenhamer RD (1985) Topology of the central nucleus of the mustache bat's inferior colliculus: correlation of single unit properties and neuronal architecture. J Comp Neurol 231:530–546.

8

The Auditory Thalamus in Bats

Jeffrey J. Wenstrup

Introduction

During the past several years, studies of the auditory cortex in bats have revealed striking examples of functional specializations in the analysis of biosonar pulses and echoes (see O'Neill, Chapter 9). By means of sharp timing or frequency selectivities, some bat cortical neurons encode particular features of sonar targets, and these features are mapped across the cortical surface. Target range is one such feature. Much of the cortical analysis of sonar signals may depend on neural interactions that occur in the medial geniculate body (MGB), the thalamic relay in the ascending pathway to the auditory cortex. For example, *combination-sensitive* neurons, selective for combinations of spectrally or temporally distinct signal elements in the sonar pulse and echo, have been well described in regions of the auditory cortex; physiological studies suggest that these responses may be created in the medial geniculate body (Olsen and Suga 1991a,b). Thus, the auditory thalamus in bats may provide new insights into the processing of complex sounds, as well as furthering our understanding of biosonar mechanisms in these animals.

Although the MGB in mammals is regarded primarily as the thalamic relay in the ascending, tonotopic pathway to the auditory cortex, its roles are multiple and diverse. It is composed of several structurally and functionally distinct nuclei, and these are believed to form the basis for functionally distinct, parallel ascending systems to the auditory cortex (Clarey, Barone, and Imig 1992; Winer 1992). The correspondence between medial geniculate nuclei in bats and their counterparts in other species is varied. How these nuclei differ among bats, and between bats and other mammals, may reveal much about their role both in the highly developed acoustic behavior of bats and in less specialized, nonchiropteran species.

Until recently, the *lateral geniculate nucleus*, the visual thalamic relay, had received more attention in bats than the MGB. Thus, despite its role in transmitting and modifying auditory information en route to the cortex, the auditory thalamus of bats is poorly understood. With a few exceptions, adequate descriptions of the structure, connections, chemical anatomy, or physiology are lacking. Most recent studies have concentrated on the mustached bat (*Pteronotus parnellii*), because its MGB has been implicated in creating novel responses for the processing of biosonar echoes. However, even the few current studies lead one to expect that other bats may differ substantially. Hence, a broader comparative approach is crucial to our understanding of the roles played by the MGB and auditory cortex in the acoustic behavior of bats.

This chapter presents studies on the mustached bat as a framework for discussion of bat medial geniculate structure and function generally. It first describes the basic organization, then examines the structural and functional properties underlying several circuits that involve different medial geniculate nuclei.

2. General Features of the Medial Geniculate Body

The MGB in bats is well developed, even among nonecholocating species (Baron 1974). However, in bats with poorly developed visual systems, such as the mustached bat (Covey, Hall, and Kobler 1987), it dominates the dorsal thalamus (Winer and Wenstrup 1994a) and forms the dorsolateral surface of the thalamus throughout the rostrocaudal extend of the MGB.

2.1 Architectonic Organization

Nuclei of the MGB are often grouped into three divisions — dorsal, ventral, and medial — although significant differences of opinion exist regarding the grouping of nuclei and their boundaries. This review follows an architectonic scheme previously identified in the cat and based on the dendritic and axonal architecture from Golgi-impregnated neurons as well as cytoarchitecture and myeloarchitecture (Morest 1964, 1965; Winer 1985). It has been extended to several other mammalian species, for example, the tree shrew (Oliver 1982), human (Winer 1984), opossum (Morest and Winer 1986; Winer, Morest, and Diamond 1988), rat (Clerici and Coleman 1990; Clerici et al. 1990), and recently the mustached bat (Winer and Wenstrup 1994a,b). Following is a brief description of some distinguishing features of each division to provide a basis for the ensuing discussion. See Winer (1992) for a more detailed treatment of the anatomical organization of the medial geniculate body in mammals.

2.1.1 The Ventral Division

The ventral division is the thalamic component of the ascending, tonotopic auditory pathway. Several features combine to distinguish it from other divisions, including (1) a tonotopic projection from the central nucleus of the

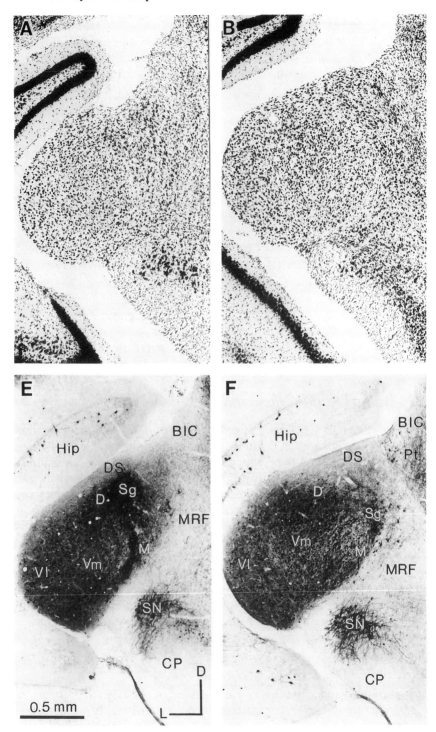

FIGURE 8.1A,B,E,F. Caption on page 372.

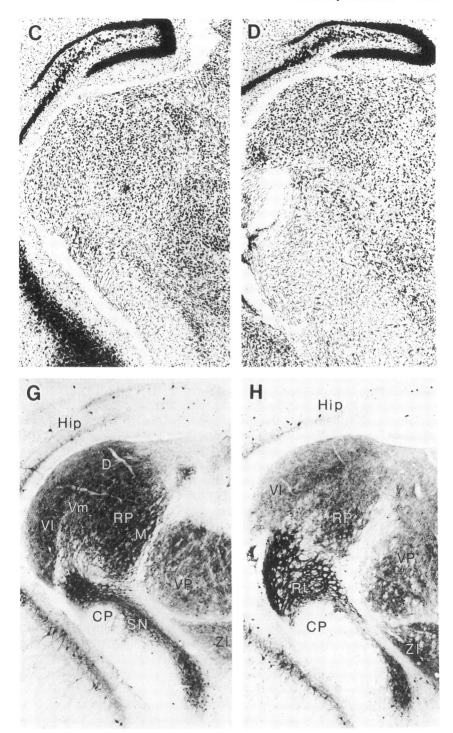

FIGURE 8.C,D,G,H. Caption on page 372.

inferior colliculus (ICC), (2) a laminar arrangement of principal cell dendrites, (3) sharply tuned, tonotopically organized responses to sound frequency, and (4) the major projection to the tonotopic, primary auditory cortical field. In many species, two major subnuclei, lateral and medial, have been identified. The lateral part (*Vl*; Figs. 8.1 and 8.2) contains the clearest laminar organization, in which the major dendrites of the principal cells lie parallel to the lateral surface of the MGB. In the medial part (*Vm*; Figs. 8.1 and 8.2), the laminar pattern may be distorted or less apparent than in the lateral part. The features of the mustached bat's ventral division generally conform to the pattern in other mammals (Olsen 1986; Wenstrup, Larue, and Winer 1994; Winer and Wenstrup 1994a,b).

2.1.2 The Dorsal Division

Several nuclei comprise the structurally and functionally diverse dorsal division. In other mammals these include superficial dorsal, dorsal, deep dorsal, suprageniculate, ventral lateral, and posterior limitans nuclei; more have been recognized in some species (Oliver 1982; Winer, Morest, and Diamond 1988). Dorsal division nuclei are united mainly in their differences from the ventral division (Winer 1992). Thus, dorsal division neurons do not have dendrites as strongly tufted as in the ventral division, and their sizes vary more among nuclei. Except for the deep dorsal nucleus, dorsal division nuclei do not receive strong ascending input from the ICC, but inputs originate instead from other inferior collicular nuclei, adjacent midbrain tegmental regions, and brainstem nuclei. The major targets of dorsal division nuclei are correspondingly varied, but all lie outside the primary tonotopic area of auditory cortex.

Most dorsal division nuclei have been identified in architectonic studies of the mustached bat (Figs 8.1 and 8.2) (Winer and Wenstrup 1994a,b). Suprageniculate neurons are among the largest in the MGB, and have radiating dendrites. Dorsal superficial neurons are bipolar or weakly tufted, while dorsal nucleus neurons are either weakly tufted or stellate. Additional

---

FIGURE 8.1A–H. Architecture of the medial geniculate body in the mustached bat, shown in Nissl-stained (A–D) and parvalbumin-immunostained (E–H) series. Each series is from a different animal, arranged in a caudal-to-rostral (left-to-right) sequence. Sections are located about 20%, 40%, 60%, and 80% through the caudal-to-rostral dimension. Both cells and neuropil are intensely parvalbumin immunoreactive in most medial geniculate subdivisions, with the exception of the superficial dorsal nucleus (DS; E,F). The rostral extreme of the medial geniculate body is only lightly or moderately labeled (H). Note also the immunostained neurons of the thalamic reticular nucleus (Rt; H). Protocol for parvalbumin immunocytochemistry: monoclonal mouse antibody (Sigma, St. Louis, MO); primary antibody dilution 1:2000; avidin-biotin-peroxidase method using heavy metal-intensified diaminobenzidine reaction.

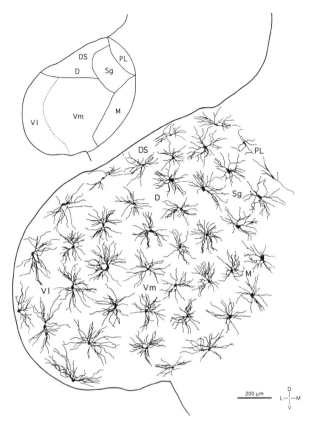

FIGURE 8.2. Illustration of Golgi preparations of the medial geniculate body (MGB) in the mustached bat. Section located near the border of the caudal and middle thirds of the MGB, corresponding roughly to Figure 8.1B. In both the lateral (Vl) and medial (Vm) parts of the ventral nucleus, principal neurons have bushy dendritic branches, but neurons in the medial part have more spherical dendritic fields. In the dorsal division, neurons with polarized dendritic fields are common in the superficial dorsal nucleus (DS), but dorsal nucleus (D) neurons have more radiate, spherical dendritic fields. Suprageniculate nucleus (Sg) neurons are larger, and their radiating branches distinguished them from the thin strip of sparsely branched, elongated neurons of the posterior limitans nucleus (PL). In the medial division (M), several neuronal varieties are impregnated, including the conspicuous magnocellular neurons. (Adapted from Winer and Wenstrup 1994b, copyright © 1994, reprinted by permission of Wiley-Liss, a division of John Wiley and Sons, Inc.)

subdivisions are also apparent, but their correspondence to other species is not always clear. For example, the rostral pole nucleus (see Fig. 8.1) was recognized by Winer and Wenstrup (1994a) as a distinctive region that has characteristics of both ventral and dorsal division neurons. However, its connections and physiology do not correspond closely to any currently known dorsal or ventral division nucleus in other mammals. Together with other parts of the dorsal division, it contains neuronal response properties that appear specialized for the analysis of sonar target features (see Section 5). These and other findings suggest that dorsal division nuclei in bats, more than the ventral division, may differ from their counterparts in other mammals and may have evolved to serve in species-specific auditory signal processing roles.

2.1.3 The Medial Division

The medial division is considered to be less closely associated with the ascending, frequency-specific auditory pathway than either the ventral or dorsal divisions (Winer and Morest 1983; but see also Rouiller et al. 1989). It is distinguished from these chiefly by its cellular population and connections. Characteristic are its large neurons (the largest in the medial geniculate) with radial dendrites, but other cell types also occur (Winer and Morest 1983; Winer 1992). The connections of the medial division reflect a broader, multisensory role, including auditory inputs from the central and external nuclei of the inferior colliculus, but also others, for example, from somatosensory and vestibular systems. Its output also differs; it projects broadly to primary and other auditory cortical regions. However, medial division projections to primary auditory cortex terminate more heavily in layer I or VI, rather than in layers III and IV (Sousa-Pinto 1973; Niimi and Naito 1974; Mitani, Itoh, and Mizuno 1987; Conley, Kupersmith, and Diamond 1991).

The medial division of the mustached bat shares a similar neuronal population with other mammals, but is much smaller in relative size than that of cats and primates; it comprises only 10% of the medial geniculate body (Figs. 8.1 and 8.2) (Winer and Wenstrup, 1994a,b). As in other mammals, the medial division in the mustached bat receives input from the ICC (see Figs. 8.4 and 8.6, later in this chapter) (Wenstrup, Larue, and Winer 1994), but its other connections are not known. In bats, functional properties of medial division neurons have not been described.

2.2 Neurochemistry

2.2.1 γ-Aminobutyric Acid (GABA)

The distribution of GABAergic neurons in the MGB differs strikingly between the two species described to date (Fig. 8.3). In the rufous horseshoe bat (*Rhinolophus rouxi*), GABAergic neurons are common in all subdivisions, but are most numerous in the ventral division (Vater, Kössl, and Horn 1992). Dendritic appendages of ventral and dorsal division GABAergic neurons are often complex and closely apposed to non-GABAergic neurons, suggesting their involvement in local circuits. GABAergic neurons are generally small, and may correspond to the small, Golgi type II cells described in other species (Morest 1975; Majorossy and Kiss 1976; Winer and Larue 1988). The mustached bat's MGB has a very different pattern (Vater,

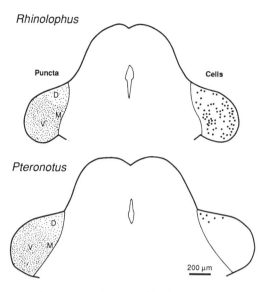

FIGURE 8.3. Schematic illustration of the distribution of GAD/GABA (glutamic acid decarboxylase/γ-aminobutyric acid)-immunoreactive puncta (*left*) and cells (*right*) in the MGB of the rufous horseshoe bat (*Rhinolophus*) and the mustached bat (*Pteronotus*). The distributions of immunopositive puncta are similar in the two species; there are fewer puncta in the dorsal division than in the other medial geniculate divisions. In contrast, the distributions of immunopositive cells are very different. In the horseshoe bat, GABAergic neurons are common in all three divisions, but are most numerous in the ventral division. GABAergic neurons in the mustached bat are rare, and most of these are located in the dorsal division. (Adapted from Vater, Kössl, and Horn, copyright © 1992, reprinted by permission of Wiley-Liss, a division of John Wiley and Sons, Inc.)

Kössl, and Horn 1992; Winer, Wenstrup, and Larue 1992). Only a very few GABAergic neurons have been found, probably less than 1% of medial geniculate neurons, and most of these were in the dorsal division.

In contrast, the form and distribution of GABAergic puncta (putative terminals) show close similarities between the two bats (see Fig. 8.3) (Vater, Kössl, and Horn 1992; Winer, Wenstrup, and Larue 1992). Thus, the dorsal and ventral divisions contain relatively fine puncta, while those in the medial division are much coarser. The dorsal division contains the fewest puncta. These patterns occurred despite differences in numbers of intrinsic GABAergic neurons, suggesting that the pattern of GABAergic terminals in the MGB may be imposed primarily by extrinsic neurons.

The dramatic difference between GABAergic neurons in the two species is very surprising, yet the results seem reliable. For example, each of the two studies (Vater, Kössl, and Horn 1992; Winer, Wenstrup, and Larue 1992) documented the scarcity of GABAergic neurons in the mustached bat using both GABA and glutamic acid decarboxylase (GAD) immunocytochemistry. These studies found few immunostained medial geniculate cells in the same histological sections where other neurons were intensely immuno-stained. Furthermore, the comparative study (Vater, Kössl, and Horn 1992), using both GABA- and GAD-immunocytochemistry, found that the numbers of GABAergic neurons differed between the two species but the distribution of puncta did not.

What is particularly surprising is that the different patterns are found in bat species that use similar biosonar signals and echo information, and that seem to use similar neuronal processing strategies. Unless some other inhibitory neurotransmitter in the mustached bat performs the same function as GABA, these results suggest major differences in the processing of ascending and descending input by medial geniculate neurons. Although MGB neurons in both species receive GABAergic input, only in the horseshoe bat do local, that is, intrageniculate, inhibitory mechanisms seem to occur. In the mustached bat, inhibitory interactions must result largely from external GABAergic input, perhaps from the thalamic reticular nucleus (Winer, Wenstrup, and Larue 1992; Wenstrup and Grose 1993), or from neurons utilizing other inhibitory neurotransmitters. However, glyci-nergic inhibition, although common in lower auditory centers, does not appear to play a role in medial geniculate processing by mammals (Aoki et al. 1988; Wenthold and Hunter 1990).

2.2.2 Calcium-Binding Proteins

Studies of the distribution of calcium-binding proteins in auditory systems have served to distinguish subsystems that process different types of information, as in the temporal processing pathways of the barn owl auditory system (Takahashi et al. 1993) or in the thalamocortical projec-tions of monkeys (Hashikawa et al. 1991). Zettel, Carr, and O'Neill (1991) examined immunoreactivity to calbindin throughout mustached bat audi-tory structures. In medial geniculate nuclei, calbindin-immunopositive neurons were abundant in dorsal and ventral divisions. Immunostaining of the neuropil was heaviest in the dorsal division, particularly the superficial dorsal nucleus. Virtually unlabeled were the suprageniculate nucleus and the posterior complex (corresponding to the caudal part of the medial division here), regions containing the largest medial geniculate neurons. In the horseshoe bat, calbindin-immunopositive cells are common in all divisions of the MGB, whereas calretinin-immunopositive cells occur only in the dorsal division and along the lateral and ventrolateral rims of the ventral division. Few axon terminals are calbindin- or calretinin-immuno-positive (Vater and Braun 1994).

Parvalbumin, another calcium-binding protein, is distributed differently in the medial geniculate body (see Fig. 8.1E–H). Both cells and neuropil are heavily labeled in several areas: the ventral division, medial division, and dorsal, rostral pole, and suprageniculate nuclei. Labeling in the most rostral part of the MGB was weaker, although present (see Fig. 8.1H). In contrast, the superficial dorsal nucleus contained virtually no immunopositive labeling (Fig. 8.1E,F). Generally, the parvalbumin-immunostained regions corresponded to those receiving input from the central nucleus of the inferior colliculus (Fig. 8.4) (Wenstrup, Larue, and Winer 1994). In the horseshoe bat, all divisions of the MGB contained parvalbumin-immunopositive cells and terminals (Vater and Braun 1994).

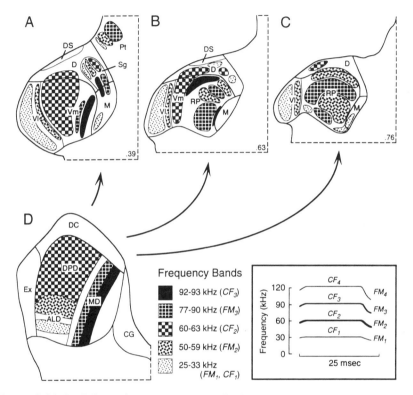

FIGURE 8.4A–D. Schematic summary of the distribution of inputs to the mustached bat's MGB (A–C) from five frequency band representations (D) in the central nucleus of the inferior colliculus (ICC) analyzing major elements of the bat's sonar signal. Three projection systems are described in the text. One system terminates in the lateral (Vl) and medial (Vm) parts of the ventral division and is tonotopically organized (A–C). A second terminates in the suprageniculate nucleus (Sg) and is also tonotopic (A). The third system was found principally in the rostral pole (RP) and dorsal (D) nuclei (B,C). These inputs may be organized according to functional biosonar components; CF_2 and CF_3 projections terminate at middle MGB levels (B), while FM_2 and FM_3 projections are extensive and terminate well into the most rostral part of MGB (B,C). Note also the projection to the pretectum from ICC regions representing frequency-modulated (FM) components of the sonar signal (A). *Inset at right*: sonogram of mustached bat biosonar pulse with signal components labeled. Thickness of lines in sonogram indicates relative intensity of harmonics. CF, constant frequency. (From Wenstrup, Larue, and Winer, copyright © 1994, reprinted by permission of Wiley-Liss, a division of John Wiley and Sons, Inc.)

Species comparisons raise questions about the functional implications of the distribution of calcium-binding proteins. For example, the distribution of parvalbumin in both the mustached bat and horseshoe bat agrees well with its distribution in monkeys (Hashikawa et al. 1991), but not in rats (Celio 1990). However, the distribution of calbindin agrees more with data in rats than monkeys (Celio 1990; Hashikawa et al. 1991). Furthermore, results differ depending on the antibody used; thus, in the mustached bat, monoclonal mouse anticalbindin labels the same medial geniculate regions (Wenstrup, unpublished data) as does polyclonal anticalbindin (Zettel, Carr, and O'Neill 1991), but it labels virtually none of the brainstem auditory regions labeled by the polyclonal antibody.

2.3 Connections of the Medial Geniculate Body

As in other mammals, the MGB in bats receives a major ascending input from the tonotopically organized central nucleus of the inferior colliculus (see Fig. 8.4). In all bats so far examined, including the mustached bat (Casseday et al. 1989; Frisina, O'Neill, and Zettel 1989; Wenstrup, Larue, and Winer 1994), the greater horseshoe bat (*Rhinolophus ferrumequinum*) (Schweizer 1981), the pallid bat (*Antrozous pallidus*) (Wenstrup and Fuzessery, unpublished data), and the big brown bat (*Eptesicus fuscus*) (Covey, unpublished data), the ICC projection is exclusively ipsilateral. In contrast, the contralateral ICC-MGB projection in the cat is significant, although smaller than the ipsilateral projection (Kudo and Niimi 1978; Rouiller and de Ribaupierre 1985).

Perhaps because bats have a special reliance on acoustic information, the MGB may receive broader input from the ICC than is the case in other mammals. This is clear in the mustached bat, where the ICC forms distinct projections to three parallel MGB systems (see Fig. 8.4), each probably serving different functional roles in acoustic orientation. These systems, considered next, include (1) the tonotopically organized system through the ventral division, (2) systems involving the suprageniculate nucleus, and (3) regions in the dorsal and rostral MGB containing specialized responses to biosonar signals. What is apparent is that these systems, and their affiliated MGB nuclei, differ considerably in the degree of their correspondence to other mammals.

3. Ventral Division: The Tonotopically Organized Medial Geniculate Body

3.1 Functional Architecture of the Ventral Division

The distinctive architectonic feature of the ventral division in the mustached bat, as in other mammals, is its laminar organization of principal cell dendrites (see Fig. 8.2) (Winer and Wenstrup 1994b). Dendrites of principal cells arise in tufts from the two somatic poles and fill a more or less planar or sheetlike expanse. The laminar arrangement is particularly evident in the lateral part, where the dendrites of tufted neurons run parallel to the lateral surface of the medial geniculate. In the medial part, the laminar organization is more variable. It is not apparent in the region, located caudally and dorsally, that receives input from the hypertrophied 60- to 63-Hz representation of the ICC. It is clearer more medially, however, where dendritic laminae lie along a ventrolateral-to-dorsomedial axis.

The laminar dendritic organization in the ventral division is matched by the arrangement of ICC afferents (Fig. 8.5A–C) (Wenstrup, Larue, and Winer 1994). Thus, in the lateral part, horseradish peroxidase-filled axons from low-frequency parts of the ICC terminate in sheets that parallel the lateral surface of the MGB (Fig. 8.5B). In the medial part, 60-kHz ICC axons do not terminate in a laminar pattern, even though the size of individual terminal fields is highly restricted (Fig. 8.5A). In contrast, ICC axons tuned above 63 kHz form terminal fields more medially, which extend along a ventrolateral-to-dorsomedial axis (Fig. 8.5C). The correspondence between dendritic arborization patterns and axon terminal fields is likely to preserve the segregation and topographic arrangement of frequencies that occurs in lower auditory nuclei. Thus, this organization closely corresponds to that in the mustached bat ICC (Zook et al. 1985), where three architectonic patterns characterize regions representing frequencies less than 60 kHz (the anterolateral division), 60–63 kHz (the dorsoposterior division), and greater than 63 kHz (the medial division). In the ICC as in the MGB the 60- to 63-kHz representation lacks a clear laminar pattern and distorts the overall laminar organization by its large size.

In other bats, the neuronal architecture of the ventral division is unknown. However, the organization of ICC input suggests somewhat different laminar patterns. Thus, in the big brown bat (Covey, unpublished data) and pallid bat (Wenstrup and Fuzessery, unpublished data), deposits in restricted ICC frequency bands result in ventral division label which extends in a dorsolateral-to-ventromedial pattern. In the little brown bat (*Myotis lucifugus*), restricted ICC deposits are reported to result in laminar patterns of labeling (Shannon and Wong 1987).

In cats, small stellate cells having only local axonal projections contribute significantly to the architecture of the ventral division (Morest 1975). Their dendrites receive synaptic contacts from tectal and cortical axons, and form axodendritic and dendrodendritic contacts with principal neurons. Many of these cells are probably GABAergic (Winer 1992), and may contribute a variety of inhibitory influences regulating the output of the larger principal cells. In the mustached bat's ventral division, small stellate cells are scarce in Golgi-impregnated material (Winer and Wenstrup 1994b), and there are very few GABAergic neurons (Winer, Wenstrup, and Larue 1992). Thus,

380   Jeffrey J. Wenstrup

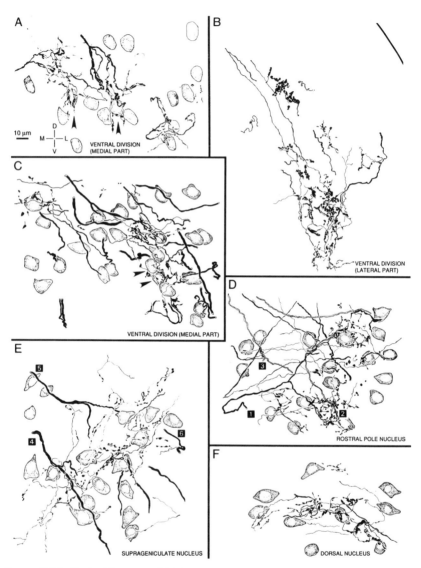

FIGURE 8.5A–F. Architecture of ICC axons in subdivisions of the medial geniculate body. (A) Labeling in the medial part of the ventral division after ICC deposits in regions tuned near 62 kHz. Boutons ended both in neuropil and near perikarya. (B) Labeling in the lateral part of the ventral division after deposits in regions tuned to 30–33 kHz. Axons and terminals formed an arc recapitulating the arrangement of principal cell dendrites in Golgi material (see Fig. 8.2). No perikarya are visible in this unstained section. *Heavy curved line*, dorsolateral surface of MGB. (C) Labeling in the medial part of the ventral division after deposits in regions tuned to 93 kHz. (D) Labeling in the rostral pole nucleus after 77- to 82-kHz deposits. Many axons (1–3) formed two branches separated by at least 100 μm. Two axons, (1) and (2), entered ventrolaterally, and each had two separate terminal plexuses that were

what is considered to be an important local inhibitory circuit in the cat's MGB apparently does not exist in the mustached bat. Whether some other circuit or some other inhibitory neurotransmitter serves the same function is unknown. As noted previously, this circuit may well exist in other bats, because horseshoe bats have many ventral division GABAergic neurons. These anatomical differences suggest that physiological properties, such as discharge patterns, tuning curves, or rate-level functions, may differ between ventral division neurons in the two bat species. Further effects may be revealed in neuropharmacological studies of ventral division neurons.

3.2 Connections of the Ventral Division

3.2.1 Inputs from the Inferior Colliculus

The ventral division receives a topographically organized projection from the central nucleus of the inferior colliculus. This projection has been studied most extensively in the mustached bat, where ICC tracer deposits were placed within regions representing each of the biosonar components of the first three harmonic elements (Frisina, O'Neill, and Zettel 1989; Wenstrup, Larue, and Winer 1994). There are three main topographic features of the ICC projection to the ventral division in this bat (see Fig. 8.4). First, low-frequency representations project laterally to the ventral division, while higher frequencies project to successively more medial loci. This topographic pattern is in general agreement with that observed in cats (Andersen et al. 1980; Kudo and Niimi 1980; Calford and Aitkin 1983). Second, 60-kHz input is expanded, in agreement with its representation in the ICC and throughout the ascending auditory system (Figs. 8.4 and 8.6).

The third feature is particularly noteworthy. Some frequency band representations in ICC appear to project only lightly to the ventral division. Thus, only modest labeling occurs within the ventral division after ICC deposits in the 50- to 59-kHz and 77- to 90-kHz frequency band representations, but much stronger labeling is found in the rostral MGB (Figs. 8.4 and 8.6). This is a significant departure from the pattern established in the cochlea and preserved within the ICC (Frisina, O'Neill, and Zettel 1989), showing that the tectothalamic projection can modify the organization within the primary auditory pathway. In this case, these frequency band

---

FIGURE 8.5A–F (*continued*) nearly congruent. (E) Labeling in the suprageniculate nucleus after 62-kHz deposits. Several axons branched in the suprageniculate nucleus (4–6). Some (4) sent collaterals into the suprageniculate while the main trunk proceeded to the ventral division. (F) Labeling in the dorsal nucleus after 62-kHz deposits. Boutons were similar to those of the ventral division in the same experiment (cf. A). (From Wenstrup, Larue, and Winer, copyright © 1994, reprinted by permission of Wiley-Liss, a division of John Wiley and Sons, Inc.)

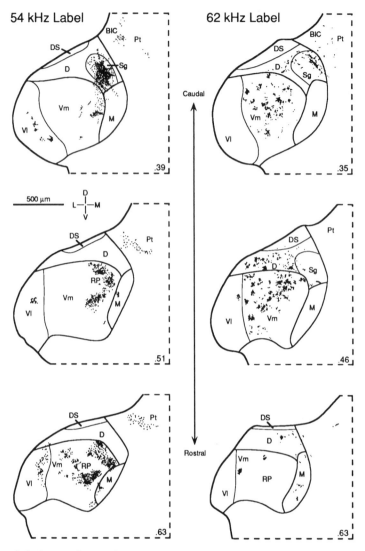

FIGURE 8.6. Comparison of tectothalamic projections from 54-kHz and 62-kHz representations of the ICC; these frequencies are contained within the FM$_2$ and CF$_2$ components, respectively, of the sonar signal. In the 54-kHz case, transport results from two large iontophoretic deposits of wheat germ agglutinin–horseradish peroxidase conjugate (WGA-HRP). In the 62-kHz case, six smaller, iontophoretic HRP deposits were placed. Note that 54-kHz regions project only lightly to the lateral part of the ventral division (Vl), but more heavily to the rostral pole nucleus (RP). In contrast, the 62-kHz ICC representation projects strongly to the medial part of the ventral division (Vm) and the adjacent dorsal division (D). Rostrally, the 62-kHz input is very weak. Both frequency representations project to the suprageniculate nucleus (Sg). (Adapted from Wenstrup, Larue, and Winer, copyright © 1994, reprinted by permission of Wiley-Liss, a division of John Wiley and Sons, Inc.)

representations correspond to higher harmonics of the frequency-modulated (FM) components of the sonar pulse, and their strong projection to the rostral MGB suggests their primary role is to carry information to specialized neurons in the rostral MGB that analyze target distance (Olsen and Suga 1991b). Furthermore, their reduced input to the ventral division suggests that they may play less of a role in other types of signal analysis, for example, elevational and azimuthal localization.

There is less information on the MGB of other species. In the pallid bat, tracer deposits in ICC frequency band representations result in terminal labeling arranged in sheets extending along a ventromedial-to-dorsolateral orientation; low-frequency label (~ 15 kHz) occurs ventrolaterally, while higher frequency label (~ 40 kHz) associated with this bat's sonar pulse occurs in more dorsal and medial sheets (Wenstrup and Fuzessery, unpublished data). A similar arrangement appears in the big brown bat (*Eptesicus fuscus*) (Covey, unpublished data). In view of the results in the mustached bat, the organization of tectothalamic input to the ventral division in other bats is of considerable interest. Does the ventral division receive more or less input from neurons tuned to frequencies within the biosonar sound? Do neurons tuned to some sonar frequencies project more heavily outside the lemniscal pathway? The pattern may be very different for bats using FM signals, because, unlike the mustached bat, they must utilize the same frequency bands to obtain several target characterizations.

3.2.2 Inputs from Other Sources

Additional inputs to the ventral division arise from the auditory cortex and the thalamic reticular nucleus (Olsen 1986; Wenstrup and Grose 1993, in press). Olsen (1986) found that deposits in AI, the tonotopically organized auditory cortex, resulted in congruent anterograde and retrograde labeling in the ventral division. Thus, corticothalamic projections follow the tonotopic organization in the ventral division. The ventral division also receives strong input from the thalamic reticular nucleus (Wenstrup and Grose 1993, in press). The auditory sector of this nucleus, located just ventral to the rostral MGB projects to most MGB regions, although its organization is not currently known.

3.2.3 Outputs

The major output of the ventral division is a topographically organized projection to tonotopically organized auditory cortex (Olsen 1986; Casseday and Pollak 1988). Thus, low frequencies in the lateral subdivision project caudally, while higher frequencies in the medial part project to more rostral cortical loci. These data agree well with the pattern of anterograde labeling in the ventral division following ICC deposits, as well as the tonotopic organization revealed in physiological studies (Olsen 1986).

3.2.4 Evidence of Intrinsic Connections

There is little evidence that the ventral division projects to or receives input from other MGB subdivisions. For example, tracer deposits in combination-sensitive regions of the dorsal division or rostral pole do not label ventral division regions outside the deposit site either retrogradely or anterogradely (Wenstrup and Grose 1993). Thus, combination-sensitive neurons do not receive low-frequency (24–30 kHz) input by way of the lateral part of the ventral division (see Section 5).

3.3 Physiological Properties of the Ventral Division

In all mammals, the ventral division is believed to maintain frequency-specific processing of acoustic information. Most ventral division neurons are sharply tuned and tonotopically organized, responding reliably and with temporal fidelity to tonal signals (Aitkin and Webster 1972; Calford 1983; Rodrigues-Dagaeff et al. 1989; Clarey, Barone, and Imig 1992). Although the evidence is limited, this generally appears to be the case for bats as well.

3.3.1 Frequency and Amplitude Tuning

Early physiological studies of MGB neurons in *Myotis oxygnathus* and the horseshoe bat (*Rhinolophus ferrumequinum*) showed that greater numbers of neurons were tuned to frequency bands within the biosonar signals of these species than to nonsonar frequencies (Vasil'ev and Andreeva 1972). In *M. oxygnathus*, a bat using broadband FM signals, nearly all neurons had moderately sharp tuning curves with Q_{10dB} values (defined as the best frequency of a neuron's tuning curve divided by the bandwidth 10 dB above threshold) less than 20; the distribution of Q_{10dB} values was similar to those in the cochlear nucleus and the auditory cortex. In the horseshoe bat, many MGB neurons were more sharply tuned, particularly at frequencies corresponding to the constant-frequency (CF) component of the bat's biosonar pulse. Ayrapet'yants and Konstantinov (1974) suggest that there is an increase in the sharpness of tuning in horseshoe bat auditory neurons from the cochlear nucleus to the MGB. In both species, some neurons were tuned in amplitude, having nonmonotonic rate-level functions and upper thresholds. Most neurons in these experiments were recorded from the "parvicellular part" of the MGB (Vasil'ev and Andreeva 1972), which includes both the ventral division and part of the dorsal division.

In the mustached bat's MGB, the best frequencies of neurons responding only to single tones ranged from 5 to 120 kHz, although very few were below 20 kHz or above 100 kHz (Olsen 1986; Olsen and Suga 1991a). Neurons tuned to frequencies in the CF sonar echoes of the second (60 kHz) and third (90 kHz) harmonic (CF_2 and CF_3, respectively) were more common than those tuned to other frequency bands in the sonar signal or to nonsonar frequencies. Many of these neurons probably lay within the ventral division, although their location by division was not reported.

Medial geniculate neurons tuned to the CF_2 and CF_3 sonar components were more sharply tuned in frequency than other MGB neurons (Fig. 8.7) (Olsen and Suga 1991a). Sharper tuning was evident near threshold (Fig. 8.7A) and well above threshold (Fig. 8.7B, C). Q_{10dB} values agree with studies on other auditory nuclei in the mustached bat (Suga, Simmons, and Jen 1975; Pollak and Bodenhamer 1981; Suga and Manabe 1982), and they reflect sharp cochlear tuning in these frequency bands, particularly at the frequencies of the CF_2 component (see Kössl and Vater, Chapter 5, this volume). However, neural inhibitory mechanisms also play a role. Thus, medial geniculate neurons display broadly tuned (i.e., broader than the excitatory tuning curve) inhibition that both sharpens the tuning curve at higher levels and restricts the range of levels to which the neuron responds (Olsen and Suga 1991a). These effects have also been documented among ventral division neurons in the cat (Aitkin and Webster 1972; Whitfield and Purser 1972; Rouiller et al. 1990). In the mustached bat, these inhibitory effects result in narrow, level-tolerant tuning curves that may contribute to the fine frequency analysis of sonar echoes (Suga and Tsuzuki 1985). This may have several functions, including (1) improved sensitivity to the magnitude of frequency (i.e., Doppler) shifts; (2) improved sensitivity to frequency-shifted CF echoes while reducing sensitivity to louder, temporally overlapping outgoing CF signals; and (3) increased sensitivity to small, periodic frequency modulations in echoes from fluttering insects.

Do circuits within the MGB contribute to inhibitory shaping of tuning curves? No studies to date have examined such effects via iontophoretic application of inhibitory transmitter antagonists. In the inferior colliculus, many of the inhibitory effects found among medial geniculate neurons have also been observed, some the result of GABAergic inhibition (Yang, Pollak, and Resler 1992; Pollak and Park, Chapter 7). While it is reasonable to suspect that inhibitory mechanisms in the cat MGB contribute to tuning sharpness (because there are many GABAergic neurons), the possibility is much less clear in the mustached bat with its lack of such neurons. If tuning sharpness is increased in the MGB, it must depend on other sources of GABAergic input or other neurotransmitters.

Some ventral division neurons may display more complex tuning than has been described thus far. In the primary auditory cortex, Fitzpatrick et al. (1993) found that many neurons tuned to the 60-kHz (CF_2) component are combination-sensitive; they are facilitated by a preceding signal tuned to frequencies in the fundamental FM biosonar component (FM_1). These have been recorded in the ventral division of the MGB (Wenstrup and Grose, in press). The origin of these responses within the ascending tonotopic system is unclear. Because preliminary studies (Mittmann and Wenstrup 1994) have reported that combination-sensitive neurons are common in the ICC, many ventral division neurons, including those responding to 60-kHz sounds,

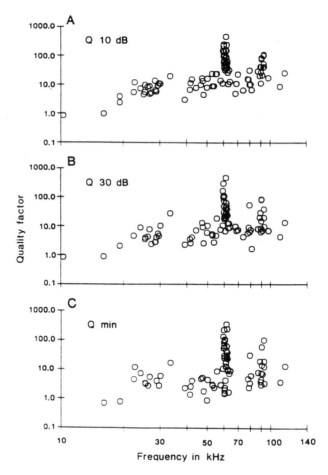

FIGURE 8.7. Sharpness of tuning among medial geniculate neurons that are not combination sensitive. Three measures of tuning were used. (A) Q_{10dB} and (B) Q_{30} dB values were calculated as the best frequency divided by the bandwidth of the tuning curve 10 dB and 30 dB above the neuron's threshold, respectively. (C) Q_{min} values were calculated as the best frequency divided by the bandwidth of the tuning curve at its widest point between threshold and 100 dB SPL (Sound Pressure Level). Although Q_{10dB} values probably reflect peripheral tuning mechanisms, Q_{30dB} and Q_{min} values also reflect the degree of neural sharpening of tuning curves. Neurons with best frequencies near 60 kHz and 90 kHz were most sharply tuned by all three measures. Data include singly tuned neurons from all parts of the medial geniculate body, not only the ventral division. (From Olsen and Suga 1991a, copyright © 1991, reprinted by permission of the American Physiological Society.)

may display sensitivity to multiple spectral elements in signals as the result of combination-sensitive ICC input.

3.3.2 Tonotopic Organization

Very few data exist on the frequency organization of the ventral division in bats. In the mustached bat, physiological mapping of best frequencies is generally consistent with data from anterograde and retrograde transport studies. Olsen (1986) recorded low-frequency responses in the lateral part of the ventral division, with higher frequencies more medially. There is a relatively large representation of the 60-kHz frequency band, corresponding to its large representation elsewhere. Also recorded are pure tone responses in the 48- to 59-kHz and 72- to 90-kHz bands, corresponding to the second and third harmonic FM components (FM_2 and FM_3, respectively) of the sonar signal (Olsen 1986). This supports connectional evidence that these frequency band representations do indeed project to the ventral division. However, their relative sizes appear to be reduced compared to the ICC. There are no published physiological studies of the frequency organization in other bats.

3.3.3 Modulation Sensitivity

Periodic modulations of frequency and amplitude have particular salience to insectivorous bats, since these modulations improve the detectability and identification of fluttering insects (see Schnitzler, Chapter 3; Pollak and Park, Chapter 7). In many other vertebrates, the ability of auditory neurons to code modulations by their temporal response pattern decreases at higher levels in the auditory pathway, and this also seems to be true in bats. Thus, cochlear nucleus/auditory nerve units (Suga and Jen 1977) lock to much higher modulation rates than do ICC units (Schuller 1979; Bodenhamer and Pollak 1983). Only one study has described modulation sensitivity in the MGB of bats. Andreeva and Lang (1977), recording evoked potentials from the horseshoe bat's MGB, were unable to demonstrate temporally locked responses to periodic amplitude modulation, whereas they obtained strong responses to such stimuli in the brainstem and inferior colliculus. Further studies with single units are needed to establish whether and in what form sensitivity to echo modulations exists among medial geniculate neurons, and also whether such sensitivity is limited to certain medial geniculate subdivisions.

3.3.4 Binaural Responses

The tonotopically organized pathway to the auditory cortex plays an important role in the analysis of sound localization cues by other mammals (Jenkins and Masterton 1982; Jenkins and Merzenich 1984; Kavanagh and Kelly 1987). In bats, there are no published accounts of the binaural responsiveness or spatial selectivity of medial geniculate neurons, and thus it is not known what role the MGB plays in analyzing and representing sound localization cues.

A preliminary report (Wenstrup 1992a) has examined projections to the MGB from aural response regions in the ICC, that is, regions characterized by a particular response to binaural sounds (Wenstrup, Ross, and Pollak 1986). Anterograde tracers were placed in parts of the 60-kHz ICC representation containing either monaural (contralaterally excited) neurons or excitatory-inhibitory (EI) neurons (excited by contralateral sound, inhibited by ipsilateral sound). Single deposits in EI regions generally labeled two target zones; one was located ventrally, in the medial part of the ventral division, while the second was located dorsally, spanning the border between the dorsal division and the medial part of the ventral division. The 60-kHz monaural region in the ICC projects mostly between the two EI inputs, to the dorsal part of the ventral division. These results suggest that aural response-specific regions may be preserved in the projection from ICC to the ventral division. Moreover, because EI neurons in ICC are sensitive to and topographically represent the interaural intensity difference (Wenstrup, Ross, and Pollak 1986), a sound localization cue, some ventral division regions may represent this target feature and project the information to primary auditory cortex.

Several questions need to be explored concerning the representation of sound location in the auditory tectothalamocortical pathway of bats, some at the cortical level but others within the MGB. Is the topographic representation of interaural intensity differences in the ICC maintained in the projection of EI neurons to the MGB? Are target elevation and azimuth analyzed by all frequency bands within sonar echoes, or are some frequencies excluded from this analysis, for example, those within the FM sweeps in the mustached bat that provide relatively little input to the ventral division? What medial geniculate subdivisions and their related cortical areas participate in analyses and representations of azimuthal and elevational spatial information?

4. The Suprageniculate Nucleus

The suprageniculate nucleus is considered to be part of the dorsal division, although it is clearly distinct from other dorsal division nuclei; others have placed it with the posterior group of thalamic nuclei (Jones and Powell 1971; Casseday et al. 1989). In the mustached bat, recent findings concerning the connections of this nucleus suggest an important role in its acoustic behavior.

The morphology of suprageniculate neurons in the mustached bat corresponds closely to that in other mammals (Casseday et al. 1989; Winer and Wenstrup 1994a,b). The principal cell is a distinctive, large, multipolar neuron with radial dendrites, about as large as the magnocellular neurons of the medial division (see Figs. 8.1 and 8.2). Medium-sized axons originating in the inferior colliculus course ventrolaterally through the suprageniculate nucleus, imparting a distinct architecture. Some ICC axons form collaterals en route to the ventral division, and these terminate within the suprageniculate nucleus (see Fig. 8.5E) (Wenstrup, Larue, and Winer 1994).

The connections of the suprageniculate nucleus are unusual in several respects. First, it receives strong input from the nucleus of the central acoustic tract (Casseday et al. 1989), a brainstem auditory nucleus that receives contralateral or possibly bilateral input from the ventral cochlear nuclei (Casseday et al. 1989) (Fig. 8.8). Physiological responses in this nucleus of the mustached bat have not been well studied, but in horseshoe bats these neurons were monaurally responsive to input from the contralateral ear (see Casseday and Covey, Chapter 5). Thus, the suprageniculate nucleus is the only thalamic auditory nucleus in the mustached bat known to receive direct auditory brainstem input. Similar, although weaker, inputs have been described in other species (Papez 1929; Morest 1965; Henkel 1983).

The suprageniculate nucleus in the mustached bat is also unusual in the input it receives from the tectum. As in other species, it receives input from the superior colliculus, but it also is the target of strong input from the central and external nuclei of the inferior colliculus (Casseday et al.

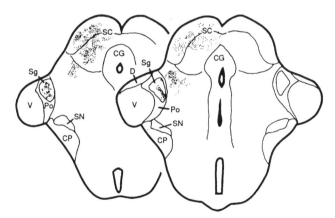

FIGURE 8.8. Anterograde labeling in the mustached bat's suprageniculate nucleus (Sg) and superior colliculus (SC) after a deposit of WGA-HRP near the nucleus of the central acoustic tract. This projection is unusual because it bypasses the inferior colliculus, proceeding to the medial geniculate body and superior colliculus by way of the central acoustic tract. (Modified from Casseday et al., copyright © 1989, reprinted by permission of Wiley-Liss, a division of John Wiley and Sons, Inc.)

1989; Wenstrup, Larue, and Winer 1994). In anterograde transport studies, Wenstrup, Larue, and Winer (1994) showed that all frequency band representations in ICC that received tracer deposits projected strongly to the suprageniculate nucleus in a frequency-specific pattern (see Figs. 8.4–8.6). Thus, lower frequency neurons (25–33 kHz; 50–59 kHz) terminate in the dorsolateral part of the nucleus, while 60-kHz inputs terminate dorsomedially (Fig. 8.6). Inputs from 80 and 90 kHz are placed more ventrally (Fig. 8.4). It is unclear how this topographic ICC projection is organized relative to inputs from the nucleus of the central acoustic tract. Moreover, because dendrites of suprageniculate neurons fill much of the nucleus, they may sample across a broad range of ICC inputs. Thus, it is unclear whether the ICC tonotopic projection will confer a physiological tonotopy onto this nucleus. The projection from the ICC is particularly interesting because it is not known to occur in any nonchiropteran species (Winer 1992).

The suprageniculate nucleus projects to broad regions of auditory cortex, but it also provides input to a limited region of frontal cortex containing robust auditory responses (Fig. 8.9) (Kobler, Isbey, and Casseday 1987). This frontal cortical region in turn projects to the superior colliculus (Kobler, Isbey, and Casseday 1987). Thus, the suprageniculate forms part of an extralemniscal (mostly) auditory subsystem involving the nucleus of the central acoustic tract, inferior collicular nuclei, the superior colliculus, and frontal cortex. Casseday et al. (1989) have suggested that the suprageniculate nucleus–frontal cortex–superior colliculus pathway serves a "priming" purpose for the activation of acousticomotor responses mediated by the superior colliculus. Inactivation of the frontal cortex during superior colliculus recordings could address possible roles of this suprageniculate nucleus–frontal cortical pathway.

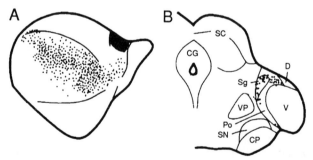

FIGURE 8.9A,B. Connection of the suprageniculate nucleus (Sg) with frontal cortex in the mustached bat. (A) WGA-HRP deposit site (*blackened region*) in frontal cortex and retrograde labeling in auditory cortical regions (*dots*). (B) Retrograde labeling in the suprageniculate nucleus and dorsal division (D). Suprageniculate neurons project to auditory cortex and to regions of frontal cortex. (Modified by permission from Kobler, Isbey, and Casseday, copyright © 1987 by the AAAS.)

5. The Dorsal and Rostral Medial Geniculate: Specialized Responses to Biosonar Signals

Dorsal and rostral areas of the MGB are perhaps the least understood of auditory thalamic regions. Their neurons often display longer latencies, broader tuning, and more variable responses to acoustic stimuli (Calford 1983), suggesting more complex analyses of sounds. In guinea pigs, dorsal division neurons display changes in frequency-receptive fields after the animals learned conditioned responses to tonal signals (Edeline and Weinberger 1991). In monkeys, dorsal division neurons may respond selectively to some social communication sounds (Olsen and Rauschecker 1992). In bats, they may participate in specialized analyses of biosonar echoes (see following) and social communication signals. Such findings suggest greater species differences in the structure and function of these areas than in the ventral division. Among bats, extensive studies have been reported only in the mustached bat. However, a preliminary study (Shannon and Wong 1987) suggested that the dorsal division in the little brown bat may also play a specialized role in biosonar information processing.

5.1 Architecture

The region considered here encompasses much of the dorsal and rostral MGB in the mustached bat (see Fig. 8.1C,D,G,H). In the present architectonic scheme, its major constituents are the dorsal nucleus and rostral pole nucleus, but not the lateral part of the ventral division. The architecture of the dorsal nucleus is similar to other mammals; it is distinguished from the ventral division by its population of stellate neurons and radiate neurons with weakly tufted dendrites (Winer and Wenstrup 1994a,b). The rostral pole nucleus, described by Winer and Wenstrup (1994a), constitutes a large part of the rostral MGB, replacing the medial part of the ventral division at levels into the rostral half (see Fig. 8.1). Although the architecture changes somewhat, neurons are similar to those in the ventral division, having relatively small somata ($\sim 8 \times 10 \ \mu$m). No somatic orientation is distinct. The neuronal architecture is less clear, because neurons in Golgi material are not well stained (Winer and Wenstrup 1994b).

Much of this region is considered by others (Olsen 1986; Olsen and Suga 1991a) to form the deep dorsal nucleus, and like the deep dorsal nucleus in the cat (Andersen et al. 1980; Calford and Aitkin 1983), it receives mainly high-frequency input from the ICC (Frisina, O'Neill, and Zettel 1989; Wenstrup 1992b; Wenstrup, Larue, and Winer 1994). However, the physiological properties and cortical targets appear quite different. Further studies are needed to determine the similarity of these regions to the deep dorsal nucleus of other mammals.

5.2 Connections of the Dorsal and Rostral Medial Geniculate Body

5.2.1 Inputs

Major inputs to the dorsal and rostral MGB are from the inferior colliculus, thalamic reticular nucleus, and auditory cortex. Input from the ICC is very strong, covering much of the rostral half of the medial geniculate body (see Fig. 8.4B,C) (Frisina, O'Neill, and Zettel 1989; Wenstrup, Larue, and Winer 1994). Axons from the ICC terminating caudally in the dorsal nucleus are similar to those in the underlying ventral division (see Fig. 8.5F), but the axons targeting the rostral pole nucleus and the rostral part of the dorsal nucleus are different, often diverging widely within these nuclei (see Fig. 8.5D) (Frisina, O'Neill, and Zettel 1989; Wenstrup, Larue, and Winer 1994). These axonal patterns may underlie the patchy organization of combination-sensitive responses observed in these regions (Olsen 1986).

The organization of ICC inputs differs from the ventral division (see Fig. 8.4B,C) (Wenstrup, Larue, and Winer 1994). Axons terminating in the rostral MGB originate in each of the ICC frequency band representations analyzing components in the first three harmonic elements. However, their organization does not follow a tonotopic pattern. Instead, they may be grouped according to their role in echolocation (and possibly social communication). Thus, ICC neurons tuned to CF_2 and CF_3 sonar components terminate in the dorsal and rostral pole nuclei of the middle third of the MGB. Input from frequency representations associated with FM_2 (48–60 kHz) and FM_3 (72–90 kHz) sonar components are very strong and cover much of the rostral third. Their projections appear to interdigitate, and it is these axons that bifurcate extensively. ICC neurons tuned to the FM_4 sonar component (96–120 kHz) terminate in a pattern similar to FM_2 and FM_3 inputs (Wenstrup, unpublished data). Responses to FM_4 frequencies are recorded in the same parts of the rostral MGB as are those to FM_2 and FM_3 frequencies (Olsen 1986; Olsen and Suga 1991a).

Although ICC neurons tuned to the fundamental of CF and FM sonar calls project to the rostral MGB, they terminate in different regions, generally located along the margins of the rostral MGB: the dorsal nucleus, the medial division, and the ventromedial extreme of the rostral pole nucleus (see Fig. 8.4). As described here, these results have significant implications for understanding the neuronal interactions that create combination-sensitive neurons.

The strong ICC input to the dorsal and rostral MGB differs from other mammals in both its organization and extent. In most mammals, only the ventral division receives such strong ICC input. However, some studies in other mammals have reported ICC input to the deep dorsal nucleus (Andersen et al. 1980; Calford and Aitkin 1983). Thus, to the extent that this region in the mustached bat corresponds to the deep dorsal nucleus, the connections are not entirely unconventional.

Inferior collicular input to combination-sensitive regions also originates in the external nucleus and the adjacent pericollicular tegmentum (Wenstrup and Grose 1993). These regions receive input from ICC frequency representations tuned to higher harmonics of the sonar signal (Wenstrup, Larue, and Winer 1994). Thus, combination-sensitive neurons in the MGB receive high-frequency input from several midbrain sources, including direct and indirect projections of the ICC.

Descending projections from the auditory cortex form a major input to combination-sensitive neurons of the rostral MGB. The strongest cortical input originates in layer VI of the dorsal auditory cortex (Wenstrup and Grose 1993, in press), an area examined in many physiological studies (see O'Neill, Chapter 9). Its combination-sensitive neurons are topographically arranged in separate CF/CF and FM-FM areas. The heaviest corticothalamic projection appears to connect MGB and cortical areas having similar functional properties. Thus, deposits of anterograde tracer placed in the dorsolateral extreme of auditory cortex, where neurons respond to FM-FM combinations (O'Neill and Suga 1982; Suga et al. 1983; Suga and Horikawa 1986), label the rostral half of the medial geniculate, where similar FM-FM neurons have been recorded (Olsen 1986). The converse is also true; deposits of retrograde tracer in FM-FM regions of the medial geniculate result in heavy labeling in the dorsolateral auditory cortex (Wenstrup and Grose 1993). A similar conclusion applies to corticothalamic input to CF/CF regions of the MGB.

The caudal part of the thalamic reticular nucleus provides strong input to the dorsal and rostral MGB, (Wenstrup and Grose 1993, in press). It is unclear whether specific thalamic reticular regions project to specific medial geniculate regions.

In the little brown bat (*Myotis lucifugus*), the dorsal division has been implicated in echo–delay-sensitive responses of auditory cortical neurons. Retrograde tracer, placed at cortical sites where delay-sensitive neurons were recorded, labeled neurons in the dorsal division of the medial geniculate body, outside the tonotopic axis of the ventral division (Shannon and Wong 1987). This finding reinforces the view that the dorsal division may be involved in the specialized processing of biosonar information in bats.

5.2.2 Outputs

The primary output is to restricted zones in the dorsolateral auditory cortex (Olsen 1986). These MGB regions appear to project to the appropriate CF/CF or FM-FM cortical combination-sensitive area. For example, tracer deposits at MGB sites showing CF/CF responses target a more lateral region than do deposits in MGB FM-FM regions.

394   Jeffrey J. Wenstrup

5.3 Physiological Properties of Combination-Sensitive Neurons

One of the most distinctive features of the dorsal and rostral MGB in the mustached bat is the presence of large numbers of combination-sensitive neurons (Olsen and Suga 1991a,b). These neurons respond best when signals contain two spectral elements, generally corresponding to the fundamental component from the emitted biosonar pulse and a higher harmonic from the sonar echo (Fig. 8.10). (Recent evidence, however, suggests that some neurons also respond to social communication sounds [Ohlemiller, Kanwal, and Suga 1992]. There is a large body of data concerning the physiological properties of cortical combination-sensitive neurons in the mustached bat (see O'Neill, Chapter 9). This section focuses

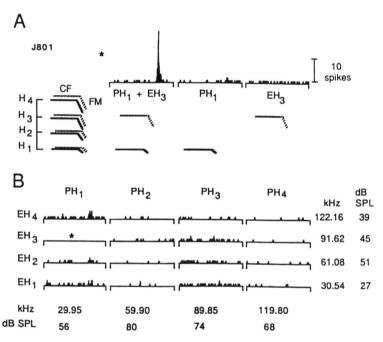

FIGURE 8.10A,B. Selectivity of an FM-FM neuron in the medial geniculate body to combinations of simulated pulse–echo pairs, shown in poststimulus time histograms. (A) For the maximum facilitated response, the neuron requires the fundamental of the pulse (PH$_1$) and the third harmonic of the echo (EH$_3$). The neuron responds poorly to either component presented alone. *Solid lines* indicate sonograms of pulse harmonics; *dashed lines*, sonograms of echo harmonics. (B) Response of same neuron to other pulse–echo pairs. The neuron did not respond well to any other combination. *Numbers at bottom* indicate constant frequency and intensity of test pulse harmonic, while *numbers at right* indicate constant frequency and intensity of the test echo harmonic. (From Olsen and Suga 1991b, copyright © 1991, reprinted by permission of the American Physiological Society.)

on the reports concerning thalamic combination-sensitive neurons and how they may differ from cortical neurons.

Two major types of combination-sensitive neurons are found in the rostral MGB and cortical areas outside the primary auditory cortex. CF/CF neurons respond best to combinations of the CF_1 component and a higher harmonic CF component (CF_n, near 60 or 90 kHz), with the frequencies of the two components being a critical stimulus feature. CF/CF neurons are believed to encode the velocity of sonar targets (Suga et al. 1983; Olsen and Suga 1991a). The second class, FM-FM neurons, responds best to a combination of spectral elements from the fundamental FM sweep and a higher harmonic FM sweep; the delay between the two components is an critical stimulus feature here. FM-FM neurons are believed to encode the distance of sonar targets (O'Neill and Suga 1982; Suga et al. 1983; Olsen and Suga 1991b). Muscimol-induced inactivation of cortical FM-FM regions decreases performance in temporal (i.e., delay) discrimination tasks by behaving bats (Riquimaroux, Gaioni, and Suga 1991).

5.3.1 CF/CF Neurons

Olsen (1986) and Olsen and Suga (1991a) examined the physiological properties of CF/CF neurons in the MGB. Only two CF combinations were recorded: CF_1/CF_2 (70%) and CF_1/CF_3 (30%). MGB CF/CF neurons typically show both a lower threshold and greater response magnitude when presented with the appropriate signal combination than in response to single tones (Fig. 8.11). The degree of facilitation varied among CF/CF neurons by 110%–5000% of the best single tone response. Reductions in threshold were correspondingly varied among units.

The facilitated response of CF/CF neurons was very sharply tuned in frequency, particularly to the higher harmonic. For the neurons shown in Figure 8.11, shifting the frequency of the CF_2 component by ± 1 kHz eliminated the response. The tuning of CF/CF neurons to CF_2 and CF_3 components was at least as sharp as singly tuned neurons responding only to these frequencies, in either the MGB or inferior colliculus. Sharpness of tuning to the CF_1 component was more variable; Olsen and Suga distinguished two populations (Fig. 8.11). Group I CF/CF neurons were relatively sharply tuned to the CF_1 (mean Q_{10dB} = 38). Their CF_1 response was always tuned within 0.7 kHz of the bat's resting frequency, and their CF_1 response area always included the resting frequency. Moreover, the best frequencies of facilitation for the CF_1 and CF_n components were in a near-exact harmonic relationship. Group II neurons differed in each of these features: they had much lower average Q_{10dB} values (9.6); they were tuned to frequencies greater than 0.7 kHz below the CF_1 resting frequency; their response areas never included the resting frequency; their best facilitation frequencies were not in an exact harmonic relationship. These two groups also differed in other ways; group I neurons did not respond

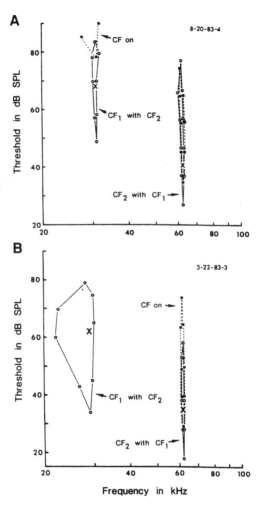

FIGURE 8.11A,B. Frequency tuning curves of CF_1/CF_2, combination-sensitive neurons, representing group I (A) and group II (B). Single tone tuning curves (*filled squares, dashed lines*); facilitative tuning curves (*open circles, solid lines*). Facilitative tuning curves were obtained by fixing the tone for one component at its best facilitative amplitude and frequency (**X**) and obtaining a tuning curve for the other component. (A) Tuning curves for group I CF/CF neuron. These neurons had sharp tuning to the CF_1 component, with the best facilitative frequency near the CF_1 resting frequency. (B) Tuning curves for group II CF/CF neuron. These neurons had broader tuning curves that excluded the CF_1 resting frequency. In both neurons, CF_2 tuning curves were very narrow. (From Olsen and Suga 1991a, copyright © 1991, reprinted by permission of the American Physiological Society.)

well when the two components were presented simultaneously, while group II neurons did.

Olsen and Suga (1991a) pointed out two significant functional aspects of the differing responses of group I and II neurons. First, each group is unresponsive to combinations of CF_1 and CF_n components within the emitted pulse or within echoes. Group I neurons are unresponsive because of their temporal sensitivity; they are inhibited when the two components are presented simultaneously. Group II neurons are unresponsive because of their spectral sensitivity; exact harmonic relationships in the pulse or in the echo do not elicit facilitation. The second feature regards the echolocation conditions under which these neurons are active; group I neurons respond to pulse–echo combinations when the bat is at rest, while group II neurons will respond when the flying bat receives Doppler-shifted CF_n echoes.

CF/CF neurons in the MGB are qualitatively similar to their counterparts in the auditory cortex, although cortical neurons may display a larger degree of facilitation (Olsen and Suga 1991a). It is unknown whether cortical neurons form populations similar to group I and II medial geniculate neurons. Nevertheless, anatomical and physiological evidence strongly suggests that medial geniculate CF/CF neurons are the source of CF/CF responses in the auditory cortex.

5.3.2 FM-FM Neurons

Medial geniculate FM-FM neurons display three response selectivities — frequency, amplitude, and delay tuning — that are well suited to encode target distance by the delay between an emitted FM_1 pulse and a returning higher harmonic FM echo (Olsen and Suga 1991b). Both the frequency and delay tuning distinguish medial geniculate FM-FM neurons from CF/CF neurons. Thus, although nearly all FM-FM neurons respond equally well to combinations of pure tone bursts as they do to FM sweeps, they are clearly tuned to frequencies within the range of the FM_1 and FM_n signal components. Moreover, they are considered to be more sharply tuned to the timing of two pulses than are CF/CF neurons.

Olsen and Suga (1991b) recorded facilitated responses to three FM combinations: FM_1-FM_2 (38%), FM_1-FM_3 (34%), and FM_1-FM_4 (28%). The frequency tuning of medial geniculate FM-FM neurons was much broader than that of CF/CF neurons, further distinguishing these two classes of medial geniculate combination-sensitive neurons (Fig. 8.12). The results of Olsen and Suga (1991b) suggest that FM-FM frequency tuning is level tolerant; facilitative tuning curves do not broaden significantly from about 10 to 50 dB or more above the threshold for facilitation. Above that level, the degree of facilitation often decreases, and many neurons have upper limits on their tuning curves. To elicit the facilitated response, FM

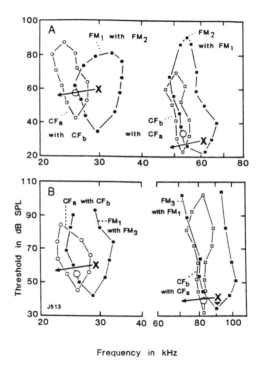

FIGURE 8.12A,B. Facilitative frequency tuning curves of FM_1-FM_2 (A) and FM_1-FM_3 (B) neurons, obtained using tone bursts and FM downsweeps. Facilitative tuning curves defined by tone bursts (*open symbols*) were obtained as described in Fig. 8.11. CF_a and CF_b refer to the test frequencies of tone bursts, where $CF_a <$ CF_b. *Large open circles* indicate best facilitative frequencies and amplitudes using tone burst stimuli. FM facilitative tuning curves (*filled symbols*) and best facilitative frequencies and amplitudes (**X**, *arrow*) are plotted using the initial frequency of the test FM sweep. To elicit a facilitated response, FM sweeps must pass into the tuning curves obtained with tone burst stimuli. Note the broader tuning of FM-FM neurons compared to CF/CF neurons (cf. Fig. 8.11). (From Olsen and Suga 1991b, copyright © 1991, reprinted by permission of the American Physiological Society.)

sweeps must pass through the facilitative tuning curves measured using combinations of tone bursts. The two facilitative tuning curves of FM-FM neurons are not in an exact harmonic relationship; generally, the higher harmonic FM curve is shifted upward. This suggests that FM-FM neurons are designed to respond to the combination of an emitted FM_1 signal and a Doppler-shifted higher harmonic FM echo. Moreover, their broader frequency tuning enables them to respond in a Doppler-tolerant fashion, that is, insensitive to the relative velocity between the bat and its target.

FM-FM neurons are also tuned in amplitude. On average, the facilitated response of FM-FM neurons was greatest about 20 dB above the threshold of their facilitated response. Significantly, both the thresholds and the best amplitudes of the facilitated response of the FM_1 component averaged 10–20 dB greater than those of the higher frequency components. Thus, the appropriate signal requires the combination of a strong FM_1 component with a weaker FM_n component. This suggests that the neurons are primarily responding to the bat's outgoing FM_1 component and the weaker FM_n component in the returning echo.

The sensitivity of FM-FM neurons to echo delay is one of their most significant features, and one closely related to the mustached bat's perception of target distance (Moss and Schnitzler, Chapter 3; Simmons et al., Chapter 4). These neurons typically display a peak response as the FM_n signal is delayed beyond the presentation of the FM_1 component (Fig. 8.13). Neurons can be characterized by their *best delay*, the delay eliciting the strongest facilitated response. In the MGB, FM-FM neurons have best delays ranging from 0 to 23 msec, although most are between 1 and 10 msec. Delay tuning curves are broader among neurons with longer best delays. Delay tuning remains relatively stable with changes in signal amplitude; significant shifts were noted mainly when amplitudes were changed near the lower and upper thresholds of the two signal components.

Olsen and Suga found several differences in stimulus locking and inhibition between medial geniculate FM-FM neurons having short best delays (< 4 msec) versus long best delays (≥ 4 msec). For example, neurons with short best delays respond poorly to single tones or FM sweeps. The latency of their facilitated response to an FM_1-FM_n combination is more closely time locked to the FM_1 signal than to the delayed FM_n signal. Third, they show little evidence of inhibition immediately following the FM_1 pulse, either tested by the introduction of a second FM_1 pulse at variable delays,

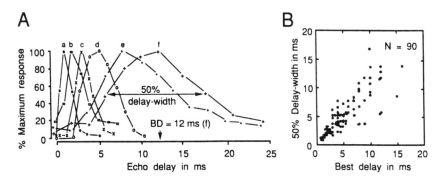

FIGURE 8.13A,B. Delay tuning of FM-FM neurons in the medial geniculate body of the mustached bat. (A) Rate versus delay functions for six representative FM-FM neurons, obtained with FM sweeps at the best facilitative amplitudes and frequencies. Best delays ranged from 0 to 23 msec. (B) Distribution of best delays and widths of delay tuning curves (50% delay width) among FM-FM neurons. Best delays were positively correlated with delay widths. (From Olsen and Suga 1991b, copyright © 1991, reprinted by permission of the American Physiological Society.)

or by the suppression of spontaneous activity. These results suggest that inhibitory mechanisms do not play a large role in the delay-sensitive, facilitated response of such neurons. One interpretation of these results is that excitatory FM_1 input initiates a brief, subthreshold facilitative period, during which the activation of an excitatory FM_n input will cause the discharge of action potentials.

FM-FM neurons with long best delays are significantly different in each of these features, suggesting fundamental differences in the mechanisms underlying their delay sensitivity. First, most such neurons respond to one or both signals presented singly, particularly to the FM_n signal (Fig. 8.14). Second, the latency of the facilitated response to FM-FM combinations is rigidly time locked to the delayed FM_n signal. Third, the FM_1 signal has an inhibitory effect. Thus, about half of these neurons show a reduction in spontaneous activity when presented with an FM_1 signal alone. In these, it appears that an early inhibitory period is triggered by the onset of the fundamental pulse, and is independent of its duration (Fig. 8.14). By examining the effect of a second FM_1 signal presented after the first, Olsen and Suga demonstrated that the second FM_1 signal could either suppress or reset the delay response. For long best-delay neurons, Olsen and Suga hypothesized that facilitation occurs when the excitation produced by a delayed FM_n signal outlasts the inhibitory period evoked by FM_1 onset, and coincides with the long-latency, FM_1-evoked excitation.

The shorter latencies of medial geniculate (compared to cortical) FM-FM neurons (Suga and Horikawa 1986; Olsen and Suga 1991b) and their input to FM-FM regions of the auditory cortex (Olsen 1986) indicate that the MGB neurons are the source of FM-FM neurons in the cortex. Olsen and Suga noted that medial geniculate neurons are very similar to cortical neurons in their delay tuning (i.e., best delays, width of delay tuning curves). However, other features differ. Nearly all medial geniculate neurons (96%) respond as well to combinations of pure tone bursts (within the appropriate frequency bands) as they do to FM sweeps, and most (83%) respond to one or both components when presented individually. In contrast, about 40% of cortical FM-FM neurons do not respond to combinations of pure tone bursts (Taniguchi et al. 1986), and most do not respond to components presented individually (Suga and Horikawa 1986). Furthermore, cortical neurons may show stronger and longer lasting facilitation. These comparisons suggest that additional processing may occur in the auditory cortex, but not with respect to delay tuning.

5.3.3 Organization of Combination-Sensitive Neurons

In the auditory cortex, distinct areas of FM-FM and CF/CF neurons occur. In each, two organizational features have been found. First, each region is topographically organized according to a particular response feature; best

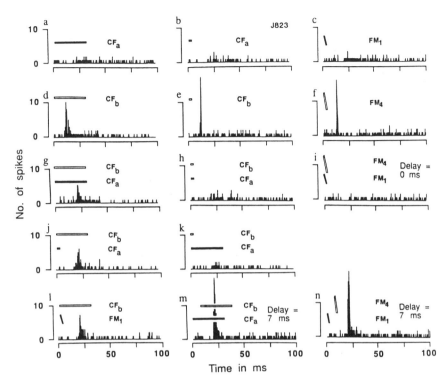

FIGURE 8.14. Inhibition in the response of an FM$_1$-FM$_4$ neuron having long best delay (7 msec). Peristimulus time histograms are shown in response to tone burst (*horizontal bars*) or FM (*oblique bars*) stimuli. *Dark bars* represent stimuli within the frequencies of the FM$_1$ component, while *open bars* are frequencies within the FM$_4$ component. CF$_a$ and CF$_b$ stimuli as in Fig. 8.12. (a–c) FM$_1$ and long or short CF$_a$ stimuli each evoke early inhibition of spontaneous firing, followed by later increased excitability. (d–f) FM$_4$ and long or short CF$_b$ stimuli evoke a short-latency, excitatory, phasic response. (g–l) When combinations of stimuli were presented at 0-msec delay, the neuron only responds if CF$_b$ is sufficiently long to outlast a period of early inhibition evoked by FM$_1$ or CF$_a$ signals (g, j, l). Moreover, the periods of inhibition and excitation are independent of the duration of FM$_1$ or CF$_a$ stimuli (g, j, l). Note that the response under these conditions occurs at longer latency than the response to FM$_4$ or CF$_b$ alone. The best response of this neuron is obtained when the FM$_4$ or CF$_b$ signal is delayed by 7 msec. Olsen and Suga (1991b) concluded that, among long best-delay neurons, delay sensitivity is determined by a period of inhibition followed by period of increased excitability, produced by the onset of the fundamental component. (From Olsen and Suga 1991b, copyright © 1991, reprinted by permission of the American Physiological Society.)

delay for FM-FM neurons and frequency shift for CF/CF neurons. Second, the topographic organization occurs in parallel for each existing signal combination. For example, the best-delay axis of FM_1-FM_2 neurons in a cortical area is aligned with an adjacent group of FM_1-FM_4 neurons, which is in turn aligned with adjacent FM_1-FM_3 neurons (Suga et al. 1983; O'Neill, Chapter 9).

In the MGB, the organization of best-delay and frequency-shift response features is poorly understood. However, CF/CF and FM-FM neurons are segregated, with the former placed more caudally, dorsally, or laterally. Thus, Olsen and Suga (1991a) reported that CF/CF neurons were only recorded in the deep dorsal division of the central one-third of the MGB, whereas FM-FM neurons lay more rostrally in the deep dorsal division. In the architectonic scheme used here, these areas correspond to the dorsal nucleus and the adjacent rostral pole nucleus, both considered parts of the dorsal division.

This agrees well with the distribution of medial geniculate inputs from CF_n and FM_n ICC representations terminating in the dorsal and rostral MGB (Wenstrup 1992b; Wenstrup, Larue, and Winer 1994). As described previously, CF_2 and CF_3 inputs terminate dorsally in the middle one-third of the MGB, while FM_2 and FM_3 (and probably FM_4) inputs terminate more rostrally (see Fig. 8.4). Surprisingly, in neither FM-FM nor CF/CF regions of MGB is there a good correspondence between the distribution of these response properties and ICC input from the fundamental harmonic element (30–24 kHz). The implications of these results are discussed next.

5.4 Mechanisms for Constructing Combination-Sensitive Neurons

Mechanisms for constructing combination-sensitive neurons are of particular interest because these neurons represent one of the best examples of the convergence of information across frequency channels, a process that appears necessary for the analysis of spectrally complex sounds. Although facilitated CF/CF and FM-FM neurons have been recorded in the MGB and auditory cortex, they were not found in the ICC (O'Neill 1985). Thus, one possibility is that such responses arise in the MGB as the result of a convergence of fundamental and higher harmonic inputs from separate parts of the tonotopically organized ICC. To date, however, such convergence has not been supported by anatomical studies.

5.4.1 Connectional Evidence

Tracer studies have failed to demonstrate a significant direct projection from ICC representations of the fundamental sonar component to combination-sensitive areas of the MGB. Wenstrup, Larue, and Winer

(1994) examined the anterograde label resulting from deposits of tritiated leucine at ICC loci responding to frequencies in the FM_1 or CF_1 components; they found very little overlap with the anterograde label obtained in other experiments involving deposits in higher frequency, FM_n or CF_n representations. Wenstrup and Grose (Wenstrup, 1992b, in press), in dual anterograde tracer experiments, placed deposits of wheat germ agglutinin conjugated to horseradish peroxidase (WGA-HRP) into FM_1 or CF_1 representations and deposits of biocytin into FM_2, CF_2, or FM_3 representations. There was little overlap between the projections from the fundamental and the higher harmonic representations in the rostral MGB. In further experiments, retrograde tracers placed in medial geniculate combination-sensitive regions strongly labeled higher harmonic CF_n or FM_n representations in the ICC, but labeled few or no cells in representations tuned to the fundamental sonar component (Wenstrup and Grose 1993, in press). These retrograde experiments also showed little evidence that other parts of the MGB, including the ventral division and other MGB regions receiving low-frequency input, projected to combination-sensitive regions.

These studies raise doubts that ICC neurons tuned to the fundamental sonar component project directly or via another MGB subdivision onto combination-sensitive neurons in the rostral and dorsal MGB. Other possibilities must be investigated. For example, low-frequency input may arrive via an indirect pathway, possibly involving a combination of two or more of the following: the lateral part of the ventral division in the MGB, the thalamic reticular nucleus, or the auditory cortex. Alternatively, combination-sensitive responses may already exist among some neurons providing input to these MGB regions. For example, the external nucleus of the inferior colliculus and the pericollicular tegmentum both provide input to combination-sensitive MGB regions (Wenstrup and Grose 1993), but their physiological properties have not been explored in the mustached bat. Recently, Mittmann and Wenstrup (1994) reported that combination-sensitive neurons occur in the ICC. This finding may explain the surprising lack of CF_1 or FM_1 inputs from the ICC to combination-sensitive regions of the MGB, because the necessary frequency convergence may occur at auditory levels below the MGB. These results suggest that high frequency representations of the ICC supply combination-sensitive response properties to MGB neurons.

5.4.2 Pharmacological Evidence

Preliminary studies by Butman and Suga (1989, 1990; Butman 1992) examined receptor mechanisms involved in the response of FM-FM neurons. For some MGB neurons having long best delays, local application of bicuculline to block $GABA_A$ receptors had the effect of shifting the neuron's best delay from long to short. This suggests that both GABAergic and non-GABAergic FM_1 inputs synapse onto combination-sensitive medial geniculate neurons.

Butman and Suga (1990; Butman 1992) also examined the excitatory amino acid sensitivity of onset and later burst components in the response of delay-sensitive neurons. Local application of APV, an N-methyl-D-aspartate (NMDA) receptor channel antagonist, eliminated the burst response, but left the delay-tuned onset response unaffected. In contrast, CNQX, a non-NMDA glutamate receptor antagonist, eliminated the onset response but left the delay-tuned burst response unaffected. A nonspecific excitatory amino acid antagonist, kyenurenic acid, eliminated both features of the component. Thus, different excitatory amino acid receptor mechanisms mediate different components to the delay-tuned response. It is not clear whether these results apply specifically to short best-delay FM-FM neurons or to the entire population.

5.4.3 Origin of Delay Tuning in Combination-Sensitive Neurons

Delay-tuned FM-FM neurons in the MGB function as coincidence detectors (Suga, Olsen, and Butman 1990; Olsen and Suga 1991b). Facilitation occurs when excitatory FM_1 and FM_n influences overlap temporally. Because the emitted FM_1 signal precedes the echo FM_n signal by several milliseconds (5.8 msec per meter of bat-target distance), the FM_1 excitatory influence must be delayed neurally to coincide with the FM_n excitatory influence.

Several mechanisms have been proposed to explain aspects of the delayed FM_1 excitation. Kuwabara and Suga (1993) reported evidence suggesting that FM_1 neural delays are created at levels below the MGB. Recording from the brachium of the inferior colliculus, presumably among axons exiting the ICC, they found a broad distribution of FM_1 latencies (3.5–15.0 msec) and a restricted distribution of FM_n latencies (3.8–6.5 msec). They concluded that mechanisms below the MGB create FM_1 delays that may account for much of the delay tuning observed among MGB neurons. However, others have reported a broader distribution of latencies among ICC neurons responding to FM_n signals (Hattori and Suga 1989; Park and Pollak 1993; Pollak and Park, Chapter 7). These studies suggest that the distribution of latencies is broad among both FM_1 and FM_n neurons in the ICC, and that delay sensitivity results from the match of appropriately timed FM_1 and FM_n inputs at the level of the MGB.

Physiological differences between short and long best-delay neurons suggest different mechanisms in their creation. Short best-delay neurons show no evidence that inhibition contributes to their delay sensitivity (Olsen and Suga 1991b); these may be constructed by a facilitating convergence that matches FM_1 and FM_n inputs having the appropriate latencies. Because these latencies need to differ by no more than 4 msec for short best-delay neurons, variations in FM_1 and FM_n latencies observed among ICC neurons could easily account for the delay sensitivity.

FM-FM neurons having long best delays seem to require a different mechanism. In physiological (see Fig. 8.14) (Olsen and Suga 1991b) and pharmacological (Butman and Suga 1989) studies, these neurons clearly showed the effects of an FM_1-elicited inhibitory mechanism. The source of the inhibition is not clear. It is unlikely that an intrageniculate inhibitory pathway exists, because so few intrinsic GABAergic neurons can be found in the MGB (Vater, Kössl, and Horn 1992; Winer, Wenstrup, and Larue 1992), and because there is little evidence of a projection to the rostral and dorsal MGB from other low-frequency MGB regions (Wenstrup and Grose 1993). If this inhibition occurs in the MGB, one possibility is that it originates from the thalamic reticular nucleus, a nucleus that is GABAergic in the mustached bat (Winer, Wenstrup, and Larue 1992), contains neurons tuned between 24 and 30 kHz (Olsen, unpublished data; Wenstrup, unpublished data), and projects to FM-FM regions of the MGB (Olsen 1986; Wenstrup and Grose 1993). In these long best-delay neurons, delay sensitivity may depend on different FM_1 input latencies, on the kinetics of GABA-activated membrane channels, or on both.

The selectivity of CF/CF neurons for the delay between the CF_1 and CF_n components is considered to be relatively broad (Olsen and Suga 1991a), perhaps reflecting differences in the mechanisms which form CF/CF versus FM-FM neurons (Park and Pollak 1993). However, delay sensitivity of CF/CF neurons is generally tested using simulated biosonar signals with long CF components or relatively long (30 msec) tone bursts, not the brief tone bursts or FM sweeps used to evaluate delay sensitivity in FM-FM neurons. Such long CF stimuli, although biologically appropriate, probably broaden the apparent delay tuning of CF/CF neurons.

For each of the mechanisms thought to underlie delay tuning, the assumption has been that MGB neurons are the coincidence detectors. However, the lack of FM1 input to combination-sensitive MGB regions (Wenstrup and Grose 1993, in press) and the finding of combination-sensitive neurons in the inferior colliculus (Mittmann and Wenstrup 1994) suggest that integration may occur lower than the MGB, and that mechanisms that create delay lines may operate on brain stem inputs to the ICC and on the ICC itself. Thus, several questions remain concerning the mechanisms of the delay-tuned facilitation. These include the site(s) at which delay-tuned responses are constructed, the sources of FM_1 input, and the mechanism(s) by which FM_1 input is delayed relative to FM_n inputs.

6. The Thalamic Reticular Nucleus

The thalamic reticular nucleus in mammals is a sheet of neurons located along the lateral and rostral margins of the thalamus, just medial to the internal capsule. It receives major inputs from nuclei of the dorsal thalamus and from corresponding cortical regions, and its main output is to the nuclei of the dorsal thalamus (Scheibel and Scheibel 1966; Jones 1975). The thalamic reticular nucleus is composed of sectors related to specific dorsal thalamic nuclei (Jones 1975). All or nearly all the neurons are GABAergic (Houser et al. 1980), thus providing a negative feedback circuit to regulate the activity of specific thalamic relay neurons. It has been implicated in several aspects of dorsal thalamic activity (see review by Shosaku et al. 1989): (1) in the generation of electroencephalogram oscillations that appear in states of reduced alertness, (2) in setting excitation levels of thalamic relay neurons, and (3) in postexcitatory and surround inhibition. In cats, thalamic reticular neurons of the auditory sector display frequent bursts of spontaneous discharge. In response to auditory stimuli, they have longer latencies and broader and more complex frequency tuning than neurons in the medial geniculate body (Simm et al. 1990).

In the mustached bat, the thalamic reticular nucleus may provide the strongest GABAergic influence on neurons of the MGB, including combination-sensitive neurons. In view of its potential role in auditory thalamic signal processing in bats, this review considers its structure and possible functions.

6.1 Anatomy

The thalamic reticular nucleus in bats is well developed, although its location and shape differ somewhat from other species. Caudally, the thalamic reticular nucleus is located just ventral to the rostral MGB, lying directly over the internal capsule and rostral cerebral peduncle (see Fig. 8.1H). Its shape is unconventional at this point, being several cell layers thick, and the neurons are scattered among fascicles of the auditory radiation. More rostrally, the nucleus assumes its more conventional appearance, a laminated shell of neurons placed along the lateral and rostral borders of the thalamus (Fig. 8.15). In the mustached bat, thalamic reticular neurons are GABAergic (Fig. 8.15) (Winer, Wenstrup, and Larue 1992) and are intensely labeled by antibodies to parvalbumin (see Fig. 8.1H).

Connectional evidence suggests that the caudal, unconventionally shaped part of the thalamic reticular nucleus is the auditory sector (Wenstrup and Grose 1993, in press). Thus, the caudal region alone is labeled by medial geniculate deposits of retrograde tracer. Whether there is a finer organization of connections with the MGB is unclear, although preliminary evidence suggests that the ventral division of the MGB receives input from somewhat different parts than does the dorsal division (Wenstrup and Grose 1993).

In bats, inputs to the thalamic reticular nucleus have not been examined in detail, although Olsen (1986) reported a projection from the MGB. In other species, both the MGB and auditory cortical fields provide input to the thalamic reticular nucleus (Jones 1975; Conley, Kupersmith, and Diamond 1991). The connections are of particular interest in the mustached bat, because the nucleus is a possible source of low-frequency input to combination-sensitive neurons of the MGB.

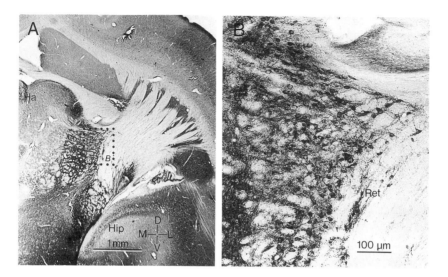

FIGURE 8.15A,B. GAD-immunolabeling of the thalamic reticular nucleus. (A) Overview of the rostral thalamus in transverse section. *Dotted square* frames the region shown at higher power in (B). (B) GAD-immunopositive neurons of the thalamic reticular nucleus (Ret) are evident. (From Winer, Wenstrup, and Larue, copyright © 1992, reprinted by permission of Wiley-Liss, a division of John Wiley and Sons, Inc.)

6.2 Physiological Responses

Very little is known about the physiology of neurons in the thalamic reticular nucleus of bats. Unpublished studies in the mustached bat (Olsen; Wenstrup) have recorded auditory responses, predominantly to frequencies in the 24- to 30-kHz band of the fundamental sonar component. These results, while preliminary, are consistent with the possible role of thalamic reticular neurons as a source of low-frequency input to combination-sensitive regions of the MGB. Such results suggest that the thalamic reticular nucleus may be involved directly in signal processing mechanisms of the MGB rather than in a modulatory role alone. Further studies clearly are necessary to address the role of this nucleus in signal processing by medial geniculate neurons in bats.

7. Summary and Conclusions

The mustached bat's medial geniculate body is composed of three major divisions distinguished by neuronal populations, connections, and physiological responses. The ventral division, part of the ascending tonotopic system, contains sharply tuned, tonotopically organized neurons. The dorsal division contains several subdivisions; most noteworthy are the dorsal and rostral parts whose combination-sensitive neurons are specialized for the analysis of species-specific, complex sounds. Suprageniculate neurons, also of the dorsal division, participate in separate pathways involving the frontal cortex and the superior colliculus. The medial division, although clearly different from the others, is not well understood. The studies in the mustached bat demonstrate in a striking way what different functional roles are played by the parallel subsystems that involve medial geniculate nuclei.

These subsystems — their mechanisms and their behavioral roles — need further study. For instance, the functional properties of the suprageniculate–frontal cortex system are poorly understood. What are the physiological responses within these areas, and how do they contribute to the acoustic behavior of the mustached bat? What processing of biosonar information occurs in the ventral division? Do frequencies in the echoes of FM components participate in the analysis of target elevation and azimuth? What mechanisms are responsible for the frequency convergence and temporal selectivity of combination-sensitive neurons? If combination-sensitive neurons are constructed at levels below the MGB, how are these responses modified in the MGB by inputs from the auditory cortex and thalamic reticular nucleus.

Necessarily, this review has focused on the mustached bat. A major gap is our lack of knowledge about medial geniculate processing in other bat species. Even the limited data from other species demonstrate sharp contrasts with the mustached bat, for example, in GABAergic inhibition and in responses encoding target distance. Basic studies of structure and function, as well as specific investigations of processing mechanisms in these other bats are likely to demonstrate different but equally interesting mechanisms of complex signal processing and neural correlates of echolocation.

Most medial geniculate subdivisions in bats correspond well to those described in other species, yet each shows modifications that may be related to the demanding acoustic behavior of bats. Even the tonotopic ventral division shows these effects in the mustached bat, whereby certain frequency bands within the biosonar signal may be underrepresented. Other parts, including the suprageniculate nucleus and rostral medial geniculate body may have undergone significantly greater modification, with different connections and response properties. These observations suggest how diverse are the roles of cell groups in the medial geniculate body of different species. However, much further work on bats is needed to understand whether these modifications are specific adaptations or part of broader phylogenetic trends.

Acknowledgments. The author thanks E. Covey for sharing unpublished data and Ms. Carol Grose for assistance in experiments and preparing the figures. This work was supported by the National Institute for Deafness and Other Communication Disorders (DC00937).

Abbreviations

Anatomical Names

ALD	anterolateral division of the inferior colliculus
BIC	brachium of the inferior colliculus
CG	central gray
CP	cerebral peduncle
D	dorsal nucleus *or* dorsal division of the medial geniculate body
DC	dorsal cortex of inferior colliculus
DNLL	dorsal nucleus of the lateral lemniscus
DPD	dorsoposterior division of the inferior colliculus
DS	superficial dorsal nucleus of the medial geniculate body
Ex	external nucleus of the inferior colliculus
Ha	habenula
Hip	hippocampus
ICC	central nucleus of the inferior colliculus
M	medial division of the medial geniculate body
MD	medial division of the inferior colliculus
MGB	medial geniculate body
MRF	mesencephalic reticular formation
PL	posterior limitans nucleus
Po	posterior thalamic nuclear group
Pt	pretectum
Pyr	pyramid
RP	rostral pole nucleus of the medial geniculate body
Rt, Ret	thalamic reticular nucleus
SC	superior colliculus
Sg	suprageniculate nucleus of the medial geniculate body
SN	substantia nigra
V	ventral division of the medial geniculate body
Vl	lateral part of the ventral division of the medial geniculate body
Vm	medial part of the ventral division of the medial geniculate body
VP	ventroposterior nucleus
ZI	zona incerta

Planes of Section

D	dorsal
L	lateral
M	medial
V	ventral

Other Abbreviations

CF	constant frequency
CF/CF	combination-sensitive neuron responding to CF components of the sonar signal
CF_a, CF_b	tone burst test frequencies used for FM-FM neurons
CF_n	nth harmonic of constant-frequency biosonar component
FM	frequency modulated
FM-FM	combination-sensitive neuron responding to FM components of the sonar signal
FM_n	nth harmonic of frequency modulated biosonar component
GABA	γ-aminobutyric acid
GAD	glutamic acid decarboxylase
HRP	horseradish peroxidase
$Q_{10\ dB}$	tuning sharpness expressed as best frequency divided by the bandwidth 10 decibels above threshold
$Q_{30\ dB}$	tuning sharpness expressed as best frequency divided by the bandwidth 30 decibels above threshold
Q_{min}	tuning sharpness expressed as best frequency divided by the maximum bandwidth of the tuning curve
WGA-HRP	wheat germ agglutinin conjugated to horseradish peroxidase

References

Aitkin LM, Webster WR (1972) Medial geniculate body of the cat: organization and responses to tonal stimuli of neurons in ventral division. J Neurophysiol (Bethesda) 35:365–380.

Andersen RA, Roth GL, Aitkin LM, Merzenich MM (1980) The efferent projections of the central nucleus and the pericentral nucleus of the inferior colliculus in the cat. J Comp Neurol 194:649–662.

Andreeva NG, Lang TT (1977) Evoked responses of the superior olive to amplitude-modulated signals. Neurosci Behav Physiol 8:306–310.

Aoki E, Semba R, Keino H, Kato K, Kashiwamata S (1988) Glycine-like immuno-reactivity in the rat auditory pathway. Brain Res 442:63–71.

Ayrapet'yants ES, Konstantinov AI (1974) Echolocation in Nature. Arlington, VA: Joint Publications Research Service.

Baron G (1974) Differential phylogenetic development of the acoustic nuclei among chiroptera. Brain Behav Evol 9:7–40.

Bodenhamer RD, Pollak GD (1983) Response characteristics of single units in the inferior colliculus of mustache bats to sinusoidally frequency modulated signals. J Comp Physiol 153:67–79.

Butman JA (1992) Inhibitory and excitatory mechanisms of coincidence detection in delay tuned neurons of the mustache bats. In: Proceedings of the 3d International Congress in Neuroethology. McGill University, Montreal.

Butman JA, Suga N (1989) Bicuculline modifies the delay-tuning of FM-FM neurons in the mustached bat. Soc Neurosci Abstr 15:1293.

Butman JA, Suga N (1990) NMDA receptors are essential for delay-dependent facilitation in FM-FM neurons in the mustached bat. Soc Neurosci Abstr 16:795.

Calford MB (1983) The parcellation of the medial geniculate body of the cat defined by the auditory response properties of single units. J Neurosci 3:2350-2364.

Calford MB, Aitkin LM (1983) Ascending projections to the medial geniculate body of the cat: evidence for multiple, parallel auditory pathways through thalamus. J Neurosci 3:2365-2380.

Casseday JH, Pollak GD (1988) Parallel auditory pathways: I. Structure and connections. In: Nachtigall PE, Moore PWB (eds) Animal Sonar: Processes and Performance. New York: Plenum, pp. 169-196.

Casseday JH, Kobler JB, Isbey SF, Covey E (1989) Central acoustic tract in an echolocating bat: an extralemniscal auditory pathway to the thalamus. J Comp Neurol 287:247-259.

Celio MR (1990) Calbindin D-28k and parvalbumin in the rat nervous system. Neuroscience 35:375-475.

Clarey JC, Barone P, Imig TJ (1992) Physiology of thalamus and cortex. In: Popper AN, Fay RR (eds) Springer Handbook of Auditory Research, Vol. 2, The Mammalian Auditory Pathway: Neurophysiology. New York: Springer-Verlag, pp. 232-334.

Clerici WJ, Coleman JR (1990) Anatomy of the rat medial geniculate body: I. Cytoarchitecture, myeloarchitecture and neocortical connectivity. J Comp Neurol 297:14-31.

Clerici WJ, McDonald AJ, Thompson R, Coleman JR (1990) Anatomy of the rat medial geniculate body: II. Dendritic morphology. J Comp Neurol 297:32-54.

Conley M, Kupersmith AC, Diamond IT (1991) The organization of projections from subdivisions of the auditory cortex and thalamus to the auditory sector of the thalamic reticular nucleus in *Galago*. Eur J Neurosci 3:1089-1103.

Covey E, Hall WC, Kobler JB (1987) Subcortical connections of the superior colliculus in the mustache bat, *Pteronotus parnellii*. J Comp Neurol 263:179-197.

Edeline J-M, Weinberger NM (1991) Subcortical adaptive filtering in the auditory system: associative receptive field plasticity in the dorsal medial geniculate body. Behav Neurosci 105:154-175.

Fitzpatrick DC, Kanwal JS, Butman JA, Suga N (1993) Combination-sensitive neurons in the primary auditory cortex of the mustached bat. J Neurosci 13:931-940.

Frisina RD, O'Neill WE, Zettel ML (1989) Functional organization of mustached bat inferior colliculus: II. Connections of the FM_2 region. J Comp Neurol 284:85-107.

Hashikawa T, Rausell E, Molinari M, Jones EG (1991) Parvalbumin- and calbindin-containing neurons in the monkey medial geniculate complex: differential distribution and cortical layer specific projections. Brain Res 544:335-341.

Hattori T, Suga N (1989) Delay lines in the inferior colliculus of the mustached bat. Soc Neurosci Abstr 15:1293.

Henkel CK (1983) Evidence of sub-collicular auditory projections to the medial geniculate nucleus in the cat: an autoradiographic and horseradish peroxidase study. Brain Res 259:21-30.

Houser CR, Vaughn JE, Barber RP, Roberts E (1980) GABA neurons are the major cell type of the nucleus reticularis thalami. Brain Res 200:345-354.

Jenkins WM, Masterton RB (1982) Sound localization: effects of unilateral lesions in central auditory system. J Neurophysiol (Bethesda) 47:987–1016.

Jenkins WM, Merzenich MM (1984) Role of cat primary auditory cortex for sound-localization behavior. J Neurophysiol (Bethesda) 52:819–847.

Jones EG (1975) Some aspects of the organization of the thalamic reticular complex. J Comp Neurol 162:285–308.

Jones EG, Powell TPS (1971) An analysis of the posterior group of thalamic nuclei on the basis of its afferent connections. J Comp Neurol 143:185–216.

Kavanagh GL, Kelly JB (1987) Contribution of auditory cortex to sound localization by the ferret (*Mustela putorius*). J Neurophysiol (Bethesda) 57:1746–1766.

Kobler JB, Isbey SF, Casseday JH (1987) Auditory pathways to the frontal cortex of the mustache bat, *Pternontus parnellii*. Science 236:824–826.

Kudo M, Niimi K (1978) Ascending projections of the inferior colliculus onto the medial geniculate body in the cat studied by anterograde and retrograde tracing techniques. Brain Res 155:113–117.

Kudo M, Niimi K (1980) Ascending projections of the inferior colliculus in the cat: an autoradiographic study. J Comp Neurol 191:545–556.

Kuwabara N, Suga N (1993) Delay lines and amplitude selectivity are created in subthalamic auditory nuclei: the brachium of the inferior colliculus of the mustached bat. J Neurophysiol (Bethesda) 69:1713–1724.

Majorossy K, Kiss A (1976) Specific patterns of neuron arrangement and of synaptic articulation in the medial geniculate body. Exp Brain Res 26:1–17.

Mitani A, Itoh K, Mizuno N (1987) Distribution and size of thalamic neurons projecting to layer I of the auditory cortical fields of the cat compared to those projecting to layer IV. J Comp Neurol 257:105–121.

Mittmann DH, Wenstrup JJ (1994) Combination-sensitive neurons in the inferior colliculus of the mustached bat. In: Proceedings of the 17th Midwinter Meeting of the Association for Research in Otolaryngology, p 93. Association for Research in Otolaryngology, Des Moines.

Morest DK (1964) The neuronal architecture of the medial geniculate body of the cat. J Anat (Lond) 98:611–630.

Morest DK (1965) The laminar structure of the medial geniculate body of the cat. J Anat (Lond) 99:143–160.

Morest DK (1975) Synaptic relationships of golgi type II cells in the medial geniculate body of the cat. J Comp Neurol 162:157–193.

Morest DK, Winer JA (1986) The comparative anatomy of neurons: homologous neurons in the medial geniculate body of the opossum and the cat. Adv Anat Embryol Cell Biol 97:1–96.

Niimi K, Naito F (1974) Cortical projections of the medical geniculate body in the cat. Exp Brain Res 19:326–342.

Ohlemiller KK, Kanwal JS, Suga N (1992) Responses of cortical and thalamic FM-FM and CF/CF neurons of the mustached bat to species-specific communication sounds. Soc Neurosci Abstr 18:883.

Oliver DL (1982) A Golgi study of the medial geniculate body in the tree shrew (*Tupaia glis*). J Comp Neurol 209:1–16.

Olsen JF (1986) Processing of biosonar information by the medial geniculate body of the mustached bat, *Pteronotus parnellii*. Ph.D. dissertation, Washington University, St. Louis, MO.

Olsen JF, Rauschecker JP (1992) Medial geniculate neurons in the squirrel monkey sensitive to combinations of components in a species-specific vocalization. Soc Neurosci Abstr 18:883.

Olsen JF, Suga N (1991a) Combination-sensitive neurons in the medial geniculate body of the mustached bat: encoding of relative velocity information. J Neurophysiol (Bethesda) 65:1254–1274.

Olsen JF, Suga N (1991b) Combination-sensitive neurons in the medial geniculate body of the mustached bat: encoding of target range information. J Neurophysiol (Bethesda) 65:1275–1296.

O'Neill WE (1985) Responses to pure tones and linear FM components of the CF-FM biosonar signal by single units in the inferior colliculus of the mustached bat. J Comp Physiol A 157:797–815.

O'Neill WE, Suga N (1982) Encoding of target range and its representation in the auditory cortex of the mustached bat. J Neurosci 2:17–31.

Papez JW (1929) Central acoustic tract in cat and man. Anat Rec 42:60.

Park TJ, Pollak GD (1993) GABA shapes a topographic organization of response latency in the mustache bat's inferior colliculus. J Neurosci 13:5172–5187.

Pollak GD, Bodenhamer RD (1981) Specialized characteristics of single units in inferior colliculus of mustache bat: frequency representation, tuning, and discharge patterns. J Neurophysiol (Bethesda) 46:605–620.

Riquimaroux H, Gaioni SJ, Suga N (1991) Cortical computational maps control auditory perception. Science 251:565–568.

Rodrigues-Dagaeff C, Simm G, de Ribaupierre Y, Villa A, de Ribaupierre F, Rouiller EM (1989) Functional organization of the ventral division of the medial geniculate body of the cat: evidence for a rostro-caudal gradient of response properties and cortical projections. Hear Res 39:103–126.

Rouiller EM, de Ribaupierre F (1985) Origin of afferents to physiologically defined regions of the medial geniculate body of the cat: ventral and dorsal divisions. Hear Res 19:97–114.

Rouiller EM, Rodrigues-Dagaeff C, Simm G, DeRibaupierre Y, Villa A, DeRibaupierre F (1989) Functional organization of the medial division of the medial geniculate body of the cat: tonotopic organization, spatial distribution of response properties and cortical connections. Hear Res 39:127–142.

Rouiller EM, Capt M, Hornung JP, Streit P (1990) Correlation between regional changes in the distributions of GABA-containing neurons and unit response properties in the medial geniculate body of the cat. Hear Res 49:249–258.

Scheibel ME, Scheibel AB (1966) The organization of the nucleus reticularis thalami: a Golgi study. Brain Res 1:43–62.

Schuller G (1979) Coding of small sinusoidal frequency and amplitude modulations in the inferior colliculus of 'CF-FM' bat, *Rhinolophus ferrumequinum*. Exp Brain Res 34:117–132.

Schweizer H (1981) The connections of the inferior colliculus and the organization of the brainstem auditory system in the greater horseshoe bat (*Rhinolophus ferrumequinum*). J Comp Neurol 201:25–49.

Shannon S, Wong D (1987) Interconnections between the medial geniculate body and the auditory cortex in an FM bat. Soc Neurosci Abstr 13:1469.

Shosaku A, Kayama Y, Sumitomo I, Sugitani M, Iwama K (1989) Analysis of recurrent inhibitory circuit in rat thalamus: neurophysiology of the thalamic reticular nucleus. Prog Neurobiol 32:77–102.

Simm GM, de Ribaupierre F, de Ribaupierre Y, Rouiller EM (1990) Discharge properties of single units in auditory part of reticular nucleus of thalamus in cat. J Neurophysiol (Bethesda) 63:1010–1021.

Sousa-Pinto A (1973) Cortical projections of the medial geniculate body of the cat. Adv Anat Embryol Cell Biol 48:1–42.

Suga N, Horikawa J (1986) Multiple time axes for representation of echo delays in the auditory cortex of the mustached bat. J Neurophysiol (Bethesda) 55:776–805.

Suga N, Jen PH-S (1977) Further studies on the peripheral auditory system of the CF-FM bats specialized for fine frequency analysis of Doppler-shifted echoes. J Exp Biol 69:207–232.

Suga N, Manabe T (1982) Neural basis of amplitude-spectrum representation in auditory cortex of the mustached bat. J Neurophysiol (Bethesda) 47:225–254.

Suga N, Olsen JF, Butman JA (1990) Specialized subsystems for processing biologically important complex sounds: cross-correlation analysis for ranging in the bat's brain. Cold Spring Harbor Symp Quant Biol 55:585–597.

Suga N, Tsuzuki K (1985) Inhibition and level-tolerant frequency tuning in the auditory cortex of the mustached bat. J Neurophysiol (Bethesda) 53:1109–1145.

Suga N, Simmons JA, Jen PH-S (1975) Peripheral specialization for fine analysis of doppler-shifted echoes in the auditory system of the "CF-FM" bat *Pteronotus parnellii*. J Exp Biol 63:161–192.

Suga N, O'Neill WE, Kujirai K, Manabe T (1983) Specificity of combination-sensitive neurons for processing of complex biosonar signals in auditory cortex of the mustached bat. J Neurophysiol (Bethesda) 49:1573–1626.

Takahashi TT, Carr CE, Brecha N, Konishi M (1993) Calcium binding protein-like immunoreactivity labels the terminal field of nucleus laminaris of the barn owl. J Neurosci 7:1843–1856.

Taniguchi I, Niwa H, Wong D, Suga N (1986) Response properties of FM-FM combination-sensitive neurons in the auditory cortex of the mustached bat. J Comp Physiol A 159:331–337.

Vasil'ev AG, Andreeva NG (1972) Characteristics of electrical responses by the medial geniculate bodies in vespertilionidae and rhinolophidae to ultrasonic stimuli with different frequencies. Neurophysiology 3:104–109.

Vater M, Braun K (1994) Parvalbumin, calbindin D-28k, and calretinin immunoreactivity in the ascending auditory pathway of horseshoe bats. J Comp Neurol 341:534–558.

Vater M, Kössl M, Horn AKE (1992) GAD- and GABA-immunoreactivity in the ascending auditory pathway of horseshoe and mustached bats. J Comp Neurol 325:183–206.

Wenstrup JJ (1992a) Monaural and binaural regions of the mustached bat's inferior colliculus project differently to targets in the medial geniculate body and pons. In: Proceedings of the 16th Midwinter Meeting of the Association for Research in Otolargnology, p. 77.

Wenstrup JJ (1992b) Inferior colliculus projections to the medial geniculate body: a study of the anatomical basis of combination-sensitive neurons in the mustached bat. Soc Neurosci Abstr 18:1039.

Wenstrup JJ, Grose CD (1993) Inputs to combination-sensitive neurons in the medial geniculate body of the mustached bat. Soc Neurosci Abstr 19:1426.

Wenstrup JJ, Grose CD Inputs to combination-sensitive neurons in the medial geniculate body of the mustached bat: the missing fundamental. J Neurosci (in press).

Wenstrup JJ, Ross LS, Pollak GD (1986) Binaural response organization within a frequency-band representation of the inferior colliculus: implications for sound localization. J Neurosci 6:962–973.

Wenstrup JJ, Larue DT, Winer JA (1994) Projections of physiologically defined subdivisions of the inferior colliculus in the mustached bat: targets in the medial geniculate body and extrathalamic nuclei. J Comp Neurol 346:207–236.

Wenthold RJ, Hunter C (1990) Immunocytochemistry of glycine and glycine receptors in the central auditory system. In: Ottersen OP, Storm-Mathisen J (eds) Glycine Neurotransmission. Chichester: Wiley, pp. 391–416.

Whitfield IC, Purser D (1972) Microelectrode study of the medial geniculate body in unanaesthetized free-moving cats. Brain Behav Evol 6:311–322.

Winer JA (1984) The human medial geniculate body. Hear Res 15:225–247.

Winer JA (1985) The medial geniculate body of the cat. Adv Anat Embryol Cell Biol 86:1–98.

Winer JA (1992) The functional architecture of the medial geniculate body and the primary auditory cortex. In: Webster DB, Popper AN, Fay RR (eds) Springer Handbook of Auditory Research, Vol. 1, The Mammalian Auditory Pathway: Neuroanatomy. New York: Springer-Verlag, pp. 222–409.

Winer JA, Larue DT (1988) Anatomy of glutamic acid decarboxylase (GAD) immunoreactive neurons and axons in the rat medial geniculate body. J Comp Neurol 278:47–68.

Winer JA, Morest DK (1983) The medial division of the medial geniculate body of the cat: implications for thalamic organization. J Neurosci 3:2629–2651.

Winer JA, Wenstrup JJ (1994a) Cytoarchitecture of the medial geniculate body in the mustached bat (Pteronotus parnellii). J Comp Neurol (346:161–182).

Winer JA, Wenstrup JJ (1994b) The neurons of the medial geniculate body in the mustached bat (Pteronotus parnellii). J Comp Neurol (346:183–206).

Winer JA, Morest DK, Diamond IT (1988) A cytoarchitectonic atlas of the medial geniculate body of the opossum, Didelphys virginiana, with a comment on the posterior intralaminar nuclei of the thalamus. J Comp Neurol 274:422–448.

Winer JA, Wenstrup JJ, Larue DT (1992) Patterns of GABAergic immunoreactivity define subdivisions of the mustached bat's medial geniculate body. J Comp Neurol 319:172–190.

Yang L, Pollak GD, Resler C (1992) GABAergic circuits sharpen tuning curves and modify response properties in the mustache bat inferior colliculus. J Neurophysiol (Bethesda) 68:1760–1774.

Zettel ML, Carr CE, O'Neill WE (1991) Calbindin-like immunoreactivity in the central auditory system of the mustached bat, Pteronotus parnellii. J Comp Neurol 313:1–16.

Zook JM, Winer JA, Pollak GD, Bodenhamer RD (1985) Topology of the central nucleus of the mustache bat's inferior colliculus: correlation of single unit response properties and neuronal architecture. J Comp Neurol 231:530–546.

9

The Bat Auditory Cortex

Wᴵʟʟɪᴀᴍ E. O'Nᴇɪʟʟ

1. Introduction

Situated as it is at the top of the hierarchy of nuclei comprising the auditory pathway, the auditory cortex should provide a wealth of information about higher order analysis of acoustic signals. The auditory cortex in bats is arguably the most intensively studied and best understood of all mammals. One species in particular, the mustached bat *Pteronotus parnelli*, has provided a wealth of detailed information about neuronal specialization and cortical organization. Despite these advances, many auditory neuroscientists tend to dismiss bat auditory neurobiology as "out of the mainstream" because they view bats as highly specialized mammals, with only one form of acoustic behavior, echolocation (but see Pollak, Winer, and O'Neill, Chapter 10). Some authors (e.g., Clarey, Barone, and Imig 1992) have questioned the applicability of the bat model of cortical organization because of the perception that bat cortex is highly specialized to analyze only the few stereotyped sounds used in echolocation. Bat researchers have inadvertently reinforced this bias by focusing their efforts only on echolocation, while ignoring the rich acoustic signal structures that bats use for communication. Only recently has there been any attention paid to the processing of the astonishingly rich variety of communication sounds used by colonial bats in social interactions (Kanwal, Ohlemiller and Suga 1993; Kanwal et al. 1994; Ohlemiller, Kanwal, and Suga 1993).

Just as there is no one typical rodent, there is no one stereotypical bat. By taking advantage of the diversity of the chiropteran order, one might find clues to the fundamental "plan" of cortical organization, if there is such a thing. Indeed, this chapter shows that, while cortical neurons may have similar properties in different bats, these neurons are not organized into a single "bat-typical" cortical pattern. That cortical organization can differ markedly even within a single mammalian order sharing a specialized acoustic behavior is a sobering discovery. It implies that there may not be an archetypal model of mammalian auditory forebrain organization upon which we can base theories about the neural bases of auditory perception.

416

Unfortunately, only 5 of more than 900 known bat species representing only four genera have been studied at the single-unit electrophysiological level, although anatomical studies have been carried out on the cortices of some other species. All the species in which the cortex has been studied are aerial insectivores: there are so far no studies published on other microchiropteran bats with different feeding habits, let alone their nonecholocating megachiropteran cousins. Generalizations from this small sample of chiropteran life should be judged carefully, indeed.

This chapter by necessity focuses solely on these five species, with incidental references to a few others as needed. These species are divided bioacoustically into two groups, the "CF/FM" and "FM" bats, based on their biosonar signals (see list of abbreviations). These two groups express rather divergent adaptations for hunting flying insects in very different habitats (see Fenton, Chapter 2). In very general terms, FM bats are known to hunt in open areas, relatively free from the acoustic clutter of vegetation. By contrast, CF/FM bats generally hunt in wooded habitats near vegetation, and it is thought that their more complex sonar signals reflect acoustic adaptations to permit them to operate in high levels of clutter (Simmons, Fenton, and O'Farrell 1979).

Regarding the auditory cortex, the best studied CF/FM bat is the mustached bat, a common neotropical species belonging to the family Mormoopidae. The best-studied FM bat is the little brown bat, *Myotis lucifugus*, a common species found throughout most of temperate North America, belonging to the largest family of "microbats," the Vespertilionidae (see Fenton, Chapter 2). Studies of two other CF/FM bats, the horseshoe bats *Rhinolophus ferrumequinum* and *R. rouxi*, and most recently, the FM bat *Eptesicus fuscus*, have provided contrasting views of cortical organization that suggest that there is no single typical plan common to all bats.

2. Defining the Auditory Cortex

A thorough definition of the boundaries of any species' auditory cortex requires detailed anatomical study of cytoarchitecture and the pattern of thalamocortical connections, combined with electrophysiological mapping of cortical cell responses to acoustic stimulation. Unfortunately, no definitive cytoarchitectonic studies have yet been published on any of the five bats mentioned here, and there have been only a few attempts at defining cortex by the pattern of thalamocortical afferents (Shannon and Wong 1987; Wenstrup and Grose 1993; see Wenstrup, Chapter 8).

We are currently forced to rely mainly on neurophysiologically defined maps of the cortex. This technique has the limitation that only those cortical areas containing cells responsive to arbitrary search stimuli are discovered and characterized. Because such stimuli might not elicit responses from cells with complex response properties, one can never rule out the possibility that cortical boundaries are underestimated. Thus, it is not surprising to find a rough correlation between the number of studies published about a particular species and the total known area of its auditory cortex.

2.1 Cytoarchitecture

The cortex in all bats studied is predominantly lissencephalic, and it therefore has no pattern of gyri and sulci by which one can immediately recognize regions of neocortex. The temporal cortex in most bats is bounded dorsally only by a shallow groove along which blood sinuses are found (Henson 1970); only in the mustached bat does one see a prominent fissure homologous to the Sylvian sulcus (Fig. 9.1A–C). Studies of cortical cytoarchitecture have concluded that the chiropteran cortex is primitive (i.e., phylogenetically old), resembling that of insectivores (Sanides 1972; Sanides and Sanides 1974; Ferrer 1987; Fitzpatrick and Henson 1994). Like the neocortex in primitive mammals, the cortex in bats is not strongly laminated (Fig. 9.1B,D). In mustached bat, nonauditory areas in the temporal cortex range in thickness from 600 to 800 μm. However, the auditory cortex is about 900–1000 μm thick (Fig. 9.1B). In both *Rhinolophus* and *Pteronotus*, fibers within the auditory cortex are more heavily myelinated than that of surrounding cortex (Fig. 9.1C).

Recent studies employing the metabolic marker [³H]2-deoxyglucose (2-DG) provide a functional view of auditory cortex in the mustached bat (Fig. 9.1D; Duncan and Henson in press; Duncan, Jiang, and Henson in manuscript). The results show that auditory cortex, bounded by the piriform cortex ventrally and the bottom of the temporal sulcus dorsally, is metabolically very active in echolocating, flying bats. This is in stark contrast to the lack of cortical 2-DG labeling in restrained, nonecholocating mustached bats passively exposed to sonar-like stimuli (O'Neill, Frisina, and Gooler 1989), and in resting bats (Duncan and Henson in press). Although quantitative analyses are not available, by inspection of the stained tissue sections it is clear that the auditory cortex is disproportionately large, occupying nearly the entire dorsoventral extent of the temporal cortex and from one-third to one-half the entire cortical hemisphere (Fig. 9.1D).

The cytoarchitecture of mustached bat primary auditory cortex is shown from Nissl-stained material in Figure 9.1E. Note the thick layer 1, thin and densely packed layers 2 and 4, and the low density of cells in layers 3 and 5. Ferrer (1987) has described the cortical cytoarchitecture of the echolocating bat *Miniopterus sthreibersi* from Nissl and Golgi material (Fig. 9.1F). In this species, there is a thick, cell-sparse molecular layer 1, bounded mainly by triangular (extraverted pyramidal) cells densely packed in a thin layer 2 (Fig. 9.1F). Unlike the occipital or central cortex in this species, the temporal cortex has a distinct granular layer 4 that in Nissl material is composed mainly of densely packed, small spherical cells lying about halfway down the depth of cortex (e.g., Fig. 9.1E). This layer in Golgi material contains spiny stellate cells (Fig. 9.1F). In *Miniopterus*, both layer 3, composed of a mixture of medium-sized triangular and pyramidal cells, and layer 5, containing large triangular and globular cells, are packed less densely than layers 2 and 4 (Fig. 9.1F). Layer 6 is more varied, including medium-sized triangular, polygonal, and globular cells.

2.2 Thalamocortical Connections

Traditionally, studies of auditory cortex have focused on its cochleotopic or tonotopic organization. All bats studied to date show some form of tonotopic organization in at least one region of cortex, and because the representation of frequencies follows the plan seen in the tonotopic primary area (AI) of other species, it has been assumed that this region is homologous. Although there is abundant neurophysiological evidence, there is scant hard anatomical evidence in favor of this homology. Olsen (1986), and more recently Wenstrup and Grose (1993), showed that the tonotopic region of mustached bat cortex is reciprocally connected to the ventral division of the medial geniculate body (MGB), a pattern typical of other mammalian species. Shannon and Wong (1987) also showed this to be true for *Myotis*. Wenstrup and Grose (1993) have demonstrated that the medial division of the mustached bat MGB is connected to all auditory cortical regions, including areas showing poor or no tonotopic organization surrounding the putative AI region (see Section 4). In some cases, the frequency tuning of neurons is poorly expressed in both the nontonotopic cortical fields and the divisions of MGB with which they are connected. Two regions dorsal to AI are connected to the unique "rostral pole" of the dorsal division of the MGB (Olsen 1986; Wenstrup and Grose 1993). As is discussed in Section 4.1, these two regions contain "combination-sensitive" neurons (cells tuned to more than one acoustic element in complex signals). Such cells first arise in the dorsal division of the MGB (Olsen 1986; see Wenstrup, Chapter 8).

The suprageniculate nucleus of the thalamus receives auditory input from the inferior colliculus (Wenstrup, Larue, and Winer 1994) and nucleus of the central auditory tract (anterolateral periolivary nucleus; Kobler, Isbey, and Casseday 1987). The suprageniculate connects broadly to the entire neocortex, auditory and nonauditory (Kobler, Isbey, and Casseday 1987).

3. Tonotopically Organized Cortex

The auditory cortex of all mammals studied so far is dominated by a tonotopic topographical organization, and bats are no exception to this rule. In fact, there are two general patterns of organization, tonotopic and

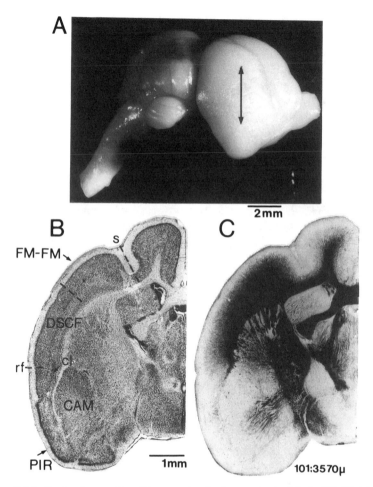

FIGURE 9.1A–E. Anatomy of auditory cortex in the mustached bat. (A) *Right side*, lateral view of forebrain. *Arrow* shows location of cross sections in B and C. Sulcus homologous to Sylvian fissure is visible (S. Radtke-Schuller, unpublished data). (B) Nissl-stained cross section through the temporal cortex at the level shown by *arrow* in A. Auditory area is bounded by Sylvian sulcus (s) dorsally and rhinal fissure (rf) ventrally. Locations of DSCF region of AI and FM-FM area are indicated. cl, claustrum; CAM, amygdala; PIR, piriform cortex (S. Radtke-Schuller, unpublished data). (C) Myelin-stained section at same level as B. Note heavy myelination in auditory cortex (S. Radtke-Schuller, unpublished). (D) Autoradiographs of serial cross sections through most of the auditory cortex from a flying bat, stained for [³H]2-deoxyglucose (2-DG) (from Duncan, Jiang, and Henson, in manuscript, reprinted by permission of the authors). *Arrows* indicate rostrocaudal direction.

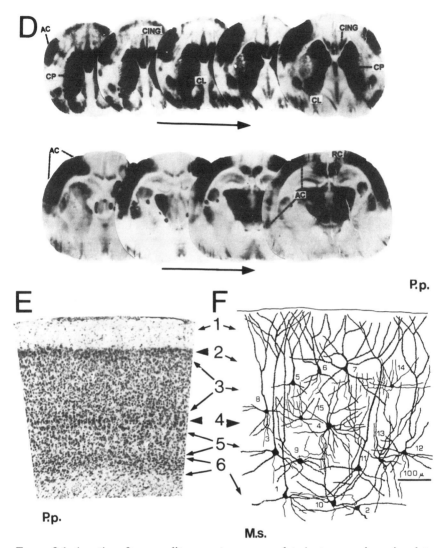

FIGURE 9.1. (*continued*) ac, auditory cortex; cp, caudate/putamen; cing, cingulate cortex; rc, retrosplenial cortex; cl, claustrum. (E) Nissl-stained cross section shows cortical layers in mustached bat AI (S. Radtke-Schuller, unpublished data). (F) Camera lucida drawings of representative Golgi-stained cortical neurons in *Miniopterus sthreibersi*, with approximate cortical layers indicated to *left* (from Ferrer 1987, copyright © 1987, J. Hirnforschung 28(2): 237–243, reprinted by permission of Academie Verlag GmbH). *1, 3, 5, 8, 11, 12*: pyramidal cells; *6,7*: extraverted pyramidal cells in layer 2; *4*: large multipolar cell with smooth dendrites; *13*: large multipolar neuron; *14*: chandelier cell; *15*: spiny stellate cell in layer 4.

nontonotopic. Primary auditory cortex is traditionally described as tono-topically organized, with neurons showing sharp tuning and relatively short latencies to tones at a single characteristic, or best, frequency (BF), arrayed by BF along the cortical surface (Woolsey 1960). The secondary auditory cortex (AII) is described as containing cells responding with longer latencies that are more broadly tuned and are often not well driven by tones (Schreiner and Cynader 1984). Some mammalian species express more than one tonotopic primary field (for review, see Clarey, Barone, and Imig 1992). Modern studies in the cat (Reale and Imig 1980) have shown that the primary auditory cortex is actually composed of four tonotopically orga-nized fields (AAF, AI, P, PES), flanked by three secondary areas (AII, DP, and V). In macaque monkey, AI is also surrounded by two other primary fields and four secondary fields, in a "core-belt" organization (Merzenich and Brugge 1973). By contrast, many species, including the ferret, only express a single primary field (Kelly, Judge, and Phillips 1986; Phillips, Judge, and Kelly 1988).

In CF/FM bats, there is typically one major tonotopic primary area, akin to AI, bordered by as many as seven secondary fields containing cells with complex response properties that typically do not respond well to single tones. Unlike neuronal responses in cat AII, which are often described as sluggish, habituating, broadly tuned, and otherwise enigmatic, cells in the nonprimary fields of CF/FM bats respond vigorously and have relatively similar latencies to neurons in AI, when presented with the appropriate complex stimulus. In mustached bat, one of these bordering fields (the CF/CF area) is actually tonotopically organized for a very narrow range of frequencies. However, it represents frequency along *two* axes, thereby forming a *bicoordinate* frequency representation (Suga et al. 1983; see Section 4.1.3).

In FM bats, cells with both simple and complex response properties are intermingled within one large tonotopic field, and there is usually one or two additional fields where tonotopy is absent or reversed in direction along the cortical surface (see Sections 3.3.2 and 3.4.2). The details of these arrangements are described next.

3.1 Mustached Bat

Auditory cortex in the mustached bat has been by far the most intensively studied. In fact, one could argue that it is the most studied of any single mammalian species, save perhaps the domestic cat. Nearly all the research on mustached bat cortex has been done in the lab of Nobuo Suga, and he and his colleagues have reviewed this work in numerous articles (Suga, Kujirai, and O'Neill 1981; Suga 1981, 1982, 1988a, b, c, 1990a, b; Suga, Niwa and Taniguchi 1983; Suga, Olsen, and Butman 1990). The mustached bat cortex is distinguished by a number of unusual (some might say "remarkable") features. The most important of these include an enormous overrepresentation of a very narrow frequency band centered around the second harmonic of the sonar pulse, the existence of two novel classes of neurons specialized to encode information related either to target ranging or velocity/motion, and a profusion of separate fields in temporal cortex where the different neurons are segregated by type, forming "processing areas" for specific types of acoustic analyses.

The terminology used to define different neuronal types and regions in mustached bat cortex is complex, but it has become somewhat of a standard for other species as well. Therefore, a short primer is necessary to aid the reader in deciphering abbreviations (see Table 9.1). Mustached bats emit signals consisting of a long constant-frequency (CF) component followed, and often preceded by, a short frequency-modulated (FM) sweep (Novick and Vaisnys 1964; Schnitzler 1970; Suga and Shimozawa 1974; Gooler and O'Neill 1987). Each cry contains up to four or five harmonics, designated H_1–H_4, and the second harmonic is usually dominant (although the bat can vary the energy in the harmonics; Gooler and O'Neill 1987). Each CF or FM component is addressed by a subscripted number (e.g., FM_1, CF_3, etc.). Cell types are assigned these labels if they are most sensitive to frequencies related to the sonar pulse. Thus, a "CF_2" neuron would be tuned to a tone at a frequency near the second harmonic CF component of its sonar pulse. "Combination-sensitive" cells (Section 4) require two or more components, such as FM_1–FM_3. Table 9.1 summarizes the properties and distribution of different cortical neurons in the mustached bat.

The duration of sonar pulses varies from as long as 30 msec during cruising or search phase to as short as 5–7 msec during terminal phase (Novick and Vaisnys 1964; Henson et al. 1987). Repetition rates increase from about 10/sec to about 100/sec as the bat progresses through search, approach, and terminal phases of target pursuit (see Fenton, Chapter 2; Moss and Schnitzler, Chapter 3; and Simmons et al., Chapter 4 for further details on sonar signals).

3.1.1 Response Properties in Tonotopic Cortex

Discharge patterns. Response properties of neurons in the tonotopically organized region of the mustached bat cortex are dependent on BF. The tonotopic region is noted for a large central region overrepresenting frequencies associated with the predominant CF_2 component of the sonar signal (i.e., the bat's "resting frequency") called the "Doppler-shifted CF" (DSCF) processing area (Suga and Jen 1976). Neurons in the DSCF have been studied most intensively. Cells in this area respond at a latency of 9–10 msec, and show two basic discharge patterns to tonal stimuli, phasic-on and sustained-on. However, discharge patterns vary widely dependent on stimulus amplitude and frequency (Suga and Manabe 1982). At BF within about 20 dB of minimum threshold (MT), most cells are transient-on

TABLE 9.1. Cortical units in *Pteronotus parnelli*

Type	Frequency tuning	Delay dependency	Key parameters	Field(s)	Organization	Information encoded
CF_2	Extremely sharp	None	Frequency, amplitude	DSCF ("fovea")	Bicoordinate: frequency vs. amplitude	Target flutter; velocity
FM_1-FM_2 FM_1-FM_3 FM_1-FM_4	Broad	Sharp	Pulse–echo time interval	FM-FM DF VF	Clustered by type Chronotopic (delay axis)	Target range
CF_1/CF_2 CF_1/CF_3	CF_1: broad; CF_2, CF_3: extremely sharp	Broad	Frequency of CF components	CF/CF	Clustered by type Bicoordinate: tonotopic	Relative target velocity (Doppler-shift magnitude)
FM_1-CF_2 (H_1-H_2)	FM_1: broad; CF_2: sharp	Broad	Pulse–echo delay; echo frequency/ amplitude	DSCF VA	As DSCF above	Target detection

CF, constant frequency; FM, frequency-modulated; DSCF, Doppler-shifted CF region of mustached bat; DF, dorsal fringe area of bat cortex; VA, area H_1-H_2) of mustached bat cortex; VF, ventral fringe of bat cortex.

responders. Sustained-on responders are comparatively rare. At frequencies below the bat's individual resting frequency, or amplitudes more than about 30 dB above threshold, transient on-off or pure off-discharge patterns occur. These periods of elevated activity at stimulus onset or offset are typically followed by periods of suppression, and rebound-off responses also occur in many cells up to 30–40 msec after cessation of the tone burst stimulus.

On-off discharge patterns are common in neurons tuned to the CF_2 frequencies throughout the mustached bat's auditory pathway. This pattern is presumably caused by prolonged "ringing" of the basilar membrane when stimulated near the bat's resting frequency (see Kössl and Vater, Chapter 5). The cochlear resonator is responsible for extremely sharp tuning of neurons tuned to this frequency band. Earlier work in the auditory periphery showed that off-discharges in auditory nerve fibers could be abolished by adding a downward sweeping FM to the end of the tone burst, mimicking the bat's sonar pulse (Suga, Simmons, and Jen 1975). However, in the inferior colliculus, about 14% of CF_2 cells show on-off responses not just at high amplitudes or lower frequencies, but at all frequencies and amplitudes within their response areas (Lesser 1987). Off-responses in these cells do not disappear with addition of an FM sweep at the end of a tone burst. Collicular on-off cells show interesting full-wave rectification properties that suggest specialization for detecting sudden but infrequent acoustic transients (Lesser et al. 1990). Whether these "pure" on-off cells occur in cortex remains to be seen.

Frequency tuning (spectral domain) properties. As is true for CF_2-tuned cells at lower levels (Suga and Jen 1977; Pollak and Bodenhamer 1981; O'Neill 1985; see Casseday and Covey, Chapter 6; Pollak and Park, Chapter 7), excitatory frequency response areas (tuning curves) are extremely sharply tuned in DSCF cells. Suga and Manabe (1982) reported Q_{10dB}* values ranging from 15 to 225, with outliers near 300. Q_{10}s this high are extraordinary, and have only been found in CF/FM bats. By contrast, Q_{10dB} of AI neurons in squirrel monkey do not exceed 20, and cells with Q_{10}s greater than 4 are considered "sharply tuned" (Shamma and Symmes 1985). Moreover, as Suga and Manabe (1982) demonstrated, central processing further sharpens frequency tuning beyond that seen at the auditory periphery. They compared tuning curve bandwidths of auditory nerve fibers and cortical cells not just near threshold with Q_{10}, but also at 30 dB and (where possible) 50 dB above threshold. They found that whereas Q_{10} values in cortical cells largely overlapped those of auditory nerve fibers, further sharpening of the response area was evident from Q_{30}

---

*Q_{10dB} equals the bandwidth of tuning curve at 10 dB above minimum threshold, divided by the best frequency. Q_{30} and Q_{50dB} are calculated from bandwidths at 30 and 50 dB above threshold, respectively.

and Q_{50dB} values at the cortical level. Therefore, cortical neurons show evidence of pronounced narrowing of the excitatory areas at moderate to high stimulus levels, and inhibition, both on- and off-BF, plays a critical role in shaping these response areas.

"Amplitude tuning" properties. Response areas can be thought of as neuronal activity maps plotted along two domains or axes, frequency and amplitude. Inhibition can sculpt the response area of a cortical neuron not only in the frequency domain, but also at both extremes of the amplitude domain. One can describe the amplitude domain behavior of a cell with an "intensity," or rate-level (input-output) function. When presented with a free-field BF stimulus (i.e., one exciting both ears equally), most *Pteronotus* cortical cells have bell-shaped, or nonmonotonic, intensity functions (Suga and Manabe 1982). This effectively "tunes" the cell to a particular "best amplitude" (BA). In a minority of this population, high stimulus levels may suppress activity to zero, producing an "upper threshold." At low stimulus levels, there is recent evidence to suggest that inhibition may also play a direct role determining a cell's threshold. Yang, Pollak, and Ressler (1993) showed in mustached bat inferior colliculus (IC) neurons that response areas, including thresholds, are shaped by γ-aminobutyric acid (GABA)-ergic inhibition (see Pollak and Park, Chapter 7). Whole-cell patch recordings in intact *Eptesicus* preparations of IC neurons with high thresholds have shown that inhibitory synaptic potentials can occur at subthreshold stimulus levels, and may therefore be important in setting threshold (Covey et al. 1993; Casseday and Covey, Chapter 6).

Inhibitory response areas. Suga and Manabe (1982) employed a modified forward masking paradigm to measure inhibition in cortical cells. Inhibitory areas were found to be extensive, in some cells covering more than an octave above and below the BF. In the extreme, cortical cells show completely circumscribed, spindle-shaped excitatory response areas. Suga and Manabe (1982) called such cells "level tolerant", that is, their frequency selectivity was essentially unchanged over the entire range of effective stimulus amplitudes. In the vast majority of DSCF neurons the discharge rate is maximized at particular values of both frequency and amplitude (i.e., at BF and BA). In this sense, DSCF neurons are tuned in both the frequency and intensity domains.

As in other mammals, the proportion of cells with nonmonotonic rate-level functions increases as one ascends from the cochlear nucleus to the cortex. At the cortical level, the net effect of the interaction of excitatory and inhibitory circuitry produces level-tolerant units functioning as frequency detectors. Such a mechanism might underlie the high acuity of frequency discrimination shown by human subjects when stimulated with moderate to high intensity signals (Hawkins and Stevens 1950; Scharf and Meiselman 1977).

3.1.2 Organization of Primary Tonotopic Area

The organization of the auditory cortex in the mustached bat is shown in Figure 9.2. By making penetrations perpendicular to the cortical surface, it was found that BFs, BAs, MTs, and response areas were roughly similar for cells within cortical columns in AI (Suga and Jen 1976; Suga and Manabe 1982). Along the cortical plane, BFs of the columns are tonotopically organized, high frequencies being represented rostrally and low frequencies caudally, resembling cat primary auditory cortex. However, the tonotopic sequence is interrupted by the DSCF area (Fig. 9.2A–C). The DSCF area accounts for fully one-third of the tonotopic region, yet it represents only a very narrow frequency band, from about 61 kHz to 63 kHz, encompassing the bat's resting and reference frequencies. The mustached bat's auditory system is especially sensitive within this frequency range, and just below this band, between about 58 and 60 kHz, the bat's ear has a pronounced insensitive region, or notch (Pollak, Henson, and Novick 1972; Suga, Simmons, and Jen 1975; Pollak, Henson, and Johnson 1979; see Kössl and Vater, Chapter 5). During Doppler-shift compensation, the bat lowers its pulse frequency to stabilize the echo CF_2 within the narrow frequency band represented within the DSCF area, and the emitted signal enters the insensitive notch, thereby reducing the difference in sensation level between the pulse and echo.

Schuller and Pollak (1979) dubbed the frequency band of extremely sharp tuning the "acoustic fovea" (see also Section 3.2). At the cortical level, the "fovea" is "personalized," in that the DSCF area represents frequencies centered around the particular cochlear resonance of the individual bat's ears. Moreover, there are sexual differences in the representation: males emit lower sonar frequencies than females, and their ears are tuned lower than females as well (Suga et al. 1987).

Within the DSCF area, Suga and Jen (1976) found that isofrequency contours were decidedly circular, suggesting a tonotopic organization with radial symmetry (Fig. 9.2B,C). Outside the DSCF area, isofrequency contour lines are slightly curved, but are mainly dorsoventrally oriented, similar to cat AI (Fig. 9.2A). Asanuma, Wong, and Suga (1983) showed that rostral to the DSCF area, the CF_3 frequency band between about 89 and 95 kHz (especially 92–94 kHz) was also overrepresented (Fig. 9.2A). Thus, the two strongest harmonics of the mustached bat's sonar cry are both represented disproportionately in AI. By contrast, the dominant FM_2 (sweeping from 60 to 50 kHz) and FM_3 (89 to 74 kHz) components of the species' biosonar signals were poorly represented along this strip of cortex. These frequency bands are displaced dorsally into what is called the "FM-FM" area, described in Section 4.1.1 (Suga and Jen 1976; Suga, O'Neill, and Manabe 1978; O'Neill and Suga 1982). Similarly underrepresented are frequencies between the dominant harmonics of the sonar signal

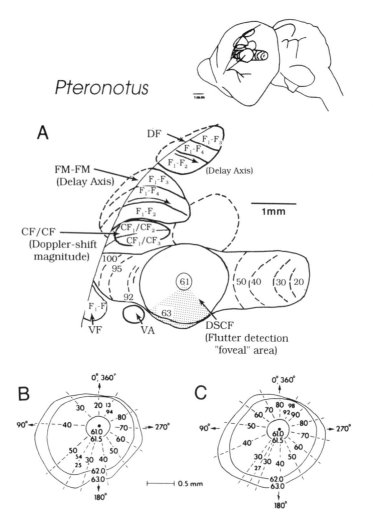

Pteronotus

1mm

A

DF

F_1-F_3
F_1-F_4
F_1-F_2
(Delay Axis)

FM-FM
(Delay Axis)

F_1-F_3
F_1-F_4

F_1-F_2

1mm

CF/CF
(Doppler-shift
magnitude)

CF_1/CF_2
CF_1/CF_3

100
95

61

92

50 40 30 20

F_1-F

63

VF   VA

DSCF
(Flutter detection
"foveal" area)

B

0°, 360°

30 20 13
94
.80
90°   40   -70-   270°
61.0
61.5   60
50   30 40
54
25
62.0
63.0

180°

0.5 mm

C

0°, 360°

80 98
70 92 90
60
-80-
90°   -50-   -70-   270°
61.0
40   61.5   60
30   50
30 40
27
62.0
63.0

180°

FIGURE 9.2A–C. Functional organization of mustached bat (*Pteronotus*) auditory cortex. (A) Cortical fields include tonotopic zone with Doppler-shifted CF region (DSCF) area in center, three dorsal fields including the FM-FM, dorsal fringe (DF), and CF/CF areas, and two ventral fields including ventral fringe (VF) and ventral anterior area (VA). *Shaded area* in DSCF indicates region containing excitatory-excitatory (EE) cells; remaining part of DSCF is predominantly excitatory-inhibitory (EI). Cells in DSCF area are exquisitely sensitive to minor frequency modulations caused by fluttering targets (Suga 1990a, reprinted from Neural Networks 3, N. Suga, "Cortical computational maps for auditory imaging", pp. 3–21, © 1990, with kind permission of Elsevier Science Ltd.). (B,C) Organization of DSCF area. Overrepresentation of frequencies near second harmonic of sonar pulse is indicated by concentric isofrequency contours. Iso-best amplitude contours are radially organized, forming a bicoordinate frequency-amplitude map with two different patterns, N type (B), where the BA *increases* in the counterclockwise direction, and V type (C), where BA *decreases* dorsoventrally in each half of the area (from Suga 1982, © 1982, reprinted by permission of Humana Press).

428

(e.g., 35–50 kHz, 65–74 kHz). It is clear from these data that the mustached bat primary tonotopic cortex is heavily weighted toward the CF components in the biosonar signal, and the FM components are displaced into a completely different region dorsal to the tonotopically mapped area (see Section 4.1).

Most importantly, Suga (1977) showed that not only frequency, but also amplitude, was systematically represented within the DSCF area. Unit thresholds within the DSCF area were all quite similar, but BA varied systematically (Fig. 9.2B,C). Suga and Manabe (1982) showed a systematic "amplitopic" representation, with iso-BA contours arrayed roughly orthogonal to the iso-BF contours, much like spokes in a wheel (Fig. 9.2B, C). Frequency and amplitude are thereby distributed in the DSCF area in a bicoordinate map.

Tunturi (1952) found evidence of a systematic difference in threshold along isofrequency contours in the dog cortex, and recent evidence from cat suggests that there is a systematic dorsoventral variation in both threshold and BA along isofrequency contours (Schreiner, Mendelson, and Sutter 1992). Neither of these examples matches the level of systematic representation of these parameters found in the mustached bat. As pointed out by the latter authors, the implication of this type of representation is that signals with different intensities will activate spatially distinct subdivisions of the cortex, and the pattern of activation will shift with intensity changes in a fairly predictable manner. Patterns of activation might then be associated with stimuli with different amplitude spectra.

3.1.3 Binaural Properties of DSCF Neurons

Despite a wealth of data concerning binaural interactions in the *Pteronotus* midbrain (see Pollak and Park, Chapter 7), only two studies have investigated binaural properties in the auditory cortex (Manabe, Suga, and Ostwald 1978; Suga, Kawasaki, and Burkhard 1990). Using a quasi-dichotic stimulation technique, Manabe, Suga, and Ostwald (1978) studied binaural interactions in DSCF neurons. Three general classes of response were found. These included binaurally excited (EE), binaurally suppressed (EI), and monaural (EO) response types. Excitatory input came from the contralateral ear in the majority of EI neurons. They were found dorsally, and EE neurons ventrally, in the DSCF area (Fig. 9.2A). Thus EE neurons summing responses to the two ears are found in the region with the lowest BAs, and EI neurons sensitive to interaural intensity differences in the region with higher BAs. This pattern is analogous to the banded, patchy organization of EE and EI cells in cat cortex (Imig and Adrian 1977; Middlebrooks, Dykes, and Merzenich 1980). Manabe, Suga, and Ostwald (1978) asserted that EE neurons are best suited for target detection because they presumably have large receptive fields and low BAs. EI neurons, on the other hand, are better suited to target localization, as they have higher BAs and smaller receptive fields than EE cells.

Beyond this segregation of binaural types, little further information exists concerning the encoding of azimuth and elevation cues in the tonotopic part of auditory cortex. Suga, Kawasaki, and Burkhard (1990) examined the spatial receptive fields of neurons in the FM-FM area of mustached bat cortex, as noted in Section 4.1.1.4. However, a systematic study akin to those on IC cells by Fuzessery and Pollak (1985), in which binaural interactions and free-field spatial receptive fields are both investigated, has not yet been attempted in the mustached bat cortex.

3.1.4 Facilitation of DSCF Neurons by Tone or FM Pairs

Recent studies using pairs of tones and frequency-modulated signals, mimicking the mustached bat's sonar signal, have surprisingly revealed facilitation of activity to stimulus combinations in DSCF neurons (Fitz-patrick et al. 1993). Specifically, DSCF cells are facilitated by pairing an FM_1 with a CF_2 signal, and this facilitation is dependent on a time delay between the two signals. The properties of these cells are best understood after considering the response properties of cells in the FM-FM area of mustached bat cortex, and discussion is therefore deferred to Section 4.1.4.

3.2 Horseshoe Bat

Bats in the family Rhinolophidae occupy an ecological niche similar to the mustached bat, and show similar echolocation behavior, including Doppler-shift compensation (Schnitzler 1968). Yet horseshoe bats are not sympatric with the mustached bat, and phylogenetically they are not closely related (see Fenton, Chapter 2). The remarkable behavioral and physiological similarities have apparently evolved independently, converging presumably from similar selective pressures.

The search phase biosonar signals of *Rhinolophus* are much longer (about 60 msec) than those of mustached bats. However, the cries are less complex, having only a very weak fundamental (H_1 component) and dominant second harmonic (H_2 component). Unlike mustached bats, the sensitivity of the ear of *Rhinolophus* does not extend to frequencies much beyond that of the second harmonic of the sonar pulse spectrum (Neuweiler 1970; Bruns 1976; Bruns and Schmieszek 1980; see Kössl and Vater, Chapter 5).

Most of what is actually known about echolocation by long CF/FM bats comes from work with horseshoe bats. However, although there is much known about auditory processing at and below the level of the inferior colliculus, there are only two published studies on the auditory cortex, one from the European greater horseshoe bat *Rhinolophus ferrumequinum*, and the other from the rufous horseshoe bat *R. rouxi*, of southern India and Sri Lanka. The cortex in these bats provides revealing contrasts to the organization of the mustached bat auditory forebrain.

3.2.1 Response Properties of Neurons in Tonotopic Cortex

Ostwald (1984) reported that more than half the units recorded in the primary auditory cortex of *R. ferrumequinum* showed only transient-on responses to tonal stimuli at all excitatory frequencies and intensities above threshold. Eighteen percent showed sustained-on responses at all stimulus levels, and 4% showed on-off responses exclusively. The remaining 24% changed their discharge patterns with changes in either intensity, frequency, or both. Like the mustached bat, on-off patterns typically occurred at intensities 20–30 dB or more above threshold, or at frequencies usually below, but sometimes higher than, BF.

Table 9.2 summarizes the response properties of cortical cells in the horseshoe bat. Similar to the mustached bat, neurons in the central part of primary cortex have BFs near or just above the frequency of the CF_2 component of the bat's signal. Excitatory areas of of units within this region are typically V shaped and very sharply tuned (Q_{10} dB values from 20 to >400), whereas those elsewhere are relatively broad (Q_{10} lower than 20). Thresholds are also lowest within the CF_2 frequency range, and although very few units show upper thresholds, most have nonmonotonic intensity functions similar to the mustached bat.

3.2.2 Frequency Representation

The representation of the cochlear partition in the horseshoe bat cortex has gross similarities to that of the mustached bat. Ostwald (1980, 1984) found a tonotopic axis in AI (defined by units with short latency responses to tones) running roughly rostrocaudally (Fig. 9.3A,B, unshaded region). Rostral and ventral to the tonotopic area, units were difficult to excite with tones and had longer latencies, and there was no clear frequency organization. In the rostral half of the tonotopic area, the frequency map is distorted by a large "CF" region, where the CF_2 component of the individual bat's sonar signal (ranging between 81 and 85 kHz in *R. ferrumequinum*) is overrepresented (Fig. 9.3B). This region is therefore functionally analogous to the DSCF area of mustached bat cortex in that it encompasses the individual bat's resting and reference frequencies. For the sake of consistency in nomenclature, the CF region shall be hereafter referred to as the "CF_2" region, because it represents the second harmonic CF component of the horseshoe bat signal. Outside the CF_2 area, isofrequency contours are generally oriented dorsoventrally. Inside the CF_2 area, contours are strongly curved, but are not quite round or ellipsoid like those in the mustached bat DSCF area. The sequence of frequency representation is nearly reversed in the CF_2 area relative to the rest of AI, with lower frequencies ventral and rostral and higher frequencies dorsal and caudal. A sharp discontinuity occurs at the dorsal border of the CF_2 area, above which frequencies found mainly in the FM_2 component (70–83 kHz) of the sonar pulse are overrepresented. Ostwald's reconstructions of the FM_2 area show a rostro-

TABLE 9.2. Cortical Units in *Rhinolophus* spp.

Type	Frequency tuning	Delay dependency	Key parameters	Field(s)	Organization	Infomation encoded
CF_2	Very sharp (Q_{10}:20~400)	None	Frequency Amplitude	CF_2 area of AI	Tonotopic	Echo frequency Target flutter
FM_1-FM_2	Broad	Sharp	Echo delay, amplitude	FM-FM	Chronotopic	Target range
CF_1/CF_2	CF_1: broad CF_2: sharp	Broad	Doppler shift	CF/CF	Unknown	Doppler-shift magnitude
CFM_1-CFM_2	CF_1:broad CF_2: sharp FM_1: broad FM_2: broad	Sharp	Echo delay Doppler shift	Mixed	Unknown	Target range; Doppler-shift magnitude

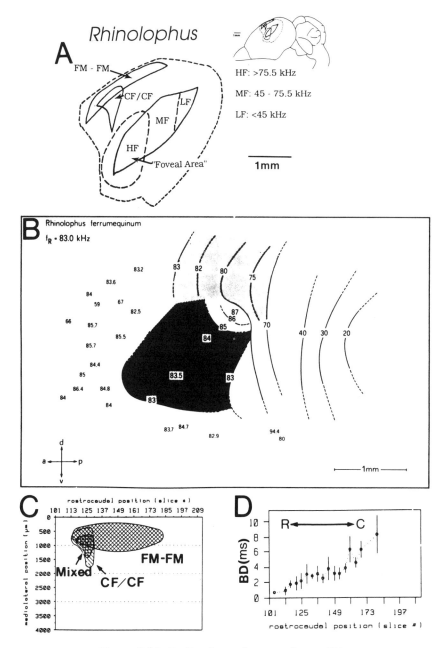

FIGURE 9.3A–D. Caption at bottom of page 434.

caudal, high-to-low frequency organization similar to that found caudal to the CF_2 area. Thus, as in *Pteronotus*, the FM_2 components of the *Rhinolophus* sonar pulse are displaced to a separate dorsal field (more detail is in Section 4.2). Ostwald (1984) showed that, relative to the innervation of receptors on the cochlear partition, there is a significant magnification of the CF_2 frequencies in the cortex.

3.3 Little Brown Bat

The cortical plan of two species whose echolocation behavior is typical of the majority of aerial insectivorous bats employing predominantly FM signals is now described. As with any classification scheme, it is easy to oversimplify the acoustical environment of the "FM bats" by ignoring the many circumstances in which these species use other types of signals (for a review, see Pye 1980; and Fenton, Chapter 2). For example, many species commonly included in this group (e.g., the Mexican free-tailed bat, *Tadarida brasiliensis*) use CF signals during search phase. Nevertheless, FM sounds are the predominant signals used for obstacle avoidance and prey capture, and research on the auditory cortex has accordingly been focused on the processing of FM sweeps.

The little brown bat, *Myotis lucifugus*, is a mosquito specialist often found foraging over water. The typical search-phase sonar pulse consists of a brief FM sweep descending over an octave from about 80 to 40 kHz in about 2–5 msec. These signals shorten to less than 1 msec, and a second harmonic appears as well, during the terminal phase.

3.3.1 Response Properties

Suga (1965a,b) recorded cortical unit responses in barbiturate anaesthetized *Myotis* to brief (4-msec) tones and FM sweeps. He described their discharge

---

FIGURE 9.3A–D. Functional organization of horseshoe bat (*Rhinolophus*) cortex. (A) Cortical fields (drawn to same scale as Figs. 9.2A, 9.4A, and 9.5A) include ventral tonotopic area and nontonotopic dorsal fields containing combination-sensitive cells. *"Foveal Area"* indicates overrepresented CF_2 region, homologous to DSCF area in mustached bat. *Inset:* Position of cortical fields in lateral view of brain (courtesy S. Radtke-Schuller). (B) Frequency representation within AI. CF_2 region shows expanded representation, reversal of frequency trend (from Ostwald 1984, © 1984, J. Comp. Physiol A 155:821–834, reprinted by permission of Springer-Verlag). (C) Functional organization of nontonotopic dorsal fields. FM-FM and CF/CF areas overlap rostrally, and CFM_1-CFM_2 ("mixed") cells are found in the region of overlap. (D) Systematic map of best delays is demonstrated within FM-FM area (C and D, adapted from Schuller, O'Neill, and Radtke-Schuller 1991, © 1991, Eur. J. Neurosci. 3:1165–1181, reprinted by permission of Oxford University Press).

patterns as "very phasic," although one might argue that the stimulus durations employed were too short to reveal more tonic response patterns. Four
general types of response areas were found, including "narrow" (sharp V-
shaped), "closed" (spindle-shaped), "double-peaked" (two regions of sensitivity), and "wide" (broad V- or U-shaped). In response to FM sweeps, both
double-peaked and wide response area cells showed about the same thresholds regardless of sweep direction. In cells showing narrow or closed receptive fields, excitatory response areas to tones were often flanked by inhibitory
areas on one or both sides. Cells with these response areas could show good
responses to downward but poor responses to upward-sweeping FM, or vice
versa. Smaller percentages of cells had upper thresholds for FM sweeps,
showed no responses to FM sweeps ("FM insensitive" or "CF specialized"),
or had thresholds to FM sweeps 10 dB or more lower than that to pure tones
("FM sensitive"). In the extreme, some cells only responded to FM sweeps,
showing no response to tonal stimuli ("FM specialized"). Two-thirds of the
cells exhibited nonmonotonic (peaked) rate-level functions.

Shannon-Hartman, Wong, and Maekawa (1992) have reexamined the
FM sweep selectivity of *Myotis* cortical neurons. Nearly every neuron (96%)
in their sample showed significantly stronger responses to FM sweeps
compared to pure tones. Eighty-three percent, dubbed "type I FM-sensitive
units," responded to both tones and FM signals that swept through the
unit's pure tone response area. Similar to Suga's (1965a) earlier study, 13 %
of the sample, termed "type II" cells, were specialized for FM sweeps,
showing little or no response to tones. The remaining 4% showed little or no
response to FM sweeps, only responding to tones (equivalent to Suga's "FM
insensitive" units). The latter "CF specialized" cells were segregated in the
cortex, as described later. The predominance of neurons preferring complex
signals (i.e., FM) over simple pure tones is a hallmark of cortical organization in *Myotis*. Table 9.3 summarizes the response properties of *Myotis*
cortical cells.

3.3.2 Organization of *Myotis* Cortex

By contrast to the CF/FM bats, the auditory cortex in *Myotis* is not as
clearly subdivided into fields containing cells with different physiological
properties. Two fields have been recognized. The larger field contains a
single tonotopic map with the AI-typical rostrocaudal progression of high
to low frequencies, formed by type I FM-sensitive cells (Fig. 9.4) (Suga
1965a; Wong and Shannon 1988). Type II cells are scattered about the
tonotopic field. The much smaller rostral field contains CF-specialized
neurons tuned to low (30–40 kHz) frequencies (Shannon-Hartman, Wong,
and Maekawa 1992). Caudally, most FM-sensitive cells are delay tuned.
Dear et al. (1993) have argued that the presence of delay-tuning properties
within a tonotopic AI provides for an efficient integration of temporal and
spectral domain processing (see Section 5).

TABLE 9.3 Cortical units in *Myotis lucifugus*

Type	Frequency tuning	Delay dependency	Key parameters	Field(s)	Organization	Information encoded
CF specialized	Sharp	Unknown	Pure tone	AI, AAF	Tonotopic	Stimulus spectrum
FM specialized	Broad	Unknown	Frequency modulation	AI only	Tonotopic (center frequency?)	FM sweeps Sonar signals
FM-FM	Broad		Paired FM Echo delay	AI only	Tonotopic No delay axis	Target range
P type		Sharp	Pulse and echo			
E type		Moderately broad	amplitude difference Repetition rate			

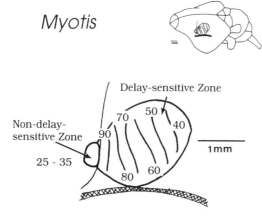

FIGURE 9.4. Organization of *Myotis* auditory cortex. By contrast to the CF/FM bats, there are only two cortical fields, both tonotopically organized. The tonotopic gradient in the larger caudal field is rostral to caudal, high to low frequency, and overrepresents the band of the sonar pulse from 80 to 40 kHz. The rostral field's tonotopic gradient is reversed, and only represents frequencies from 25 to 35 kHz. Cells in the caudal, but not the rostral, field are delay sensitive. Scale of drawings is same as in Figs. 9.2A, 9.3A, and 9.5A. (Courtesy of D. Wong.)

3.4 Big Brown Bat

The auditory behavior of this species is by far the most intensively studied, and there has been much research on the midbrain and auditory brainstem (see Covey and Casseday, Chapter 6). However, there is a paucity of data on the auditory cortex of big brown bats. The search/approach phase sonar pulse is a short FM sweep with two harmonics (FM$_1$ and FM$_2$). FM$_1$ sweeps downward over about an octave from about 50 kHz to 25 kHz. One or two additional harmonics appear, and the entire signal decreases in frequency, during late approach and terminal phase.

3.4.1 Response Areas and Intensity Functions

Jen, Sun, and Lin (1989) described the basic response properties of *Eptesicus* cortical neurons. They reported that all units discharged phasically at the onset of tonal stimuli. However, as in Suga's (1965a) study, the tonal stimulus they used was too short to draw conclusions about the relative abundance of phasic versus tonic response patterns. The lowest thresholds occurred in the 50- to 70-kHz range of BFs, and the highest thresholds were found in the 20- to 40-kHz BF range. Tuning curves were all V shaped, and Q$_{10dB}$ values ranged from 2 to 30. However, despite the open shape of the tuning curves, nearly every unit tested had nonmonotonic rate-level functions.

3.4.2 Frequency Representation

Jen, Sun, and Lin (1989) described the representation of both frequency and space within the *Eptesicus* cortex. With regard to frequency, they reported columnar organization of BFs and a tonotopic axis in the cortical plane along which frequency decreased systematically from anterior to posterior. This tonotopic region spans the frequency range between 20 and 90 kHz, but there is an overrepresentation of frequencies between 30 and 75 kHz. Similarly, Dear et al. (1993) reported an overrepresentation of the 20- to 60-kHz range in the tonotopic region (their "area A"; Fig. 9.5A). Both laboratories describe a "variable" region anterior to the tonotopic zone ("area B") where frequency is not tonotopically represented either within or across animals. Dear et al. (1993) also recognized a third area "C" rostral to area B where a tonotopic pattern also occurs (Fig. 9.5A). In area C, the axis of tonotopic organization is reversed from that seen in area A. Dear et al.

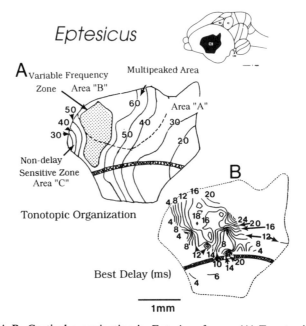

FIGURE 9.5A,B. Cortical organization in *Eptesicus fuscus*. (A) Tonotopic organization, showing three functional divisions. Tonotopic zone *(Area "A")* is caudal to the variable frequency zone *(Area "B", stippled),* and rostral tonotopic zone *(Area "C"),* which shows a frequency representation reversal. *"Multipeaked Area"* (bounded by *dashed line*) in dorsorostral part of area A contains cells tuned to two frequencies. (B) Iso-best delay (BD) contour map. Delay-tuned cells are found in both area A and especially, variable-frequency area B; they are absent in area C. In area B, there is a rostrocaudal, short-to-long BD progression, but in tonotopic area A, no clear delay map is visible. (From Dear et al. 1993, © 1993, J. Neurophysiol. 70:1988–2009, by permission of the American Physiological Society.)

reported that the dominant FM_1 frequencies of the *Eptesicus* sonar pulse (20–50 kHz) occupy about 74% of auditory cortex. As mentioned earlier, specialized delay-sensitive neurons are intermingled with unspecialized neurons in the tonotopic zone. As is discussed later, delay-tuned neurons have been found throughout areas A and B.

By contrast to Jen, Sun, and Lin (1989), Dear et al. (1993) found many instances where BFs did not show columnar organization in penetrations that were apparently perpendicular to the cortical surface. They have ruled out the possibility that this result is simply caused by improperly aligned penetration angles. One is forced to conclude that frequency representation in *Eptesicus* is not two dimensional within the cortical plane, as in other mammals, but includes a third depth dimension *along* cortical columns.

3.4.3 "Multipeaked" Neurons

Although the measurement of frequency response areas was not done systematically in their study, Dear et al. (1993) found that a number of neurons showed two threshold minima. They dubbed these cells "multipeaked neurons." BFs varied widely in multipeaked cells, and did not often coincide with frequencies of the sonar signals. The ratio of the two BFs was typically 1:3, and in most cells, the lower BF was below 20 kHz. Multipeaked cells were found in the anterior half of auditory cortex, mainly in the high-frequency part of the tonotopic area A and throughout area B. This is somewhat analogous to the distribution of delay-tuned cells (see Fig. 9.5B and Section 4.4.3).

Dear et al. (1993) pointed out that complex targets reflect multiwavefront echoes with spectral peaks and notches, the notches being caused by wavefront interference from the spatially separated surfaces (Beuter 1980; Habersetzer and Vogler 1983; Kober 1988; Kober and Schnitzler 1990; Simmons and Chen 1989; see Simmons et al., Chapter 4). Bats apparently use these spectral features to discriminate complex targets (Habersetzer and Vogler 1983; Mogdans and Schnitzler 1990; Schmidt 1988; Simmons et al. 1989). Dear et al. suggested that multipeaked neurons could decode spectral notches, as the notch frequencies are odd harmonics of each other (e.g., 1, 3, 5, etc.), similar to the 1:3 ratio of BFs in these cells. The problem with this hypothesis is that multipeaked cells respond best to spectral *peaks*, not notches. Dear et al. are content with the proposition that notched echoes would actually *decrease* the firing rate of multipeaked neurons. However, they provide no evidence that these cells have sufficient levels of background activity against which decreases in firing rate might be detected.

3.4.4 Auditory Space Representation

Jen, Lin and Sun (1989) determined the spatial receptive fields of cortical neurons using a roving free-field loudspeaker. All receptive field centers were located in the contralateral hemifield, and grew larger with amplitude. Receptive fields also depended on BF, such that higher BF units showed smaller receptive fields. The response centers of units with higher BFs tended to be located more toward the midline than those with lower BFs. Thus, the representation of auditory space systematically changes within the tonotopic axis of the cortex. Receptive field azimuths shift from frontal to lateral along the rostrocaudal axis of decreasing BF. By contrast, the receptive field elevations show no such dependency on location/BF. Contrary to the results of Dear et al. (1993), Jen, Lin and Sun (1989) found very little variation of BF, MT, tuning-curve shape, or azimuth of the spatial receptive field center within cortical columns.

The dependence of receptive field location on BF is predictable from the directionality of the pinna (Jen and Chen 1988). Pinna directionality improves at higher frequencies, and may be optimized at certain locations (Fuzessery, Hartley, and Wenstrup 1992). As frequency increases, the area of best sensitivity moves toward the midline, and often below the horizontal axis of the head, in every bat species studied so far (Grinnell and Grinnell 1965; Shimozawa et al. 1974; Grinnell and Schnitzler 1977; Fuzessery and Pollak 1984; Makous and O'Neill 1986).

3.5 Summary

In summary, each species of bat so far studied shows at least one tonotopically organized cortical field, representing frequency along a rostrocaudal gradient similar to AI in the cat. There is an overrepresentation of the biosonar frequencies in each species that is especially pronounced in CF/FM bats. However, in CF/FM bats the representation of the FM components of the sonar pulse is poor within AI, and is instead segregated into separate, nontonotopic fields. FM bats, on the other hand, have a tonotopically organized cortex composed of cells that are typically more responsive to FM stimuli than they are to pure tones. These fundamental distinctions are further explored next.

4. "Nontonotopic" Cortical Organization

The most intriguing feature of the auditory cortex in all bats so far studied is that new types of organization beyond tonotopy emerge from populations of cells with complex response properties. In this context, "complex" means that cells are specialized in some way for complex signals. "Specialization" can refer, for example, to a cell that responds to a complex, time-varying signal, such as an FM sweep, but does not respond to pure tones. As mentioned in Section 3, such cells are relatively common in the *Myotis* cortex (Suga 1965b; Shannon-Hartman, Wong, and Maekawa 1992). Specialization can also refer to a cell that shows facilitation or a markedly lower threshold to two or more signal elements ("combinations"), but is unresponsive to these signal elements presented alone, or in the wrong temporal order. In the microbat cortex, the most common type of facilitation occurs when pairs of FM signals are presented with particular time delays between the two sweeps. Combination-sensitive cells responsive to this stimulus are called "delay sensitive" or "delay tuned," and they occur in some form in the cortex of all bats studied so far. They also are known to occur in the MGB in at least one species, the mustached bat (Olsen 1986; Olsen and Suga 1991a,b; see Wenstrup, Chapter 8), and were actually first recorded in the midbrain intercollicular (pretectal) area of *Eptesicus* (Feng, Simmons, and Kick 1978). The properties of these cells are well suited for temporal and spectral pattern recognition. Moreover, their arrangement has revolutionized the traditional view of auditory cortex, emphasizing functional (task-oriented) rather than anatomical (cochleotopic) organizational principles.

4.1 Mustached Bat

Dorsal and ventral to the tonotopically organized area of mustached bat cortex are additional fields containing neurons with response properties that are much more complex than cells in AI (Table 9.2; see however Section 4.1.4 concerning the DSCF area). The first two of these fields to be discovered, and still by far the most intensively studied, are the FM-FM and CF/CF areas. It is within these regions that combination sensitive neurons were first found in cortex.[†] The typical combination-sensitive neuron is facilitated by two stimuli that alone do not elicit any response. As such, they are relatively "invisible" to tonal or wideband noise search stimuli routinely used to explore auditory centers. Such neurons combine two signal elements, and in the case of the mustached and horseshoe bats, these elements are always (and rather unexpectedly) derived from *different* harmonics of the bat's sonar signal.

4.1.1 FM-FM Area

Facilitation and delay tuning. The FM-FM area (Fig. 9.2B, Fig. 9.7) dorsal and rostral to the tonotopic area in the anterior auditory cortex of the mustached bat is notable because it contains a pure population of one type of combination-sensitive neuron. These neurons are specialized to respond to time-separated *pairs* of frequency-modulated sweeps (Suga, O'Neill, and Manabe 1978; Fig. 9.6A). The initial essential component of the pair

---

[†]Delay tuning and combination sensitivity were concurrently reported in both the auditory midbrain, and also, apparently, the auditory cortex in *Eptesicus* by Feng, Simmons, and Kick (1978). However, this paper only reported the response properties of midbrain cells. Details of *Eptesicus* cortical cells have only recently been described (see Section 4.4).

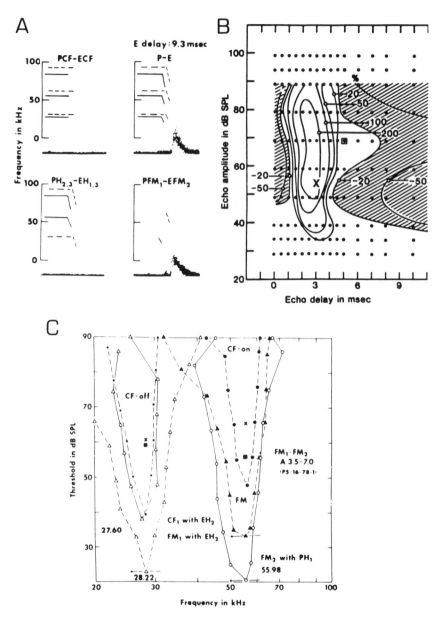

FIGURE 9.6A–C. Response properties of FM-FM cells in mustached bat. (A) Peristimulus time (PST) histograms and schematic spectra demonstrating selectivity of delay-sensitive FM-FM cells for combinations of a first-harmonic FM and (in this case) a second-harmonic FM stimulus. On *top right* is shown the strong facilitation produced by stimuli mimicking first three harmonics of a natural CF-FM sonar pulse *(solid lines)* and a 9.3-msec delayed echo *(dashed lines)*, compared to lack of response to the CF components alone *(top left)*. On the *bottom*, the facilitated

mimics the FM_1 (25–30 kHz). The delayed second sweep must mimic either the FM_2, FM_3, or FM_4 component (depending on the neuron), but never the FM_1 (Fig. 9.6C; Suga, O'Neill, and Manabe 1978). The majority are therefore classified as FM_1-FM_2, FM_1-FM_3, or FM_1-FM_4 cells. Each cell type is dorsoventrally segregated in the FM-FM area into three separate bands (O'Neill and Suga 1979, 1982). At the borders between the bands, cells facilitated by more than one FM component in the delayed sweep can be found (O'Neill and Suga 1982). These include FM_1-$FM_{2,3}$, FM_1-$FM_{3,4}$ and even FM_1-$FM_{2,3,4}$ neurons.

The most striking attribute of FM-FM neurons is that they are "delay tuned." Most cells respond best at a specific delay between the FM sweeps, called the "best delay" (BD). When presented with FM pairs at different repetition rates mimicking the different stages of target pursuit, the majority of cells show stable delay-tuning functions ("delay tuning curves"; Figs. 9.6B, 9.7A). Echo delay is the behavioral cue used to estimate target range (Simmons 1971, 1973). If one assumes that these cells are stimulated by the FM_1 component in the emitted pulse, and one (or more) FM components in returning echoes, then the probable role of delay-sensitive cells is to encode target range.

A small number of FM-FM neurons had best delays that were not stable, but rather became shorter at higher repetition rates. Therefore, Suga, O'Neill, and Manabe (1978) called cells with stable best delays *delay tuned*, and those with labile best delays *tracking* neurons. The delay-tuning curves of tracking neurons effectively "retune" themselves to shorter echo delays at higher repetition rates, such as when a bat increases its emission rate during

---

FIGURE 9.6A–C. (*continued*) response to a pair of FM_1 and delayed FM_2 components *(right)* contrasts to lack of discharges in response to multiharmonic stimulus lacking the first harmonic in the pulse and second harmonic in the delayed echo *(left)* (adapted from Suga et al. 1983, © 1983, J. Neurophysiol. 49:1573–1626, by permission of American Physiological Society). (B) Delay-tuning curve of area VF FM_1-FM_3 cell showing temporal inhibition. Contour values are expressed as a percentage relative to the sum of spike counts to FM_1 and FM_3 presented alone. Facilitation occurs when FM_1 is paired with FM_3 at delays and intensities indicated by *unshaded contours*. Inhibition is produced when a second FM_3 echo is presented with FM_1-FM_3 stimulus, indicated by *shaded areas* (FM_3 level and delay in the pair shown by *small square* at 6-msec delay, 70 dB sound pressure level [SPL]). Inhibition sharpens the delay-tuning curve, but does not alter BD (adapted from Edamatsu and Suga 1993, © 1993, J. Neurophysiol. 69:1700–1712, by permission of American Physiological Society). (C) Frequency-tuning curve of FM_1-FM_2 neuron in FM-FM area. Response areas for pure tones are shown as *filled circles, dashed lines*; for FM sweeps alone as *filled triangles* and *dashed lines*; and for various pairs (combinations) of CF and FM components, as labeled. Lowest thresholds were for FM_1-FM_2 pairs. *Arrows* indicate frequency range of FM sweeps (from Suga et al. 1983; © 1983, American Physiological Society).

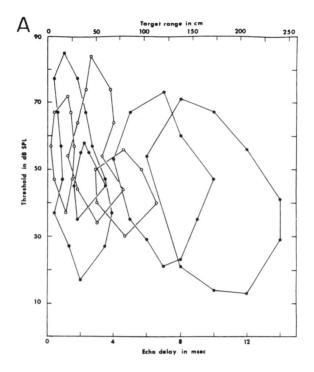

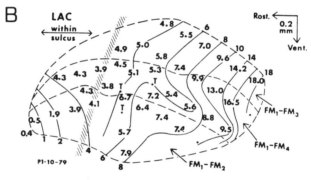

FIGURE 9.7A,B. Organization of the FM-FM area in mustached bat. (A) Delay-tuning curves of delay-tuned neurons show increase in delay-tuning curve widths in cells with longer best delays. Different echo delays can be encoded by different neurons. (B) Iso-best delay map reconstructed from rostrocaudally oriented penetrations parallel to cortical plane in left hemisphere (LAC). Cells recorded within banks of Sylvian fissure are displayed unfolded to *left* of *shaded line*. Three bands containing FM₁-FM₂, FM₁-FM₃, and FM₁-FM₄ cells are indicated. (From O'Neill and Suga 1982; © 1982, J. Neurosci. 2:17–31, by permission of the Society for Neuroscience.)

target approach. By having a labile temporal window, tracking neurons would in effect "lock on" to targets throughout the attack, unlike delay-tuned cells that would only discharge when the target was in a specific range "window."

Edamatsu and Suga (1993) have found that when two echoes are presented after a pulse, one in the delay-sensitive period and the other at shorter or longer delay, the delay tuning curve becomes sharper (Fig. 9.6B, shaded areas). The function of these temporal inhibitory areas is analogous to that of lateral inhibition in the spectral domain, but in this instance it improves range acuity and echo-amplitude tolerance.

Suga and colleagues have dubbed facilitation involving different harmonic components "heteroharmonic facilitation" (Suga et al. 1983). To test the hypothesis that FM-FM neurons are responding to emitted pulses and returning echoes, Kawasaki, Margoliash, and Suga (1988) recorded from vocalizing bats under anechoic conditions (i.e., they faced the bats out a window!), and played back artificial echoes time locked to the vocalizations. They found that the self-stimulation of the bat's own emissions activated FM-FM cells when $FM_{n\ (n=2,3,\ or\ 4)}$ echo stimuli were presented alone at the appropriate time delays; they were not responsive to the bat's emissions or the artificial echoes alone.

Why do FM-FM neurons require this seemingly odd set of spectral components to encode target range? One advantage of heteroharmonic facilitation that has been suggested (Suga and O'Neill 1979; O'Neill and Suga 1982; Suga et al. 1983) is that it reduces the likelihood of jamming of the target ranging system by the sonar signals of other bats. If the ranging system relied on high-amplitude signals in both the pulse and echo, the sonar cries of other bats might activate delay-tuned cells inappropriately. Because the emitted FM_1 signal is 30–40 dB weaker than the H_2 component, only the bat's own cries would be sufficiently intense to stimulate its FM-FM cells. This affords each bat its own "private line" for target ranging, and could perhaps explain the remarkable ability of these bats to echolocate successfully in the confined and crowded conditions typical of bat roosting sites.

Frequency and amplitude domain properties of FM-FM neurons. Two additional properties remain to be mentioned regarding FM-FM neurons. One is that they are rather broadly tuned in the frequency domain (Fig. 9.6C). This "frequency tolerance" enables these cells to respond to the broadband FM components of pulses and echoes regardless of the magnitude of Doppler compensation behavior or echo Doppler shift, without impairing the ability to encode target distance.

The second interesting property is that FM-FM cells are often sharply tuned in the amplitude domain. Echo amplitude is affected by both target size and distance from the bat (collectively, the subtended angle of the target). The facilitation of FM-FM cells is typically nonmonotonically related to the amplitude of the echo FM_n component, and most cells showed "best facilitation amplitudes" (BFA; Suga et al. 1983).

One question that arises is whether BFAs are matched to the stimulus levels impinging upon the bat's auditory system in natural conditions. The actual level of stimulation reaching the auditory forebrain is affected by such factors as the directionality of the ear and oral cavity, middle ear muscle contraction during vocalization, and neuronal suppression (Suga and Schlegel 1972; Suga and Shimozawa 1974). Kawasaki, Margoliash, and Suga (1988) estimated from cochlear microphonic recordings that the vocal self-stimulation by the FM_1 component is equivalent to about 70 dB sound pressure level (SPL), somewhat higher than the mean BFA for the FM_1 component in cortical cells (63 dB SPL). Middle ear muscle activation during vocalization (not accounted for by their calibration) would further attenuate the signal and might make up for the difference observed here.

For the echo FM components, the system is adapted to a fairly wide range of moderate amplitudes. However, there is behavioral evidence that mustached bats compensate echo amplitude by adjusting the intensity of their pulses, and thereby stabilize the stimulation of the ears at a nearly constant level regardless of target distance (Kobler et al. 1985). Estimates from *Eptesicus* by Kick and Simmons (1984) of the level of echo stimulation reaching the bat during approach phase range from 25 to 30 dB SPL, near the lower end of the range of BFAs for echo FM_2 components (56.7 ± 14.5 dB SPL; Suga et al. 1983).

Functional organization of FM-FM area. The best delays represented in the FM-FM area range from approximately 0.4 to 18 msec, but the majority of cells have BDs between 3 and 8 msec (Fig. 9.7A) (Suga and O'Neill 1979; O'Neill and Suga 1982). In terms of target range, these BDs span the biologically relevant time delays encoding distances between a few centimeters and about 3 m (Griffin 1958, 1971). Cells recorded in penetrations perpendicular to the cortical surface show similar delay tuning curves, suggesting columnar organization of this response property. Moreover, there is an orderly "chronotopic" representation of BD along the rostrocaudal axis of the FM-FM area, with an overrepresentation of echo delays of 3–8 msec (Fig. 9.7B; Suga and O'Neill 1979). Iso-best delay contours drawn on the cortical surface traverse the boundaries of the FM_1-FM_2, FM_1-FM_3, and FM_1-FM_4 cortical strips. Thus the FM-FM area contains a single map of echo delay, and thereby target range, that is congruent across echo harmonics. Because targets will reflect different echo spectra depending on their size, having delay-tuned cells tuned to all three upper harmonics enables the system to encode distance for a wide range of target sizes.

Suga and O'Neill (1979) made a rough calculation of target-range acuity based upon the distance along the cortical surface occupied by the range map, the range of BDs expressed along the range axis, and the estimated width of cortical columns. Assuming a column occupies about 20 μm, maximum range resolution in the linear region from 3 to 8 msec BD calculated to be about 2 cm, in agreement with range discrimination data for real (as opposed to jittered phantom) targets (Simmons 1971, 1973; see Chapter 4).

The discovery of these unusual cells, their clustering into separate fields, and their orderly arrangement along a temporal rather than a spectral axis, was the first clear demonstration in the mammalian auditory system of an organizational principle not based on the tonotopy of the cochlear partition. This place map representation for target distance followed closely the remarkable discovery of a "space map" in a division of the barn owl midbrain containing cells selective for both stimulus azimuth and elevation (Knudsen and Konishi 1978), and provided independent evidence that computational properties of the auditory pathways can create data structures in neural space that mimic spatial structures in extrapersonal space.

Directional selectivity of FM-FM neurons. Delay-sensitive FM-FM neurons are able to encode spatial information along one dimension related to target distance. But might they also be able to encode the location of a target in all three dimensions? Suga, Kawasaki, and Burkhard (1990) investigated the spatial tuning of range-tuned neurons in the FM-FM area to sound azimuth and elevation, to see whether they might encode target location in three-dimensional space. They found that the free-field spatial receptive fields determined from FM-FM best-delay stimuli were very large, more than 70° in both azimuth and elevation, even for amplitudes near threshold. Receptive fields in some cells ended at the midline, suggesting that they are suppressed by signals in the ipsilateral hemifield. In other cells, activity could be elicited in both hemifields, suggesting binaural excitation. Regardless of the type of cell, the receptive field centers were typically within the contralateral hemifield, but their centers differed according to the essential harmonic for the echo FM stimulus in a pair. Mean best azimuths were confined to the zone between midline and about 25° contralateral, nearly the same as more peripheral measures of ear directionality for these frequency bands (Fuzessery and Pollak 1984; Makous and O'Neill 1986). Best delays remained more or less constant within the receptive fields. Suga, Kawasaki, and Burkhard (1990) concluded that delay-tuned cortical cells are unsuited for target localization, but can provide reliable estimates of target range in a large region of frontal auditory space.

4.1.2 Dorsal Fringe, Ventral Fringe, and "H_1-H_2" Areas

Three other fields of mustached bat cortex contain delay-sensitive cells. The three regions are ipsilaterally connected "in series," and are also connected to the corresponding contralateral fields via the anterior commissure (Fritz et al. 1981). The dorsal fringe (DF) area lies just dorsal to the FM-FM area, rather close to the midline and straddling the terminus of the temporal sulcus (see Fig. 9.2A) (Suga and Horikawa 1986). Although smaller, this region is almost a carbon copy of the FM-FM area, in that there are the same three rostrocaudally elongated bands of cells tuned to the different echo harmonics, and there is a map of BD. However, the range of BDs found along the rostrocaudal axis extends out only to 9 msec (Suga and Horikawa 1986). Anterograde tracing studies have shown that DF receives a projection from the ipsi- and contralateral FM-FM areas (Fritz et al. 1981), but surprisingly there is no indication of a thalamocortical projection to this field. In this sense, the DF area could be considered secondary auditory cortex.

Another even smaller field containing FM-FM neurons is the ventral fringe (VF) area (Edamatsu, Kawasaki, and Suga 1989). The VF lies beneath the rostral end of the tonotopic AI region and receives a projection from the DF area (see Fig. 9.2A) (Fritz et al. 1981). The distribution of BDs in VF cells is truncated even more than that in the DF area, extending only to 5–6 msec. The small size of VF makes it difficult to determine whether there are discrete bands, clusters, or randomly intermingled FM-FM neurons tuned to different echo harmonics.

Smaller still is the H_1-H_2 region (also referred to as area VA), lying caudal to VF and just underlying the rostral border of the DSCF area. Area VA receives a weak intracortical projection from the FM-FM area (Suga 1984) and the CF/CF area (Olsen 1986). In this region, most neurons are facilitated by any combination of first and second harmonic CF and FM components: CF_1-FM_2, CF_1/CF_2, FM_1-FM_2, and FM_1/CF_2. They are somewhat similar to "CFM_1-CFM_2" neurons described recently in the horseshoe bat cortex (Schuller, O'Neill, and Radtke-Schuller 1991; see Section 4.2). In many respects, they are also similar to DSCF area cells recently found to be facilitated by FM_1-CF_2 combinations (Fitzpatrick et al. 1993; see Section 4.1.4). Because of this similarity in response properties, Fitzpatrick et al. (1993) have suggested that the H_1-H_2 region may in fact not constitute a separate field, but rather might be simply a previously unrecognized part of the DSCF area.

Differences in response properties exist between the FM-FM, DF, and VF areas, but the differences are small and do not clearly indicate a different function for each area. Edamatsu and Suga (1993) hypothesized that neurons in the three areas would show increasing specialization, based on the chainlike interconnection of the FM-FM → DF → VF pathway. In fact, what they found was that only rather subtle differences distinguish the FM-FM and VF fields. In both areas, for example, the delay-tuning curves of about 80% of the cells in both areas show temporal inhibition for pulse–echo delays both shorter and longer than the best delay (Edamatsu and Suga 1993). On the other hand, VF cells are strongly adapted by rapidly repeated FM stimulus pairs, and consequently can only function during search phase, whereas most FM-FM area cells can follow up to at least 100 pairs/sec (terminal-phase rates) with little change in either BD or delay-tuning curve shape. This result suggests that different cortical fields may take part in target range processing only during particular phases of echolocation. As is discussed in Section 4.3.3, an analogous situation occurs in the *Myotis* auditory cortex. To what end this strategy is applied by the brain is still unknown.

4.1.3 CF/CF Area

Response properties. A second type of combination-sensitive neuron discovered in the mustached bat is the CF/CF[‡] cell. CF/CF cells are facilitated by combinations of two constant-frequency tones (Fig. 9.8A) (Suga, O'Neill and Manabe 1979; Suga et al. 1983), and there are only two types, CF_1/CF_2 and CF_1/CF_3. As with FM-FM neurons, CF/CF cells are confined to a separate cortical field, overlying the rostral end of the tonotopic area, just ventral to the FM-FM area (see Fig. 9.2A). Another gross similarity to the organization of the FM-FM area is that CF_1/CF_2 and CF_1/CF_3 cells are segregated into two narrow strips running rostrocaudally (Fig. 9.8C). That there are no CF_1/CF_4 cells is not unexpected, as cells with BFs greater than about 112 kHz are rare in the mustached bat auditory system.

The similarities between CF/CF and FM-FM cells end at that point. The spectral and temporal tuning properties of CF/CF neurons are almost diametrically opposite to FM-FM cells, as summarized in Table 9.2. Whereas FM-FM cells are broadly frequency tuned and sharply delay tuned, CF/CF cells are sharply frequency tuned and broadly delay tuned.

The facilitation tuning curves for CF/CF cells (Fig. 9.8B) resemble a plot combining the tuning curves of CF_1-tuned cells with CF_2 or CF_3 tuned cells from the tonotopic field of auditory cortex (Suga, O'Neill, and Manabe 1979; Suga et al. 1983; Suga and Tsuzuki 1985) or lower centers like the IC (O'Neill 1985). The tuning for the CF_3 component in CF_1/CF_3 cells is especially sharp, with Q_{10dB} values of 65–70 versus 40–50 for CF_3-tuned cells in AI or the periphery (Suga and Jen 1977; Asanuma, Wong, and Suga 1983; Suga and Tsuzuki 1985). This is significant because it is the only example so far reported of central neurons showing better Q_{10} values than peripheral neurons. Suga and Tsuzuki (1985) showed that inhibitory sidebands make important contributions to the sharp tuning of these cells even at very high stimulus levels, rendering the cells level tolerant.

Suga et al. (1983) examined the relationship between the best facilitation frequencies (BFFs) for the essential components. Some CF/CF cells have harmonically related BFFs (i.e, the CF_n is an integer multiple of CF_1),

---

[‡]The use of the "/" in the abbreviations of the CF components facilitating these cells is deliberate, and indicates that there must be temporal *overlap* between the two components. This is in contrast to abbreviations for FM-FM cells, where "-" indicates that there must be temporal *separation* between the FM components of the stimulus.

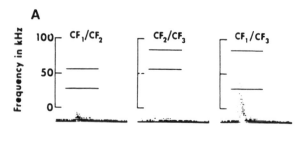

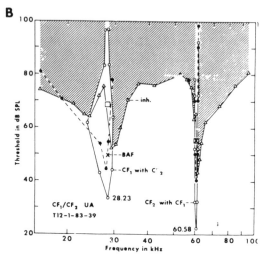

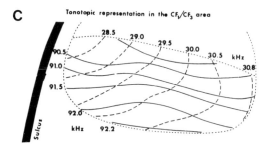

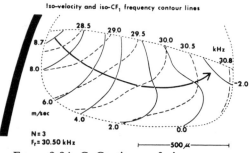

FIGURE 9.8A–C. Caption on facing page.

implying that they could respond to pulse or echo alone, although response to the latter is unlikely given the weak first harmonic. However, the majority of CF/CF cells are slightly "mistuned", that is, the best frequency for the CF_n component is not harmonically related to CF_1. Consequently, these cells cannot respond to harmonically related elements in pulses or echoes alone. Instead, they can respond only to signals with different fundamental frequencies, such as an emitted pulse and an overlapping Doppler-shifted echo. CF/CF cells could thereby encode the magnitude of Doppler shifts by comparing the fundamental frequency of the pulse to the higher harmonics of the echo. Because of their extraordinary tuning sharpness, each cell in effect represents a single Doppler-shift magnitude. Because echo Doppler shifts are caused by relative velocity differences between the bat and its surroundings, each CF/CF cell could represent a particular relative velocity.

By and large, CF/CF neurons are not delay tuned. Although some cells prefer a particular echo delay between the two CF components for maximum response, strong facilitation occurs at nearly any delay, as long as there is some temporal overlap. This temporal overlap can be as short as 1 msec and still produce noticeable facilitation. In a behavioral context, this broad "delay tolerance" enables CF/CF cells to encode relative velocity over a wide range of target distances.

Olsen (Olsen 1986; Olsen and Suga 1991a) found this type of neuron in the medial geniculate body of the mustached bat, but found that thalamic CF/CF cells responded much better to individual CF components presented alone than did cortical cells (see Wenstrup, Chapter 8, this volume).

Organization of the CF/CF area. The CF/CF area is divided such that CF_1/CF_2 cells are found dorsally and CF_1/CF_3 are found ventrally (see Fig. 9.2A) (Suga et al. 1983). Cells responding to either the CF_2 or the CF_3 component ($CF_1/CF_{2,3}$ cells) are located at the border between the two

←

FIGURE 9.8A–C Response properties of cells in CF/CF area of mustached bat cortex. (A) PST histograms and schematic spectra show responses to various pairs of CF signals mimicking sonar components. Cell shown is from the ventral half of the CF/CF area, selective for combinations of CF_1 and CF_3 (from Suga et al. 1983, © 1983, by permission of American Physiological Society). (B) Excitatory *(open)* and inhibitory *(shaded)* response areas of a CF_1/CF_3 neuron, measured with tones alone *(dashed lines)* or pairs of CF stimuli *(solid lines)*. Contrast these areas to those for the FM-FM cell in Figure 9.7. x indicates the BAF, best amplitude for facilitation (from Suga and Tsuzuki 1985, © 1985, J. Neurophysiol. 53:1109–1145, by permission of American Physiological Society). (C) Tonotopic representation *(top)* and isovelocity contours of Doppler-shift magnitude *(bottom)* in the CF_1/CF_3 region. CF_1 and CF_3 are represented roughly orthogonally, and the axis of increasing velocity is shown by the *arrow* (from Suga, Kujirai and O'Neill 1981; © 1981, in Syka J., Aitkin L. (eds), Neuronal Mechanisms of Hearing, pp. 197–219, by permission of Plenum Press.)

strips. CF/CF cells recorded in single cortical columns have essentially identical BFFs, that is, they are tuned to the same Doppler shift. Suga et al. (1983) showed that there is a bicoordinate representation of BFFs, with iso-BFF contours for the CF_1 component oriented more or less orthogonally to those for the CF_2 or CF_3 components (Fig. 9.8C, top). Although the map only covers a limited range of frequencies around the CF components of the sonar signal, this was the first demonstration of a *bicoordinate* tonotopic representation in the auditory system. This map encodes frequency differences ranging from 0 ventrocaudally to about 3 kHz rostrodorsally (Fig. 9.8C).

The relatively large and constant Doppler shift caused by a flying bat's motion relative to its surroundings has been termed the "DC" component of the Doppler-shifted echo, whereas the small, periodic frequency modulations caused by the local movement of insect targets can be considered the "AC" component (Suga and Manabe 1982). Because each CF/CF cell could precisely encode a specific Doppler shift, Suga et al. (1983) proposed that the CF/CF area systematically encodes flight velocity, that is, the DC component of the Doppler-shift, within a maplike representation on the cortical surface. In the CF/CF area, the represented Doppler shifts compute to velocities ranging from about -1 or 0 to about 9 m/sec, with an overrepresentation of the range from 0 to 5 m/sec (Fig. 9.8C, bottom).

However, Schuller and Pollak (1979) have argued that evolution has not selected for an exquisitely tuned ear and Doppler-shift compensation behavior simply for the purpose of velocity perception per se. They pointed out that velocity is readily available to all bats from the rate of change in echo delay across successive echo presentations. In their view, these extraordinary adaptations have evolved to enable the detection of small frequency and amplitude modulations (i.e., the echo AC component), caused by the beating of an insect's wings (Schnitzler et al. 1983). Indeed, Goldman and Henson (1977) have shown that mustached bats will only attack fluttering targets. Suga, Niwa, and Taniguchi (1983) have also shown that CF/CF cells show exquisite phase-locking to sinusoidal FM of the echo CF_n component, a stimulus that crudely approximates the AC component of the Doppler-shifted echo from a fluttering target. This suggests a possible role for the CF/CF area in local target motion analysis.

Gooler and O'Neill (1987, 1988) and O'Neill and Basham (1992) have proposed an additional possible function of the CF/CF area: encoding the magnitude of the Doppler shift not for measuring relative velocity, but rather for computation of the error signal needed by the vocal motor system to control Doppler-shift compensation behavior itself. For further discussion of vocal motor control systems and Doppler-compensation circuitry, see Schuller (1974, 1977), Suga, Simmons, and Shimozawa (1974), Gooler and O'Neill (1987), Schuller and Radtke-Schuller (1990), and Metzner (1993).

It seems possible that the CF/CF area could in fact represent both velocity and local target movement. Trappe and Schnitzler (1982) showed that flying horseshoe bats Doppler-shift compensate their flight velocity relative to their surroundings, and not to that of insect prey they were pursuing. Thus, echoes from the surroundings would stimulate a specific "iso-velocity" contour in the CF/CF area encoding the bat's flight velocity, whereas the insect echo, modulated by the movement of the wings, would stimulate other parts of the CF/CF area, within which cells would encode the echo modulation patterns by phase-locking their discharges.

4.1.4 Facilitation in the DSCF Area

Fitzpatrick et al. (1993) have found that about 75% of neurons in the DSCF area can also exhibit delay-dependent facilitation to paired stimuli mimicking biosonar components. These "FM_1-CF_2" cells constitute a third type of facilitation neuron, similar to those originally described in the H_1-H_2 area (Table 9.2; see Section 4.1.2). The best delays of DSCF cells are relatively long, from 5 to 30 msec, with a mean at about 21 msec corresponding to a target distance of 3.6 m. Thus, facilitation improves the sensitivity of DSCF neurons to faint echoes from distant targets. Fitzpatrick et al. (1993) pointed out that the target range at which facilitation first occurs in this population corresponds to the distance at which mustached bats initiate the pursuit of flying insects, and they suggest that facilitation might play a role in initiating approach-phase behavior. They point out that this result has important implications for the function of primary auditory cortex, namely that even AI neurons can exhibit complex response properties.

4.1.5 Role of Cortical Fields in Mustached Bat Echolocation Behavior

Recent experiments have examined the possible involvement of the DSCF and FM-FM areas in frequency and target range discrimination. Riquima-roux, Gaioni, and Suga (1991, 1992) attempted to inactivate the DSCF or the FM-FM area selectively with muscimol (a GABA receptor agonist), and then determined the bat's acuity for frequency and distance discriminations using a conditioned avoidance procedure. When muscimol was applied to the DSCF area, subject bats could discriminate large, but not small, frequency differences. By contrast, target distance discriminations remained unimpaired. When muscimol was instead applied to the FM-FM area, then fine, but not coarse, target range discrimination was impaired, and frequency discrimination remained normal. Thus, as predicted from the characteristics of cells and their functional organization, the FM-FM area is involved in the perception of distance, but has apparently little to do with frequency discrimination. The opposite is true for the DSCF area. Individual cortical fields appear to play pivotal roles in specific aspects of perception related to particular features of acoustic signals. In no other species so far investigated has there been as clear a demonstration of the function of specific cortical fields in perception.

4.2 Horseshoe Bat

Schuller and colleagues (Schuller, Radtke-Schuller, and O'Neill 1988; Schuller, O'Neill, and Radtke-Schuller 1991) carried out an extensive study of the FM region of the horseshoe bat cortex, and compared the results with data from the FM-FM, DF, and CF/CF areas in mustached bats gathered using identical experimental techniques in the same laboratory. In general, the cells in the FM region (as defined by Ostwald 1984) preferred hetero-harmonic combinations like those in mustached bat (cf. Tables 9.2 and 9.3). However, facilitation in horseshoe bats was less striking than in mustached bats, and many cells in the horseshoe bat showed good responses to CF or FM sounds presented alone. The fields containing combination-sensitive cells, although distinct, are also smaller in horseshoe bats than those in mustached bats, and so far there is no indication that more than one field of cortex contains neurons of each type (see Fig. 9.3A).

Consistent with the simpler biharmonic structure of the horseshoe bat sonar signal, only three types of combination-sensitive neurons were found in the horseshoe bat: FM_1-FM_2, CF_1/CF_2, and CFM_1-CFM_2. This latter class of "mixed" cells is facilitated by both CF_1/CF_2 and FM_1-FM_2 combinations, and is functionally equivalent to H_1-H_2 cells in area VA of the mustached bat (see Section 4.1.2 and 4.1.4). Mixed cells are also found in the FM-FM region of mustached bat cortex (Suga et al. 1983; Schuller, O'Neill, and Radtke-Schuller 1991), but they are not as prevalent, and have not been previously recognized as a separate class.

4.2.1 Response Areas of Combination-Sensitive Neurons

The tuning properties of horseshoe bat combination neurons resemble those in mustached bat, with some exceptions. As in mustached bat, horseshoe bat CF_1/CF_2 cells are typically broadly tuned for the CF_1 component and very sharply tuned for the CF_2 component. For FM_1-FM_2 cells, tuning for both components was broad, but there were cases in which the response area for the FM_1 component was narrower than that to the CF_1 component in a typical CF_1/CF_2 cell.

4.2.2 Delay Dependency

As in the mustached bat, horseshoe bat FM-FM cells are sharply delay tuned, whereas CF/CF cells were more often broadly tuned. Unlike the mustached bat, the majority of horseshoe bat FM-FM cells were tuned to short best delays between 1.5 and 4.5 msec (longest, 9.5 msec). However, there is a chronotopic representation of BD within a rostrocaudally elongated strip in the horseshoe bat cortex, distorted by an overrepresentation of BDs from 2 to 4 msec (see Fig. 9.3B,C). Within this strip, facilitated neurons are intermingled with nonfacilitated neurons, resembling FM bats (see following) but unlike the mustached bat. CF/CF and mixed cells are found in a mediolaterally elongated strip overlapping the rostral end of the FM-FM zone (see Fig. 9.3B).

Interestingly, facilitated cells were recorded only within cortical layer 5, at depths of 400–800 μm. This is similar to the preferred depth for recording such cells in the mustached bat (Schuller, O'Neill, and Radtke-Schuller 1991). Cells near the surface responded to FM_1 signals alone, whereas those below layer 5 often responded best to FM_2 signals alone. Thus, facilitation is restricted to layer 5 neurons sandwiched between nonfacilitated neurons in the superficial and deep layers tuned to the individual harmonics.

One of the more puzzling results to come from the study by Schuller, O'Neill, and Radtke-Schuller (1991) was that most horseshoe bat combination-sensitive neurons were tuned to frequencies that were appropriate for *negative* rather than positive Doppler shifts. In the case of CF/CF cells, rather than being tuned to frequencies equal to or higher than the second harmonic of the best CF_1 frequency for facilitation (i.e., positive Doppler shifts), the BFF for CF_2 was usually *lower* than twice the CF_1. The same was true for FM_1-FM_2 cells. Tuning for negative Doppler shifts also occurs in mustached bat, but to a lesser extent. Because only positive Doppler shifts can occur when the bat approaches a target, tuning to negative Doppler shifts would seem to be maladaptive. Of course, FM_1-FM_2 cells are broadly tuned to the lower FM component. However, CF/CF neurons are sharply tuned for the upper frequency for facilitation, making them less able to tolerate any mismatches between the stimulus and their response area. Thus, what exact role these cells play in perception is clouded by this rather puzzling finding.

One possible explanation comes from field observations of horseshoe bats (Link, Marimuthu, and Neuweiler 1986). These bats often adopt a strategy of ambushing prey from a hanging perch. In this instance, negative Doppler shifts would occur when the flying prey passed by the bat's position. This situation might elicit responses from CF/CF and mixed-type neurons. Unfortunately, whereas the echo frequency might be optimal when the prey has passed the bat's position, the pulse-echo time delay would likely be too long to excite the majority of FM-FM cells, whose BDs are 2–4 msec (34–68 cm target range).

Aside from the foregoing rather weak counterarguments, for the horseshoe bat it is difficult to reconcile the apparent selectivity for negative Doppler shifts with the hypothesis that combination-sensitive cells encode information about Doppler-shift magnitude or target range. Either the hypothesis needs to be revised, or something vital about the behavior of these bats is being overlooked (hardly surprising) that would reconcile the dilemma.

In general, then, despite many similarities in basic response properties and functional organization, it is clear that the horseshoe bat cortex is less differentiated than mustached bat cortex. Although the mustached bat has multiple large cortical fields and partial redundancy of representation of target range information, the horseshoe bat apparently makes due with only one relatively small ranging area, and the chronotopic axis heavily favors nearby targets. If one conceives of cortical maps as "data representations" that are involved in some way with perception, then one could argue that, because it has more maps, the mustached bat's perceptual world must be richer than that of the horseshoe bat, despite close similarities in echoloca-tion behavior. Do redundant or even partially redundant maps in cortex subserve different percepts, or control different behaviors? This exciting and important question remains mostly unanswered, despite the relative accessibility for experimentation of these cortical fields. As noted in Section 4.1.5, Riquimaroux, Gaioni, and Suga (1991, 1992) have shown that one type of percept can be selectively impaired by inactivation of the appro-priate cortical field, leaving other percepts intact. Does this imply that more widespread disruption of perception would occur in bats with less cortical differentiation in a similar experiment?

4.3 Little Brown Bat

4.3.1 Delay Tuning

Following closely on the discovery of delay-sensitive neurons in *Eptesicus* midbrain (Feng, Simmons, and Kick 1978) and *Pteronotus* cortex (Suga, O'Neill and Manabe 1978), Sullivan (1982a,b) investigated the response of cortical neurons in *Myotis* to paired FM stimuli. He found that fully 84% of the recorded cells were indeed facilitated by pairs of identical FM sweeps (*Myotis* sonar pulses typically have only one harmonic), and most of these were delay tuned as well (Sullivan 1982a). Unlike the CF/FM bats in which facilitation requires heteroharmonic combinations, delay-tuned facilitation in this species was expressed only when there was an *amplitude* difference between the two FM stimuli.

Delay-sensitive neurons fall into two categories, P-type and E-type (see Table 9.4). P-type (Fig. 9.9A, left) cells have short best echo delays, narrow delay tuning curves, and response latencies to FM pairs similar to that for loud (70–80 dB SPL) FM pulses delivered alone (Fig. 9.9A, top left). E-type cells (Fig. 9.9A, right), by contrast, have long best delays, broader delay tuning curves, and response latencies time locked to the echo FM (Fig. 9.9A, top right). That is, P-type cells discharge at a fixed latency to the pulse, and are facilitated only over a narrow range of echo delays. The latency of E-type units is locked to the echo, and their response is simply facilitated at particular delays by the presence of the earlier pulse FM. In this regard, P-type cells respond more like mustached and horseshoe bat FM-FM cells than do E-type cells.

On the basis of these observations, Sullivan (1982a) suggested that range might be encoded in *Myotis* by two distinct mechanisms operating at different target distances. Targets far away would excite E-type units, and distance might be encoded by the temporal pattern of facilitated discharge,

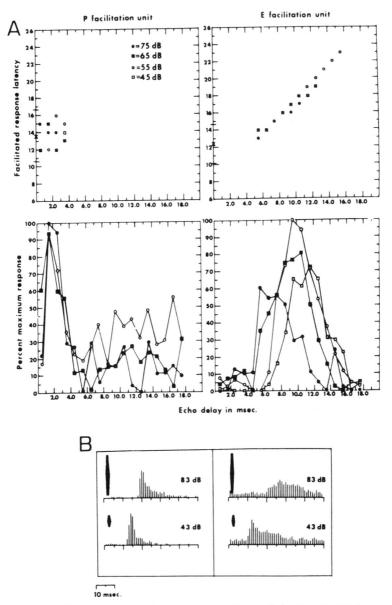

FIGURE 9.9A,B. Response properties of delay-sensitive cells in *Myotis*. (A) Response latency *(top)* and magnitude *(bottom)* as a function of echo delay for a P-type *(left)* and an E-type *(right)* cell. P-type cells were tuned to short-echo FM delays, and the response latencies were locked in time to the pulse FM component. E-type cells were tuned for longer echo delays, and the responses are locked to the echo component, rather than the pulse (from Sullivan 1982a; © 1982, J. Neurophysiol. 48:1011–1032, by permission of American Physiological Society). (B) Two examples of paradoxical latency shift. In each, the latency in response to a brief, high-amplitude FM stimulus *(top)* is longer than that to a low-amplitude stimulus *(bottom)*. Best delay is related to the *difference* in latency (from Sullivan 1982b; © 1982, J. Neurophysiol. 48:1033–1047, by permission of American Physiological Society).

rather than by a delay-tuned place mechanism. Targets nearby would excite
P-type cells, and these might encode distance by a place mechanism (as in
CF/FM bats), by virtue of their stable delay tuning.

However, Sullivan made an intriguing observation that suggested that
these response properties might be somewhat interchangeable. He found
that by lowering the amplitude of the pulse FM stimulus, an E-type cell
could be made to respond like a P-type unit. He suggested that the bat
might be able to control the type of range coding used by the system, simply
by altering the amplitude of its sonar emissions.

4.3.2 Paradoxical Latency Shift

Sullivan (1982b) investigated further the stimulus parameters essential for
facilitation of delay-sensitive cells. He found that in many units there was
a "paradoxical" relationship between response latency and stimulus amplitude (Fig. 9.9B). That is, rather than latency decreasing with increased
amplitude, he showed that latency *increased* in delay-sensitive cells. Units
showing such a "paradoxical latency shift" had level-response functions
with two peaks, one at low and the other at high amplitudes. Responses to
stimuli in the low-amplitude peak had shorter latencies than those in the
high-amplitude peak. Consequently, the latency to vocalized pulses would
be longer than that for most echoes, because pulses are normally much
more intense than echoes. In addition, Sullivan discovered that for P-type
cells, the latency to high-amplitude FM sweeps was related to the best delay,
and that best delay was related to the *difference* between the latencies for
strong and weak FM sweeps presented singly. Olsen later found that a
similar relationship existed for FM-FM delay-tuned cells in the mustached
bat medial geniculate body (Olsen 1986; Olsen and Suga 1991b; see also
Wenstrup, Chapter 8). Thus, in *Myotis* it appears that pulses and echoes are
differentiated simply on the basis of amplitude, rather than frequency as
found in CF/FM bats.

4.3.3 Effect of Repetition Rate and Duration on Delay Tuning

In a series of recent papers, Wong and colleagues have identified novel and
important additional factors that affect target range coding by delay-
sensitive neurons. Pinheiro, Wu, and Jen (1991) had already shown that IC
neurons in the big brown bat (*Eptesicus*) were selective for repetition rate.
Wong, Maekawa, and Tanaka (1992) and Teng and Wong (1993) found
that *Myotis* delay-sensitive cells also preferred certain repetition rates
between 5 and 50/sec. As in mustached bat, delay-tuned neurons showed
little change in delay tuning, whereas tracking neurons changed their delay
tuning, with increases in repetition rate.

Another parameter affecting delay sensitivity is pulse duration. *Myotis*
biosonar signals range from about 2–4 msec in search phase to as short as
0.2 ms in terminal phase (Griffin, Webster, and Michael 1960; Sales and

Pye 1974). Tanaka, Wong, and Taniguchi (1992) found that shortening the stimulus either extended delay-dependent facilitation to higher repetition rates, decreased the range of repetition rates over which which a cell was delay sensitive, or abolished facilitation altogether.

The combined effects of repetition rate and duration of biosonar signals on facilitation can thereby shift the number and location of active range-tuned cells in the cortex. During each phase of target pursuit, different but overlapping sets of delay-sensitive neurons are activated. The largest number of cells would be recruited to encode delay at repetition rates between 10 and 20 Hz, that is, during early approach phase. Moreover, at these repetition rates virtually the entire range of best delays is represented in the active population. This means that during search phase neurons are available that can encode target distances from about 17 cm to about 300 cm, that is, just about the entire range physically perceivable by an echolocating bat (Griffin 1958, 1971). In addition, neurons preferring higher repetition rates, those presumably active during mid-approach and terminal phase, have shorter BDs.

It is somewhat puzzling that there is such a diminution of the active range-encoding population during the later phases of target approach, because intuitively it seems that precise range encoding would be needed at that time. Perhaps, as has recently been hypothesized by Dear, Simmons, and Fritz (1993), these bats develop a detailed acoustic "image" of the objects arrayed in front of them only during the search phase, before the start of target pursuit. In *Myotis* this image would seem to deteriorate during the approach phase, assuming that the number of range encoding neurons impacts on image "quality." However, it must be remembered that despite their fewer numbers, delay acuity is maximum among neurons tuned to short delays.

4.3.4 Effect of Frequency on Delay Tuning

There is considerable evidence that, in all bats so far studied, target distance information is not carried over one spectral channel. Instead, at least two spectrally distinct channels separately encode pulses and echoes. Berkowitz and Suga (1989) have shown that, unlike CF/FM bats, delay-sensitive neurons in *Myotis* do not rely on different harmonics of the FM sweep for facilitation, but this is not to say that different frequencies were preferred in the pulse and echo. Using pure tone pairs (CF-CF stimuli), they found that the BFFs for pulses were about 8 kHz *higher* than for echoes. Maekawa, Wong, and Paschal (1992) divided the idealized *Myotis* FM sweep into four spectral quartiles, and found that most delay-tuned neurons preferred different spectral quartiles in pulses and echoes. Most units required the lower quartile in pulse FM sweeps; in echoes, most preferred one or more of the three lower quartiles.

In summary, most neurons in *Myotis* auditory cortex are delay tuned and presumably contribute to the processing of target range information. The mechanism for creating delay tuning is quite different from that seen in CF/FM bats. Rather than relying on spectral features (different harmonics) to differentiate pulses from echoes, the *Myotis* system utilizes amplitude differences. Moreover, delay tuning is strongly dependent on the duration and repetition rate of sonar signals, and the availability of range-encoding neurons is heavily influenced by the vocal behavior of the bat. As such, the ranging mechanism is more labile and dependent on the behavioral situation than it is in CF/FM bats.

4.4 Big Brown Bat

Dear et al. (1993) have completed the first detailed investigation of facilitation and delay tuning in *Eptesicus* cortical cells (Table 9.4). They found that, like *Myotis*, there is no apparent segregation of delay-sensitive neurons outside the tonotopically organized primary field. However, by contrast to *Myotis* and the CF/FM bats, they reported that only a small proportion of cortical neurons (about one in six) were delay tuned. Like delay-tuned neurons in *Myotis*, those in *Eptesicus* showed facilitation to paired FM stimuli, and only weak or no response to FM stimuli presented singly. Facilitation occurred only over a delimited set of pulse–echo delays, measured in this study with a single (5/sec) stimulus repetition rate. Regarding CF/CF-type facilitation, although there are a significant number of multipeaked cortical cells (Section 3.4.3), Dear et al. did not test these cells with simultaneously presented CF stimuli.

4.4.1 Response Properties and Latency Variation in Delay-Tuned Cells

Dear et al. (1993) found that nearly all delay-tuned cells had brief onset discharge patterns (average 1 spike/stimulus pair). Pulse and echo amplitude were significant for facilitation of delay-sensitive cells, and there is a reasonable, if not perfect, match between level of self-stimulation and pulse BFA. Not unexpectedly, echo BFAs were lower than pulse BFAs. Both the average echo BFA and the lack of correlation between echo BFA and BD are properties consistent with observations in *Pteronotus* delay-tuned cells (Taniguchi et al. 1986).

Dear et al. (1993) measured pulse and echo facilitation latencies in delay-tuned cells. The range of pulse facilitation latencies measured was enormous, from 9 to 42 msec. Echo facilitation latencies also varied widely, from about 5 to 35 msec. The authors make the important point that for any given echo delay (i.e., target range), the neuronal discharges signaling the occurrence of a target at a particular distance would be temporally dispersed over many milliseconds within the subpopulation of cells tuned to that delay.

Dear, Simmons, and Fritz (1993) addressed the consequences of such a

TABLE 9.4 Cortical Units in *Eptesicus fuscus*

Type	Frequency tuning	Delay dependency	Key parameters	Field(s)	Organization	Information encoded
CF	Sharp	Unknown	Frequency	A,C B	Tonotopic Variable	Stimulus spectrum
FM-FM	Broad	Sharp	Paired FM; echo delay; pulse and echo amplitude difference	A,B only	Tonotopic; delay axis in area B?	Target range

temporal dispersion of responses in range-tuned neurons. The authors pointed out that in a complex environment, each pulse emitted by a bat would generate multiple echoes from targets located at different distances. This array of echoes constitutes the "acoustic scene" perceived by the bat at a given moment in time. Of course, the echoes from each object in the scene actually arrive at different times, and were they to be encoded in real time by delay-tuned neurons responding at a single, fixed latency, perceptual "binding" of the elements in the acoustic scene would have to be accomplished by linking neural responses distributed in time. However, because delay-tuned cells exhibit a wide array of response latencies for any given echo delay, targets located at various distances from the bat could elicit responses *simultaneously* from subpopulations of neurons with differing best delays (recall the single-spike, stimulus-evoked discharge pattern and lack of spontaneous discharge). In the words of the authors, this process " . . . transform(s) the sequential arrival times of echoes with different delays into a concurrent, accumulating neural representation of multiple objects at different ranges. . . ." (Dear, Simmons, and Fritz 1993).

Consider, for example, the simple case of two targets, one near and the other far from the bat. At any given moment following the arrival of both echoes, two subpopulations of delay-tuned cells will be *simultaneously* active: (1) a group of *short*-BD cells with *long* latencies, encoding the nearby target, and (2) a group of *long*-BD cells with *short* response latencies, encoding the distant target. At each successive moment in time, new subpopulations encoding each target are recruited in the order of their increasing response latencies. Ensembles of cells with different best delays discharging simultaneously can thus evoke "snapshots" of the entire acoustic scene. These "snapshots" are updated continuously, adding the images of progressively more distant objects until the next pulse is emitted. This process is analogous to viewing visual scenes with stroboscopic illumination. Thus, the array of latencies in delay-tuned cells could be considered as yet another example of *delay lines* by which behaviorally related events occurring at different times (representing in this case targets at different distances) can be encoded at the neuronal level by coincident activity in a population of cells.

An additional feature of the Dear, Simmons, and Fritz (1993) scene-analysis schema is that for at least some delay-tuned neurons, range acuity improves with time after vocalization. They showed that delay-tuning curve sharpness is better in cells with longer facilitation latencies, and claim that successive acuity improvements are consistent with a computational algorithm called "multiresolution decomposition," an image-analysis technique useful for such things as edge detection and object recognition. This view of cortical range processing is provocative and worthy of further consideration, because all bats studied so far show similar facilitation latency distributions for their delay-tuned populations.

4.4.2 "Amplitude-Shift" Delay-Tuned Neurons

About 13% of the delay-tuned neurons recorded by Dear et al. (1993) changed their BDs with changes in pulse amplitude. These cells were dubbed "amplitude-shift" neurons, and were nearly equally divided between cortical areas A and B. Most of these cells exhibited a fairly large decrease in BD with a decrease in pulse amplitude, but a couple showed an increase in BD under similar conditions. Small changes in the effective pulse amplitude (i.e., the amplitude stimulating the auditory system versus that actually impinging upon the ears) could thereby have drastic effects on the range tuning of this small population of cells.

As mentioned in Section 4.1.1 and 4.3.3, cells in *Pteronotus* and *Myotis* called "tracking neurons" also have labile delay-tuning properties (O'Neill and Suga 1979, 1982; Wong, Maekawa, and Tanaka 1992). Like amplitude-shift cells in *Eptesicus*, tracking neurons are few in number and are found scattered among delay-tuned neurons. While amplitude-shift cells in *Eptesicus* behave somewhat similarly, it is not clear that they are the equivalent of tracking neurons in *Pteronotus* and *Myotis*, because tracking neurons change their delay tuning not with changes in pulse *amplitude*, but rather with changes in pulse–echo *repetition rate*. Only one study of *Pteronotus* directly investigated whether delay tuning changed with pulse or echo amplitude (Taniguchi et al. 1986). They found that BD shortened with decreased amplitude in about one-third of their sample, but unlike amplitude-shift cells, the shift in BD was very small. Dear et al. (1993) also point out that while the delay tuning curves of tracking neurons become narrower with higher pulse repetition rates, those of amplitude-shift cells are unaffected by amplitude. Other disimilarities in the response properties of these cells bring into question the functional homology between amplitude-shift cells and tracking neurons, and this issue requires further study.

4.4.3 Organization of Delay-Tuned Neurons in *Eptesicus*

Similar to *Myotis*, delay-tuned neurons were found mainly in the anterior half of the tonotopic area A and throughout the nontonotopic area B of auditory cortex in *Eptesicus*, but none were found in the anterior tonotopic area C (Dear et al. 1993). Contour plots of BD averaged from eight bats (Fig. 9.5B) showed no single overall place representation of echo delay like that seen in CF/FM bats. However, it could be argued that BD is systematically represented in smaller patches of cortex, especially within the dorsorostral part of area B. Area B, in turn, is a region superficially homologous to the nontonotopic cortical fields in CF/FM bats where delay-tuned cells are found. Whether these small regions constitute a map of target range in *Eptesicus* is debatable. Not debatable, however, is the fact that delay-tuned cells in *Eptesicus* are intermingled with cells not sensitive to delay, as they are in both *Rhinolophus* and *Myotis*. They are also tonotopically organized in primary cortex, like *Myotis* but quite different from *Pteronotus* and *Rhinolophus*.

Based on the results from other species, one might expect to find an overrepresentation of certain BDs related to behaviorally significant target ranges. Unique to *Eptesicus*, Dear et al. (1993) found a bimodal distribution of BDs, forming two groups of cells with BDs less than 9 msec and more than 12 msec. Why there is this curious gap in BD representation is not immediately obvious. Does this mean that *Eptesicus* has little or no range acuity for objects between about 175 and 200 cm away? A behavioral test of this hypothesis would seem worthwhile. Dear et al. (1993) found that the BDs of the majority of cells in the long-BD group clustered between 12 and 22 msec, although there were a few outliers with BDs greater than 30 msec. This BD range corresponds to distances between 200 and 400 cm. The prominence of long-BD cells in the population is distinctly different from the situation in the CF/FM bats, where cells with BDs longer than 12 msec are very rare. Dear et al. also found that short-BD neurons were more typically found in the tonotopic area A, whereas long-BD neurons were clustered in the nontonotopic area B (see Fig. 9.5).

No studies have as yet systematically examined either the spectral response preferences or repetition rate sensitivity of delay-tuned neurons in *Eptesicus*. These factors are critical to whether cells even show delay sensitivity in *Myotis*. Until their effects are studied, it will remain unclear whether the representations of echo delay currently published accurately reflect the organization of *Eptesicus* cortex.

5. Circuit Models of Delay Tuning

The discovery of neurons able to encode time differences between temporally separated acoustic events has important implications for the auditory perception of complex signals. From his discovery of paradoxical latency shift and its relationship to delay sensitivity in *Myotis* cortical neurons, Sullivan (1982b) suggested a model that establishes delay tuning from the convergence of two afferent pathways, one processing high-amplitude and the other low-amplitude signals. This model can also be extended to apply to CF/FM bats, if one substitutes heteroharmonic facilitation for the paradoxical latency shift mechanism. One variant of the model involves delay lines, whereby the high-amplitude (FM bat)/low-frequency (CF/FM bat) pathway is delayed to produce longer latencies than the low-amplitude (high-frequency) pathway (Fig. 9.10A). The delay-tuned cell acts as a coincidence detector (logical AND gate), responding to the echo only when it is delayed acoustically by an amount equal to the difference in latencies of the two pathways. This delay-line model resembles that first put forth by Jeffress (1948) to explain detection of interaural time differences, for which

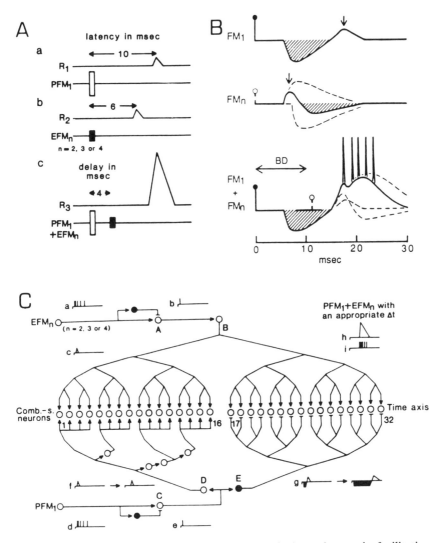

FIGURE 9.10A–C Circuit models for delay tuning via heteroharmonic facilitation mechanism, as used by CF/FM bats. (A) *Delay line model*, in which facilitation by a pulse FM and an echo FM at the best delay (c) is related to the difference between latencies to the pulse (a) and the echo (b) presented alone. In FM bats, amplitude differences between pulse and echo produce paradoxical latency shift having a similar effect (see Fig. 9.9B). This model is thought to be valid for delay-tuned cells with short BDs, and involves delay lines in the FM_1 channel produced by pathlength differences shown on the left half of the circuit model shown in (C) (from Suga 1990a; © 1990, with permission from Elsevier Science Ltd.). (B) *Neural inhibition* model for long BD cells. The response to the FM_1 alone *(top)* is inhibitory. At the end of the inhibitory period, there is a slight rebound in response that might give rise to a weak, long-latency response. The response to the echo FM component alone *(middle)* is excitatory, with a short latency, and may or may not be followed by

there is strong experimental support from work in both the avian (Konishi et al. 1988) and mammalian (Yin and Chan 1988) auditory system.

The other model suggested by Olsen (1986) and elaborated by Suga (1990a,b) involves similar coincidence detection, but the delay lines are created from the interplay of excitation and inhibition (Fig. 9.10B). In this model, the high-amplitude (low-frequency) afferent pathway is inhibitory, and the low-amplitude (high-frequency) pathway is excitatory, on the target delay-tuned cell. In this case, the best delay is related to the difference between the latency for recovery from inhibition (technically, the pulse-alone latency), which presumably varies across afferents to cells with different BDs, and the latency for the echo, which may be relatively similar across afferents. Olsen (1986) and Olsen and Suga (1991b) showed that, in delay-tuned neurons in the medial geniculate body, a delay-line mechanism might account for delay tuning for short delays (<4 msec), while the inhibition mechanism seems responsible for long best delays(>4 msec). The complete model (Suga 1990b) combining both delay-line mechanisms is shown in Figure 9.10C.

One difficulty facing the paradoxical latency shift theory is its vulnerability to disturbance by the many factors influencing pulse and echo amplitudes. There are many mechanisms, both behavioral and physiological, that reduce the amplitude differences between pulses and echoes. In FM bats, attenuation of self-stimulation during pulse emission amounts to about 35–40 dB, attributable to the combined effects of ear directionality (Grinnell and Grinnell 1965), vocal tract directionality (Hartley and Suthers 1989), middle ear muscle activation (Suga and Jen 1977), and central suppression at the level of the midbrain (Suga and Schlegel 1972; Suga and Shimozawa 1974). Moreover, the attenuation of the auditory system wanes with time following pulse emission, reducing the effective amplitude of echoes from nearby, but not distant, targets. This resembles an "automatic gain control," and its effect is to disassociate echo amplitude from target

---

FIGURE 9.10A–C. (*continued*) inhibition (tonic *[dashed line]* vs. phasic *[solid line]* response). When pulse is combined with an echo delayed by an amount equal to the latency difference between pulse and echo alone (i.e., at BD), the excitation in the echo afferent channel combines nonlinearly with the inhibitory rebound in the pulse afferent channel, resulting in strong facilitation and high discharge rates (*dashed lines* show underlying membrane responses to pulse and echo alone, and to pulse paired with echo) (from Suga, Olsen, and Butman 1990, © 1990, by permission of Cold Spring Harbor Laboratory Press). (C) Circuit model proposed by Suga (1990b) for delay tuning. *Circles* in the center represent delay-tuned neurons. Input from the echo (EFMn) afferent channels *(top)* arrives with little or no delay. Input from the pulse (PFM$_1$) afferent channel passes through delay lines established by pathlength differences (short best delay cells, *left half*) from pathway labeled "D", or inhibition (long best delay cells, *right half*) from pathway labeled "E" (from Suga 1990b, © 1990, Sci. Am. 262:60–68, by permission of Scientific American, Inc.).

range (Kick and Simmons 1984; Hartley and Suthers 1990). In CF/FM bats there is also evidence for "amplitude compensation" whereby the bat decreases pulse amplitude as it approaches obstacles, effectively stabilizing the amplitude difference between pulses and echoes (Kobler et al. 1985). The net effect of both these mechanisms is to attenuate the pulse amplitude and thereby reduce the amplitude differences between pulses and echoes. As Sullivan (1982b) noted, it is possible under certain conditions that echo amplitude might exceed vocal self-stimulation. How then can the *Myotis* auditory system discriminate pulses from echoes?

6. Discussion and Summary

It should be clear from the foregoing that, while there are many similarities in the response properties of cortical cells in different bats, there are also important species differences in the organization of those cells in the auditory cortex. These differences seem to be most pronounced between FM and CF/FM bats, and it is intriguing to speculate about which type of cortical organization is more representative. Common features include at least one tonotopically organized cortical field, overrepresentation of sonar signal frequencies, and specialized delay-sensitive neurons predominantly or exclusively responsive to paired FM sweep stimuli. In both *Myotis* and *Eptesicus*, the primary tonotopic field dominates the cortex, and columns of FM-specialized, delay-sensitive neurons are embedded among columns of unspecialized, frequency-tuned cells. In *Eptesicus*, neurons with multi-peaked tuning curves also lie within this matrix of columns. The intermingling of delay-tuned, multipeaked, and single-frequency tuned columns may well have some significance in terms of information processing. Dear et al. (1993) have put forth the idea that this type of cortical organization better subserves the processing of echo information, nearly all of which they feel is expressed in the time domain. In their opinion, it is logical for multipeaked cells that possibly encode spectral notches to be located near delay-tuned cells reporting overall target range. This is because spectral notches are caused by small-scale time differences in the multiple wave-fronts reflected from a complex target (Beuter 1980), and Simmons, Moss, and Ferragamo (1990) have provided evidence suggesting that bats perceive notches as if they were time-domain encoded. In this view, spectral and temporal information is encoded along a single temporal axis, and it would be logical to find cells encoding this information intermingled in the same field of auditory cortex. Whether in fact FM bats encode spatial information exclusively in the time domain is currently the topic of considerable debate (see Pollak 1993).

As discussed in Section 4.4.1, Dear, Simmons, and Fritz (1993) have pointed out that the delay-sensitive cells respond with a variety of latencies, and could provide the bat a type of "acoustic scene analysis" analogous to the way visual scenes are represented. While it has not yet been determined in *Eptesicus*, in *Myotis* many delay-tuned neurons only reveal delay sensitivity at certain stimulus rates, suggesting that different subsets of neurons are recruited into the analysis of range at different stages of target approach. This imparts an adaptive lability to the information processing in *Myotis* cortex. By contrast, if one ignores the small population of tracking neurons, there is little or no dependence of range tuning on vocal emission rate in the mustached bat. Mustached bat cortical organization is thereby more machine-like, providing a stable representation of echo information regardless of the behavioral state. It is not yet known whether other CF/FM bats such as *Rhinolophus* also show a stable cortical range representation under different conditions of stimulation.

The two CF/FM species have other links to the FM bats. In the mustached bat, multipeaked and delay-sensitive neurons have recently been found in the DSCF area of AI, similar to the organization of AI in FM bats (Fitzpatrick et al. 1993). However, delay-sensitive DSCF cells are not well suited for target ranging, in that their delay tuning is broad, variable, and favors very distant targets. The role played by these cells may be simply to amplify the probability of target detection at the physical limits of the bat's echo detection envelope. FM-FM cells with short BDs that are well suited for ranging are completely segregated in the mustached and horseshoe bats into distinct cortical fields. CF/CF cells, which seem especially suited to signal processing in species with long CF components in their sonar signals, also are segregated both from the tonotopic and the range-tuned areas in the CF/FM bats. However, in at least one respect, *Rhinolophus* may represent a "missing link" between FM bat and mustached bat cortical organization, because nearly half the cells recorded in the *Rhinolophus* combination-sensitive zones were tuned to tonal stimuli and were not combination sensitive. Like FM bats, combination-sensitive cells are intermingled in *Rhinolophus* cortex with cells tuned to simpler stimuli. However, in *Rhinolophus* this area is separate from AI and is not tonotopically organized.

Other differences between the two groups of species are more clearly definable. First, there is a major difference in the mechanism producing delay sensitivity. In CF/FM bats, the two channels bringing range information to the forebrain are tuned to different harmonic components that identify the pulse and the echo stimulus. O'Neill and Suga (1982) named this "heteroharmonic" facilitation. The hypothesis for this mechanism is that the pulse is encoded by first-harmonic-tuned cells, and routed through an array of delay lines, while the echo is encoded by higher harmonic-tuned cells and presumably routed through a fast pathway. In the FM bats, pulse and echo FM signals are apparently differentiated by amplitude: high-amplitude pulse information is captured by a yet-unknown mechanism and passed via an array of delay lines to the forebrain (paradoxical latency shift; Section 4.3.2), whereas low-amplitude echoes are routed through a fast pathway. Interestingly, even in the amplitude-based system of *Myotis*, the delay-tuned neurons are facilitated by different FM sweep parcels in the pulse and echo. It is interesting to speculate whether the heteroharmonic facilitation seen in CF/FM bats that use multiharmonic signals is simply an evolutionary development from the FM bat system that utilizes different parts of the broadband single-harmonic FM sweep in pulses and echoes.

The second big difference between CF/FM and FM bat cortex is in the expression of a neural map of target range, and the segregation of that map to a discrete cortical field. Maplike representations of sensory information are widespread and prominent in the visual, somatosensory, and auditory systems. Typically, these maps simply recapitulate the spatial arrangement of receptors in the sensory periphery, accompanied by overrepresentation of particular receptive fields. However, maps representing the spatial location of sound sources, including target range information in echolocating bats, must be computed by neural circuitry in the auditory system. The circuitry for computing target range clearly involves both midbrain (inferior colliculus and nuclei of the lateral lemniscus) and forebrain structures (auditory thalamus and cortex). In all bats so far studied, this circuitry culminates in delay-tuned forebrain neurons that can encode target range. However, the expression of a systematic target range map is highly developed only in the mustached bat, where it extends into at least three separate cortical fields, while such a map is distinct but less well developed in the horseshoe bat. By contrast, even though there are many range-encoding neurons in both the FM bats studied, there is no clearly defined map of echo delay (although it has been argued here that such an organization might occur on a small scale in *Eptesicus*).

This major difference in the organization of delay-tuned cells brings up the question of the role played by the mapped fields in CF/FM bats. No one would argue that target-range processing is any less sophisticated in FM bats because they lack a map of target range. Indeed, there is behavioral evidence that FM bats have better range acuity than CF/FM bats (Simmons 1973), as predicted from the advantage their broadband signal structure gives them in temporal resolving power (Simmons, Howell, and Suga 1975; Simmons and Stein 1980).

Might the answer to this question lie in the difference in habitats preferred by these species for hunting insects? CF/FM bats have evolved their elaborate sonar specializations to enable them to hunt in cluttered habitats (Neuweiler et al. 1987), where they must parse out many conflicting echoes vying for their attention with those from their intended prey. FM bats typically hunt over open ground or water, where there are presumably fewer objects cluttering their sonar "screen," and local target motion is not a critical identifying factor necessary to spot prey. When FM bats increase the pulse emission rate during a pursuit, they bring groups of range-tuned cells "on line," and the activation pattern in the cortex is complex and unsystematic.

CF/FM bats, on the other hand, put a premium on detecting and tracking a moving target in a background of echo clutter. Having a maplike representation of the acoustic scene that segregates the targets systematically along a range axis seems intuitively easier to interpret than having a more random excitation pattern like that in FM bats. The well-conceived experiments of Riquimaroux, Gaioni, and Suga (1991) showed that fine, but not coarse, range discrimination was disrupted by lesions of the FM-FM area in the mustached bat. Further, Altes (1989) has argued that having a maplike representation provides the bat with a data structure well suited for computation of target range on a finer scale than is possible just from considering the granularity of BD representation or delay-tuning-curve widths alone (sort of a range hyperacuity phenomenon). This may permit mustached and horseshoe bats to overcome some of the range acuity limitations inherent in their narrow-bandwidth sonar signals (Simmons 1973; Simmons, Howell, and Suga 1975), as well as provide a better neural substrate to assist in the spatial localization of small targets in highly cluttered habitats.

Despite the progress made in the last decade or so, much remains to be discovered about the auditory cortex in bats. For example, there have only been three, fairly small-scale, studies done on spatial sensitivity or binaural interaction in the bat cortex (Manabe, Suga, and Ostwald 1978; Jen, Sun, and Lin 1989; Suga, Kawasaki, and Burkhard 1990). Efforts need to be made to link bat cortical organization more clearly with that in other mammals. For example, it would be worthwhile to know whether such properties as tuning-curve sharpness, FM sweep rate selectivity, minimum threshold, or best amplitude of cells in the tonotopic fields are organized along isofrequency contours as they are in cat cortex (see Clarey, Barone, and Imig 1992 for review). Also, the current experiments of Ohlemiller, Kanwal, and Suga (1993), examining how cells tuned to range and Doppler-shift magnitude respond to the enormously rich variety of mustached bat communication sounds, should inspire further work on this important and long-neglected area of the bat acoustic behavior. The structural similarity of acoustic elements in these sounds to syllables in human speech is often striking (Kanwal et al. 1994), and understanding how these elements are processed by specialized cortical neurons in bats might give rise to useful models of speech sound analysis at higher levels of the auditory system.

Finally, it is hoped that the wealth of information already gleaned from study of the cortex in bats will inspire others to combine biologically relevant complex and simple stimuli in their attempts to understand the organization of auditory cortex in other species.

Acknowledgments.   I gratefully acknowledge the contribution of G. Schuller, S. Radtke-Schuller, G. E. Duncan, O. W. Henson and D. Wong for anatomical material on bat cortex, Martha Zettel for assistance with figures, Nobuo Suga, George Pollak, Donald Wong, and Jeff Wenstrup for valuable discussions about the auditory forebrain, and Willard Wilson, Art Popper, and Dick Fay for helpful comments on the manuscript. Support for the author was provided in part by the National Institute for Deafness and Communicative Disorders (R01-DC00267) and the National Institute on Aging (P01-AG09524).

Abbreviations

AI	primary auditory cortex
AII	secondary auditory cortex
BA	best amplitude
BD	best delay
BF	best frequency
BFA	best amplitude for facilitation
BFF	best frequency for facilitation
CF	constant frequency (signal component)
CF/CF	CF/CF area of mustached bat cortex; stimulus pair consisting of two CF components
CF/FM	sonar signal consisting of CF and FM components
CF_1/CF_2	combination-sensitive neuron facilitated by a first- and second-harmonic CF signal presented simultaneously
CF_1/CF_3	combination-sensitive neuron facilitated by a first- and third-harmonic CF signal presented simultaneously
CF_1, CF_2, CF_3, CF_4	CF components of the four harmonics of mustached bat sonar signal
CFM-CFM	combination-sensitive neuron facilitated by both CF and FM components ($= H_1 - H_2$ [below])
DF	area DF (dorsal fringe) of mustached bat cortex
DSCF	Doppler-shifted CF region of mustached bat AI
FM	frequency-modulated signal; a type of sonar signal.
FM-FM	region of mustached bat cortex containing delay-sensitive neurons tuned to pairs of FM signals
FM_1, FM_2, FM_3, FM_4	FM components of the four harmonics of mustached bat sonar signal
FM_1-CF_2	combination of signal elements that facilitates neurons in mustached bat DSCF area
FM_1-FM_2	delay-sensitive neuron facilitated by pairs of first- and second-harmonic FM signals
FM_1-FM_3	delay-sensitive neuron facilitated by pairs of first- and third-harmonic FM signals
FM_1-FM_4	delay-sensitive neuron facilitated by pairs of first- and fourth-harmonic FM signals
H_1-H_4	four harmonics of the mustached bat sonar signal (includes both CF and FM components)

H_1-H_2	delay-sensitive neuron facilitated by pairs of either CF or FM components (see "CFM-CFM" above); region of mustached bat cortex (= area VA)
IC	inferior colliculus
MGB	medial geniculate body
MT	minimum threshold
$Q_{10\,dB}$, $Q_{30\,dB}$, $Q_{50\,dB}$	best frequency divided by the tuning-curve bandwidth 10, 30, or 50 dB above MT
VA	area VA of mustached bat cortex (= area H_1-H_2)
VF	area VF (ventral fringe) of mustached bat cortex

References

Altes RA (1989) An interpretation of cortical maps in echolocating bats. J Acoust Soc Am 85:934–942.

Asanuma A, Wong D, Suga N (1983) Frequency and amplitude representations in anterior primary auditory cortex of the mustached bat. J Neurophysiol (Bethesda) 50:1182–1196.

Berkowitz A, Suga N (1989) Neural mechanisms of ranging are different in two species of bats. Hear Res 41:255–264.

Beuter KJ (1980) A new concept of echo evaluation in the auditory system of bats. In: Busnel RG, Fish JF (eds) Animal Sonar Systems. New York: Plenum, pp. 747–764.

Bruns V (1976) Peripheral auditory tuning for fine frequency analysis of the CF-FM bat, *Rhinolophus ferrumequinum*. II. Frequency mapping in the cochlea. J Comp Physiol A 106:87–97.

Bruns V, Schmieszek E (1980) Cochlear innervation in the greater horseshoe bat: demonstration of an acoustic fovea. Hear Res 3:27–43.

Clarey JC, Barone P, Imig TJ (1992) Physiology of thalamus and cortex. In: Popper AN, Fay RR (eds) Springer Handbook of Auditory Research, Vol 2, The Mammalian Auditory Pathway: Neurophysiology. New York: Springer-Verlag, pp. 232–334.

Covey E, Johnson BR, Ehrlich D, Casseday JH (1993) Neural representation of the temporal features of sound undergoes transformation in the auditory midbrain: evidence from extracellular recording, application of pharmacological agents, and in vivo whole cell patch clamp recording. Soc Neurosci Abstr 19:535.

Dear SP, Simmons JA, Fritz J (1993) A possible neuronal basis for representation of acoustic scenes in auditory cortex of the big brown bat. Nature 364:620–623.

Dear SP, Fritz J, Haresign T, Ferragamo M, Simmons JA (1993) Tonotopic and functional organization in the auditory cortex of the big brown bat, *Eptesicus fuscus*. J Neurophysiol (Bethesda) 70:1988–2009.

Duncan GE, Henson OW (1994) Brain activity patterns in flying, echolocating bats *(Pteronotus parnellii)*: assessment by high resolution autoradiographic imaging with ³H-2-deoxyglucose. Neurosci 59:1051–1070.

Edamatsu H, Suga N (1993) Differences in response properties of neurons between two delay-tuned areas in the auditory cortex of the mustached bat. J Neurophysiol (Bethesda) 69:1700–1712.

Edamatsu H, Kawasaki M, Suga N (1989) Distribution of combination-sensitive neurons in the ventral fringe area of the auditory cortex of the mustached bat. J Neurophysiol (Bethesda) 61:202–207.

Feng AS, Simmons JA, Kick SA (1978) Echo detection and target-ranging neurons in the auditory system of the bat *Eptesicus fuscus*. Science 202:645–648.

Ferrer I (1987) The basic structure of the neocortex in insectivorous bats (*Miniopterus sthreibersi* and *Pipistrellus pipistrellus*). A Golgi study. J Hirnforsch 28:237–243.

Fitzpatrick DC, Henson OW (1994) Cell types in the mustached bat auditory cortex. Brain Behav Evol 43:79–91.

Fitzpatrick DC, Kanwal JS, Butman JA, Suga N (1993) Combination-sensitive neurons in the primary auditory cortex of the mustached bat. J Neurosci 13:931–940.

Fritz JB, Olsen J, Suga N, Jones EG (1981) Connectional differences between auditory fields in a CF-FM bat. Soc Neurosci Abstr 7:391.

Fuzessery ZM, Pollak GD (1984) Neural mechanisms of sound localization in an echolocating bat. Science 225:725–728.

Fuzessery ZM, Pollak GD (1985) Determinants of sound location selectivity in bat inferior colliculus: a combined dichotic and free-field stimulation study. J Neurophysiol (Bethesda) 54:757–781.

Fuzessery ZM, Hartley DJ, Wenstrup JJ (1992) Spatial processing within the mustache bat echolocation system: possible mechanisms for optimization. J Comp Physiol A 170:57–71.

Goldman LJ, Henson OW (1977) Prey recognition and selection by the constant frequency bat, *Pteronotus p. parnellii*. Behav Biol Sociobiol 2:411–419.

Gooler DM, O'Neill WE (1987) Topographic representation of vocal frequency demonstrated by microstimulation of anterior cingulate cortex in the echolocating bat, *Pteronotus parnelli parnelli*. J Comp Physiol A 161:283–294.

Gooler DM, O'Neill WE (1988) Central control of frequency in biosonar emissions of the mustached bat. In: Nachtigall PE, Moore PWB (eds) Animal Sonar: Processes and Performance. New York: Plenum, pp. 265–270.

Griffin DR (1958) Listening in the Dark. New Haven: Yale University Press.

Griffin DR (1971) The importance of atmospheric attenuation for the echolocation of bats. Anim Behav 19:55–61.

Griffin DR, Webster FA, Michael C (1960) The echolocation of flying insects by bats. Anim Behav 8:141–154.

Grinnell AD, Grinnell VS (1965) Neural correlates of vertical localization by echolocating bats. J Physiol 181:830–851.

Grinnell AD, Schnitzler H-U (1977) Directional sensitivity of echolocation in the horseshoe bat *Rhinolophus ferrumequinum*. II. Behavioral directionality of hearing. J Comp Physiol A 116:63–76.

Habersetzer J, Vogler B (1983) Discrimination of surface structured targets by the echolocating bat *Myotis myotis* during flight. J Comp Physiol A 152:275–282.

Hartley DJ, Suthers RA (1989) The sound emission pattern of the echolocating bat, *Eptesicus fuscus*. J Acoust Soc Am 85:1348–1351.

Hartley DJ, Suthers RA (1990) Sonar pulse radiation and filtering in the mustached bat, *Pteronotus parnellii rubiginosus*. J Acoust Soc Am 87:2756–2772.

Hawkins JE, Stevens SS (1950) The masking of pure tones and speech by white noise. J Acoust Soc Am 22:6–13.

Henson OW (1970) The central nervous system of Chiroptera. In: Wimsatt WA (ed) Biology of Bats. New York: Academic Press, pp. 57–152.

Henson OW, Bishop A, Keating A, Kobler J, Henson M, Wilson B, Hansen R

(1987) Biosonar imaging of insects by *Pteronotus p. parnellii*, the mustached bat. Natl Geogr Res 3:82–101.

Imig TJ, Adrian HO (1977) Binaural columns in the primary field (AI) of cat auditory cortex. Brain Res 138:241–257.

Jeffress (1948) A place theory of sound localization. J Comp Psychol 41:35–39.

Jen PHS, Chen D (1988) Directionality of sound pressure transformation at the pinna of echolocating bats. Hear Res 34:101–118.

Jen PHS, Sun X, Lin PJJ (1989) Frequency and space representation in the primary auditory cortex of the frequency modulating bat *Eptesicus fuscus*. J Comp Physiol A 165:1–14.

Kanwal JS, Ohlemiller KK, Suga N (1993) Communication sounds of the mustached bat: classification and multidimensional analyses of call structure. Assoc Res Otolaryngol Abstr 16:111.

Kanwal JS, Matsumura S, Ohlemiller KK, Suga N (1994) Analysis of acoustic elements and syntax in communication sounds emitted by mustached bats. J Acoust Soc Amer 96:1229–1254.

Kawasaki M, Margoliash D, Suga N (1988) Delay-tuned combination-sensitive neurons in the auditory cortex of the vocalizing mustached bat. J Neurophysiol (Bethesda) 59:623–635.

Kelly JB, Judge PW, Phillips DP (1986) Representation of the cochlea in primary auditory cortex of the ferret. Hear Res 24:111–115.

Kick SA, Simmons JA (1984) Automatic gain control in the bat's sonar receiver and the neuroethology of echolocation. J Neurosci 4:2725–2737.

Knudsen EI, Konishi M (1978) Space and frequency are represented seperately in auditory midbrain of the owl. J Neurophysiol (Bethesda) 41:870–884.

Kober R (1988) Echoes of fluttering insects. In: Nachtigall PE, Moore PW (eds) Animal Sonar: Processes and Performance. New York: Plenum, pp. 477–482.

Kober R, Schnitzler H-U (1990) Information in sonar echoes of fluttering insects available for echolocating bats. J Acoust Soc Am 87:882–895.

Kobler JB, Isbey SF, Casseday JH (1987) Auditory pathways to the frontal cortex of the mustache bat, *Pteronotus parnellii*. Science 236:824–826.

Kobler JB, Wilson BS, Henson OW Jr, Bishop AL (1985) Echo intensity compensation by echolocating bats. Hear Res 20:99–108.

Konishi M, Takahashi TT, Wagner H, Sullivan WE, Carr CE (1988) Neurophysiological and anatomical substrates of sound localization in the owl. In: Edelman GM, Gall WE, Cowan WM (eds) Auditory Function: Neurobiological Bases of Hearing. New York: Wiley, pp. 721–745.

Lesser HD (1987) Encoding of amplitude-modulated sounds by single units in the inferior colliculus of the mustached bat, *Pteronotus parnelli*. Ph.D. Thesis, University of Rochester, Rochester, NY.

Lesser HD, O'Neill WE, Frisina RD, Emerson RC (1990) ON-OFF units in the mustached bat inferior colliculus are selective for transients resembling "acoustic glint" from fluttering insect targets. Exp Brain Res 82:137–148.

Link A, Marimuthu G, Neuweiler G (1986) Movement as a specific stimulus for prey catching behavior in rhinolophid and hipposiderid bats. J Comp Physiol A 159:403–413.

Maekawa M, Wong D, Paschal WG (1992) Spectral selectivity of FM-FM neurons in the auditory cortex of the echolocating bat, *Myotis lucifugus*. J Comp Physiol A 171:513–522.

Makous JC, O'Neill WE (1986) Directional sensitivity of the auditory midbrain in the mustached bat to free-field tones. Hear Res 24:73–88..

Manabe T, Suga N, Ostwald J (1978) Aural representation in the Doppler-shifted-CF processing area of the auditory cortex of the mustache bat. Science 200:339–342.

Merzenich MM, Brugge JF (1973) Representation of the cochlear partition on the superior temporal plane of the macaque monkey. Brain Res 50;275–296.

Metzner W (1993) An audiovocal interface in echolocating horseshoe bats. J Neurosci 13:1862–1878.

Middlebrooks JC, Dykes RW, Merzenich MM (1980) Binaural response-specific bands in primary auditory cortex (AI) of the cat: Topographical organization orthogonal to isofrequency contours. Brain Res 181:31–48.

Mogdans J, Schnitzler H-U (1990) Range resolution and the possible use of spectral information in the echolocating bat, *Eptesicus fuscus*. J Acoust Soc Am 88:754–757.

Neuweiler G (1970) Neurophysiologische Untersuchungen zum Echoortungssystem der Grossen Hufeisennase *Rhinolophus ferrumequinum* Schreber. J Comp Physiol A 67:273–306.

Neuweiler G, Metzner W, Heilmann U, Rubsamen R, Eckrich M, Costa HH (1987) Foraging behavior and echolocation in the rufous horseshoe bat (*Rhinolophus rouxi*) of Sri Lanka. Behav Ecol Sociobiol 20:53–67.

Novick A, Vaisnys JR (1964) Echolocation of flying insects by the bat, *Chilonycteris parnellii*. Biol Bull 127:478–488.

Ohlemiller KK, Kanwal JS, Suga N (1993) Do cortical auditory neurons of the mustached bat have a dual function for processing biosonar signals and communication sounds? Assoc Res Otolaryngol Abstr 16:111.

Olsen JF (1986) Processing of biosonar information by the medial geniculate body of the mustached bat, *Pteronotus parnellii*. Ph.D. Thesis, Washington University, St. Louis, MO.

Olsen JF, Suga N (1991a) Combination-sensitive neurons in the medial geniculate body of the mustached bat: Encoding of relative velocity information. J Neurophysiol (Bethesda) 65:1254–1274.

Olsen JF, Suga N (1991b) Combination-sensitive neurons in the medial geniculate body of the mustached bat: encoding of target range information. J Neurophysiol (Bethesda) 65:1275–1296.

O'Neill WE (1985) Responses to pure tones and linear FM components of the CF-FM biosonar signal by single units in the inferior colliculus of the mustached bat. J Comp Physiol A 157:797–815.

O'Neill WE, Basham M (1992) Pulse-echo stimulus combinations can facilitate sonar signal vocalizations elicited by electrical stimulation of the anterior cingulate cortex in the mustache bat. In: Proceedings of Third International Congress of Neuroethology, Montreal, Quebec, CA, Aug 9–14, 1992. Soc for Neuroethology: p. 272.

O'Neill WE, Suga N (1979) Target-range sensitive neurons in the auditory cortex of the mustached bat. Science 203:69–73.

O'Neill WE, Suga N (1982) Encoding of target range and its representation in the auditory cortex of the mustached bat. J Neurosci 2:17–31.

O'Neill WE, Frisina RD, Gooler DM (1989) Functional organization of mustached bat inferior colliculus: I. Representation of FM frequency bands important for target ranging revealed by ^{14}C-2-deoxyglucose autoradiography and single unit mapping. J Comp Neurol 284:60–84.

Ostwald J (1980) The functional organization of the auditory cortex in the CF-FM bat *Rhinolophus ferrumequinum*. In: Busnel RG, Fish JF (eds) Animal Sonar Systems. New York: Plenum, pp. 953–956.

Ostwald J (1984) Tonotopical organization and pure tone response characteristics of single units in the auditory cortex of the greater horseshoe bat. J Comp Physiol A 155:821–834.

Phillips DP, Judge PW, Kelly JB (1988) Primary auditory cortex in the ferret (*Mustela putorius*): neural response properties and topographic organization. Brain Res 443:281–294.

Pinheiro AD, Wu M, Jen PH-S (1991) Encoding repetition rate and duration in the inferior colliculus of the big brown bat, *Eptesicus fuscus*. J Comp Physiol A 169:69–85.

Pollak GD (1993) Some comments on the proposed perception of phase and nanosecond time disparities by echolocating bats. J Comp Physiol A 172:523–531.

Pollak GD, Bodenhamer RD (1981) Specialized characteristics of single units in inferior colliculus of mustache bat: frequency representation, tuning, and discharge patterns. J Neurophysiol (Bethesda) 46:605–620.

Pollak GD, Henson OW Jr, Johnson R (1979) Multiple specializations in the peripheral auditory system of the CF-FM bat, *Pteronotus parnellii*. J Comp Physiol 131:255–266.

Pollak GD, Henson OW Jr, Novick A (1972) Cochlear microphonic audiograms in the pure tone bat *Chilonycteris parnellii parnellii*. Science 176:66–68.

Pye JD (1980) Echolocation signals and echoes in air. In: Busnel R-G, Fish JF (eds) Animal Sonar Systems. New York: Plenum, pp. 309–354.

Reale RA, Imig TJ (1980) Tonotopic organization of auditory cortex in the cat. J Comp Neurol 192:265–291.

Riquimaroux H, Gaioni SJ, Suga N (1991) Cortical computational maps control auditory perception. Science 251:565–568.

Riquimaroux H, Gaioni SJ, Suga N (1992) Inactivation of DSCF area of the auditory cortex with muscimol disrupts frequency discrimination in the mustached bat. J Neurophysiol (Bethesda) 68:1613–1623.

Sales G, Pye D (1974) Ultrasonic Communication by Animals. London: Chapman and Hall.

Sanides F (1972) Representation in the cerebral cortex and its areal lamination patterns. In: Bourne GH (ed) The Structure and Function of the Nervous System. New York: Academic Press, pp. 329–453.

Sanides D, Sanides F (1974) A comparative Golgi study of the neocortex in insectivores and rodents. Z Mikrosk Anat Forsch (Leipz) 88:957–977.

Scharf B, Meiselman CH (1977) Critical bandwidth at high intensities. In: Evans EF, Wilson JP (eds) Psychophysics and Physiology of Hearing. London: Academic Press, pp. 221–232.

Schmidt S (1988) Evidence for spectral basis of texture perception in bat sonar. Nature 331:617–619.

Schnitzler H-U (1968) Die Ultraschall-Ortungslaute der Hufeisen-Fledermause (Chiroptera-Rhinolophidae) in verschiedenen Orientierungssituationen. Z Vgl Physiol 57:376–408.

Schnitzler H-U (1970) Comparison of the echolocation behavior in *Rhinolophus ferrum-equinum* and *Chilonycteris rubiginosa*. Bijdr Dierkd 40:77–80.

Schnitzler H-U, Menne D, Kober R, Heblich K (1983) The acoustical image of fluttering insects in echolocating bats. In: Huber F, Markl H (eds) Neuroethology and Behavioral Physiology, Berlin: Springer-Verlag, pp. 235–250.

Schreiner CD, Cynader MS (1984) Basic functional organization of second auditory cortical field (AII) of the cat. J Neurophysiol (Bethesda) 51:1284–1305.

Schreiner CE, Mendelson JR, Sutter ML (1992) Functional topography of cat primary auditory cortex: representation of tone intensity. Exp Brain Res 92:105–122.

Schuller G (1974) The role of overlap of echo with outgoing echolocation sound in the bat *Rhinolophus ferrumequinum*. Naturwissenshaften 61:171–172.

Schuller G (1977) Echo delay and overlap with emitted orientation sounds and Doppler-shift compensation in the bat, *Rhinolophus ferrumequinum*. J Comp Physiol A 114:103–114.

Schuller G, Pollak GD (1979) Disproportionate frequency representation in the inferior colliculus of Doppler-compensating greater horseshoe bats: evidence for an acoustic fovea. J Comp Physiol 132:47–54.

Schuller G, Radtke-Schuller S (1990) Neural control of vocalization in bats: mapping of brainstem areas with electrical microstimulation eliciting species-specific echolocation calls in the rufous horseshoe bat. Exp Brain Res 79:192–206.

Schuller G, O'Neill WE, Radtke-Schuller S (1991) Facilitation and delay sensitivity of auditory cortex neurons in CF-FM bats, *Rhinolophus rouxi* and *Pteronotus p. parnellii*. Eur J Neurosci 3: 1165–1181.

Schuller G, Radtke-Schuller S, O'Neill WE (1988) Processing of paired biosonar signals in the cortices of *Rhinolophus rouxi* and *Pteronotus parnellii*. In: Nachtigall PE, Moore PWB (eds) Animal Sonar: Processes and Performance. New York: Plenum, pp. 259–264.

Shamma SA, Symmes D (1985) Patterns of inhibition in auditory cortical cells in awake squirrel monkeys. Hear Res 19:1–13.

Shannon S, Wong D (1987) Interconnections between the medial geniculate body and the auditory cortex in an FM bat. Soc Neurosci Abstr 13:1469

Shannon-Hartman S, Wong D, Maekawa M (1992) Processing of pure-tone and FM stimuli in the auditory cortex of the FM bat, *Myotis lucifugus*. Hear Res 61:179–188.

Shimozawa T, Suga N, Hendler P, Schuetze S (1974) Directional sensitivity of echolocation system in bats producing frequency-modulated signals. J Exp Biol 60:53–69.

Simmons JA (1971) Echolocation in bats: signal processing of echoes for target range. Science 171:925–928.

Simmons JA (1973) The resolution of target range by echolocating bats. J Acoust Soc Am 54:157–173.

Simmons JA, Chen L (1989) The acoustic basis for target discrimination by FM echolocating bats. J Acoust Soc Am 86:1333–1350.

Simmons JA, Stein RA (1980) Acoustic imaging in bat sonar: echolocation signals and the evolution of echolocation. J Comp Physiol A 135:61–84.

Simmons JA, Fenton MB, O'Farrell MJ (1979) Echolocation and pursuit of prey by bats. Science 203:16–21.

Simmons JA, Howell DJ, Suga N (1975) Information content of bat sonar echoes. Am Sci 63:204–215.

Simmons JA, Moss CF, Ferragamo M (1990) Convergence of temporal and spectral information into acoustic images of complex sonar targets perceived by the echolocating bat, *Eptesicus fuscus*. J Comp Physiol A 166:449–470.

Simmons JA, Freedman EG, Stevenson SB, Chen L, Wohlgenant TJ (1989) Clutter interference and the integration time of echoes in the echolocating bat, *Eptesicus fuscus*. J Acoust Soc Am 86:1318–1332.

Suga N (1965a) Functional properites of auditory neurones in the cortex of echolocating bats. J Physiol 181:671–700.

Suga N (1965b) Responses of cortical auditory neurones to frequency-modulated sounds in echo-locating bats. Nature 206:890–891.

Suga N (1977) Amplitude spectrum representation in the Doppler-shifted-CF processing area of the auditory cortex of the mustache bat. Science 196:64–67.

Suga N (1981) Neuroethology of the auditory system of echolocating bats. In: Katsuki Y, Norgren, Sato (eds) Brain Mechanisms of Sensation. New York: Wiley, pp. 45–60.

Suga N (1982) Functional organization of the auditory cortex: Representation beyond tonotopy in the bat. In: Woolsey CN (ed) Cortical Sensory Organization, Vol 3, Multiple Auditory Areas. Clifton, NJ: Humana, pp. 157–218.

Suga N (1984) The extent to which biosonar information is represented in the bat auditory cortex. In: Edelman GM, Gall WE, Cowan WM (eds) Dynamic Aspects of Neocortical Function. New York: Wiley, pp. 315–373.

Suga N (1988a) Auditory neuroethology and speech processing: Complex sound processing by combination-sensitive neurons. In: Edelman GM, Gall WE, Cowan WM (eds) Auditory Function: Neurobiological Bases of Hearing. New York: Wiley, pp. 679–719.

Suga N (1988b) What does single-unit analysis in the auditory cortex tell us about information processing in the auditory system? In: Rakic P, Singer W (eds) Neurobiology of Neocortex. New York: Wiley, pp. 331–349

Suga N (1988c) Parallel-hierarchical processing of biosonar information in the mustached bat. In: Nachtigal PE, Moore PWB (eds) Animal Sonar: Processes and Performance. New York: Plenum, pp. 149–159.

Suga N (1990a) Cortical computational maps for auditory imaging. Neural Networks 3:3–21.

Suga N (1990b) Biosonar and neural computation in bats. Sci Amer 262:60–66

Suga N, Horikawa J (1986) Multiple time axes for representation of echo delay in the auditory cortex of the mustached bat. J Neurophysiol (Bethesda) 55:776–805.

Suga N, Jen PH (1976) Disproportionate tonotopic representation for processing CF-FM sonar signals in the mustache bat auditory cortex. Science 194:542–544.

Suga N, Jen PH (1977) Further studies on the peripheral auditory system of 'CF-FM' bats specialized for fine frequency analysis of Doppler-shifted echoes. J Exp Biol 69:207–232.

Suga N, Manabe T (1982) Neural basis of amplitude-spectrum representation in auditory cortex of the mustached bat. J Neurophysiol (Bethesda) 47:225–255.

Suga N, O'Neill WE (1979) Neural axis representing target range in the auditory cortex of the mustached bat. Science 206:351–353.

Suga N, Schlegel P (1972) Neural attenuation of responses to emitted sounds in echolocating bats. Science 177:82–84.

Suga N, Shimozawa T (1974) Site of neural attenuation of responses to self-vocalized sounds in echolocating bats. Science 183:1211–1213.

Suga N, Tsuzuki K (1985) Inhibition and level-tolerant frequency tuning in the auditory cortex of the mustached bat. J Neurophysiol (Bethesda) 53:1109–1145.

Suga N, Kawasaki M, Burkard RF (1990) Delay-tuned neurons in auditory cortex of mustached bat are not suited for processing directional information. J Neurophysiol (Bethesda) 64:225–235.

Suga N, Kuzirai K, O'Neill WE (1981) How biosonar information is represented in the bat cerebral cortex. In: Syka J, Aitkin L (eds) Neuronal Mechanisms of Hearing. New York: Plenum, pp. 197–219.

Suga N, Niwa H, Taniguchi I (1983) Representation of biosonar information in the auditory cortex of the mustached bat, with emphasis on representation of target velocity information. In: Ewert J-P, Capranica RR, Ingle DJ (eds) Advances in Vertebrate Neuroethology. New York: Plenum, pp. 829–867.

Suga N, Olsen JF, Butman JA (1990) Specialized subsystems for processing biologically important complex sounds: Cross-correlation analysis for ranging in the bat's brain. Cold Spring Harbor Symp Quant Biol 55:585–597.

Suga N, O'Neill WE, Manabe T (1978) Cortical neurons sensitive to combinations of information-bearing elements of biosonar signals in the mustached bat. Science 200:778–781.

Suga N, O'Neill WE, Manabe T (1979) Harmonic-sensitive neurons in the auditory cortex of the mustache bat. Science 203:270–274.

Suga N, Simmons JA, Jen PH-S (1975) Peripheral specialization for fine analysis of Doppler-shifted echoes in the auditory system of the "CF-FM" bat, *Pteronotus parnellii*. J Exp Biol 69:207–232.

Suga N, Simmons JA, Shimozawa T (1974) Neurophysiological studies on echolocation systems in awake bats producing CF-FM orientation sounds. J Exp Biol 61:379–399.

Suga N, Niwa H, Taniguchi I, Margoliash D (1987) The personalized auditory cortex of the mustached bat: adaptation for echolocation. J Neurophysiol (Bethesda) 58:643–654.

Suga N, O'Neill WE, Kujirai K, Manabe T (1983) Specificity of "combination sensitive" neurons for processing complex biosonar signals in the auditory cortex of the mustached bat. J Neurophysiol (Bethesda) 49:1573–1626.

Sullivan WE (1982a) Neural representation of target distance in auditory cortex of the echolocating bat *Myotis lucifugus*. J Neurophysiol (Bethesda) 48:1011–1032.

Sullivan WE (1982b) Possible neural mechanisms of target distance coding in auditory system of the echolocating bat, *Myotis lucifugus*. J Neurophysiol (Bethesda) 48:1033–1047.

Tanaka H, Wong D, Taniguchi I (1992) The influence of stimulus duration on the delay tuning of cortical neurons in the FM bat, *Myotis lucifugus*. J Comp Physiol A 171:29–40.

Taniguchi I, Niwa H, Wong D, Suga N (1986) Response properties of FM - FM combination-sensitive neurons in the auditory cortex of the mustached bat. J Comp Physiol A 159:331–337.

Teng H, Wong D Temporal and amplitude tuning of delay-sensitive neurons in the auditory cortex of *Myotis lucifugus*. J Neurophysiol (Bethesda) (in press).

Trappe M, Schnitzler H-U (1982) Doppler-shift compensation in insect-catching horseshoe bats. Naturwissenshaften 69:193–194.

Tunturi AR (1952) A difference in the representation of auditory signals for the left and right ears in the iso-frequency contours of the right middle ectosylvian auditory cortex of the dog. Am J Physiol 168: 712–727.

Wenstrupp JJ, Grose CD (1993) Inputs to combination-sensitive neurons in the medial geniculate body of the mustached bat. Soc Neurosci Abstr 19:1426.

Wenstrup JJ, Larue DT, Winer JA (1994) Projections of physiologically defined subdivisions of the inferior colliculus in the mustached bat: targets in the medial geniculate body and extrathalamic nuclei. J Comp Neurol 346:207–236.

Wong D, Shannon SL (1988) Functional zones in the auditory cortex of the echolocating bat, *Myotis lucifugus*. Brain Res 453:349–352.

Wong D, Maekawa M, Tanaka H (1992) The effect of pulse repetition rate on the delay sensitivity of neurons in the auditory cortex of the FM bat, *Myotis lucifugus*. J Comp Physiol A 170:393–402.

Woolsey CN (1960) Organization of cortical auditory system: A review and a synthesis. In: Rasmussen G, Windle W (eds) Neural Mechanisms of the Auditory and Vestibular Systems. Springfield: Thomas, pp. 165–180.

Yang L, Pollak GD, Ressler C (1993) GABAergic circuits sharpen tuning curves and modify response properties in the mustache bat inferior colliculus. J Neurophysiol (Bethesda) 68:1760–1774.

Yin TCT, Chan JCK (1988) Neural mechanisms underlying interaural time sensitivity to tones and noise. In: Edelman GM, Gall WE, Cowan WM (eds) Auditory Function: Neurobiological Bases of Hearing. New York, Wiley: pp. 385–430.

10

Perspectives on the Functional Organization of the Mammalian Auditory System: Why Bats Are Good Models

GEORGE D. POLLAK, JEFFERY A. WINER,
AND WILLIAM E. O'NEILL

1. Introduction

Echolocating bats, more than most other mammals, rely on their sense of hearing for obtaining information about their external world (Griffin 1958; also see Fenton, Chapter 2, and Moss and Schnitzler, Chapter 3). In keeping with their reliance on hearing, their auditory systems are not only well developed, but are also proportionately much larger than are the auditory systems of other mammals. Nevertheless, bats are rarely used as models to illustrate basic features of the mammalian auditory system. The reasons for this are partially historical. The cat has traditionally been employed in studies of the central auditory system, and thus the studies of other mammals are frequently overshadowed by the large number of reports on the cat. However, we believe there is also another reason. This reason stems from a notion that echolocation, the ability to "see" objects in the external world with ultrasonic echoes, required fundamental modifications of the auditory system. These modifications changed the nature of acoustic processing, and thus separated the bat auditory system from that of other mammals.

The purpose of this chapter is to frame the issue of the generality of the auditory systems of echolocating bats more explicitly by considering the functional organization and mechanisms of the brainstem and then the forebrain auditory systems of several species of bats, and contrasting their features with those acknowledged as common in other mammals. In the following sections we outline the reasons that led us to two main hypotheses. The first hypothesis is that the information conveyed from the cochlea to the central nervous system is processed in a similar manner with similar circuitry by the brainstem auditory system in most, if not all, mammals.

The second is that the forebrain auditory system differs considerably among mammals and expresses species-specific features that are of adaptive value to the particular species. Furthermore, we conclude by proposing that both the brainstem and forebrain auditory systems of bats are good models of the mammalian auditory system, but for very different reasons.

2. The Anatomy and Functional Organization of the Auditory Brainstem Is Conserved

Here we propose that the information that the cochlea conveys to the brain is processed in fundamentally the same way by similar mechanisms and structures in the brainstem auditory systems of mammals. The hypothesis of a common mammalian processing strategy is supported by a large number of studies which show that the anatomical and functional features of the brainstem auditory system, from cochlear nucleus to inferior colliculus, appear to be conserved throughout mammalian evolution. The homologies among the auditory systems of bats and other mammals have been discussed in a number of previous reports (e.g., Zook and Casseday 1982a,b, 1985, 1987; Zook et al. 1985, Zook and Leake 1989; Ross, Pollak, and Zook 1988; Pollak and Casseday 1989; Kuwabara and Zook 1991, 1992; Grothe et al. 1992, 1994; Pollak 1992; Winer, Larue, and Pollak 1995).

This does not imply that the brainstem auditory systems among mammals are identical, because they are not. The size and complexity of some cell groups, for example, are more highly elaborated in certain echolocating bats than they are either in other bats or in other mammals. A case in point is the marginal cell group in the mustache bat's anteroventral cochlear nucleus (Zook and Casseday 1982a). Other cell groups are more highly developed in all the echolocating bats that have been studied than they are in other mammals. The most prominent example is the intermediate nucleus of the lateral lemniscus (Zook and Casseday 1982a; Covey and Casseday 1991; Schwartz 1992). Such differences, however, are not limited only to bats. Considerable variations exist among brainstem structures in all mammals. These variations are especially evident in some periolivary nuclei (Casseday, Covey, and Vater 1988; Schofield 1991; Kuwabara and Zook 1992; Schofield and Cant 1992; Grothe et al. 1994), in the arrangement of the cell groups comprising the olivocochlear feedback system (Bishop and Henson 1987, 1988; Ostwald and Aschoff 1988; Warr 1992; Grothe et al. 1994), and in the variable expression of the medial superior olive (Harrison and Irving 1966; Casseday, Covey, and Vater 1988; Covey, Vater, and Casseday 1991; Grothe et al. 1992; Kuwabara and Zook 1992; Grothe et al. 1994). Nevertheless, the differences in the brainstem auditory nuclei among mammals appear to be largely in the degree to which one or another cell group is developed (Harrison and Irving 1966; Zook and Casseday 1982a; Pollak and Casseday 1989; Covey and Casseday 1991; Kuwabara and Zook 1992; Schwartz 1992), or in the relative position of a homologous cell group (Brown and Howlett 1972; Pollak and Casseday 1989; Schwartz 1992; Warr 1992), rather than in the emergence of new nuclei or novel processing strategies. The global view of the mammalian auditory brainstem, then, is one of striking similarity and fundamental continuity across many dimensions.

3. The Medial Geniculate Body Has Species-Specific Features

In contrast to the auditory brainstem, certain aspects of forebrain organization differ markedly among mammals. These differences are especially apparent in the auditory thalamus, where disparate patterns of γ-aminobutyric acid (GABA)ergic and calbindin-like immunoreactivity have been found in the medial geniculate body of the rat (Celio 1990), chinchilla (Kelley et al. 1992), bat (Zettel, Carr, and O'Neill 1991), and macaque monkey (Jones and Hendry 1989). In the rat (Winer and Larue 1988) and mustache bat (Vater, Kössl, and Horn 1992; Winer, Wenstrup, and Larue 1992; Winer, Larue, and Pollak 1995), for example, about 1% of the medial geniculate cells are GABAergic. In contrast, 25%–30% of the cells in the medial geniculate body of the cat (Rinvik, Ottersen, and Storm-Mathisen 1987) and squirrel monkey (Smith, Séguela, and Parent 1987) are GABAergic. Such immunocytochemical diversity occurs not only among orders of mammals, but also between species of bats. This is illustrated by the marked difference in the proportion of GABAergic medial geniculate cells in the mustache bat (Winer, Wenstrup, and Larue 1992; Winer, Larue, and Pollak 1995) compared to the horseshoe bat (Vater, Kössl, and Horn 1992): the horseshoe bat has an abundance of GABAergic cells, especially in the ventral division of the geniculate, whereas the ventral division of the mustache bat's geniculate is virtually devoid of GABAergic cells.

The disparate patterns of GABA and calbindin in the medial geniculate body imply that the neurochemical adaptations in the medial geniculate body are pivotal evolutionary features subserving some important facet of species-specific signal processing. This proposition is supported by recent neurophysiological studies of the medial geniculate in both the mustache bat and squirrel monkey. Turning first to the mustache bat, Olsen and Suga (1991a) found a prominent tonotopic organization in the ventral division, where cells are sharply tuned to only one frequency. The tonotopic organization and tuning features of neurons in the ventral division correspond closely to the patterns in other mammals. However, novel features emerge in portions of the medial and dorsal divisions of the medial geniculate body. In those divisions there is a convergence of projections from two or more isofrequency contours to produce combinatorial response properties that are not present in the mustache bat's inferior colliculus (O'Neill 1985). As explained next, the combinatorial neurons integrate information across frequency contours, corresponding to the harmonic frequencies of the mustache bat's orientation calls, to create emergent features that are of importance to the bat. More recently, Olsen and Rauschecker (1992) also found combination-sensitive neurons in the squirrel monkey medial geniculate. However, the combinatorial properties are for features of the communication calls of the squirrel monkey, and are therefore different from the combinatorial properties in the mustache bat geniculate.

3.1 The Medial Geniculate Body of the Mustache Bat Illustrates That Combinatorial Properties Are Species Specific

One of the primary factors that render combinatorial neurons species specific is the specificity with which the response properties of neurons from one frequency contour are matched with the response properties of neurons from another frequency contour. This is clearly illustrated by the specific acoustic features that are required to drive the various types of combinatorial neurons in the forebrain of the mustache bat. Because the relevance of those features are only apparent when related to the structure of the bat's biosonar signals, we first briefly consider the composition of the mustache bat's orientation calls and the types of information conveyed in components of the calls, and then turn to the acoustic signals that best drive combinatorial neurons.

The calls emitted by the mustache bat are characterized by an initial long constant-frequency (CF) component, and a terminal brief, downward sweeping frequency-modulated (FM) component (Fig. 10.1; also see Moss and Schnitzler, Chapter 3). Each call is emitted with a fundamental frequency and four harmonics, but the second harmonic always contains the most energy. This species relies to an inordinate extent on the second harmonic of the CF component. The importance of this CF component is reflected in Doppler-shift compensation, whereby a flying bat continually adjusts the frequency of its emitted CF component to compensate for the upward Doppler shifts that are imposed on the echo by the flight speed of the bat relative to background objects (Schnitzler 1970; Trappe and Schnitzler 1982). The CF and FM components convey different categories of information. The fluttering wings of a flying insect impose frequency and amplitude modulations on the echo CF component, and the modulation patterns are used for target recognition (see Moss and Schnitzler, Chapter 3). In addition, the difference in frequency between the emitted CF component and the Doppler-shifted echo CF component may be used to determine target flight speed (Suga 1988; Olsen and Suga 1991a; see O'Neill,

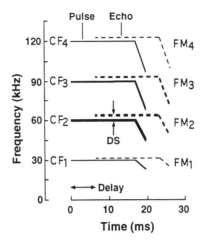

FIGURE 10.1. Schematic illustration of the time-frequency structure (sonogram) of the mustache bat's orientation call and its echo. Each call is emitted with four harmonics, and each harmonic has both a constant-frequency (CF) and frequency-modulated (FM) component. The second harmonic is always the most intense. The echo is higher in frequency than the emitted pulse because of Doppler shifts (DS). There is also a delay between the onset of the pulse and the onset of the echo due the distance between the bat and its target.

Chapter 9). In contrast, the FM component is used to evaluate target distance, which is conveyed by the time separating the emitted and echo FM components (O'Neill and Suga 1982; Olsen and Suga 1991b).

The combinatorial neurons are selective both for particular harmonic combinations and for one of the components of the bat's orientation call; one type of cell is selective for the frequencies in the CF components, the so-called CF/CF neurons, whereas other cells are selective for frequencies in the terminal FM components and are called FM-FM neurons (Olsen and Suga 1991a,b; also see Wenstrup, Chapter 8, and O'Neill, Chapter 9). The CF/CF neurons are characterized by their weak discharges to individual tone bursts and the vigorous firing when two tone bursts having different frequencies are presented together. They are also distinguished by their relative insensitivity to the temporal interval separating the presentation of the two frequencies. CF/CF neurons are tuned to two of the harmonics of the CF components of the bat's orientation calls; thus, CF_1-CF_2 neurons are driven only when the fundamental and second harmonic of the CF components are combined, and CF_1-CF_3 neurons by the fundamental and third harmonic. The exact frequency requirements for optimally driving each type of CF/CF neuron are slightly different among the population, in that each neuron is tuned to a small deviation from an exact harmonic relationship. Thus, the population of CF_1-CF_2 neurons are optimally driven by the range of CF_1 frequencies the bat would emit during Doppler-shift compensation, combined with the CF_2 frequencies that correspond to the stabilized echoes it receives from Doppler-shift compensation. Similar precise frequency pairings are expressed by the population of CF_1-CF_3.

FM-FM neurons are also most effectively driven by two harmonically related frequencies that correspond to the terminal FM components of the bat's orientation calls. Thus, there are FM_1-FM_2 neurons as well as FM_1-FM_3 and FM_1-FM_4 neurons. For these neurons the frequency specificity is not as strict as in the CF/CF neurons, but rather the neurons require that the two signals be presented in a specific temporal order, indicative of a target at a certain distance. Moreover, each neuron is tuned to a particular delay, its best delay, and the population expresses the range of pulse-echo delays that the mustache bat would normally receive during echolocation.

The foregoing features show the striking concordance between the highly specified spectral and temporal requirements of the signals that drive these neurons optimally and the spectral and temporal features of the biosonar signals that the mustache bat emits and receives. These combinatorial properties, then, are tailored to the mustache bat's orientation signals, and thus are unique to that animal.

4. The Functional Organization of the Mustache Bat's Auditory Cortex Is Species Specific

The various types of combinatorial and noncombinatorial neurons are elegantly arranged in the mustache bat cortex, and thereby illustrate the conceptually important point that the functional organization of the auditory cortex can be species specific. The auditory cortex of the mustache bat encompasses several cortical fields (Fig. 10.2, top left panel). One field has a pronounced tonotopic organization, and corresponds to the primary auditory cortices of other mammals. Within the tonotopically organized field, the 60-kHz contour, which Suga and Jen (1976) termed the "Doppler-shifted constant frequency" (DSCF) area, is greatly enlarged, as it is in all lower nuclei. Until recently, it was thought that the functional organization of the 60-kHz contour was similar to the frequency contours in the primary auditory cortices of other mammals (Suga and Jen 1976; Manabe, Suga, and Ostwald 1978). However, a recent study showed that combination sensitivity and delay tuning are also characteristics of neurons in the 60-kHz contour, but that these features are not present in the other frequency contours of the tonotopically organized field (Fitzpatrick et al. 1993). The properties of the combination sensitivity of neurons in the 60-kHz cortical contour also differ in a number of significant respects from the combinatorial properties of the CF/CF and FM-FM neurons described previously, and further illustrate the species-specificity of the mustache bat's auditory

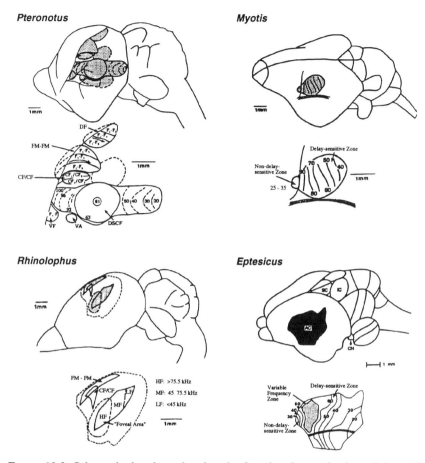

FIGURE 10.2. Schematic drawings showing the functional organization of the auditory corticies of four species of bats. On the *left* are shown two long CF/FM bats,
the mustache bat (*Pteronotus parnellii, top*) and the horseshoe bat (*Rhinolophus*
spp., *bottom*). On the *right* are shown two FM bats, the little brown bat (*Myotis
lucifugus, top*) and the big brown bat (*Eptesicus fuscus, bottom*). For each species,
the top drawing in each panel shows the location of the known auditory cortical
fields in a side view of the brain and the bottom drawing showns an expanded view
of the functional organization. Note that for the two CF/FM species, there is a
segregation of combination-sensitive neurons into harmonically tuned (CF/CF) and
delay-tuned (FM-FM) areas, distinct from the tonotopically organized zones of the
primary auditory cortex. For the mustache bat, several additional combinatorial
regions are shown: dorsal fringe area (DF), ventral fringe area (VF), and ventral area
(VA). FM-FM areas contain a systematic map of best delays. By contrast, except for
the extreme rostral part of the auditory cortex, delay-tuned cells are not completely
segregated in FM bats: most cells within the tonotopically organized cortical field are
also delay sensitive, and there is no clear map of best delays. CF/CF neurons are not
found in FM bats, probably because they lack a long CF component in their biosonar
signals. (Adapted from O'Neill, Chapter 9.)

cortex. Whether these features are formed in the cortex or are a consequence of processing in the geniculate has not been investigated.

Other cortical fields contain only CF/CF neurons or FM-FM neurons (Fig.10.2, top left panel) (Suga, O'Neill, and Manabe 1978, 1979; O'Neill and Suga 1982; Suga 1988; also see O'Neill, Chapter 9). Each of the combination-sensitive fields has additional levels of organization in which neurons that combine the first and second harmonics are arrayed in parallel with neurons that combine first and third, and first and fourth, harmonics. The CF/CF field is further characterized by an orderly change in the best frequencies of the neurons that is repeated in each of the harmonically related regions. The FM-FM region, on the other hand, is characterized by the orderly arrangement of neurons having different best delays, which creates a map of best delays that spans the harmonic regions. Finally, several fields that border the primary fields further elaborate the FM-FM delay-tuned features, and each field also expresses an orderly map of best delays (Suga and Horikawa 1986; Suga 1990).

In summary, both the response properties of neurons and their topographic arrangement in the mustache bat cortex are highly ordered. In addition, the functional organization corresponds to both the structure of the mustache bat's orientation calls and the types of information that the bat needs to extract from those signals. These features then appear to be tailored to the requirements of the mustache bat, which suggests that the functional organization of its cortex is species specific. Next, we show that in other bats the response properties of cortical neurons and their functional organization are different from that of the mustache bat. These data not only provide additional support for the hypothesis that mustache bat cortex is species specific, but suggest that the functional organization of the auditory cortices of other bats is also species specific.

5. Auditory Cortices of Other Bats Have Different Functional Organizations

The cortices of other bats also express combinatorial features, although no other species has been as thoroughly studied as the mustache bat. The general picture emerging from the studies of the auditory cortex of various species of bats is that the combinatorial requirements of their neurons, as well as their functional arrangement, are different, and sometimes markedly so, from that of the mustache bat.

Horseshoe bats, as one example, emit long CF/FM biosonar calls, similar to the orientation calls of the mustache bat. Horseshoe bats, like mustache bats, also have CF/CF and FM-FM cortical neurons that require harmonically related frequencies. The combinatorial neurons are topographically segregated from each other and from the tonotopically organized field (see Fig.10.2, lower left panel) (Schuller, O'Neill, and Ratke-Schuller 1991), features that are also comparable to the arrangement of the mustache bat cortex. There are, however, several notable differences between the horseshoe and mustache bats in the properties of the combinatorial cells and their arrangement in the cortex. One example is the properties of CF/CF neurons in the two species. The optimal frequency combinations in the horseshoe bat are not as strict as they are in the mustache bat, nor are the frequency combinations that best drive each of the cells arranged in an orderly manner, as they are in the mustache bat. Another example concerns the FM-FM neurons. Although the best delays of the neurons in the FM-FM field have an orderly arrangement, comparable to the arrangement in the mustache bat, in the horseshoe bat FM-FM neurons are represented in only one field. In the mustache bat, on the other hand, delay-tuned maps are present in several cortical fields.

The auditory cortices of both the big brown bat, *Eptesicus fuscus* (Jen, Sun, and Lin 1989; Dear et al. 1993) and the little brown bat, *Myotis lucifugus* (Sullivan 1982a,b; Wong and Shannon 1988; Berkowitz and Suga 1989; Maekawa, Wong, and Paschal 1992) have a very different functional organization from that of either the mustache or horseshoe bats (Fig. 10.2, right panels). While the cortices of the brown bats have delay-tuned neurons, the presence of CF/CF neurons has not been reported, and for reasons explained in a later section, it is unlikely that they would have such neurons. Delay-tuned neurons in the brown bats have some of the same combinatorial requirements as the delay-tuned neurons in the mustache bat forebrain. For instance, these cells respond best to different spectral components of the first and second signals when presented with a particular temporal delay (Sullivan 1982a,b; Berkowitz and Suga 1989; Maekawa, Wong, and Paschal 1992). However, the harmonic relationship is not required, as it is in the horseshoe and mustache bats, and the arrangement of neurons with different best delays are distributed in portions of the primary auditory cortex without apparent order (Wong and Shannon 1988; Berkowitz and Suga 1989; Dear et al.).

These studies show that although there are some shared properties of combinatorial neurons in the cortices of bats, there are also properties that are not shared and thus are different among species. In addition, these studies also show that the functional organization of the combinatorial regions is unique to each species.

6. Combinatorial Properties Reflect Adaptations for Some Ecological Niche and Are Selectively Advantageous

The studies reviewed here suggest that the combinatorial features expressed in the forebrain reflect adaptations for some ecological niche, and therefore may have evolved in response to selective pressures. This proposition is supported by the comparable combinatorial features in the forebrains of bats that exploit similar habitats. One example is the presence of CF/CF neurons in the forebrains of mustache and horseshoe bats. These bats, unlike most other bats, hunt for insects under the forest canopy (Bateman and Vaughan 1974; Neuweiler 1983, 1984). This habitat provides a rich source of food for which there is probably little other competition, but it is also an environment that requires the bat to detect and perceive prey in the midst of the echoes from background objects. To deal with the acoustic clutter, horseshoe and mustache bats evolved similar biosonar systems that emphasize a long constant-frequency component coupled with Doppler-shift compensation. Such a biosonar system greatly enhances the bat's ability to detect and recognize fluttering insects against a background of vegetation (Pollak and Casseday 1989; also see Moss and Schnitzler, Chapter 3, and Pollak and Park, Chapter 7). Concurrently, the bats also evolved numerous specialized features in their peripheral and central auditory systems for processing the CF component. Of interest in this regard is that horseshoe bats are found only in the Old World whereas mustache bats are confined to the tropics of the New World. Consequenly, the specialized constant-frequency sonar systems evolved independently in the two species. It follows, then, that there must also have been an independent genesis of CF/CF neurons in their forebrains, and that in both cases the genesis was driven by the selective advantages that accrue to a constant-frequency sonar system.

The proposition that combinatorial properties are adapted for an ecological niche is also supported by the studies of the forebrains of the brown bats. The brown bats employ a biosonar system that is very different from the long CF-FM bats and exploit a different ecological niche. These bats mainly use a brief, downward-sweeping FM chirp for echolocation, and generally hunt for insects in the open skies (Griffin 1958; Neuweiler 1983, 1984). While at times big brown bats add a short CF component to the FM chirp, the CF component is far less important to this bat and is used differently than it is by long CF-FM bats. This is evident because the CF component is only emitted when the bat is searching for prey and is eliminated once a target has been detected (Simmons, Fenton, and O'Farrell 1979). Furthermore, the brown bats do not compensate for Doppler shifts in the echo CF component. It is, therefore, not surprising that there have been no reports of CF/CF combinatorial neurons in the forebrains of the brown bats, and it is unlikely that these bats would have such neurons. Rather, the delay-tuned combinatorial neurons in their cortices emphasize the importance of extracting information from the FM chirp. Furthermore, the acoustic requirements of their delay-tuned combinatorial neurons, as well as the way they are functionally organized in the cortex, are different from the delay-tuned neurons of either the horseshoe bat or the mustache bat.

6.1 Other Mammals Should Also Have Combinatorial Neurons That Are Selectively Advantageous for the Particular Species

If portions of the auditory forebrains of bats evolved in response to selective pressures, then the forebrain of other mammals should also be adapted for processing specific classes of acoustic cues that confer selective advantages for the acoustic environment in which they evolved. This hypothesis gains additional credence from several recent reports concerned with the forebrain auditory systems of other mammals. One is the report by Olsen and Rauschecker (1992), mentioned previously, which showed that neurons in the medial geniculate of the squirrel monkey are sensitive to particular combinatorial features of the animal's communication calls. Others are recent studies of the auditory cortices of the cat and mouse, which have revealed specialized regions that may be analogous, if not homologous, to the combinatorial cortical regions in the long CF-FM bats. Sutter and Schreiner (1991), for example, reported that the dorsal region in the cat primary auditory cortex has populations of cells tuned to two, or in some cases three, frequencies. The response properties of these neurons are in some ways similar to those of CF/CF neurons, but in other ways differ considerably from the features that characterize those combinatorial neurons in the mustache bat. The authors point out that these neurons may be involved in complex sound processing, and suggest that the region of the cat cortex in which these neurons are found may be comparable to the CF/CF region in the cortex of the mustache bat. Similarly, Stiebler (1987) and Hoffstetter and Ehret (1992) reported that a separate cortical field in the mouse, adjacent to the primary auditory cortex, represents only high ultrasonic frequencies and lacks the prominent tonotopic organization characteristic of primary auditory cortex. These authors emphasize that this region is probably important for the analysis of ultrasonic signals related to mother–infant and sexual interactions, and also suggest that this region may be comparable to the combinatorial regions in the mustache bat's cortex.

The studies cited appear to support the general hypothesis that portions of the forebrain are adapted to extract features of acoustic signals, and that the particular features extracted may be species specific. However, we have little understanding of exactly what signal features are important to these animals, and what combinatorial features each of these animals needs to extract from those signals. The combinatorial features in the forebrain of bats relate to the information in the pulse–echo combinations they receive while echolocating. However, these features must surely be different from those that are important to mice or cats. Furthermore, relevant combinatorial features most likely differ among mammals as distantly related as rodents, carnivores, and primates.

7. Some Unifying Themes

The studies reviewed in the previous sections lead us to propose some fundamental differences between the functional organization of the auditory brainstem and auditory forebrain. Our scenario assumes that all mammals need to process similar types of acoustic cues in the lower stages of their central auditory systems, and that these cues are processed in the same way with similar mechanisms and structures in their auditory brainstems. Among the universal cues that must be extracted are binaural cues for spatial localization, temporal and spectral cues for the recognition of communication and other signals, and intensity cues for judgments of loudness. The cues in each spectral component of the signal are coded and represented in the frequency contours of the brainstem auditory nuclei. The coded cues are then partially recombined in the auditory forebrain, resulting in the emergence of new, combinatorial properties. The novel feature of the recombination is the integration of information from two or more frequency contours, thereby partially reconstituting the spectrum of the original signal. We are, therefore, led to the view that one of the functional distinctions between the brainstem and the forebrain is the role that each region plays in the decomposition of the spectrum of an acoustic signal, and subsequently in the partial reconstitution of the original spectrum. We envision the decomposition and reconstruction as a three-stage process that entails different strategies for signal processing in the cochlea, brainstem, and forebrain.

The first stage occurs in the cochlea where the spectral components of a sound are dissembled and the constituent frequencies are arranged as a place code along the cochlear partition. The second stage occurs in the brainstem, where the spectral decomposition is preserved in the various frequency channels of the auditory pathway. Thus, each frequency is processed in a series of parallel pathways that ultimately converge in the inferior colliculus. Some of the pathways merge the information from the two ears, and thus begin to code for the spatial location of the signal. Other pathways maintain the separation of the two ears, and thus process information that is not influenced by the signals received at the other ear (e.g., Pollak and Casseday 1989). The afferents to the lateral superior olive (LSO) and its ascending projections to the inferior colliculus constitute one of the binaural parallel pathways, while the projections from the dorsal cochlear nucleus to the inferior colliculus exemplify one of the monaural parallel pathways. The significant features of the parallel pathways are that processing is accomplished on a frequency-by-frequency basis in the isofrequency contours of each successive nucleus, and that the structures, connections, and mode of processing in each nucleus are fundamentally similar among mammals.

The third stage begins in the medial geniculate, which is the first site in the auditory system where information from two or more frequencies is integrated to create combination-sensitive neurons. By partially recombining the spectral features of the original signal, combinatorial neurons represent some aspect of the signal that cannot be represented adequately by neurons tuned to a single frequency. Thus, one of the features that distinguishes forebrain from brainstem processing is the partial reconstitution of the original spectrum by integrating information across frequency contours. The processing of combinatorial properties is then further elaborated in the various fields of the auditory cortex.

One common denominator of the mammalian forebrain, then, may be the presence of combinatorial properties that are initially created in the medial geniculate body and are functionally arranged in cortical fields adjacent to the primary auditory cortex. However, it appears that the medial geniculate possesses an evolutionary plasticity which allows for the construction of a diversity of combinatorial properties in different species. One indication of such a plasticity is the disparate patterns of neurochemical indicators in the medial geniculate body, which suggests that different forebrain strategies for processing information evolved among mammals. These considerations lead us to hypothesize that the medial geniculate does not construct a standard set of combinatorial response features that are present, with minor variation, in the forebrains of all mammals. Rather, what is constructed are species-specific combinatorial properties that are tailored to the needs of the species. The unique properties emerge from the ways in which various response properties of neurons from different frequency contours are integrated in the medial geniculate body. Furthermore, the disparate neurochemical patterns, particularly the differences in the population of GABAergic medial geniculate neurons, suggest that the means by which integration is achieved may also be different among species. The net result, however, is that the emergent combinatorial properties provide selective advantages to that animal for the perception of its communication signals or, in the case of echolocating bats, for the representation of objects in the external world.

8. The Brainstem and Forebrain Auditory Systems of Bats Are Both Good Models of the Mammalian Auditory System

We now return to the original topic of this chapter: whether the bat auditory system is an appropriate model for the mammalian auditory system. We propose that because the brainstem auditory system is highly conserved, the bat brainstem auditory system is a very good model of the mammalian auditory system. The contrary assertion, that the bat auditory brainstem possesses unique attributes that are fundamentally different from those of other, more generalized mammals, is inconsistent with the wealth of data showing a fundamental continuity in its physiological and anatomical arrangement.

Auditory forebrains, on the other hand, are more variable in their functional organization and apparently have species-specific features, which suggests that there is no prototypical model of the mammalian auditory forebrain. It follows that the functional organization of the forebrain is revealed only when probed with the "appropriate" stimuli. Furthermore, knowledge of what constitutes "appropriate" stimulation comes from an appreciation of the acoustic cues on which a particular species relies, and which features of the signals its auditory forebrain is designed to reconstruct. One advantage that accrues to using echolocating bats for neurophysiological investigations of the auditory system is that the way in which they manipulate their orientation calls is known in considerable detail, as are the acoustic cues from which these animals derive information about objects in the external world. This knowledge allows investigators to generate signals that mimic the types of signals that bats emit and receive during echolocation, as well as the sorts of acoustic cues that are contained in the echoes reflected from flying insects. The auditory forebrain of bats in general, and the mustache bat in particular, are excellent examples of species-specific adaptation in the mammalian auditory forebrain, in that they illustrate how a particular species solves the problem of constructing and uniquely organizing combinatorial properties of acoustic attributes that are of clear importance for that animal's perception of its external world. Deriving the rules that govern such processes in the forebrain will provide insights into the biological strategies of information processing that evolved in species with diverse acoustic behaviors.

Acknowledgments. We thank Evan Balaban, Mike Ferrari, David McAlpine, Lynn McAnelly, Tom Park, Carl Resler, Mike Ryan, Brett Schofield, Wesley Thompson, Harold Zakon, and John Zook for their helpful comments on this manuscript. Supported by NIH grant DC 20068.

References

Bateman GC, Vaughan TA (1974) Nightly activities of mormoopid bats. J Mammal 55:45–65.

Berkowitz A, Suga N (1989) Neural mechanisms of ranging are different in two species of bats. Hear Res 5:317–335.

Bishop AL, Henson OW Jr (1987) The efferent projections of the superior olivary complex in the mustached bat. Hear Res 31:175–182.

Bishop AL, Henson OW Jr (1988) The efferent auditory system in Doppler-shift compensating bats. In: Nachtigall PE, Moore PWB (eds) Animal Sonar: Processes and Performance. New York: Plenum, pp. 307–310.

Brown JC, Howlett B (1972) The olivo-cochlear tract in the rat and its bearing on the homologies of some constituent cell groups of the mammalian superior olivary complex: a thiochline study. Acta Anat 83:505–526.

Casseday JH, Covey E, Vater M (1988) Connections of the superior olivary complex in the rufous horseshoe bat, *Rhinolophus rouxi*. J Comp Neurol 278:313–329.

Celio MR (1990) Calbindin D- 28k and parvalbumin in the rat nervous system. Neuroscience 35:375–475.

Covey E, Casseday JH (1991) The monaural nuclei of the lateral lemniscus in an echolocating bat: parallel pathways for analyzing temporal features of sound. J Neurosci 11:3456–3470.

Covey E, Vater M, Casseday JH (1991) Binaural properties of single units in the superior olivary complex of the mustached bat. J Neurophysiol (Bethesda) 66:1080–1094.

Dear SP, Fritz J, Haresign T, Ferragamo M, Simmons JA (1993) Tontopic and functional organization in the auditory cortex of the big brown bat, *Eptesicus fuscus*. J Neurophysiol (Bethesda) 70:1988–2009.

Fitzpatrick DC, Kanwal J, Butman JA, Suga N (1993) Combination sensitive neurons in the primary auditory cortex of the mustached bat. J Neurosci 13:931–940.

Griffin DR (1958) Listening in the Dark. New Haven: Yale University Press.

Grothe B, Vater M, Casseday JH, Covey E (1992) Monaural interaction of excitation and inhibition in the medial superior olive of the mustached bat: an adapation for biosonar. Proc Natl Acad Sci USA 89:5108–5112.

Grothe B, Schweizer H, Pollak GD, Schuller G, Rosemann C. (1994) Anatomy and projection patterns of the superior olivary complex in the Mexican free-tailed bat, *Tadarida brasiliensis mexicana*. J Comp Neurol 343:630–646.

Harrison JM, Irving R (1966) Visual and nonvisual auditory systems in mammals. Science 154:738–742.

Hoffstetter KM, Ehret G (1992) The auditory cortex of the mouse: connections of the ultrasonic field. J Comp Neurol 323:370–386.

Jen PH-S, Sun X, Lin PJJ (1989) Frequency and space representation in the primary auditory cortex of the frequency modulating bat, *Eptesicus fuscus*. J Comp Physiol A 165:1–14.

Jones EG, Hendry SHC (1989) Differential calcium binding proten immunoreactivity distinguishes classes of relay neurons in monkey thalamic nuclei. Eur J Neuroci 1:222–246.

Kelly PE, Frisina RD, Zettel ML, Walton JP (1992) Differential calbindin-like immunoreactivity in the brain stem auditory system of the chinchilla. J Comp Neurol 320:196–212.

Kuwabara N, Zook JM (1991) Classifying the principal cells of the medial nucleus of the trapezoid body. J Comp Neurol 314:707–720.

Kuwabara N, Zook JM (1992) Projections to the medial superior olive from the medial and lateral nuclei of the trapezoid body in rodents and bats. J Comp Neurol 324:522–538.

Maekawa M, Wong D, Paschal WG (1992) Spectral sensitivity of FM-FM neurons in the auditory cortex of the echolocating bat, *Myotis lucifugus*. J Comp Physiol A 171:513–522.

Manabe T, Suga N, Ostwald J (1978) Aural representation in the Doppler-shifted-CF processing area of the primary auditory cortex of the mustached bat. Science 200:339–342.

Neuweiler G (1983) Echolocation and adaptivity to ecological constraints. In: Huber

F, Markl H (eds) Neuroethology and Behavioral Physiology: Roots and Growing Pains. Berlin Heidelberg: Springer-Verlag, pp. 280–302.

Neuweiler G (1984) Foraging, echolocation and audition in bats. Naturwissenschaften 71:446–455.

Olsen JF, Suga N (1991a) Combination sensitive neurons in the medial geniculate body of the mustached bat: encoding of relative velocity information. J Neurophysiol (Bethesda) 65:1254–1274.

Olsen JF, Suga N (1991b) Combination sensitive neurons in the medial geniculate body of the mustached bat: encoding of target range information. J Neurophysiol (Bethesda) 65:1275–1296.

Olsen JF, Rauschecker JP (1992) Medial geniculate neurons in the squirrel monkey sensitive to combinations of components in a species-specific vocalization. Soc Neurosci Abstr 18:883.

O'Neill WE (1985) Responses to pure tones and linear FM components of the CF/FM biosonar signals by single units in the inferior colliculus of the mustached bat. J Comp Physiol A 157:797–815.

O'Neill WE, Suga N (1982) Encoding of target-range information and its representation in the auditory cortex of the mustached bat. J Neurosci 47:225–255.

Ostwald J, Aschoff A (1988) Only one nucleus in the brainstem projects to the cochela in horseshoe bats: the nucleus olivo-cochlearis. In: Nachtigall PE, Moore PWB (eds) Animal Sonar: Processes and Performance. New York: Plenum, pp. 347–352.

Park TJ, Larue DT, Winer JA, Pollak GD (1991) Glycine and GABA in the superior olivary complex of the mustache bat: projections to the central nucleus of the inferior colliculus. Soc Neurosci Abstr 17(1):300.

Pollak GD (1992) Adaptations of basic structures and mechanisms in the cochlea and central auditory pathway of the mustache bat. In: Popper AN, Fay RR, Webster DB (eds) Evolutionary Biology of Hearing. New York: Springer- Verlag, pp. 751–778.

Pollak GD, Casseday JH (1989) The Neural Basis of Echolocation in Bats. Berlin Heidelberg New York: Springer-Verlag.

Rinvik E, Ottersen OP, Storm-Mathisen J (1987) Gamma-aminobutyrate-like immunoreactivity in the thalamus of the cat. Neuroscience 21:787–805.

Ross LS, Pollak GD, Zook JM (1988) Origin of ascending projections to an isofrequency region of the mustache bat's inferior colliculus. J Comp Neurol 270:488–505.

Schofield BR (1991) Superior paraolivary nucleus in the pigmented guinea pig: separate classes of neurons project to the inferior colliculus and the cochlear nuclues. J Comp Neurol 312:68–76.

Schofield BR, Cant NB (1992) Organization of the superior olivary complex in the guinea pig: II. Patterns of projections from the periolivary nuclei to the inferior colliculus. J Comp Neurol 317:438–455.

Schnitzler H-U (1970) Comparison of echolocation behavior in *Rhinolophus ferrumequinum* and *Chilonycteris rubiginosa*. Bijdr Dierkd 40:77–80.

Schuller G, O'Neill WE, Ratke-Schuller S (1991) Facilitation and delay sensitivity of auditory cortex neurons in CF-FM bats, *Rhinolophus rouxi* and *Pteronotus parnellii*. Eur J Neurosci 3:1165–1181.

Schwartz IR (1992) The superior olivary complex and lateral lemniscal nuclei. In: Webster DB, Popper AN, Fay RR (eds) The Mammalian Auditory Pathway: Neuroanatomy, Vol. 1. New York: Springer-Verlag, pp. 117–167.

Simmons JA, Fenton MB, O'Farrell M (1979) Echolocating and the pursuit of prey by bats. Science 203:16–21.

Smith Y, Séguela P, Parent A (1987) Distribution of GABA-immunoreactive neurons in the thalamus of the squirrel monkey (Saimiri sciureus). Neuroscience 2:579–591.

Stiebler I (1987) A distinct ultrasound-processing area in the auditory cortex of the mouse. Naturwissenschaften 74:96–97.

Suga N (1988) Auditory neuroethology and speech processing: complex-sound processing by combination sensitive neurons. In: Edelman GM, Gall EW, Cowan MW (eds) Auditory Function: Neurobiological Bases of Hearing. New York: Wiley, pp. 679–720.

Suga N (1990) Cortical computational maps for auditory imaging. Neural Networks 3:3–21.

Suga N, Horikawa J (1986) Multiple time axes for representation of echo delays in the auditory cortex of the mustached bat. J Neurophysiol 55:776–805.

Suga N, Jen PH-S (1976) Disproportionate tonotopic representation for processing species-specific CF-FM sonar signals in the mustache bat auditory cortex. Science 194:542–544.

Suga N, O'Neill WE, Manabe T (1978) Cortical neurons sensitive to particular combinations of information bearing elements of bio-sonar signals in the mustache bat. Science 200:778–781.

Suga N, O'Neill WE, Manabe T (1979) Harmonic sensitive neurons in the auditory cortex of the mustached bat. Science 203:270–274.

Sullivan WE (1982a) Neural representation of target distance in auditory cortex of the echolocating bat, Myotis lucifugus. J Neurophysiol (Bethesda) 48:1011–1031.

Sullivan WE (1982b) Possible neural mechanisms of target distance coding in the auditory system of the echolocating bat, Myotis lucifugus. J Neurophysiol (Bethesda) 48:1032–1047.

Sutter ML, Schreiner CE (1991) Physiology and topography of neurons with multipeaked tuning curves in cat primary auditory cortex. J Neurophyisol (Bethesda) 65:1207–1226.

Trappe M, Schnitzler H-U (1982) Doppler-shift compensation in insect-catching horseshoe bats. Naturwissenschaften 69:193–194.

Vater M, Kössl M, Horn AKE (1992) GAD- and GABA-immunoreactivity in the ascending auditory pathway of horseshoe and mustache bats. J Comp Neurol 325:183–206.

Warr WB (1992) Organization of olivocochlear efferent systems. In: Webster DB, Popper AN, Fay RR (eds) The Mammalian Auditory Pathway: Neuroanatomy, Vol.1. New York: Springer-Verlag, pp. 410–448.

Winer JA, Larue DT (1988) Anatomy of glutamic acid decarboxylase immunoreactive-reaction neurons and axons in the rat medial geniculate body. J Comp Neurol 278:47–68.

Winer JA, Wenstrup JJ, Larue DT (1992) Patterns of GAB Aergic immunoreactivity define subdivisions of the mustached bat's medial geniculate body. J Comp Neurol 319:172–190.

Winer JA, Larue DT, Pollak GD (1995) GABA and glycine in the central auditory system of the mustache bat: Structural substrates for inhibitory neuronal organization. J Comp Neurol (in press).

Wong D, Shannon SL (1988) Functional zones in the auditory cortex of the echolocating bat, Myotis lucifugus. Brain Res 453:349–352.

Zettel ML, Carr CE, O'Neill WE (1991) Calbindin-like immunoreactivity in the central auditory system of the mustached bat, *Pteronotus parnellii*. J Comp Neurol 313:1–16.

Zook JM, Casseday JH (1982a) Cytoarchitecture of auditory systems in lower brainstem of the mustache bat, *Pteronotus parnellii*. J Comp Neurol 207:1–13.

Zook JM, Casseday JH (1982b) Origin of ascending projections to inferior colliculus in the mustache bat, *Pteronotus parnellii*. J Comp Neurol 207:14–28.

Zook JM, Casseday JH (1985) Projections from the cochlear nuclues in the mustache bat, *Pteronotus parnellii*. J Comp Neurol 237:307–324.

Zook JM, Casseday JH (1987) Convergence of ascending pathways at the inferior colliculus of the mustache bat, *Pteronotus parnellii*. J Comp Neurol 261:347–361.

Zook JM, Leake PA (1989) Connections and frequency representation in the auditory brainstem of the mustache bat, *Pteronotus parnellii*. J Comp Neurol 290:243–261.

Zook JM, Winer JA, Pollak GD, Bodenhamer RD (1985) Topology of the central nucleus of the mustache bat's inferior colliculus: correlation of single unit properties and neuronal architecture. J Comp Neurol 231:530–546.

Index

Indexing is by scientific name of the various species. When the generic name is used alone this usually is because the reference is to several different species of the same genus. Most vertebrate species are referenced under the scientific names. The only exceptions are very common species such as mouse, rat, cat, chicken, dog, and human. A glossary of scientific and common names of most bat species used in this volume can be found in the appendix to chapter 2.

Sound localization is referred to in the index as sound source localization.